THE YEASTS

Volume I

Second Edition

THE YEASTS

Edited by

ANTHONY H. ROSE

School of Biological Sciences, Bath University,
Claverton Down, Bath, UK

J. S. HARRISON

Ashley House, Upper Frog Street, Tenby,
Dyfed, UK

Volume I

Second Edition

Biology of Yeasts

1987

ACADEMIC PRESS

Harcourt Brace Jovanovich, Publishers

London Orlando New York San Diego Austin
Boston Sydney Tokyo Toronto

ACADEMIC PRESS INC. (LONDON) LTD
24/28 Oval Road
London NW1

United States Edition published by
ACADEMIC PRESS INC.
Orlando, Florida 32887

British Library Cataloguing in Publication Data

The Yeasts. — 2nd ed.
Vol. 1
1. Yeast fungi
I. Rose, A.H. II. Harrison, J.S.
589.2'33 QK617.5

ISBN 0-12-596411-0

Printed in Great Britain at the Alden Press
Oxford, London and Northampton

Contributors

D.G. Ahearn, Department of Microbiology, Georgia State University, University Plaza, Atlanta, GA 30303, USA.

A. Hagler, Department of Microbiology, Georgia State University, University Plaza, Atlanta, GA 30303, USA.

J.S. Harrison, Ashley House, Upper Frog Street, Tenby, Dyfed, UK.

R. Hurley, Department of Microbiology, Institute of Obstetrics and Gynaecology, University of London, Queen Charlotte's Maternity Hospital, Goldhawk Road, London, UK.

N.J.W. Kreger-van Rij, Laboratory of Medical Microbiology, State University of Groningen, Oostersingel 59, 9713 EZ Groningen, The Netherlands.

C.P. Kurtzman, Northern Regional Laboratory, 1815 North University Street, Peoria, IL 61604, USA.

J. de Louvois, Department of Microbiology, Institute of Obstetrics and Gynaecology, University of London, Queen Charlotte's Maternity Hospital, Goldhawk Road, London, UK.

A. Mulhall, Department of Microbiology, Institute of Obstetrics and Gynaecology, University of London, Queen Charlotte's Maternity Hospital, Goldhawk Road, London, UK.

H.J. Phaff, Department of Food Science and Technology, University of California, Davis, CA 95616, USA.

A.H. Rose, School of Biological Sciences, University of Bath, Claverton Down, Bath, UK.

W.T. Starmer, Department of Biology, Syracuse University, New York, USA.

J.P. van der Walt, Microbiology Research Group, South African Council for Scientific and Industrial Research, PO Box 395, Pretoria 0001, South Africa.

A.E. Wheals, School of Biological Sciences, University of Bath, Claverton Down, Bath, UK.

Preface

The days when a single author was able to produce a series of volumes which could satisfactorily cover the whole gamut of a general science such as chemistry, physics or biology are long past. More recently these continuously expanding disciplines, and many more including mathematics and chemical engineering, have become essential basics for the complete study of a single class of living organisms.

We have endeavoured in this multivolume treatise to provide as wide a coverage as possible of the different areas of investigation that comprise the science of zymology, the study of yeasts. Four volumes involving 50 chapters have resulted; two further volumes are in the planning stage. Inevitably there will be gaps when dealing in this very broad field; for such we apologize and trust that any deficiencies may not seriously detract from the value of the present work. In this, as in all similar cases, prominent workers in each relevant discipline present an overall account of their chosen field with particular attention to the results of the most recent advances.

The first requirement is the recognition and definition of the members of the group, that is taxonomy and classification, closely followed by such essentials as biological functions related to the external environment and intracellular mechanisms. In these fields modern research has been very active and rewarding, advances since the publication of the first edition of *The Yeasts* in 1969–1971 being quite phenomenal. Genetics also has made tremendous strides forward during the same period, the molecular aspect being especially important. The many branches of knowledge discussed in detail are too numerous to mention here, but each volume has been planned to cover a group of related subjects, and so form a useful book in its own right.

We gratefully acknowledge the debt we owe, not only to the contributors to these volumes and others concerned for their efforts, often under difficult circumstances, but also to the vast number of workers in related fields in the past whose investigations have made possible the rapid exponential advances in zymological knowledge. We hope that these volumes will provide a useful overall view of the present state of the art.

Anthony H. Rose
J. Stuart Harrison

October 1986

Contents

1 Introduction

Anthony H. Rose and J. Stuart Harrison

School of Biological Sciences, University of Bath, Bath, UK and Ashley House, Upper Frog Street, Tenby, Dyfed, UK

Yeasts hold a unique place in the long history of mankind. In the Microbial World, they can be likened to cattle among higher animals, having been cultivated by Man as a source of food and drink, albeit unwittingly until the last century, for the leavening of bread and for the production of beverages which contain what was probably the earliest soporific and anaesthetic, ethyl alcohol. More recently they have been used for the manufacture of chemical products, fuel and edible single-cell protein.

It is just over a hundred years since Pasteur (1866, 1876) demonstrated the essential participation of live yeast in fermentation processes. Since that time a small number of species of the genus *Saccharomyces* have been produced in amounts measured annually in millions of tonnes; this subject is addressed in a later volume in this Second Edition. The fact that strains of *Saccharomyces cerevisiae* are capable of reproducing both sexually and asexually in simple medium either in the presence or absence of oxygen and are non-toxic, has not only made them commercially viable, but their ready availability has provided an ideal source of microbial biomass for research into cellular behaviour. Because of the general similarity between the mechanisms operating within the cells of so many living organisms, fundamental yeast research is of much wider application than fermentation research. The past 15 years have witnessed a massive increase in the use of yeasts as model eukaryotic organisms with which to study basic cellular phenomena. Not a little of this exploitation of yeasts in cell biology studies has been prompted by the vastly increased knowledge which biologists now have of the genome in yeasts, especially in *Saccharomyces cerevisiae*, and in ways in which this genome can be engineered. In our Introduction to Volume 2 of the first edition of *The Yeasts* (Rose and Harrison, 1971), we proposed the setting up of Project Y,

The Yeasts Vol. 1, 2nd edition
ISBN 0-12-596411-0

an integrated approach to a study of cell biology using yeast as the model organism. To a considerable extent, this suggestion has been taken up by cell biologists, needless to say without any acknowledgement of our clarion call. One of us (Rose, 1981) returned recently with a further advocacy of Project Y, by which time the battle had been virtually won.

The extent of advances in various fields has, however, been highly variable. In the manufacture of fermented drinks, for example, striking changes would not be expected, partly because production methods are restricted by official regulations, but also because alcohol yields have for many years been near the theoretical maximum (Harrison and Graham, 1970) while taste and appearance, like the manufacturing methods, are largely governed by personal preference and tradition. It is natural that, where procedures have been developed empirically over hundred of years, results will have approached the optimum, at least until some entirely new approach is established. On the other hand, research into the biochemistry, physiology and genetics of yeasts has proceeded at an ever increasing pace. Techniques for isolating intracellular organelles from yeasts have been refined, with the result that our knowledge of their composition and structure is now extensive; Volume 3 in this Second Edition reviews knowledge on yeast organelles. All the major metabolic pathways operating in yeasts have now been charted, the last to yield its secrets being that used by certain yeasts to incorporate atoms from methanol into cell constituents. Volume 4 in this Second Edition covers yeast metabolism. But it is in the field of yeast genetics that progress over the past 15 years has been most dramatic. Indeed, it is in this field that yeasts have increasingly been adopted as model eukaryotic organisms, in the true spirit of Project Y. Much of this recent progress is reviewed in excellent texts edited by Broach *et al.* (1981a, b) and Spencer and his colleagues (1983).

The first volume in the Second Edition of *The Yeasts* is devoted to the biology of these organisms. The early chapters deal with yeast taxonomy and systematics. A number of taxonomic changes are listed by Kreger-van Rij (Chapter 2) who recognizes 60 genera. This contrasts with only 39 given 15 years earlier by the same author (Kreger-van Rij, 1969), the increase being largely due to improved methods of differentiation. The generic names, with the abbreviations she has allocated to them and which are used throughout this treatise, are listed at the beginning of each volume in the Second Edition. The division of micro-organisms into narrowly defined permanently established groups according to their common characteristics is inevitably becoming increasingly difficult. Originally, visual and behavioural observations were sufficient to distinguish between types but, as the study of biological phenomena has become more detailed, it is evident that finer and fundamentally more logical distinctions can be made between organisms. The important new concept of DNA relatedness as a method of classification, the delimitation of species on the basis of phenotypic discontinuity, and other modern systems are discussed in subsequent chapters. It is evident that, as more fundamental research reveals the specific functions of individual components of living cells, the more logical the classification system will become, and the significance of the extra- and intracellular interactions will be better understood.

The identification of strains is a necessary step in the study of yeasts found in environments that are of interest to man, including water, soil, plants and insects. Although the majority of species appear innocuous, there remain some which may in various ways affect higher animals including human beings, who may suffer pathogenic infections. In each of the chapters dealing with these matters, the first considerations are isolation of uncontaminated specimens and determination of their species identity. As more information becomes available on the behaviour of each type, the easier it will be to predict the influence for good or bad that the particular yeast will produce. In view of the consequences of genetic background, cross-breeding, mutation and deliberate modification of gene structure, there would seem to be no limit to the possible variations in behaviour within species. This has for some years been apparent in the breeding of industrial yeasts by hybridization techniques (Fowell, 1969a) but more recently, particularly for academic purposes, by the planned introduction of selected segments into genes in order to investigate specific enzyme mechanisms. A word of warning should be sounded, however, in case genetic modifications introduced commercially may produce undesirable effects.

An important feature in the study of yeasts as of any other group of organisms is the ease or otherwise of distribution from one environment to others. This is dependent to some degree on the characteristics of the cell cycle of the particular species, a subject discussed in a chapter in which changes in cell volume and mass are followed in relation to synthesis of protein, enzymes, DNA, RNA and other compounds and structures. In specific cases, co-ordination of growth and cell division in both mitosis and meiosis are considered. In the first edition of *The Yeasts*, Fowell (1969b) contributed a chapter on the wider question of life cycles related to yeast genera, dividing them into basic types in which the sequence of events differs considerably, some systems being more complex than others which may lack a sporulation stage. The kind of life cycle will also influence distribution phenomena in accordance with such variables as ease of sporulation and time relationships of the various stages.

It is intended that subsequent volumes of this treatise should cover the full range of disciplines applicable to yeasts, bringing together under one title contributions by leading specialists, on all aspects of their function from both practical and theoretical angles. This work is clearly necessary owing to the great advances achieved since the first edition was published.

References

Broach, J., Strathern, J. and Jones, E. (eds) (1981a). "Molecular Biology of the Yeast *Saccharomyces*. Life Cycle and Inheritance." Cold Spring Harbor Laboratory, Cold Spring Harbor, New York.

Broach, J., Strathern, J. and Jones, E. (eds) (1981b). "Molecular Biology of the Yeast *Saccharomyces*. Metabolism and Gene Expression." Cold Spring Harbor Laboratory, Cold Spring Harbor, New York.

Fowell, R.R. (1969a). *In* "The Yeasts" (A.H. Rose and J.S. Harrison, eds), vol. 1, pp. 303–383. Academic Press, London.

Fowell, R.R. (1969b). *In* "The Yeasts" (A.H. Rose and J.S. Harrison, eds), vol. 1, pp. 461–471. Academic Press, London.

Harrison, J.S. and Graham, J.C.J. (1970). *In* "The Yeasts" (A.H. Rose and J.S. Harrison, eds), vol. 3, pp. 283–348. Academic Press, London.

Kreger-van Rij, N.J.W. (1969). *In* "The Yeasts" (A.H. Rose and J.S. Harrison, eds), vol. 1, pp. 5–78. Academic Press, London.

Pasteur, L. (1866). "Études sur le Vin." Imprimeurs Impérials, Paris.

Pasteur, L. (1876). "Études sur la Bière." Gauthier-Villars, Paris.

Rose, A.H. (1981). *In* "Current Developments in Yeast Research" (G.G. Stewart and I. Russell, eds), pp. 645–652. Pergamon Press, Toronto.

Rose, A.H. and Harrison, J.S. (1971). *In* "The Yeasts" (A.H. Rose and J.S. Harrison, eds), vol. 2, pp. 1–2. Academic Press, London.

Spencer, J.F.T., Spencer, D.M. and Smith, A.R.W. (eds) (1983). "Yeast Genetics. Fundamental and Applied Aspects." Springer-Verlag, New York.

2 Classification of Yeasts

N.J.W. Kreger-van Rij

Laboratory of Medical Microbiology, State University, Groningen, The Netherlands

I. INTRODUCTION

Yeasts are unicellular fungi which reproduce by budding or fission. This definition covers a wide variety of organisms, among them ascomycetes and basidiomycetes. Yeasts in this morphological concept may be a stage in the life cycle of hyphal fungi. The yeasts discussed here are restricted to those included in the third edition of "The Yeasts, a Taxonomic Study" (N.J.W. Kreger-van Rij, ed., 1984a). In this, as in the earlier editions, not all fungi with a yeast phase have been included, but a choice has been made for

The Yeasts Vol. 1, 2nd edition
ISBN 0-12-596411-0

traditional and practical reasons. Extension is possible in future with fungi for which a combination of morphological, physiological and biochemical features is advantageous or required for species differentiation.

In the first section, the criteria used for differentiation and classification are mentioned: in the second, the principles of classification of species, genera and higher taxa are discussed. In the last section, a survey of the classification of the three groups: the ascosporogenous yeasts, the basidiosporogenous yeasts and the imperfect yeasts is given, with special emphasis on the genera.

Since the appearance of the second edition of "The Yeasts" (J. Lodder, ed., 1970), yeast taxonomy has developed in several ways. The number of genera has increased from 39 to 60 and the number of species from 349 to 500. Moreover, new criteria have been introduced and reclassifications have been proposed. There is not always consensus of opinion on these classifications among the yeast taxonomists. The third edition gives an up-to-date account of yeast taxonomy. As far as possible, the present chapter follows the classification of species as given by the authors of this work.

II. CRITERIA USED IN THE CLASSIFICATION OF YEASTS

A. Choice of criteria

Features which are different in the various yeasts may be used as criteria in classification. What is generally indicated as a feature is the response of the yeast to certain prescribed test conditions: a change in the test conditions may bring out a different response: this holds for most morphological, physiological and biochemical features. Since differences are important in systematics, it is necessary to compare the responses under identical conditions. This has led to the development of certain standardized tests which have become more or less generally accepted. However, small differences in methods used by yeast workers occur. These are not considered to be of great importance but, in the event of doubt when comparing a yeast strain with the standard description of a certain species, it is advisable to use identical methods. The important details of a test method are: the condition and amount of the inoculum, the composition of the medium, the apparatus used, the time and temperature of incubation, and the method of evaluating the results of the test.

For use in systematics it is desirable that a property of a yeast be constant under standard conditions; however, several properties show variations. Yeasts are examined as cultures on certain media, which means that selection is started from the moment of isolation. Keeping yeasts for a prolonged period on these media extends the selection and may lead to a fairly constant product which differs, however, from the original organism. The use of lyophilized cultures may considerably diminish the risk of change of properties, although selection is not entirely excluded.

B. Criteria used in the standard description

The procedures for establishing the criteria used for the description of yeast species have been described by Wickerham (1951), van der Walt (1970a) and van der Walt and Yarrow (1984a). For details the reader is referred to these works.

1. Vegetative reproduction

The manner of vegetative reproduction, including the shape and size of the cells, is examined in liquid and on solid media, the latter on slants, on slides or in Dalmau plates. Various media of standardized composition are used, for instance laboratory prepared and commercial malt extract, a medium containing glucose, yeast extract (malt extract) and peptone, corn-meal agar and potato (glucose) agar.

The shape and size of the cells are influenced by the composition of the medium and other conditions for growth, and comparison of the results of identical test conditions should be made.

There are two means of vegetative reproduction in yeasts: by the production of buds, and by the formation of cross-walls in hyphae or single cells. Buds may have a narrow or a broad base. They occur multilaterally on the cell or are confined to special regions, for instance on the shoulders of elongate cells or exclusively at one or both poles (Fig. 1a). The latter mode of budding may lead to the formation of multiple scars, i.e. concentric ridges of bud scars.

Budding cells which remain attached to each other are referred to as pseudomycelium (Fig. 1b). Very often there is a differentiation among them of cells which grow out to a greater length (pseudomycelium cells) and others which remain short and are known as blastospores or blastoconidia. Cells formed as buds on true hyphae are also named blastospores or blastoconidia; these are sometimes formed on small elevations. Another type of vegetative reproduction is to be found in the genus *Sterigmatomyces* in which buds form on stalks (Fig. 1c); these cells are not forcibly discharged. Ballistospores also arise on stalks or sterigmata, but are discharged by a water-drop mechanism. True hyphae, which may form side branches, grow at the tip: cross-walls are

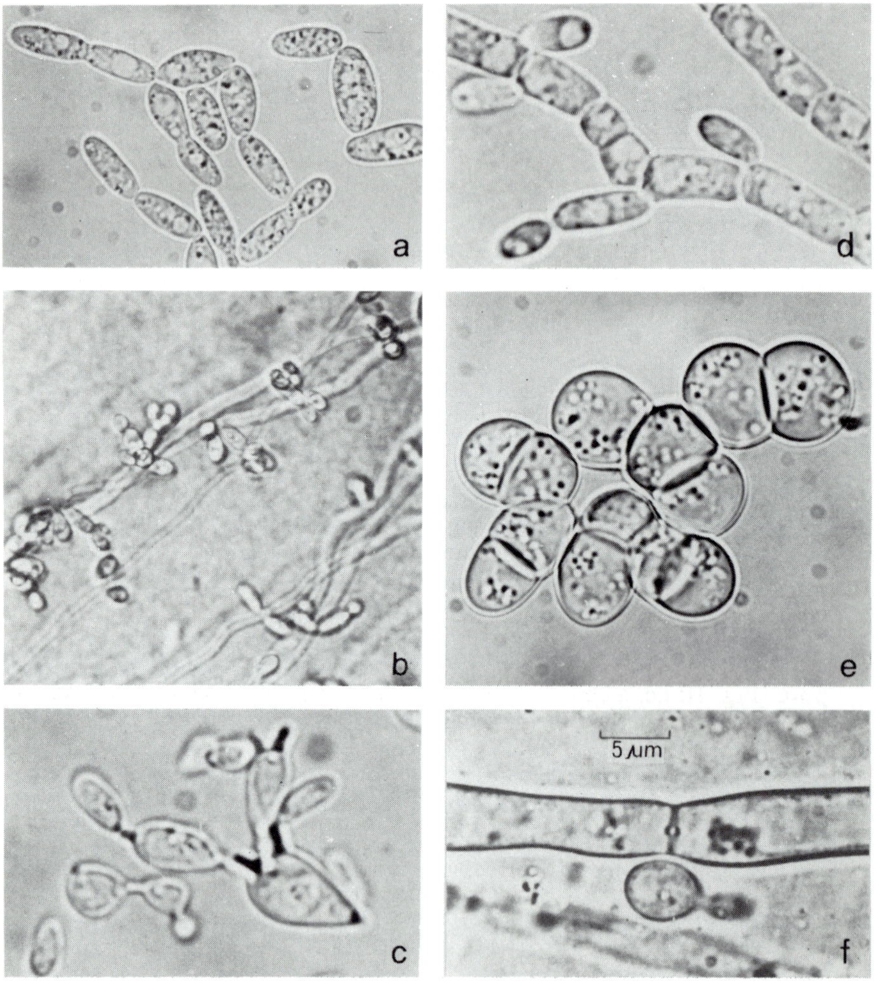

Fig. 1. Photomicrographs of vegetative reproduction in yeasts. (a) *Nadsonia commutata.* Vegetative cells with bipolar budding on a broad base. Magnification × 1250. (b) *Candida tropicalis.* Pseudomycelium. Magnification × 500. (c) *Sterigmatosporidium polymorphum*, the perfect state of a *Sterigmatomyces* species. Buds are formed on sterigmata in a haploid strain of the species. Magnification × 2600. (d) *Trichosporon cutaneum.* True mycelium with septa and blastospores. The mycelium begins to disintegrate into arthrospores by splitting along the septa. Magnification × 2100. (e) *Schizosaccharomyces octosporus.* Vegetative cells reproducing by fission. Magnification × 1550. (f) *Ambrosiozyma platypodis.* Hypha with septum and a dolipore in the middle of it, visible as a dark ring with a light centre. From Kreger-van Rij and Veenhuis (1969a).

formed without constriction. The hyphae may be split up at the cross-walls to form arthrospores (Fig. 1d). Cross-wall formation in single cells followed by fission is typical of the genus *Schizosaccharomyces* (Fig. 1e).

Ultrastructural studies of cell walls have revealed taxonomically important characteristics. First, a distinct difference is apparent in the wall of vegetative cells and in bud formation between ascomycetous and basidiomycetous yeasts. Sections of the wall fixed with osmium tetroxide or potassium permanganate show a broad electron-light inner layer and a thin electron-dense outer layer in ascomycetous yeasts, whereas a structure of several light and dark layers is found in basidiomycetous yeasts. In the latter, the thickness of the wall varies, depending on the age of the cells (Kreger-van Rij and Veenhuis, 1971).

In ascomycetous yeasts a new wall, which becomes the wall of the bud, arises from under the wall of the mother cell, penetrating it during growth (Kreger-van Rij and Veenhuis, 1969b, 1975a; Kreger-van Rij, 1978b). In basidiomycetous yeasts, the wall of the young bud is continuous with the inner layer of the wall of the mother cell (Prusso and Wells, 1967; Kreger-van Rij and Veenhuis, 1971).

A second characteristic revealed by electron microscopic studies is the structure of the cross-wall or septum in true hyphae and in pseudohyphae which appears to show variations. Septa in ascomycetous yeasts may have a dolipore (Fig. 1f; i.e. a central pore with a thickening of the wall around it), plasmodesmata (see Fig. 2f), a central micropore with blunt edges (Gow *et al.*, 1980), or a closure line (i.e. the remnants of the plasmalemma in the middle of the cross-wall after it is closed by centripetal growth).

In basidiomycetous yeasts, a simple pore with a tapered edge or a dolipore or a closure line occurs. The membranous structure around the dolipore (i.e. the cap or parenthesome in higher fungi) is in yeasts more or less developed or lacking; it may have a vesicular structure (Kreger-van Rij and Veenhuis, 1971; see Fig. 2e). The ultrastructure of the cross-wall in ascomycetous yeasts is generally different in pseudomycelium and in true mycelium; in pseudohyphae a closure line occurs, and in true hyphae plasmodesmata or a dolipore. However, there are exceptions, for instance the septa in the true mycelium of *Pichia burtonii* have a closure line, and in the cross-wall between mother cell and bud of *Ambrosiozyma platypodis* a dolipore may occur. Whereas the value of the distinction between pseudo- and true mycelium has depreciated in yeast taxonomy, great taxonomic value has been attached to the various types of septal structure, especially for generic differentiation. Although tests involving electron microscopy are not in the routine of yeast identification, ultrastructural data occur in the standard description of several species.

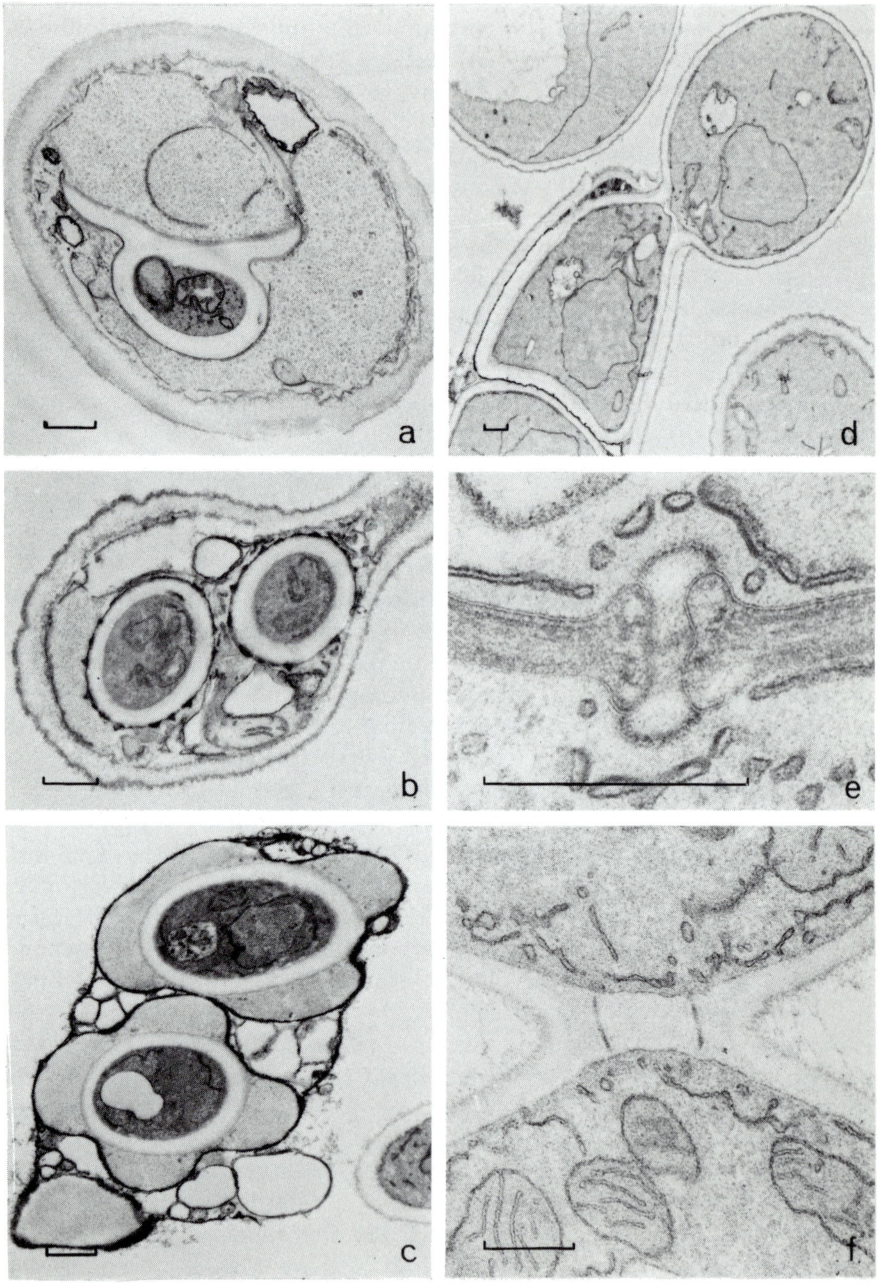
a
b
c
d
e
f

Morphological features may vary because of the occurrence of smooth and rough variations of cultures, and the change from smooth to rough in a yeast strain. This change is not only apparent from the aspect and consistency of the streak culture, but also involves the shape and association of cells as well as pellicle formation. The glistening, smooth and soft culture changes to a mat, rough, tough form; the round to short oval cells growing singly or in pairs become more elongate and produce tree-like formations; the absence of pseudomycelium gives way to a well-developed pseudomycelium. In mycelial yeasts there is a tendency to produce more hyphae and fewer budding cells, also fewer arthrospores.

Wickerham (1964b) recognizes these changes as mutations from fermentative to oxidative forms in which many genes are involved, with transitional stages. He studied the phenomena in a strain of *Hansenula petersonii*, and proposed the theory that the genes in question are not individually mutated, but controlled by other genes which regulate metabolism in accordance with the environment. The changes in metabolism involve a decreased fermentative capacity, the utilization of other metabolites instead of sugars, and the production of hydrophobic instead of hydrophilic molecular structures on the outer wall. Thus the environment strongly influences the smooth/rough (S-R) variation. For instance, the smooth form of *Hansenula petersonii* appeared to be stable on agar containing glucose, peptone, yeast extract and malt extract but on malt agar the colony form changed from smooth to rough. This change in colony form on malt agar is observed in many yeasts.

The influence of the S-R variation on the formation of pseudomycelium causes this feature to have limited value for differentiation. For instance, in the former distinction of the genera *Candida* and *Torulopsis* by the presence or absence of pseudohyphae, a few species with smooth cultures had first

---

Fig. 2. Electron micrographs of ascospores and of septa in yeasts (bar represents 0.5 μm). (a) *Candida freyschussii*. Hat-shaped ascospore formed after fusion of two cells of opposite mating type. Strains of opposite type were discovered by Rodrigues de Miranda (1980) who considers *Candida freyschussii* the imperfect state of *Pichia rhodanensis*. (b) *Clavispora lusitaniae*. Cross section through ascus with ascospores. The walls of the latter have small dark warts. (c) *Sporopachydermia cereana*. Section through ascospores liberated from the ascus. The spore wall has a light inner layer, a thin dark outer layer, and in between an irregular greyish layer. (d). *Saccharomyces cerevisiae*. Section through a germinating ascospore still within the ascus. A bud is formed which is separated from the mother cell by a thin light layer and the beginning of darker layers at both sides. (e) *Filobasidium uniguttulatum*. Section through a dolipore. The cap around the pore is partly cup-shaped. (f) *Candida ingens*. In the cross-wall between mother cell and bud two plasmodesmata are visible. The plasmodesmata are arranged in a circle. Rodrigues de Miranda (1980) discovered opposite mating types among the strains of *Candida ingens*. After mixing cells of opposite type, gametangiogamy occurs and asci with thick-walled spores are formed.

been classified as *Torulopsis* species, but after changing to the R-form, as *Candida* species. However, there are also types of pseudomycelium which occur in the smooth forms, such as pseudohyphae in *Candida albicans*.

2. Sexual reproduction

The main features of sexual reproduction in the ascosporogenous yeasts are the mode of ascus formation and the shape of the ascospores. The shape of the ascus, the number of spores per ascus, the colour of the spores, their possible release from the ascus and their mode of germination may also be typical criteria.

Ascospore formation is an important taxonomic criterion. A number of different sporulation media are used. Usually, they allow limited growth or no growth at all; for instance, when usable nitrogen is scarce or lacking. The media are inoculated with yeast in a well-nourished condition, the pre-sporulation medium being of importance. Various temperatures of incubation may be tried, generally lower than the optimum temperature for growth. Sporulation is often confined to a restricted region of the streak culture. Species of certain genera may sporulate preferentially on a special medium, for instance *Saccharomyces* species on acetate agar.

On the sporulation media, the ascus may be formed by a direct change from a vegetative cell, or sporulation may be preceded by conjugation either of two single cells or of a mother cell and a bud. In nearly all yeasts conjugation is somatic, that is, two yeast cells fuse and become the zygote in which nuclear fusion takes place. A type of conjugation which may occur in mycelial yeasts is gametangiogamy, a conjugation of two cells which act as gametangia and which are outgrowths of the hyphae. The hyphal cells are separated by cross-walls from the fused gametangia which form the ascus.

A special type of conjugation was observed in certain *Saccharomycopsis* species (Kreger-van Rij and Veenhuis, 1975b). After fusion of two hyphal cells or a hyphal cell and a blastospore, a new wall was formed between the two cells which turned into asci or on which buds were formed which became asci. In some of these *Saccharomycopsis* species ascus formation had been indicated before as parthenogenetic, because no fused cells were visible when the asci were present.

If spores are formed in a culture from a single spore or from a haploid cell, the yeast is homothallic; if a mixture of cells of opposite mating type is required for sporulation, the yeast is heterothallic. However, Kurtzman (1984d) has pointed out that ascospores of two-spored asci may be diploid, and germinate to give diploid cells which sporulate without preceding conjugation (e.g. in *Pichia amylophila*). Several yeast species are known to be both homothallic and heterothallic. In these yeasts the culture from a single haploid cell sporulates after diploidization, but it also represents one of the mating types. Very often, but not always, mixtures of cells of opposite mating type sporulate more readily than cells of a single type. Cells of opposite type may agglutinate when mixed, as for instance in the species *Saccharomyces kluyveri* (Wickerham, 1958).

Variation in the shape of the ascospores of yeasts provides a good criterion for differentiation. Round, oval, reniform, cylindrical and needle-shaped spores occur. A ledge in the middle or on one side of round or oval spores gives them a Saturn- or hat-shaped appearance. Wartiness is another typical feature. It may be visible under the light microscope, but sometimes the warts are so small that they only appear under the scanning- or transmission-electron microscope. Scanning-electron microscopy has also revealed ridges on the spores, for example in *Debaryomyces marama* (Kurtzman *et al.*, 1975) and a ledge which is not circumfluent in *Pichia dispora* (Kurtzman and Smiley, 1974). Transmission-electron microscopy of sections shows more detail than scanning-electron microscopy and also affords an insight into the internal structure of the spores. Fixation of the material with osmium tetroxide or potassium permanganate reveals a wall in which layers of different electron-density are visible. In the simplest form there is a dark outer layer and a lighter inner layer. Variations in the thickness and shape of these layers occur, for instance a dark warty outer layer or a light inner layer forming warts. The inner layer is not always of even electron density; dark and light zones may be distinguished, the latter as the inner zone, the former between this and the outer layer. A dark line occasionally separates these zones. The outer zone may be very dark, and thus not distinguishable from the outer dark layer, for instance in the spores of *Saccharomycopsis capsularis*. The ledge belongs to the outer zone of the inner layer. Observations on the germination of hat-shaped spores show that the light inner zone becomes the wall of the vegetative cell, while the rest of the spore wall, including the ledge, vanishes. Removal of the outer layers proceeds either by stretching and by a gradual disintegration, or by rupturing as the cell swells. In the latter case, the empty shell generally remains attached to the cell and observations by light microscopy lead to the conclusion that the spore had two membranes. A double inner layer, namely an inner layer with two zones of which the innermost forms the wall of the vegetative cell after germination, was not only observed in hat-shaped spores, but also in the round spores of *Saccharomyces cerevisiae*. In spores with a simple inner layer, a new wall may be formed for the vegetative cell during germination.

Since the early electron microscopy studies by Kawakami (1960), Besson (1966) and Kreger-van Rij (1969b), many others have followed, for instance Black and Gorman (1971), Kurtzman *et al.* (1980) and others mentioned in the discussion of the genera. Data on the ultrastructure of the ascospores are far from complete, and if this character is considered to be of taxonomic importance (e.g. for generic differentiation) it may be worthwhile to pursue these studies.

Most species of the basidiosporogenous yeasts are heterothallic and the haploid stage is in the form of yeast cells. The first step in the life cycle is the fusion of compatible yeast cells, namely cells of opposite mating type. This results in a dikaryotic mycelium with clamp connections at the septa. In the Filobasidiaceae, one of the groups of the basidiosporogenous yeasts, long thin basidia are formed on the hyphae where nuclear fusion and meiosis take place. Basidiospores are formed on the apex of the basidium, either sessile basidiospores arranged in a whorl, as in *Filobasidium*, or in four chains as in *Filobasidiella*. In the second group of the basidiosporogenous yeasts, the teliospore-forming yeasts, thick-walled cells (teliospores) develop on the mycelium: here, nuclear fusion occurs. During germination of the teliospores, meiosis takes place. A metabasidium is formed from which basidiospores arise by budding. In addition to a heterothallic life cycle, a homothallic one can develop from a single haploid cell to include dikaryotic hyphae and either filiform basidia, or teliospores and metabasidia.

A third type of life cycle, described in the teliospore-forming yeasts, involves a self-sporulating mechanism. In this case, from a single yeast cell a uninuclear mycelium develops with teliospores. Details of the life cycles of the Filobasidiaceae are discussed by Kwon-Chung and Fell (1984a, b) and of the teliospore-forming yeasts by Fell and Statzell Tallman (1984a, b, c).

A new basidiosporogenous genus, *Sterigmatosporidium*, described by Kraepelin and Schulze (1982) differs from the Filobasidiaceae and the teliospore-forming yeasts. It is discussed on p. 46.

3. Physiological and biochemical criteria

The principal physiological criteria used in the classification of yeasts are fermentation and assimilation. The standard description of species includes the capacity to ferment certain sugars tested in Durham tubes for a fixed period of time. Characterization of the fermentation as vigorous, good, slow or weak depends on the rate of fermentation, and is observed as the quantity of gas in the insert tube checked every two days over a period of two weeks. The usual sugars tested are glucose, galactose, sucrose, maltose, lactose and raffinose. Additional sugars, such as trehalose and melibiose and the polysaccharides inulin and starch, may also be tried. In the fermentation of the trisaccharide raffinose different enzymes are involved. Complete fermentation of this sugar may depend either on the presence of invertase and melibiase and α-glucosidase.

The value of the assimilation of carbon compounds as a character for the distinction of yeast species has been emphasized by Wickerham (1951) and Wickerham and Burton (1948). In the standard description of the species, the result of tests of 18–30 carbon compounds is recorded. These compounds have been chosen for their differentiating value, although detailed knowledge about the biochemical background of their utilization is often lacking. Barnett (1968) has given a survey of biochemical data, as far as is known, about the nutritional tests.

There are fewer carbon compounds used in the standard description given in the third edition of "The Yeasts, a Taxonomic Study" (Kreger-van Rij, ed., 1984a) than in the second edition (Lodder, ed., 1970). Compounds with a low differentiating value, and those that are often weakly and slowly assimilated and may give inconsistent results, have been omitted. The list of carbon compounds in most standard descriptions include galactose, sucrose, maltose, cellobiose, trehalose, lactose, raffinose, soluble starch, D-xylose, L-arabinose, D-ribose, L-rhamnose, erythritol, ribitol, D-mannitol, succinic acid, citric acid and inositol. However, in some genera, extra compounds are required for differentiation of the species or as useful data for characterization purposes. These are: L-sorbose, melibiose, melezitose, inulin, D-arabinose, methanol, ethanol, glycerol, galactitol (dulcitol), D-glucitol (sorbitol), DL-lactic acid, (−)-malic acid, glucuronic acid, 2-ketogluconate, 5-ketogluconate, α-methyl-D-glucoside, arbutin, salicin, glucono-δ-lactone, D-glucosamine, decane and hexadecane.

Barnett (1968) has shown that enzymic links exist between the utilization of some compounds. For instance, a single enzyme is probably responsible for hydrolysing the β-glucosides cellobiose, arbutin and salicin, and the splitting and assimilation of these compounds represents one characteristic instead of three.

The method for testing assimilating ability is of great importance since, by the presence or absence of the opportunity for adaptation, different results may be obtained. As the standard method, that prescribed by Wickerham (1951) has been accepted. It includes tests in liquid medium in stationary cultures for three weeks. By the use of shaken cultures (Ahearn *et al.*, 1960), the test period may be shortened to one week.

To obtain quick preliminary results, it is also possible to use auxanographic tests which take two to three days, and for which positive results agree with those obtained using the liquid medium test. If necessary, the compounds which have not been assimilated by this method may be tested afterwards using liquid medium. Assimilation of β-glucosides may be quickly and conveniently examined by growing the yeast on a solid medium containing arbutin, splitting of which gives a brown colour in the presence of iron salts. Assimilation of nitrogenous compounds is tested in the same way as for carbon compounds, namely in liquid medium. Potassium nitrate is the most important test compound. Others which may be tried are sodium nitrite, ethylamine, cadaverine, L-lysine, creatinine and creatine. Utilization of nitrogenous compounds may also be tested with the auxanographic method, which often gives the same results as the liquid medium test.

Apart from these fermentation and assimilation tests, the standard description includes a number of other criteria of general importance and others which are only useful for the distinction of species in certain groups or genera. The standard description of all species includes: growth in a vitamin-free medium, growth at 37°C, and growth on 50% (w/w) glucose–yeast extract agar or in 10% sodium chloride plus 5% glucose in yeast nitrogen base (Wickerham, 1951). Relatively few yeast species can grow in a vitamin-free medium and this criterion may therefore be useful for the identification of a yeast strain. For this test, and also for establishing special vitamin requirements, which are occasionally mentioned in the standard description, the conditions have to be strictly standardized if the results are to be used for comparison. Growth at 37°C or at any other specific temperature, and especially the maximum temperature of growth, may be a valuable species characteristic (Vidal-Leiria *et al.*, 1979).

Two tests used in the standard description are of importance for the distinction of basidiomycetous yeasts, the Diazonium blue B (DBB) colour test and the urease test. In the DBB test (van der Walt and Hopsu-Havu, 1976; Hagler and Ahearn, 1981), the cells grown under special conditions give a dark-red colour with the dye. The mechanism of the reaction is not known. Lachance (1982) found that an ascomycetous yeast, *Sporopachydermia quercuum*, also gives a positive DBB reaction. So far, it is the only exception. The urease test (Seeliger, 1956) on solid medium (Christensen, 1946) or in liquid medium (Roberts *et al.*, 1978), is not exclusive for basidiomycetous yeasts; a few ascomycetous yeasts are also positive (Hagler and Ahearn, 1981). However, the test has the advantage of easy performance.

Two characteristics of taxonomic importance, the G+C content of DNA and the coenzyme-Q system (Co-Q), are given in the standard description of some, but not all, species. These features are not routinely established, and the data on the Co-Q structure cited in Tables 1, 2 and 3 all come from Yamada and coworkers (Yamada and Kondo, 1972; Yamada *et al.*, 1973a, b, c, 1976a, b, 1977, 1980; Yamada and Banno, 1984a). The G+C content, or the molar percentage of guanine plus cytosine of DNA, is discussed in Chapter 3. The Co-Q or ubiquinone system of the yeasts varies in the number of isoprene units per molecule. Coenzyme-Q is an enzyme of the electron transport system. Other criteria of the standard description of secondary importance are, for instance, production of starch-like compounds, acid production, splitting of fat, cycloheximide resistance and production of esters.

C. Other criteria

Some characteristics of yeasts not mentioned in the standard description are important for taxonomy. To these belong the chemical composition of the cell wall, DNA base sequence homology and enzyme patterns. Research on the chemical composition of the cell wall and of extracellular compounds has been reviewed by Phaff (1971, 1977, 1984a). The following data are chiefly taken from the last publication. Glucan, mannan and chitin are components of the yeast cell wall. As far as is known, in ascomycetous yeasts β-glucan and chitin occur, and in basidiomycetous yeasts β-glucan, chitin, and in some cases α-glucan. The last compound was found in *Cryptococcus* species and also in the ascomycete *Schizosaccharomyces* (Bacon *et al.*, 1968). Bastide *et al.* (1979) demonstrated the difference in the composition of the cell wall between ascomycetous and basidiomycetous yeasts by treatment of the cells with different enzymes, the glucanases, in order to obtain sphaeroplasts, i.e. protoplasts liberated from the cell wall. For ascomycetous yeasts, 1,3-β-glucanase was required, and for some basidiomycetous yeasts, 1,3-β-glucanase and 1,4-α-glucanase.

Comparative studies of yeast mannans of different yeast species were made by Kocourek and Ballou (1969) and by Gorin and Spencer (1970). Kocourek and Ballou used a method involving acetolysis and analysed the resulting oligosaccharides. Gorin and Spencer made proton magnetic resonance spectra of isolated mannans. The results of these studies and their use for yeast taxonomy are discussed by Phaff (1984a).

Extracellular compounds, either present in the capsule or excreted into the medium, include polysaccharides and lipids. Phosphomannans of various compositions are produced by *Hansenula* and *Pichia* species; *Pichia mucosa* produces a mannan (Seymour *et al.*, 1976).

Slodki *et al.* (1966) found similar extracellular heteropolysaccharides containing D-glucuronic acid, D-xylose, D-mannose and *O*-acetyl groups in capsules of *Cryptococcus* and *Tremella* species, and suggested a relationship on that basis. Von Arx and Weijman (1979) extended this investigation to include many yeast species, and used whole cells for analysis. Xylose appeared to occur in the cells of a group of basidiomycetous yeasts which the authors indicated as Filobasidiaceae, and to be absent from another group of basidiomycetous yeasts, the Sporobolomycetaceae, and from ascomycetous yeasts.

Production of extracellular lipids by yeasts, reviewed by Stodola *et al.* (1967), has not yet been used in yeast systematics, although the authors point out that it may be a useful criterion. It is typical of some *Cryptococcus* and *Rhodotorula* species (Ruinen and Deinema, 1964). Immunological studies of the cell wall have been made on yeasts of medical importance (for data see Tsuchyia *et al.*, 1974). Establishment of DNA sequence homology or relatedness between yeast strains is of prime importance for the distinction of species. These tests and their use are discussed in Chapter 3.

Enzyme patterns may be studied by the method of electrophoresis, and from comparative studies conclusions about relationship may be drawn. For instance, Baptist and Kurtzman (1976) compared six different enzymes in *Cryptococcus laurentii* and its varieties and in two *Tremella* species. On the basis of data obtained and on other features, they reclassified two strains of *Cr. laurentii* var. *magnus*. Yamazaki and Komagata (1981) compared seven enzymes in *Rhodotorula* and *Rhodosporidium* strains and distinguished several groups among them.

D. Identification of yeasts

For the identification of yeasts, keys are required. With the increase in the number of yeast species, the keys have become more complex. Key characters should be clear and unambiguous and, therefore, features from tests which give weak or slow results are less desirable. Characters that vary within a species cannot be used, but they may be suitable for other species where they do not vary.

Several keys are possible for the same group of yeast species, and in all cases the criteria which make the primary divisions are very important. A key may follow the taxonomic classification and first lead to genera, and then within each genus to the species. For ascosporogenous yeasts, this generally means that characteristics of the ascospores are used and that the yeast must first be induced to sporulate. The basidiosporogenous yeasts are most frequently isolated in the haploid phase and, if they are heterothallic, identified as imperfects. In the imperfect yeasts the easily established criteria of vegetative reproduction are often decisive for generic differentiation.

Keys may also lead directly to species, not via genera. In this case it is possible to make keys without the use of features of sexual reproduction, and even exclusively with physiological criteria. In the latter case the unambiguous nature of the tests is most important. Impurities in the compounds tested, and inaccurate test conditions, may give aberrant results leading to wrong identifications. The deviations in the descriptions by different authors of the same species show that such mistakes do occur. Examples of keys made by computers leading directly to the species, using only or mainly physiological criteria are those by Barnett and Pankhurst (1974) and Barnett *et al.* (1979, 1983). In the last mentioned publication, the number of tests involved in the keys is 55 — 57 among which vitamin requirements are frequently used. The keys given in the third edition of "The Yeasts, a Taxonomic Study"

referred for further information. Although Hansen and Klöcker were already aware of different levels of relationship between the yeasts, Guilliermond (1909a,b) was the first to explain his phylogenetic views by describing the principles. He assumed, in the fungi which produce the same type of primitive ascus, an evolutionary development from mycelial to budding species, and observed stages of this behaviour in the genera *Eremascus*, *Endomyces* (= *Endomycopsis* Stelling-Dekker = *Saccharomycopsis* Klöcker) and *Saccharomyces*. Another line connected *Eremascus*, via *Endomyces*, with *Schizosaccharomyces*. Strains of *Eremascus* form only true mycelium, those of *Saccharomycopsis* true mycelium and budding cells, and strains of *Saccharomyces* only budding cells. In the second line, strains of *Endomyces* produce true mycelium that falls apart into arthrospores, and those of *Schizosaccharomyces* form single cells by fission. From this point of view vegetative reproduction, especially either budding or fission, is of primary importance.

Wickerham (1951) and Wickerham and Burton (1962) proposed a phylogenetic theory with a different approach from that of Guilliermond. They considered the shape of the ascospores to be a more important character than budding and fission. Wickerham made a very intensive study of the genus *Hansenula*, which is characterized by the shape of the spores and by nitrate assimilation. According to his theory, which is derived from observations on *Hansenula*, the fundamental evolutionary trend is a progressive change from haploid to diploid species. This is associated with a decreased dependence on external vitamins, the ability to ferment an increased number of sugars, and the capacity to form more hyphae. The last feature is in sharp contrast with Guilliermond's theory. However, it may be possible, as a result of ultrastructural studies on hyphae, to reconcile both theories to a certain extent. These studies showed various types of septal structure (see Section III.B, p. 9); for instance, a difference was found between the septa of *Saccharomycopsis* species with plasmodesmata, and of *Hansenula* species in which they were absent. This possibly fundamental difference distinguishes both types of hyphae. In the author's opinion it is possible to adhere to Guilliermond's view that the mycelial genus *Saccharomycopsis* is primitive, and that the budding genus *Hansenula* in which true hyphae may be formed after an S-R mutation is more advanced. Another primitive feature of most *Saccharomycopsis* species is that they are haploid.

C. Higher taxa

The two families of the ascosporogenous yeasts, the Spermophthoraceae and the Saccharomycetaceae are distinguished by the shape of the ascospores, needle-shaped in the former and of different shape in the latter. It is questionable whether this difference suffices for a separation of the families. For instance, the genus *Metschnikowia* of the Spermophthoraceae is very similar to the genera of the Saccharomycetaceae (Lodder, 1970). A closer examination of the species of the other genera of the Spermophthoraceae, especially of their life cycles, may show more similarities and may perhaps lead to merging of the families.

The criteria for the subdivision of the Saccharomycetaceae into four subfamilies are mentioned in Section IV.A (p. 23). Of these, bipolar budding brings together in the Nadsonioideae, genera and species that differ markedly in the method of ascus formation and in the shape of the ascospores. In view of this, the value of bipolar budding for distinguishing taxa at the level of subfamilies may be questioned. Moreover, budding from the same site, as in bipolar budding, is not confined to the Nadsonioideae but is also found, for example, in *Arthroascus javanensis* and *Ambrosiozyma cicatricosa*.

The teliospore-forming yeasts and the Filobasidiaceae are the two groups in which the basidiosporogenous yeasts are classified. Both have unique features which makes it difficult to indicate their place in the Basidiomycotina. Fell (1984) classified the teliospore-forming yeasts in the order Ustilaginales, but does not indicate a family; Kwon-Chung (1984) accepts the family of the Filobasidiaceae for the genera *Filobasidium* and *Filobasidiella*. Jülich (1981) classifies the Filobasidiaceae in a new order Filobasidiales. The new genus *Sterigmatosporidium* has not yet been classified in a higher taxon.

The imperfect yeasts are classified into two families, the Sporobolomycetaceae and the Cryptococcaceae, distinguished by the formation of ballistospores in the former. This classification is faulty. For instance, the species of the two genera of the Sporobolomycetaceae may be related to quite different perfect fungi: *Sporobolomyces* species to the teliospore-forming *Sporidiobolus* species, and *Bullera* species to certain Sirobasidiaceae of the Tremellales. The Cryptococcaceae are very heterogeneous and include all other imperfect yeasts, both ascomycetous and basidiomycetous. Although it is now possible to recognize imperfect yeasts as ascomycetous or basidiomycetous, and to establish by DNA-homology tests the relationship between perfect and imperfect yeasts, the classification of the imperfect yeasts in a system comparable to the perfect genera is still very difficult or impossible because of the paucity of data. Attempts at reclassifications, as mentioned on p. 52, may therefore lead to provisional changes, with the disadvantages of temporary changes of names.

IV. CLASSIFICATION

In the following section a survey of the classification of yeasts is given, with the emphasis on a more detailed discussion of the genera. Three groups are recognized in the yeasts: A, the ascosporogenous yeasts; B, the basidiosporogenous yeasts; and C, the imperfect yeasts.

A. Ascosporogenous yeasts

The ascosporogenous yeasts are classified in the Hemiascomycetes, which are Ascomycotina that lack ascocarps (fruiting bodies) and ascogenous hyphae (Ainsworth, 1973). In the class of the Hemiascomycetes, yeasts are classified in two families, the Saccharomycetaceae and the Spermophthoraceae of the order Endomycetales. The two families are distinguished by the shape of the ascospores: in the Spermophthoraceae the spores are needle-shaped, in the Saccharomycetaceae they have a different shape. The Saccharomycetaceae are distinguished from two other families in the Endomycetales, the Ascoideaceae, and the Endomycetaceae, by the presence of single budding cells (Kreger-van Rij, 1973).

The Saccharomycetaceae comprise four subfamilies (Table I). (1) Schizosaccharomycetoideae, in which vegetative reproduction is exclusively by fission. (2) Saccharomycetoideae, which include all ascosporogenous genera which show multilateral budding and in which the ascus is not formed in the peculiar way of the Lipomycetoideae, which contains only the genus *Lipomyces* (see Section IV.A 3, p. 38). (3) Lipomycetoideae, with the typical ascus and ascospores mentioned above. (4) Nadsonioideae, in which budding is bipolar.

1. Schizosaccharomycetoideae

The single genus of the Schizosaccharomycetoideae, *Schizosaccharomyces*, is characterized by vegetative reproduction exclusively by fission (see Fig. 1e), either of single cells or of true mycelium splitting up into arthrospores. The absence of budding distinguishes the genus from all other yeasts. The cell wall of *Schizosaccharomyces* is also exceptional in that chitin is lacking (Kreger, 1954). The wall of older cells shows multiple scars (Streiblová and Wolf, 1972). Ascus formation is preceded by somatogamous conjugation and four or eight spores are present in the ascus. The spores are round, oval (see Fig. 3a) or reniform; the ascospores of *Schiz. malidevorans* are warty. The spores are released from the ascus, and most give a positive starch test with iodine solution. Fermentation is vigorous. The four *Schizosaccharomyces* species require inositol for growth and have a maximum temperature for growth of 40–42°C (Slooff, 1970b).

The two species with four-spored asci, *Schiz. pombe* and *Schiz. malidevorans*, have a Co-Q-10 system, and *Schiz. octosporus* and *Schiz. japonicus* with eight spores per ascus, have a Co-Q-9 system (Yamada *et al.*, 1973a).

Table I Characteristics of the yeast genera in the families Saccharomycetaceae and Spermophthoraceae

Saccharomycetaceae

1. Schizosaccharomycetoidea

Schizosaccharomyces

Fission cells (true mycelium and arthrospores).
Ascospores round, oval or reniform, smooth or warty, 4–8 per ascus, liberated.
Fermentation positive.
Coenzyme Q-9 and Q-10.

2. Saccharomycetoideae

Ambrosiozyma

Budding cells, true mycelium, hyphal septa with a dolipore.
Asci on the hyphae; ascospores hat-shaped.
Fermentation positive.
Nitrate assimilation positive or negative.
Coenzyme Q-7.

Arthroascus

Budding on a broad base, true mycelium and arthrospores, hyphal septa with a closure line.
Asci generally spindle-shaped; ascospores hat- or Saturn-shaped, smooth or warty, liberated.
Fermentation negative.
Coenzyme Q-8.

Arxiozyma

Budding cells.
Asci unconjugated; ascospores round or oval, warty, not liberated.
No pellicle formation.
Fermentation vigorous.
Coenzyme Q-6.

Citeromyces

Budding cells.
Ascospores round and warty, not liberated.
Fermentation positive.
Nitrate assimilated.
Coenzyme Q-8.

Clavispora

Budding cells.
Ascospores club-shaped, warty, liberated.
Fermentation positive.
Coenzyme Q-8.

Cyniclomyces

Budding cells.
Ascospores oval to cylindrical, generally not liberated.
Fermentation positive.
Growth between 30 and 40°C.
In stomach of rabbits.

Debaryomyces

Budding cells.
Conjugation between mother cell and bud; ascospores round or oval, warty or with ridges, not liberated.
Fermentation positive or negative.
Coenzyme Q-9.

Dekkera

Budding cells.
Ascospores hat-shaped, liberated.
Fermentation positive.
Nitrate assimilation positive or negative.

Table I. Continued

Short-lived on malt agar.
Coenzyme Q-9.

Guilliermondella
Budding cells, true mycelium, hyphal septa with plasmodesmata.
Asci on the hyphae; ascospores lunate, liberated.
Fermentation positive.
Coenzyme Q-8.

Hansenula
Budding cells (true mycelium).
Ascospores hat- or Saturn-shaped, or hemispherical, generally liberated.
Fermentation positive or negative.
Nitrate assimilation positive.
Coenzyme Q-7 and Q-8.

Issatchenkia
Budding cells.
Asci unconjugated; ascospores round and warty, not liberated.
Pellicle formation.
Fermentation postive.
Coenzyme Q-7.

Kluyveromyces
Budding cells.
Ascospores reniform, crescentiform, round or oval, 1–16 (or more) per ascus, liberated.
Fermentation positive.
Coenzyme Q-6.

Lodderomyces
Budding cells.
Ascospores oval, not liberated.
Fermentation positive.
Coenzyme Q-9.

Pachysolen
Budding cells.
Ascus thick-walled with a tube; the end of the tube is thin-walled and contains the spores; ascospores hat-shaped, liberated.
Fermentation positive.
Nitrate assimilation positive.
Coenzyme Q-8.

Pachytichospora
Budding cells.
Ascospores round to oval with an eccentrically thickened wall, not liberated.
Fermentation positive.

Pichia
Budding cells (true mycelium, arthrospores).
Ascospores hat- or Saturn-shaped, round or hemispherical, smooth or warty, generally liberated.
Fermentation positive or negative.
Coenzyme Q-7, Q-8 and Q-9.

Saccharomyces
Budding cells.
Asci unconjugated; ascospores round or oval, smooth, not liberated.
No pellicle formation.
Fermentation vigorous.
Coenzyme Q-6.

Saccharomycopsis
Budding cells, true mycelium (arthrospores), hyphal septa with plasmodesmata.
Asci formed on the hyphae; ascospores round, hat- or Saturn-shaped, smooth or warty, liberated or not liberated.

Table I. Continued

Fermentation positive or negative.
Coenzyme Q-8.

Schwanniomyces
Budding cells.
Conjugation between mother cell and bud; ascospores round or oval with an equatorial ledge, warty, not liberated.
Fermentation positive.
Coenzyme Q-9.

Sporopachydermia
Budding cells.
Ascospores round or oval with a thick wall, liberated.
Fermentation negative.
Inositol assimilation positive.

Stephanoascus
Budding cells, true mycelium, hyphal septa with plasmodesmata.
Asci round with a small top cell, ascospores hemispherical with a thick wall at the convex side, not liberated.
Fermentation negative.

Torulaspora
Budding cells.
Asci formed after conjugation, usually between mother cell and bud; ascospores round or oval, warty or smooth, not liberated.
Fermentation vigorous.
Coenzyme Q-6.

Wickerhamiella
Budding cells.
Ascospores oblong and rugose, liberated.
Fermentation negative.
Nitrate assimilation positive.

Wingea
Budding cells.
Ascospores lentiform, light brown, not liberated.
Fermentation positive.
Coenzyme Q-9.

Yarrowia
Budding cells, true mycelium, hyphal septa with central micropore or closure line.
Asci on the hyphae, ascospores round or hat-shaped, the ledge may be of irregular shape, spores liberated.
Fermentation negative.
Coenzyme Q-9.

Zygosaccharomyces
Budding cells.
Conjugated asci, ascospores round or oval, not liberated.
Fermentation vigorous.
Coenzyme Q-6.

3. Lipomycetoideae

Lipomyces
Budding cells, usually capsulated.
"Active buds" may conjugate or directly change into asci; ascospores oval, amber-coloured, smooth, warty or with ridges, 1–16 or more per ascus, usually liberated.
Fermentation negative.
Coenzyme Q-9.

4. Nadsonioideae

Hanseniaspora
Budding cells, bipolar on a broad base.

Table I. Continued

Ascospores round, hat-, helmet- or Saturn-shaped, smooth or warty, liberated or not liberated.
Fermentation positive.
Coenzyme Q-6.

Nadsonia
Budding cells, bipolar on a broad base.
Conjugation between mother cell and bud; ascospores round, warty, brown, not liberated.
Fermentation positive or negative.
Coenzyme Q-6.

Saccharomycodes
Budding cells, bipolar on a broad base.
Conjugation of germinating spores in the ascus; ascospores round with a narrow ledge.
Fermentation positive.
Coenzyme Q-6.

Wickerhamia
Budding cells, bipolar on a broad base.
Ascospores cap-shaped, usually one per ascus, liberated.
Fermentation positive.
Coenzyme Q-9.

Spermophthoraceae

Coccidiascus
Budding cells.
Asci large, banana-shaped; ascospores spindle-shaped, 8 per ascus, liberated.

Metschnikowia
Budding cells.
Asci large, cylindrical or clavate; ascospores needle-shaped, 1-2 per ascus, liberated or not liberated.
Fermentation positive or negative.
Coenzyme Q-9.

Nematospora
Budding cells, true mycelium.
Asci large, cylindrical; ascospores spindle-shaped, 8 per ascus, liberated.
Fermentation positive.
Coenzyme Q-6.

2. Saccharomycetoideae

The Saccharomycetoideae include 26 genera, six of which are characterized by the production of abundant true mycelium in addition to single budding cells. They are *Saccharomycopsis, Guilliermondella, Ambrosiozyma, Arthroascus, Stephanoascus* and *Yarrowia*. In contrast, the genera *Pichia* and *Hansenula* include a few species which produce true hyphae to a limited extent, with the exception of *Pichia burtonii* which forms abundant true mycelium.

Among the species with true hyphae there have been several changes in generic names since 1970. Stelling-Dekker (1931) classified many of these yeasts in a new genus, *Endomycopsis*. Van der Walt and Scott (1971) pointed out that the name of the genus should be *Saccharomycopsis* after the oldest included species, *Saccharomycopsis capsularis*. The second cause for changes of names were reclassifications as the result of ultrastructural studies of the

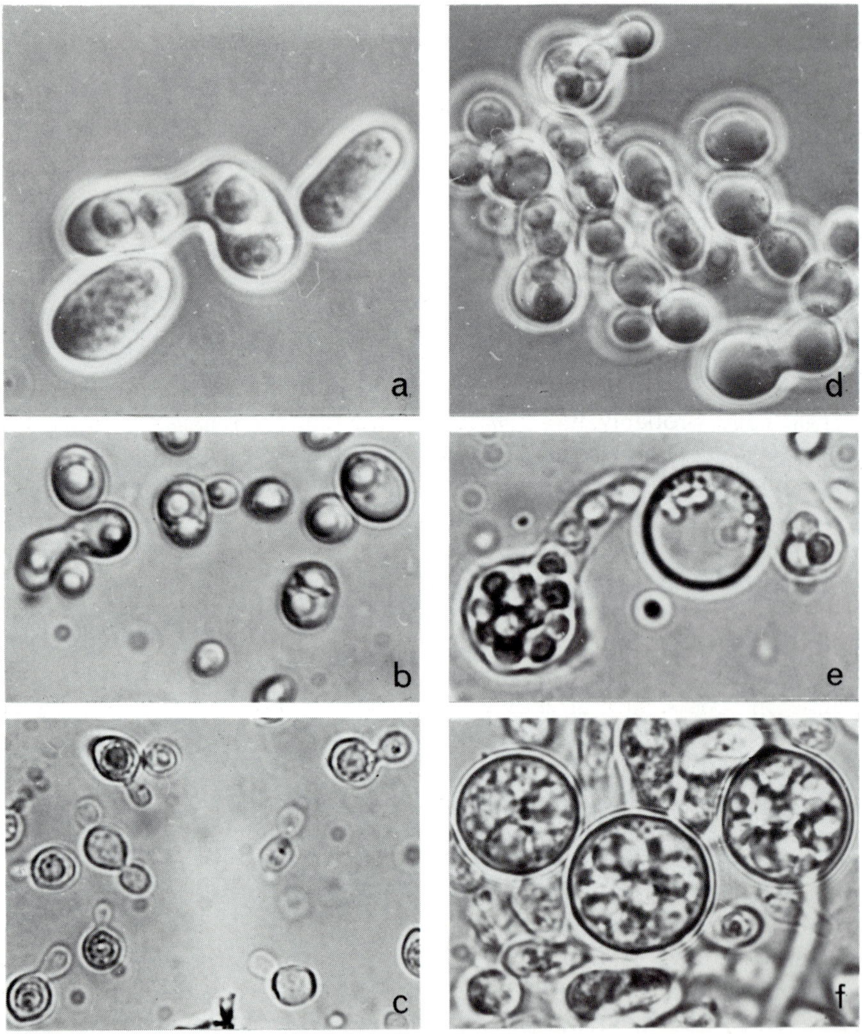

Fig. 3. Photomicrographs of asci and spores of various yeasts. (a) *Schizosaccharomyces pombe*. Ascus resulting from the fusion of two cells with four round to oval spores. Magnification × 2350. (b) *Hansenula saturnus*. Asci with Saturn-shaped spores. Magnification × 1850. (c) *Debaryomyces hansenii*. Asci with round, warty spores. Conjugation between mother cell and bud precedes ascus formation. Magnification × 1600. (d) *Zygosaccharomyces rouxii*. Round spores are present in the asci. The latter are formed after conjugation between mother cell and bud or between separate cells. Magnification × 1850. (e) *Lipomyces starkeyi*. Cell with a sac-like ascus with many spores attached to it. Magnification × 2000. (f) *Sporidiobolus ruinenii*. Large round thick-walled lipid-containing teliospores are shown in the preparation. Magnification × 1850.

hyphal septa by Kreger-van Rij and Veenhuis (1969a, 1973). These authors found various types of structures (see p. 9), which were considered to be important for generic differentiation. Other features of use in this respect are the shape of the ascospores and of the ascus, and formation of the ascus on the mycelium or not.

The six species of the genus *Saccharomycopsis* accepted by Kreger-van Rij (1984c) have plasmodesmata in the hyphal septa. One of these, *S. crataegensis* (Kurtzman and Wickerham, 1973), is heterothallic, the others are homothallic. Kreger-van Rij and Veenhuis (1975b) have shown that in *S. capsularis* fusion between hyphal cells or hyphal cells and blastospores takes place, and is followed by the formation of a new wall between the two cells. Both cells may turn into asci, or buds which later become asci may form on them. This type of conjugation also occurs in *S.fibuligera, S. vini, S. crataegensis* and, probably, in *S. synnaedendra*, but not in *S. malanga*. In the last species, two hyphal cells or blastospores fuse; buds, which turn into asci, are formed on the zygote.

The ascospores of the six species vary in shape and wall structure. All have a ledge; in *S. vini* even a second ledge is present. Some spores are warty. In *S. capsularis* the ascospores have a thick outer layer which, during germination, bursts along a crack. The ledge present in very young spores may disappear during maturation (Kreger-van Rij and Veenhuis, 1975b).

Strains of *S. fibuligera*, which ferment starch, have been isolated from starchy materials, such as chalky bread and raji, a chinese leavening agent containing mould and yeast species. *Saccharomycopsis malanga* has also been isolated from raji (Kurtzman *et al.*, 1974), *S. vini* from wine and *S. synnaedendra* from tunnels of bark beetles.

Von Arx (1972) reclassified *S. synnaedendra* in a new genus, *Botryoascus*, since he failed to find single budding cells in cultures of this species. This author (van der Walt and von Arx, 1980) transferred *S. vini* and *S. crataegensis* to the genus *Endomycopsella* Boedijn, because of the "spiny Saturn-shaped ascospores which become markedly pigmented during maturation". Unfortunately, no strains of the single species of *Endomycopsella phaeospora* described by Boedijn (1960) are available for comparison. Van der Walt and von Arx (1980) renamed *S. fibuligera* as *Endomyces fibuliger* Lindner because the ascospores of *S. fibuligera* resemble those of *Endomyces decipiens* Reess. This last species has not yet been isolated in pure culture. These reclassifications on sparse, and in yeast taxonomy quite insufficient, evidence seem to be premature and may lead to confusion.

The genus *Guilliermondella* is characterized by the shape of the ascospores which are round, oval or reniform; the wall at the two poles is thickened giving them a lunate appearance. As in most *Saccharomycopsis* species, hyphal cells fuse and are subsequently separated by a wall before they turn into asci (Kreger-van Rij and Veenhuis, 1976a). The spores are released from the ascus. The single species, *G. selenospora*, ferments glucose, and strains have been isolated from tan liquid. The hyphal septa have plasmodesmata. The genus *Guilliermondella* shows much resemblance with the genus *Saccharomycopsis* and merging it with the latter, as proposed by Stelling-Dekker (1931), might be considered.

The single species in the genus *Stephanoascus*, *Steph. ciferrii* (Smith *et al.*, 1976) has plasmodesmata in the septa. It is heterothallic and the ascus is formed after fusion of compatible hyphal cells. The ascus is spherical with a small apical cell. When the ascospores are young, they are rather flat and hat-shaped with a narrow ledge. During maturation, the wall swells at one side and the mature spores are hemispherical with a thick wall at the convex side. The two spores are not released from the ascus. Fermentation is absent from *Steph. ciferrii*. Strains have been isolated from cows and pigs, and from soil.

The genus *Ambrosiozyma* is characterized by the presence of a plugged dolipore in the hyphal septa (Kreger-van Rij and Veenhuis, 1969a) which is visible under the light microscope as a small dark dot in the middle of the septum (see Fig. 1f). The four species have hat-shaped spores. All are fermentative, and in liquid cultures produce a fragrant or fruity odour. The species have been isolated from bark beetles and tunnels of bark beetles. *Ambrosiozyma platypodis* and *A. philentoma* are very similar, but differ in the assimilation of nitrate which is positive in *A. platypodis*. Because of this, von Arx (1972) classified it in a separate genus *Hormoascus*. In the author's opinion this separation is not warranted. The species *A. monospora* and *A. cicatricosa* are also similar in both morphological and physiological characters. *Ambrosiozyma cicatricosa* is conspicuous by forming dactyloid cells, i.e. blastospores which produce a series of buds at the top and show polar elongation with multiple scars (Scott and van der Walt, 1971).

The species of the genera *Yarrowia* and *Arthroascus* have in the septum a central micropore or, when it is closed, a closure line formed by the remnants of plasmalemma. The micropore has blunt edges, in contrast with the typical ascomycete pore in higher fungi, which has a tapered edge.

The genus *Yarrowia* was described by van der Walt and von Arx (1980) for the species *Saccharomycopsis lipolytica*, the perfect state of *Candida lipolytica*. Wickerham *et al.* (1970) discovered mating types among the strains of the latter species and originally described it as *Endomycopsis lipolytica*, in which species asci are usually formed on the mycelium. The ascospores are of various shapes. Most of them have a ledge, which may be irregular. *Yarrowia lipolytica* differs from the *Saccharomycopsis* species in the Co-Q system; it has Co-Q-9 and the *Saccharomycopsis* species Co-Q-8 (Yamada *et al.*, 1976b).

In the genus *Arthroascus*, which contains only the species *Ar. javanensis*, buds are formed on a broad base and true mycelium may fall apart into arthrospores. Multiple scars may be present at the poles of the cells. Conjugation between separate or connected cells precedes ascus formation (Kreger-van Rij and Veenhuis, 1976a). One of the fused cells turns into an ascus with one to four spores and is then often swollen in the middle. The spores are round or somewhat flattened, with an equatorial or subequatorial ledge. They are smooth or warty. Strains of *Ar. javanensis* have been isolated from soil, slime-flux and rotten apples.

The genus *Pichia*, in the description given by Kurtzman (1984d) is, with 56 species, the largest ascosporogenous genus. There is a wide variation in morphological and physiological features, such as the presence or absence of pseudo- and true mycelium, the shape of the ascospores, homo- and heterothallism, ploidy, pellicle formation and fermentative ability. The genus *Pichia* differs from *Hansenula* in that the species do not assimilate nitrate. Most *Pichia* species have ascospores with a ledge, which makes them hat- (see Fig.2a) or Saturn-shaped, and form evanescent asci. The ascospores of *P. fluxuum* have a narrow ledge and a warty wall. The spores of other *Pichia* species, including those with round spores, are smooth. Certain species, previously classified as *Pichia* which form round warty spores, have been transferred to the genera *Issatchenkia* and *Debaryomyces* (see pp. 34 and 33 respectively).

Pichia species with round smooth ascospores, such as *P. carsonii* and *P. etchellsii*, are considered by some authors (van der Walt and Johannsen, 1975) to be atypical of the genus *Pichia*. However, in *P. membranaefaciens*, strains with hat-shaped ascospores and strains with round spores without a ledge occur and these strains may be mated. Therefore, it is difficult at present to separate the round-spored species from the genus *Pichia*. These *Pichia* species resemble in the shape of the spores certain *Kluyveromyces* and *Saccharomyces* species but, unlike the latter, they do not exhibit vigorous fermentation. They also differ in the Co-Q system from *Saccharomyces* and *Kluyveromyces* species; the latter have Co-Q-6, the round-spored *Pichia* species Co-Q-9.

Pichia ohmeri belongs to a group of similar haploid heterothallic species which form pseudomycelium, and some which produce true mycelium to a limited extent. However, as far as is known, the septa in the true hyphae do not have plasmodesmata, which distinguishes them from *Saccharomycopsis* species. The ascospores of *Pichia ohmeri* vary in shape: they may have a distinct ledge, or, under the light microscope, they are round without a ledge. However, the round spores appear to have a narrow ledge when viewed in sections under the electron microscope (N.J.W.Kreger-van Rij, unpublished observations). One of the *Pichia* species accepted by Kurtzman (1984d), *P. burtonii*, produces abundant true hyphae which are split up into arthrospores. The septa have no plasmodesmata. Von Arx and van der Walt (1976) had earlier reclassified this species in a new genus, *Hyphopichia*.

Most *Pichia* species may be easily distinguished by the features of the standard description. However, recently several *Pichia* strains have been isolated from cacti (Phaff *et al*., 1978b; Starmer *et al*., 1978a, b, 1979) which closely resemble *P. membranaefaciens*. They had to be distinguished from this species because of failure to mate with *P. membranaefaciens* types, and having distinct differences in G + C content. Sources of *Pichia* strains include bark beetles, *Drosophila*, exudate of trees, tanning liquid and fruit.

The genus *Hansenula* is characterized by the formation of hat- and Saturn-shaped (Fig. 3b) and hemispherical ascospores, and by the assimilation of nitrate. With the exception of the latter character, *Hansenula* species are very similar to *Pichia* species. They are also homo- or heterothallic, haploid or diploid, may or may not produce pseudomycelium (certain species form some true hyphae) and vary in fermentative ability and pellicle formation. As far as is known, true hyphae in *Hansenula* spp. have no plasmodesmata or dolipores in the septa. Cultures with many ascospores may be pink or red, just as in certain *Pichia* spp. Some species, such as *H. capsulata* and *H. holstii*, have capsulated cells, the capsule material containing mannans and phosphomannans (Gorin, 1973; Seymour *et al*., 1976); some *Hansenula* species (e.g. *H. polymorpha*) assimilate methanol.

Kurtzman (1984b) accepted 30 species in the genus. As a result of the establishment of G + C content and, especially, of DNA sequence complementarity between strains, a few changes in the species differentiation have been made in Wickerham's (1970) description of the genus. For instance, Kurtzman (1984a) found that *H. bimundalis* var. *americana* should be classified as a separate species *H. americana*, and Fuson *et al*. (1979) established 78% DNA complementarity between the strains of *H. canadensis* and *H. wingei* which made *H. wingei* a synonym of *H. canadensis*. Kurtzman *et al*. (1979) showed that *H. jadinii* was the perfect state of *Candida utilis* with 85% DNA complementarity. Finally, Kurtzman (1984b) found 68% DNA complementarity between a strain of *H. mrakii* and one of *P. sargentensis*, both species with Saturn-shaped ascospores. This discovery showed the narrow relationship between the two genera as presently defined and made readjustments necessary. Therefore, Kurtzman (1984e) transferred the hat-shaped species of *Hansenula* to *Pichia*. Following von Arx *et al*. (1977) he proposed to classify the *Hansenula* and *Pichia* species with Saturn-shaped spores in the old genus *Williopsis*.

The genus *Pachysolen* is characterized by the shape of the ascus which is a cell with a long neck. Most of the wall is very thick with the exception of the apex of the cell which includes four hat-shaped spores. The spores are easily liberated from the ascus. The single species *Pa. tannophilus* is homothallic and has been isolated from tanning liquid: it assimilates nitrate and ferments glucose.

The genus *Wickerhamiella* is another nitrate-positive genus. It has been described by van der Walt (van der Walt and Liebenberg, 1973) who found ascospores in the type strain of *Torulopsis domerqii*. The spores are oblong and rugose, and the single spore is released from the ascus, although sporulation was found to be scarce. *Wickerhamiella domerqii*, which is nonfermentative, has been isolated from a wine vat in South Africa.

In the two species of the genus *Dekkera*, the assimilation of nitrate varies. The genus is characterized by physiological features, namely strong acid production from glucose and slow growth on malt agar. On the latter medium the cells are short-lived and the cultures have a characteristic odour. The yeasts are fermentative. They produce hat-shaped ascospores and the asci are evanescent. The species are homothallic (van der Walt, 1970b). They have their imperfect state in the genus *Brettanomyces*, and have been isolated from beer and wine.

The genus *Citeromyces* is nitrate-positive and contains one species, *Cit. matritensis*. It is heterothallic (Wickerham, 1958) and has warty spores which are not released from the ascus. Haploid and diploid strains may be isolated from material with a high sugar concentration such as fruit syrup and sweetened condensed milk. *Citeromyces matritensis* ferments glucose and sucrose.

The species of the genus *Debaryomyces* have ascospores with warts (Fig. 3c), except for *Deb. marama* which has ridged spores. The species are haploid, and conjugation between mother cell and bud usually precedes ascus formation. *Debaryomyces* is nitrate-negative, but strains of certain species may assimilate nitrite.

The distinction between *Debaryomyces* and *Torulaspora*, another genus with haploid species and warty spores, lies in the structure of the spore wall (Kreger-van Rij and Veenhuis, 1975a, 1976b), in the Co-Q system (*Debaryomyces* Q-9 and *Torulaspora* Q-6) and in the fermentation (vigorous in *Torulaspora* and not vigorous in *Debaryomyces*), the last feature being important for identification. The distinction between *Debaryomyces* and the round-spored *Pichia* species lies in the presence or absence of warts on the ascospores. Kurtzman and Smiley (1974) made a study of ascospores of *Debaryomyces* and *Pichia* with the scanning electron microscope. They found that *P. pseudopolymorpha* had warty spores; this species has since been transferred to the genus *Debaryomyces* (Price and Phaff, 1979). Price *et al.* (1978) determined G + C contents and DNA base sequence relatedness in *Debaryomyces* and certain *Pichia* species. From this study, it appeared that *P.*

polymorpha was identical with *Deb. cantarellii* and *Deb. castellii*. The result was a new nomenclature combination, *Debaryomyces polymorphus* (Price and Phaff, 1978). Price *et al.* (1978) found a relatedness of 39.7–96.7% between seven species considered to be synonyms of *Deb. hansenii* (Kreger-van Rij, 1984a) and *Deb. hansenii*. The type strain of *Deb. nepalensis* showed only 15.4% relatedness with *Deb. hansenii*, and *Deb. nepalensis* and *Deb. hansenii* should, therefore, be considered as separate species. However, with the usual characteristics of the standard description it is difficult to distinguish between the two species, and Kreger-van Rij (1984a) retained *Deb. nepalensis* provisionally as a synonym of *Deb. hansenii*. Chemotaxonomic studies by Nakase and Suzuki (1985a, b) confirmed the distinction between *Deb. hansenii* and *Deb. nepalensis*, but did not provide simple criteria for differentiation either. *Debaryomyces* species can all grow on a medium with a high sugar concentration (50%, w/w, glucose) and *Deb. hansenii* strains have also been isolated from sources with a high salt concentration, such as pickled cucumbers. Other sources are cheese, sausages, tobacco, nails and skin.

The genus *Schwanniomyces* is characterized by the formation of warty ascospores with an equatorial ledge. There is usually one, seldom two, spores per ascus and they are not released. The vegetative cells are haploid and conjugation between mother cell and bud precedes ascus formation (Kreger-van Rij, 1977a). Studies on DNA base homology (Price *et al.*, 1978) of the four *Schwanniomyces* species accepted by Phaff (1970c) showed a high degree of relatedness among them and resulted in the merging of three species with *Schw. occidentalis*. The species *Schw. persoonii* was reduced to a variety of *Schw. occidentalis*. All *Schwanniomyces* strains came from soil.

Kudriavzev (1960) first described the genus *Issatchenkia* based on the species *I. orientalis*. Kurtzman *et al.* (1980) included three other species in the genus, *I. occidentalis*, *I. scutulata* (syn. *Pichia scutulata*; Phaff *et al.*, 1976) and *I. terricola* (syn. *Pichia terricola*; van der Walt, 1957). Kurtzman and Smiley (1976) discovered that *I. occidentalis* and *I. terricola* are heterothallic. These authors and Kurtzman *et al.* (1980) examined ascospores of *Issatchenkia* species, with the exception of *I. orientalis*, with scanning- and transmission-electron microscopes and found all spores to be warty.

Kurtzman *et al.* (1980) established G + C content and DNA complementarity of three of the species and of the imperfect states *Candida krusei* (of *I. orientalis*) and *C. sorbosa* (of *I. occidentalis*).

In the definition of Kurtzman (1984c) the genus is characterized by formation of warty ascospores in unconjugated asci which are persistent. Nitrate is not assimilated; fermentation is weak and slow to rapid. Pellicles are formed on liquid media. The Co-Q system is 7, which is different from *Debaryomyces* and the round-spored *Pichia* species which both have Q-9, and from *Saccharomyces* which has Q-6. In contrast with *Issatchenkia*, the

Debaryomyces and the round-spored *Pichia* species are haploid, and *Saccharomyces* species do not form pellicles. Sporulation of *I. orientalis* is poor on the usual media. The species has been isolated from fruit juices and tea beer.

The genus *Clavispora* has been described by Rodrigues de Miranda (1979) who discovered opposite mating types among the strains of *Candida lusitaniae*. The ascospores in *Cl. lusitaniae* are club-shaped, and have small warts which are visible in the electron microscope (see Fig. 2b) but not under the light microscope. The spores are released from the ascus. *Clavispora lusitaniae* has fermentative ability and grows at 37°C. It is an opportunist pathogen (Holzschu *et al.*, 1979) and has been isolated from human and animal sources, corn meal and citrus peel juice.

Sporopachydermia is another genus with heterothallic species described by Rodrigues de Miranda (1978). By mating experiments the author found the perfect states of *Cryptococcus lactativorus* and *Cr. cereanus, Sporop. lactativora* and *Sporop. cereana* respectively. The ascospores have a very thick wall (Kreger-van Rij, 1978a) (see Fig. 2c) and are released from the ascus. The two species assimilate inositol, they are non-fermentative and their cultures have a disagreeable odour. Sources of isolation are cacti, sea-water and material from a human patient.

The genus *Wingea*, with the single species *Wi. robertsii*, is characterized by the formation of lentiform, smooth, yellowish-brown ascospores. Conjugation between mother cell and bud precedes ascus formation. The ascus is persistent. *Wingea robertsii* is fermentative and has been isolated from larval feed of *Xylocopa caffra* in South Africa.

In the genus *Lodderomyces* the ascospores are oblong with obtuse ends. One, seldom two, is formed per ascus; they are not released. The vegetative cells are diploid and the single species, *Lod. elongisporus*, is homothallic. It ferments sugars, can assimilate alkanes and has been isolated from orange juice.

Cyniclomyces guttulatus (syn. *Saccharomycopsis guttulata*) is the single species of this genus. Its habitat is the stomach of the rabbit. It has large cylindrical budding cells and forms oval to cylindrical ascospores without immediately preceding conjugation. The species has outstanding physiological properties: it grows at temperatures between 35 and 40°C in a pH range of 2.0–6.5, and requires amino acids for growth (Phaff, 1984b). Richle and Scholer (1961) found that good growth was possible on a solid medium provided the carbon dioxide concentration of the atmosphere was increased to about 10% (v/v). The yeast grows well at pH 4.5 in a liquid medium containing 2% glucose, 1% proteose peptone and vitamins (Phaff, 1970b). Glucose and sucrose are fermented.

The genus *Pachytichospora* has been described by van der Walt (1978) for the species originally named *Saccharomyces transvaalensis* (van der Walt,

1956). It has typical ascospores which are oval with an eccentrically thickened wall. One or two spores are formed in each ascus, which is persistent. The vegetative cells are diploid and the species is heterothallic. *Pachytichospora transvaalensis* has fermentative ability and has been isolated from soil.

The genus *Kluyveromyces* is characterized by the ascospores which are round, oval, crescentiform or lunate, and are released from the ascus. The last feature distinguishes round-spored *Kluyveromyces* species from *Saccharomyces* and *Zygosaccharomyces* species. The internal ultrastructure of the spore wall is similar in most *Kluyveromyces* species, but a few have aberrant spore walls, namely *K. waltii, K. thermotolerans* (syn. *K. veronae*) (Kreger-van Rij, 1979) and *K. africanus* (Kreger-van Rij, 1980). The number of spores per ascus is one to four in most species, but up to 16 and more in *K. polysporus* and *K. africanus*. The species are homo- or heterothallic. Fermentation is vigorous. Several species produce a red pigment closely related to pulcherrimin (Wickerham and Burton, 1956) which may diffuse into agar medium.

Species differentiation within the genus *Kluyveromyces* has been studied by establishment of G + C content and by DNA homology tests (Martini, 1973; Martini *et al.*, 1972; Phaff *et al.*, 1978a) and by mass matings of auxotrophic mutant strains (Johannsen and van der Walt, 1978). As a result of the latter, Johannsen (1980) reduced the number of *Kluyveromyces* species accepted by van der Walt (1970c) by nine. These nine species are now partly classified as varieties of the species *K. marxianus* or are synonyms of one of the varieties. For instance, *K. fragilis*, a lactose-fermenting yeast is a synonym of *K. marxianus* var. *marxianus*. The imperfect state of this variety is *Candida pseudotropicalis*, strains of which have been isolated from milk products, and also from human sources.

The four genera *Saccharomyces, Arxiozyma, Zygosaccharomyces* and *Torulaspora* have much in common: round to short-oval ascospores (Fig. 3d) which are not released from the ascus, a vigorous fermentation, and Co-Q system 6. *Saccharomyces* is distinguished by having diploid vegetative cells. The species have smooth ascospores. Van der Walt and Yarrow (1984b) have recently reclassified *Saccharomyces telluris*, the only *Saccharomyces* species with warty spores, in a new genus *Arxiozyma. Zygosaccharomyces* and *Torulaspora* species both have haploid vegetative cells. The difference between the two genera is not distinct. In *Zygosaccharomyces* spp. conjugation between two different cells, seldom between mother cell and bud, precedes ascus formation. The ascospores are smooth, but *Zygosacch. rouxii* may have warty spores. In *Torulaspora* species conjugation between mother cell and bud precedes ascus formation, or vegetative cells with a protuberance turn into asci without apparent conjugation. The ascospores are usually warty, seldom smooth.

Lodder and Kreger-van Rij (1952) and van der Walt (1970d) had combined the species of the three genera in the genus *Saccharomyces*, but van der Walt and Johannsen (1975), von Arx *et al.* (1977) and Barnett *et al.* (1979) have since separated the haploid species from *Saccharomyces* and classified them in *Zygosaccharomyces* and *Torulaspora*. Since 1970, the number of species in each genus has notably diminished. This is the result of a new concept of the *Saccharomyces* species. As early as 1939 Winge and Laustsen showed that certain *Saccharomyces* species produced fertile hybrids, but this discovery did not lead to merging of the species. Several of them differed in a single enzyme, and by mutation one species might change into another. For instance, Scheda and Yarrow (1966) obtained mutants of *Saccharomyces* strains which had acquired the ability to ferment galactose, sucrose or maltose. The study of DNA base homology among the *Saccharomyces* species by Bicknell and Douglas (1970), and especially the more extensive study by Price *et al.* (1978), have shown relatedness in this respect of many *Saccharomyces, Zygosaccharomyces* and *Torulaspora* species. This has resulted in the combination of many of them into a lower number of species.

The genus *Saccharomyces* includes seven species (Barnett *et al.*, 1979). Sixteen of those accepted by van der Walt (1970d) have been reduced to synonymy with *Sacch. cerevisiae*. *Saccharomyces kluyveri* may occur in various degrees of ploidy (Wickerham, 1958). It is homothallic as well as heterothallic, and cells of opposite mating type agglutinate when mixed. *Arxiozyma telluris*, the perfect state of *Candida pintolopesii*, resembles *Issatchenkia* species in the formation of warty ascospores, diploid vegetative cells and in physiological features. However, there is a difference in the Co-Q system, which in *Arxiozyma telluris* is Q-6 and in *Issatchenkia* Q-7. *Saccharomyces* and *Arxiozyma* species differ from *Issatchenkia* in the absence of pellicle formation on liquid media.

Species of the genera *Zygosaccharomyces* and *Torulaspora* are all osmotolerant and can grow on 50% (w/w) glucose–yeast extract agar, some on 60% (w/w). The genus *Zygosaccharomyces* was originally described by Barker (1901) for a species which he did not name. Saccardo and Sydow (1902) named it *Zygosacch. barkeri* and this species is the type for the genus. From Barker's description von Arx *et al.* (1977) concluded that it is synonymous with *Zygosacch. rouxii*, although no type material of *Zygosacch. barkeri* was available for comparison.

From a study of the ultrastructure of the ascospore wall (Kreger-van Rij, 1979), three types appear to occur among the *Zygosaccharomyces* species accepted by Barnett *et al.* (1979), *Zygosacch. rouxii* with warty spores of the type of *Debaryomyces*, which is different from *Torulaspora* ascospores, *Zygosacch. bailii* with a spore wall composed of a single electron-light inner layer and a thin dark outer layer, and *Zygosacch. florentinus* spores with a double light inner layer and a rather thick dark outer layer, the same structure as found in *Sacch. cerevisiae* (Kreger-van Rij, 1978b; see Fig. 2d). These observations seem to point to the heterogeneity of the genus *Zygosaccharomyces*. The species *Zygosacch. bailii* and *Zygosacch. rouxii* are both important contaminant yeasts. *Zygosaccharomyces bailii* is resistant to several preservatives and may occur in products containing acetic acid (Kurtzman *et al.*, 1971). *Zygosaccharomyces rouxii* can grow in media containing 70% (w/w) glucose.

In the genus *Torulaspora* three species are accepted by Barnett *et al.* (1979). *Torulaspora delbrueckii* includes the former *Saccharomyces* species *Sacch. rosei, Sacch. fermentati, Sacch. inconspicuus, Sacch. saitoanus* and *Sacch. vafer*. As a result of this merging, assimilation and fermentation reactions with galactose, sucrose and maltose vary in the description of *T'spora delbrueckii. Torulaspora globosa* was originally described by Klöcker (1909) as *Debaryomyces globosus*. It is homo- and heterothallic. The ascospores of *T'spora pretoriensis* are round, but elongate spores may also occur. Kreger-van Rij and Veenhuis (1976b) studied the ultrastructure of *Torulaspora* ascospores and observed in sections a double light inner layer and a rather thick dark outer layer in the wall. The layering resembled that of *Sacch. cerevisiae* (Kreger-van Rij, 1978b), but the outer zone of the light layer formed warts.

3. Lipomycetoideae

The genus *Lipomyces* of the Lipomycetoideae is characterized by a typical process of ascus formation (Slooff, 1970a): (i) an active bud on a cell may change into an ascus (see Fig. 3e), (ii) an active bud and a protuberance on another cell may conjugate and the cell may turn into an ascus, (iii) the mother cell, after having produced an ascus through an active bud, may itself change into an ascus. The ascospores are amber-coloured and 2-30 or more are formed per ascus, the spores may or may not be released. They are smooth, with irregular folds (Smith and Batenburg-van der Vegte, 1984), with lengthwise ridges or, in *L. anomalus*, surrounded by an exosporium-like membrane (Babjeva and Gorin, 1975). Starkey (1946), who first described *Lipomyces* yeasts, used a sporulation medium without a nitrogen source. The cells of most species are capsulated and a starch reaction after growth in a special medium is given by all strains, although sometimes weak. Slodki and Wickerham (1966) showed galactose in the constituents of the extracellular polysaccharides in *L. starkey*, but not in *L. lipofer*, and used this feature to differentiate the species. Nieuwdorp *et al.* (1974) also found galactose in *L. kononenkoae* and *L. tetrasporus* cells. Phaff and Kurtzman (1984) accepted five *Lipomyces* species; all of them had been isolated from soil.

4. Nadsonioideae

Nadsonia is the first genus described in this chapter in which budding is generally, but not exclusively, bipolar (see Fig. 1a). Ascus formation is preceded by conjugation between mother cell and bud. The ascospores are formed in the mother cell or in the bud, or the contents of the zygote move into a second bud at the opposite side of the first, and here meiosis takes place and the spores are formed. The spores are round, brown and warty. One, seldom two, is formed per ascus; they are not liberated, but the bud containing the spores may be detached from the mother cell. Of the three *Nadsonia* species accepted by Miller and Phaff (1984), two are fermentative; the third, *N. commutata*, described by Golubev (1973) is non-fermenting. The strains of the species have been isolated from soil and slime flux in the U.S.S.R. and Japan. They do not grow at 30°C; the maximum temperature of growth for *N. commutata* being 24°C and for *N. elongata* 26°C (Miller and Phaff, 1984). The Co-Q system of *N. elongata* and *N. fulvescens* is Q-6 (Yamada *et al.*, 1976a).

The second genus with bipolar budding is *Hanseniaspora*. Meyer *et al.* (1978), after a study of DNA composition and DNA sequence relatedness, accepted six species. They are physiologically similar, all fermenting, assimilating few carbon compounds, among which are β-glucosides, and requiring inositol for growth (Phaff, 1970a). However, they show marked differences in the shape of the ascospores. *Hanseniaspora guilliermondii* has four and *H'spora valbyensis* two hat-shaped spores per ascus which are released, *H'spora uvarum* and *H'spora occidentalis* (Smith, 1974) have round spores with an equatorial or subequatorial ledge, mostly one per ascus, not liberated, warty in the former species and smooth in the latter, and *H'spora osmophila* and *H'spora vineae* have round warty spores, generally one per ascus and not released. Meyer *et al.* (1978) found that *H'spora osmophila* and *H'spora vineae* strains both had a G + C content of 40.2–40.5% and showed 38–60% DNA relatedness, which suggests that *H'spora vineae* is a synonym or a variety of *H'spora osmophila* rather than a separate species. Miller and Phaff (1958) postulated a diploidization during germination of a single spore based on nuclear stainings of the spores of *H'spora valbyensis* and *H'spora guilliermondii*. Ultrastructural studies of the ascospores have been made by Kreger-van Rij and Ahearn (1968) and Kreger-van Rij (1977b).

The genus *Saccharomycodes*, with the single species *S'codes ludwigii*, has round ascospores with a narrow ledge, usually four in each ascus. On germination, they conjugate in pairs while still in the ascus and thus give diploid vegetative cells. *Saccharomycodes ludwigii* ferments glucose and sucrose. Strains have been isolated from grape must, wines and exudate of an oak.

The genus *Wickerhamia*, with the single species *W. fluorescens* (Soneda, 1959), has cap-shaped ascospores, i.e. the ledge at one side is much wider than at the other side of the spore.There is generally one spore per ascus which splits along an equatorial plane, releasing the spore. Rooney and Moens (1973a,b) showed in an ultrastructural study of spore formation that only one lobe of the four-lobed meiosis nucleus is enveloped by the pro-spore wall. The single strain of *W. fluorescens* has been isolated from faeces of a wild squirrel. In contrast with the other three genera with bipolar budding which contain Co-Q system 6, *W. fluorescens* has Q-9 (Yamada *et al.*, 1976a).

The three genera of the Spermophthoraceae are characterized by needle-shaped ascospores. The ascus is larger than the vegetative cells. The single species of the first genus *Coccidiascus, Coccidiascus legeri*, occurs as a parasite in the intestinal cells of *Drosophila* species where it produces eight fusiform ascospores which are intertwined in a helix. The organism has not been obtained in pure culture, but Lushbaugh *et al.* (1976) deposited material containing *Coccidiascus legeri* in the New York Botanical Garden Cryptogamic Herbarium.

Nematospora coryli is the single species in the second genus. It occurs as a parasite in fruits such as tomatoes and hazelnuts, and is transmitted by certain plant-feeding insects. The eight spores are easily liberated from the ascus. They are pointed at both ends and one end is very thinly drawn, which gives the impression of a whip-like appendage (Kreger-van Rij, 1969a). *Nematospora coryli* has fermentative ability; it forms true mycelium with a few septa and pseudomycelium.

In the third genus, *Metschnikowia*, the ascospores are needle-shaped and attenuate at one or both ends. One or two spores are formed per ascus. Two lying closely together in the large ascus may be mistaken for one. Metschnikoff (1884) described the first species, *Monospora bicuspidata*, present as a parasite in *Daphnia magna.* A second species, parasitic in the brine shrimp, *Artemia salina*, was described by Kamienski (1899) who also named the genus *Metschnikowia.* Neither species had been isolated. Van Uden and Castelo-Branco (1961) were the first to obtain pure cultures of similar organisms from sea-water, fish intestines and kelp. For a historical survey of the name of the genus see Miller and van Uden (1970). Wickerham (1964a) discovered mating types in strains of *Candida pulcherrima* and *C. reukaufii*, and Pitt and Miller (1968) first found asci with needle-shaped ascospores in the species, which were described as *M. pulcherrima* and *M. reukaufii. Metschnikowia pulcherrima* has been isolated from fruits and flowers. This species forms big oil-containing chlamydospores, the pulcherrima cells; they may turn into asci. Several strains produce pulcherrimin, a red pigment which diffuses into the medium. *Metschnikowia reukaufii* closely resembles *M. pulcherrima*, but forms oval to cylindrical chlamydospores which may become asci. Strains of this species frequently occur in nectar of flowers. One of the terrestrial *Metschnikowia* species, *M. lunata* (Golubev, 1977), is homothallic.

B. Basidiosporogenous yeasts

These yeasts are a haploid phase in the life cycle of mostly heterothallic basidiomycetes; they originate, for instance, from germinated basidiospores. The yeasts are classified into three groups, (i) the teliospore-forming yeasts, (ii) the Filobasidiaceae with a yeast phase, and (iii) the genus *Sterigmatosporidium* (see Table II).

Table II. Characteristics of the genera of the basidiosporogenous yeasts

Teliospore-forming yeasts

Leucosporidium

Budding cells, true mycelium with and without clamp connections. Hyaline cultures. No ballisto spores.
Teliospores, metabasidia with basidiospores.
Fermentation positive or negative.
Coenzyme Q-8, Q-9 and Q-10.

Rhodosporidium

Budding cells, true mycelium with and without clamp connections. Yellow, orange or red cultures due to carotenoid pigments. No ballistospores.
Teliospores, metabasidia with basidiospores.
Fermentation negative.
Coenzyme Q-8, Q-9 and Q-10.

Sporidiobolus

Budding cells, true mycelium with clamp connections. Pink or red cultures due to carotenoid pigments. Ballistospores.
Teliospores, metabasidia with basidiospores.
Fermentation negative.
Coenzyme Q-10.

Filobasidiaceae

Filobasidiella

Budding cells, true mycelium with clamp connections. Hyaline cultures.
Basidiospores in four chains on the apex of the basidium.
Fermentation negative.
Coenzyme Q-10.

Filobasidium

Budding cells, true mycelium with clamp connections. Hyaline cultures.
Basidiospores sessile in a whorl on the apex of the basidium.
Fermentation positive or negative.
Coenzyme Q-9 and Q-10.

Unclassified

Sterigmatosporidium

Budding cells, formed on sterigmata and separated at the distal end of the stalk. Pseudomycelium may be formed. True mycelium with clamp connections. Hyaline cultures. No ballistospores.
Two types of chlamydospores are formed on the mycelium, round and long-oval to cylindrical ones. The latter germinate and haploid spores are formed terminally by budding.
Fermentation negative.
Nitrate assimilation negative.
Coenzyme Q-10.

For most basidiomycetous yeasts the complete life cycle is still unknown. However, it is possible to recognize these yeasts as basidiomycetous by certain characteristics. Two of the most important ones are the ultrastructure of the cell wall and of bud formation, and nuclear behaviour during mitosis (McCully and Robinow, 1972a, b). Other features of basidiomycetous yeasts are nòt uniformly present or they occur also in a few ascomycetous yeasts; examples are ballistospore formation, carotenoid pigments, high G + C content, production of deoxyribonuclease and urease, and the presence of α-glucan in the cell wall. A colour reaction with Diazonium blue B (see p.16) of the culture grown under special conditions is given by basidiomycetous, but not by ascomycetous yeasts, with a single exception.

In addition to the basidiomycetes of the three groups mentioned above, basidiomycetous yeasts may be the haploid stage of Tremellaceae and Sirobasidiaceae of the Tremellales. Compatible cells of the yeasts in these families may conjugate and produce a dikaryotic mycelium on which a basidiocarp, basidia and basidiospores develop (Bandoni, 1984). Rodrigues de Miranda and Bandoni (1984) have described the yeast state of eight *Tremella* species, of *Holtermannia corniformis, Sirobasidium magnum* and *Fibulobasidium inconspicuum*, all isolated from basidiospores by R.J. Bandoni. *Sirobasidium magnum* and *Fib. inconspicuum* strains produce ballistospores and their description agrees with the diagnosis of the genus *Bullera*. The inositol-positive species of *Tremella* agree with *Cryptococcus*, the inositol-negative species with *Candida*. The strain of *Holt. corniformis* is inositol-positive, and in this respect agrees with *Cryptococcus* sp. Although true and pseudomycelium are usually absent, a few septate hyphae may be formed.

1. Teliospore-forming yeasts

Yeasts of this type were first discovered by Nyland (1949) who described *Sporidiobolus johnsonii*, and by Banno (1967) who found a sexual cycle in strains of *Rhodotorula glutinis*. Later, Fell and coworkers (Fell *et al.*, 1969, 1973; Fell, 1970b; Fell and Statzell Tallman, 1981) have made important contributions to the elucidation of life cycles in several species of teliospore-forming yeasts. Fell and Statzell Tallman (1984a, b, c) have published a comprehensive survey of life cycles and classification of these yeasts.

A heterothallic life cycle begins with the fusion of yeast cells of compatible mating type, which results in the formation of a dikaryotic mycelium with clamp connections. Mating tests are performed by thoroughly mixing loopfuls of cells of two yeast strains on a suitable medium, for example corn-meal agar. After some time hyphae may be observed growing out from the edge of the colony, generally within the agar. Thick-walled cells, teliospores (see

Fig. 3f), develop on the hyphae. They are terminal, lateral or intercalary and of various shapes, for example round, oval or angular, in some species of a dark colour. In the teliospores nuclear fusion occurs. The teliospores germinate and form a metabasidium in which meiosis takes place. The shape of the metabasidium varies among the species; it may be one-celled, a holobasidium, or two to four-celled, a phragmometabasidium. Basidiospores are formed on the metabasidium, terminally on the holobasidium, and terminally or laterally on the phragmometabasidium. In some species primary basidiospores are formed, the sporidia, on which secondary basidiospores develop. The basidiospores reproduce by budding and thus the haploid yeast phase is re-established. Germination of the teliospores requires a special technique of soaking the spores in distilled water for some time and transferring to a solid medium, where they germinate.

Fell (1974, 1984) described three types of mating systems in the teliospore-forming yeasts: a bipolar mating system with two alleles (*A1* and *A2*); a bipolar multiallelic system (*A1, A2, A3* etc.) with the compatible combinations *A1A2, A1A3, A2A3* etc.; and a tetrapolar system (*A1B1* and *A2B2*) with the compatible combinations *A1B1A2B2* and *A1B2A2B1*.

The teliospore-forming yeasts include three genera, *Sporidiobolus, Rhodosporidium* and *Leucosporidium*. *Sporidiobolus* and *Rhodosporidium* species are orange and red, due to the presence of carotenoid pigments; *Leucosporidium* species have hyaline cultures. *Sporidiobolus* is the only genus with ballistospore formation. Of the four *Sporidiobolus* species accepted by Fell and Statzell Tallman (1984c), two are homothallic and two heterothallic. The former are *Sporid. johnsonii* and *Sporid. ruinenii*, both isolated from plant material (Nyland, 1949; Ruinen, 1963). The imperfect states of the two heterothallic species, *Sporid. pararoseus* and *Sporid. salmonicolor*, are *Sporobolomyces* species. Bandoni *et al.* (1975) first found mating types among the strains of *Sporobolomyces hispanicus, Sp. odorus* and *Sp. salmonicolor*, and Fell and Statzell Tallman (1981) described the perfect state as *Sporid. salmonicolor*. Fell and Statzell Tallman (1980) found that the strain described as *Aessosporon salmonicolor* and considered to be the perfect state of *Sp. salmonicolor* by van der Walt (1970e) mated with one of the types of *Sporid. salmonicolor*, and they reduced the name *Aessosporon salmonicolor* to a synonym of *Sporid. salmonicolor*.

Holzschu *et al.* (1981) made DNA studies of *Sporid. ruinenii* and *Sporid. johnsonii* and found a distinct difference in G + C contents and DNA complementarity between them. The two species are physiologically similar, but *Sporid. johnsonii* grows at 37°C which *Sporid. ruinenii* fails to do. All four *Sporidiobolus* species produce a dikaryotic mycelium with clamp connections and teliospores. The cultures of the homothallic species may be dark-coloured due to the presence of the teliospores. The heterothallic species have only been isolated in the haploid state. They come from various sources, for example air, soil and sea-water.

The genus *Rhodosporidium* in the description of Fell and Statzell Tallman (1984b) includes nine species. They show variations in mating system, shape of teliospores, and metabasidium and basidiospore formation. Most species are heterothallic and strains have been isolated in the haploid yeast state. In a few species, self-sporulating strains occur in which uninucleate hyphae without clamp connections develop from a single yeast cell. In *Rhodosp. toruloides* a heterothallic and homothallic cell cycle may occur. In the latter, dikaryotic mycelium with clamp connections develops from a single yeast cell. The eight species with a heterothallic life cycle have their imperfect states in the genera *Rhodotorula* and *Cryptococcus*, but only the type strain of *Cryptococcus infirmo-miniatus* mates with one of the types of *Rhodosp. infirmominiatum*. The type strains of the *Rhodotorula* species do not mate with haploid *Rhodosporidium* species although they are very similar. Three *Rhodosporidium* species, *Rhodosp. toruloides, Rhodosp. sphaerocarpum* and *Rhodosp. diobovatum*, have a haploid state similar to *Rhodotorula glutinis*. The three *Rhodosporidium* species are distinguished by the shape of the teliospores and by their G + C content. Inositol assimilation and nitrate assimilation vary among the *Rhodosporidium* species. Fermentative ability is absent.

Fell and Statzell Tallman (1984a) accept six *Leucosporidium* species, all described by Fell *et al.* (1969). They form cream-coloured cultures which are often mucoid. The yeast phase may form pseudomycelium, and four species have their imperfect state in the genus *Candida*. Assimilation of inositol varies among the species, but all are nitrate-positive. The following data on the life cycles of *Leucosporidium* species have been taken from Fell and Statzell Tallman (1982, 1984a). Two species, *Leu. scottii* and *Leu. antarcticum* have a heterothallic life cycle, *Leu. scottii* with a tetrapolar, multiallelic mating system and *Leu. antarcticum* with a bipolar mating system. After fusion of compatible cells mixed, for instance, on corn meal agar, a dikaryotic mycelium with clamp connections and teliospores develops. The latter germinate to form a four-celled metabasidium with lateral and terminal basidiospores. In addition to the heterothallic cycle, self-sporulating strains occur in both species. The four other *Leucosporidium* species are all self-sporulating. They produce mycelium without clamp connections and teliospores which germinate to form a one- to two-celled metabasidium. These species have fermentative ability, in contrast to the two heterothallic species, and they do not form starch-like compounds. The Co-Q system varies among the *Leucosporidium* species: *Leu. antarcticum* has Q-10, strains of *Leu. scottii* Q-9 or Q-10, and *Leu. frigidum, Leu. gelidum, Leu. nivalis* and *Leu. stokesii* Q-8 (Yarmada and Kondo, 1972). *Leucosporidium* species, with the exception of some *Leu. scottii* strains, do not grow at temperatures higher than 20°C. They have been isolated from soil, snow and marine waters in Antarctic regions (di Menna, 1966; Fell *et al.*, 1969).

2. Filobasidiaceae

Olive (1968) established the family Filobasidiaceae as a new genus with the species *Filobasidium floriforme* isolated from plant material. The family is characterized by the formation of slender, non-septate basidia with terminal sessile basidiospores. Blastospores may be formed on the hyphae, but teliospores are not produced. A second genus of the family, *Filobasidiella*, was described by Kwon-Chung (1975) who discovered mating types among the strains of *Cryptococcus neoformans*.

In the heterothallic life cycle of the species in these genera, yeast cells of opposite mating type fuse and a dikaryotic mycelium with clamp connections develops. A slender basidium is produced in which karyogamy and meiosis take place. In *Filobasidium* 4–10 sessile basidiospores are formed on the apex of the basidium in a petal-like whorl; in *Filobasidiella* four long chains of basidiospores are formed by basipetal budding from four sites on the apex of the basidium.

Most *Filobasidium* and *Filobasidiella* strains have been isolated in the haploid yeast state. Exceptions are the original strains of *F. floriforme* and a strain of *F. capsuligenum* described by Rodrigues de Miranda (1972). Haploid strains of *F. floriforme* resemble *Cryptococcus albidus*, but do not give mating reactions with the type strain of this species (Kwon-Chung and Fell, 1984b). Rodrigues de Miranda found that yeast strains obtained from a sporulating culture of *F. capsuligenum* can be mated with the types discovered among *Torulopsis capsuligenum* strains by Fell *et al.* (1969) and originally classified by these authors in the genus *Leucosporidium*.

A third species in the genus *Filobasidium, F. uniguttulatum*, was described by Kwon-Chung (1977) who found opposite mating types among the strains of *Cryptococcus uniguttulatus*. Mating reactions in the three *Filobasidium* species were obtained by mixing yeast cells of compatible types on agar, for example corn-meal or hay-infusion agar. The dikaryotic hyphae formed after fusion grow mostly within the agar and the basidia are formed on the surface. *Filobasidium capsuligenum* is the only *Filobasidium* species with fermentative ability. It has Co-Q system 10, whereas *F. uniguttulatum* has Q-9.

The genus *Filobasidiella* has a single species, *Fil. neoformans*, with the varieties *neoformans* and *bacillispora*. They differ in the shape of the basidiospores and in a few physiological features such as assimilation of (−)-malic acid, glycine and creatinine. The variety *neoformans* includes serotypes A and D; variety *bacillispora* includes serotypes B and C. The two varieties showed a 55.2–64.0% DNA relatedness (Aulakh *et al.*, 1981). *Filobasidiella neoformans* var. *neoformans* is the perfect state of *Cryptococcus neoformans* var. *neoformans*, and var. *bacillispora* of *Cr. neoformans* var. *gattii*.

Several media have been used in mating experiments, for instance malt-extract agar, V8 agar, hay-infusion agar, sucrose–yeast agar (Kwon-Chung and Fell, 1984a), a pyrimidine medium (Ahearn and Schlitzer, 1981), and a medium containing pigeon faeces (Staib, 1981). Basidia with basidiospores are often formed at the edge of the mixed colony on dikaryotic mycelium that is partly within the agar. Kwon-Chung (1980) found that the mother nuclei formed after meiosis in the basidium divided mitotically. The daughter nuclei migrated into the basidiospores, of which there may be up to 20 per chain. The basidiospores in one chain may vary in mating type. Some strains of *Cr. neoformans* have a homothallic life cycle (Erke, 1976).

3. Sterigmatosporidium

Kraepelin and Schulze (1982) described the perfect state of a new *Sterigmatomyces* species, *Sterigmatosporidium polymorphum* (see Fig. 1c). It is heterothallic and on the dikaryotic mycelium with clamp connections, two types of chlamydospores are formed, round and oval cylindrical ones. In the latter, karyogamy takes place and during germination meiosis occurs and haploid spores are terminally formed as buds. Kraepelin and Schulze (1982) hesitated to indicate the oval to cylindrical hyaline chlamydospores as teliospores and did not yet classify the genus *Sterigmatosporidium* in a higher taxon. The dikaryotic hyphae of the species have a dolipore in the septa (N.J.W.Kreger-van Rij, unpublished work) which also distinguishes the genus from the teliospore-forming yeasts.

C. Imperfect yeasts

This group includes yeasts that do not form ascospores or basidiospores (Table III). Sporulation may be lacking, for instance, because suitable conditions are unknown, or because the yeast is heterothallic and haploid, and the opposite mating type has not yet been discovered. Several imperfect yeast species closely resemble perfect species, and only differ in the ability to form sexual spores. The sporulating species is indicated as the perfect state (teleomorph) of the non-sporulating species, and the latter as the imperfect state (anamorph) of the sporulating one.

Most imperfect genera are either ascomycetous or basidiomycetous; in a few, both ascomycetous and basidiomycetous species are classified. Generic

Table III. Characteristics of the genera of the imperfect yeasts

Sporobolomycetaceae

Bullera

Budding cells, true mycelium may be formed. Cream to yellow cultures.
Ballistospores rotationally symmetrical.
Fermentation negative.

Sporobolomyces

Budding cells, true mycelium. Red or pink cultures due to carotenoid pigments.
Ballistospores bilaterally symmetrical.
Fermentation negative.
Coenzyme Q-10.

Cryptococcaceae

Aciculoconidium

Budding cells, true mycelium and arthrospores, terminal blastospores needle-shaped.
Fermentation positive.

Brettanomyces

Budding cells, non-septate true mycelium may be formed.
Fermentation positive.
Nitrate assimilation positive or negative.
Short-lived on malt agar.
Coenzyme Q-9.

Candida

Budding cells, pseudo- and true mycelium may be formed.
Fermentation positive or negative.
Nitrate assimilation positive or negative.
Inositol assimilation positive or negative. Inositol-positive strains form pseudomycelium.
Coenzyme Q-6, Q-7, Q-8, Q-9 and Q-10.

Cryptococcus

Budding cells, often capsulated. No pseudo- or true mycelium. Hyaline or red cultures due to carotenoid pigments; seldom brown or black cultures.
Fermentation negative.
Nitrate assimilation positive or negative.
Inositol assimilation positive.
Coenzyme Q-8, Q-9 and Q-10.

Eeniella

Budding cells, bipolar on a broad base.
Fermentation positive.
Nitrate assimilation negative.
Short-lived on malt agar.
Coenzyme Q-9.

Fellomyces

Budding cells, formed on sterigmata and separated at the distal end of the stalk, true mycelium may be formed. Hyaline or salmon-coloured to orange cultures due to carotenoid pigments. No ballistospores.
Fermentation negative.
Nitrate assimilation positive or negative.
Inositol assimilation positive or negative.
Coenzyme Q-10.

Kloeckera

Budding cells, bipolar on a broad base.
Fermentation positive.
Coenzyme Q-6.

Malassezia

Budding cells, monopolar on a broad base, true mycelium may occur.
Fermentation negative.
From human and warm-blooded animal skin.

Table III. Continued

Oosporidium
- Budding cells on a broad base, true mycelium. Cultures pink or orange-yellow due to non-carotenoid pigment.
- Fermentation negative.
- Nitrate assimilation positive.

Phaffia
- Budding cells, chlamydospores. Orange or red cultures due to carotenoid pigments.
- Fermentation positive.
- Inositol assimilation negative.

Rhodotorula
- Budding cells, often capsulated, true mycelium may occur. Red or yellow cultures due to carotenoid pigments.
- Fermentation negative.
- Nitrate assimilation positive or negative.
- Inositol assimilation negative.
- Starch not produced.
- Coenzyme Q-9 and Q-10.

Schizoblastosporion
- Budding cells, bipolar on a broad base.
- Fermentation negative.

Sterigmatomyces
- Budding cells, formed on sterigmata and separated at a septum in the mid-region of the stalk, true mycelium may be formed. Hyaline cultures. No ballistospores.
- Fermentation negative.
- Nitrate assimilation positive or negative.
- Inositol assimilation negative.
- Coenzyme Q-9.

Sympodiomyces
- Cells budding from a conidiophore, (true mycelium).
- Fermentation negative.

Trichosporon
- Budding cells, true mycelium and arthrospores.
- Fermentation positive or negative.
- Nitrate assimilation positive or negative.
- Inositol assimilation positive or negative.

Trigonopsis
- Budding cells, oval or triangular.
- Fermentation negative.

differentiation of the imperfect yeasts is based on morphological characteristics and on a few physiological characters, such as acid production and assimilation of inositol.

The eight ascomycetous genera will be discussed first. Three of them are characterized by bipolar budding: *Kloeckera, Eeniella* and *Schizoblastosporion. Kloeckera*, according to Meyer *et al.* (1978), includes six species which have their perfect state in six *Hanseniaspora* species, all with fermentative ability and Co-Q-6 (Yamada *et al.*, 1976a). The new genus *Eeniella* described by Smith *et al.* (1981) is also fermentative, but distinguished from *Kloeckera* by strong acid formation from glucose. In this characteristic and in the Co-Q system 9 (Yamada *et al.*, 1980), the single species, *Eeniella nana* resembles *Brettanomyces* species. The strain of *Eeniella nana* came from beer. *Schizoblastosporion* with one species, *Schizobl. starkeyi-henricii*, is not fermentative: strains have been isolated from soil.

The genus *Trigonopsis* is characterized by the occurrence of triangular cells. Ellipsoidal buds are formed on slight elevations on the cell and these protuberances give the typical shape to the cells which may also be tetra- or rhombohedral (Slooff, 1970d). The single species, *Trig. variabilis*, has been isolated from beer.

In *Sympodiomyces*, described by Fell and Statzell (1971), buds are formed sympodially on a conidiophore at the end of a yeast cell. After production of the first bud, the conidiophore grows out sideways and produces a second bud, and so on. True mycelium with septa may also be formed. Strains of the single species, *Symp. parvus*, have been isolated from Antarctic waters. Its maximum temperature of growth is 25°C.

Two ascomycetous genera produce true mycelium abundantly: *Oosporidium* and *Aciculoconidium*. The single *Oosporidium* species, *O. margaritiferum*, forms buds on a broad base and true septate hyphae, but no arthrospores. Growth is very slow, and the cultures are coloured pink to orange-yellow by a non-carotenoid pigment (do Carmo-Sousa, 1970a). Strains have been isolated from slime fluxes from elm, oak and lime. The genus *Aciculoconidium* has been established by King and Jong (1976) for the species *Trichosporon aculeatum*, first described by Phaff *et al.* (1956). It produces arthrospores and needle-shaped blastospores in addition to round and oval forms. King and Jong (1976) did not consider the disarticulation of the hyphae as arthrospore formation and therefore removed the species from the genus *Trichosporon*. If this concept is not accepted, the genus *Aciculoconidium* may be retained because of the typical shape of the blastospores. The species ferments glucose weakly and has been isolated from *Drosophila*.

The last ascomycetous genus, *Brettanomyces*, is characterized by formation of acetic acid from glucose, and by some features which are associated with this characteristic and which have been mentioned in the discussion of the genus *Dekkera* (p. 33). Two of the nine *Brettanomyces* species have their perfect state in the genus *Dekkera*. All *Brettanomyces* species have fermentative ability, although a few ferment only slowly. In the latter species, acid formation is also low. Sources of isolation of *Brettanomyces* strains are beer (lambic beer), wine and grape must.

Of the eight basidiomycetous genera, *Bullera* and *Sporobolomyces* are both distinguished by formation of ballistospores. Derx (1930) classified these genera in the Sporobolomycetaceae. In the descriptions of Fell and Statzell Tallman (1984d) and Rodrigues de Miranda (1984a), *Sporobolomyces* species produce pink, red or orange cultures, and *Bullera* species cream to yellow cultures. The ballistospores in the former genus are usually bilaterally symmetrical, in the latter rotationally symmetrical, but this character is no longer decisive for the distinction of the genera (Phaff, 1970d,e). Of the seven *Sporobolomyces* species, two have their perfect state in the genus *Sporidiobolus* (Fell and Statzell Tallman, 1981). Some species produce true mycelium. Strains have been isolated, for instance, from plant material and from the air.

The strains of the six *Bullera* species mostly have mucoid cultures. Their perfect states are unknown, but haploid strains of *Sirobasidium magnum*, belonging to the Tremellales, agree with the diagnosis of *Bullera* (Bandoni, 1984). Inositol assimilation varies among the *Bullera* species. Inositol-positive species which have lost the ability to form ballistospores agree with the diagnosis of the genus *Cryptococcus*.

In the genus *Sterigmatomyces* described by Fell (1966), buds are formed on stalks on the mother cell. The stalks vary in length from 0.5 to 26 μm (Fell, 1970a). The cultures are generally dry, but mucoid variants occur. In the latter, the sterigmata are short or lacking. Kurtzman *et al.* (1983) established DNA relatedness among the species. This resulted in the merging of the nitrate-negative species *St. indicus* with the nitrate-positive species *St. halophilus*. More recently, Yamada and Banno (1984a,b) found two different Co-Q systems among the strains of *Sterigmatomyces* spp. They retained the species with Co-Q-9 in the genus *Sterigmatomyces* and classified those with Co-Q-10 in a new genus *Fellomyces*. The genera *Sterigmatomyces* and *Fellomyces* also differ in the place on the stalk where the bud is separated from the mother cell: in *Sterigmatomyces* at the mid-region of the stalk, in *Fellomyces* at the distal end. One of the *Fellomyces* species, *Fellomyces nectairii*, has salmon-coloured colonies, the other *Fellomyces* and *Sterigmatomyces* spp. are white to cream coloured. *Sterigmatomyces* now includes the species *St. halophilus* and *St. elviae*.

The two species of the genus *Malassezia* are characterized by monopolar budding. The cells are round to bottle-shaped with a collarette at the site of bud formation. One of the species, *Mal. furfur*, may form hyphae which fall apart. The wall structure of both species is typical with a spiral groove in the inner layer (Keddie, 1966; Gordon, 1979). These yeasts have been isolated from the healthy and diseased skin of man and warm-blooded animals. *Malassezia furfur* occurs in the hyphal state in the skin disease pityriasis versicolor. It requires vegetable oil or fatty acids in the medium; *Mal. pachydermatis* can grow without these additives. The genus has also been named *Pityrosporum* (Sabouraud, 1904), but the name *Malassezia* (Baillon, 1889) has priority (Slooff, 1970c; Gordon, 1979). *Malassezia pachydermatis* gives a positive Diazonium blue B colour reaction (Hagler and Ahearn, 1981) and is, therefore, considered to be basidiomycetous.

Rhodotorula and *Phaffia* include basidiomycetous yeasts with red or orange cultures, *Rhodotorula* non-fermenting and *Phaffia* fermenting. An important characteristic of *Rhodotorula* species is the inability to assimilate inositol which distinguishes *Rhodotorula* from *Cryptococcus*. Another distinguishing feature is the inability to form starch-like compounds. Most strains have mucoid cultures owing to capsulated cells. One of the eight species accepted by Fell *et al.* (1984), *Rh. glutinis*, has perfect states in the genus *Rhodosporidium*. Sources of isolation are, for instance, plant material, the air and marine waters.

The single species of the genus *Phaffia, Ph. rhodozyma* (Miller *et al.*, 1976) is also inositol-negative, but produces starch-like compounds. Chlamydospores may be formed. Strains have been isolated from slime fluxes from beech, birch and elder (Miller, 1984).

The genus *Cryptococcus* shows similarities with *Rhodotorula*, but is distinguished by the ability to assimilate inositol. Three of the 19 *Cryptococcus* species accepted by Rodrigues de Miranda (1984b) have red cultures, the others are hyaline or yellowish. Most species have capsulated cells and mucoid cultures; fermentative ability is absent and pseudo- and true mycelium is not formed. Formation of starch-like compounds occurs in most species, but not in all. All *Cryptococcus* species are basidiomycetous with the exception of *Cr. melibiosum* (Rodrigues de Miranda, 1978) which is one of the few inositol-positive ascomycetous yeasts. Van der Walt *et al.* (1981) classified *Cr. melibiosum* in a new genus *Myxozyma*. Golubev *et al.* (1974) found that glucuronic acid is assimilated only by basidiomycetous *Cryptococcus* species and would, therefore, be more suitable than inositol assimilation for the characterization of *Cryptococcus* as a basidiomycetous genus.

Cryptococcus species have perfect states in the genera *Filobasidium*, *Filobasidiella* and *Rhodosporidium*. Moreover, haploid strains of several *Tremella* species and of *Holtermannia corniformis* agree with the diagnosis of *Cryptococcus* (see Section IV.B, p. 42). Kurtzman (1973) discovered mating types among the strains of *Cr. laurentii* which after fusion produced a dikaryotic mycelium with clamp connections and septa with a dolipore (Kreger-van Rij and Veenhuis, 1974; Rhodes *et al.*, 1981), but no sexual spores. One of the few *Cryptococcus* species growing at 37°C is the pathogen *Cr. neoformans*.

The last two imperfect genera, *Trichosporon* and *Candida*, in the descriptions of Kreger-van Rij (1984d) and Meyer *et al.* (1984) respectively, include both ascomycetous and basidiomycetous species. *Trichosporon* is characterized by the formation of loose budding cells, true mycelium and arthrospores (see Fig.1d). The six basidiomycetous species are non-fermenting. Their perfect state is yet unknown. One of them is *Tr. cutaneum* (*Tr. beigelii*), the description of which has wide limits and may well include more than one species. For instance, the maximum temperature of growth ranges from 29 to 41°C (do Carmo-Sousa, 1970b). Strains have been isolated from hair and skin, and also from soil, fresh water, wood-pulp and mushrooms. King and Jong (1975) transferred *Tr. inkin* (do Carmo-Sousa and van Uden, 1967) to a new genus, *Sarcinosporon*. They considered the sarcina-like conglomerates of cells formed by septation in different planes as sporangia. The author cannot agree with this opinion, and has retained the species in the genus *Trichosporon*.

The ascomycetous *Trichosporon* species probably have their perfect state in various genera, such as *Pichia* (*Hyphopichia*) and *Endomyces* (*Dipodascus*). Von Arx (von Arx *et al*., 1977) and Weiman (1979), in an attempt to restrict *Trichosporon* to basidiomycetous species, reclassified several ascomycetous species in the genus *Geotrichum*. The genus *Geotrichum* has become heterogeneous by these reclassifications, for instance by the inclusion of *Tr. terrestre*, a nitrate-positive species that produces loose budding cells. Barnett *et al.* (1983) followed von Arx and Weijman in these reclassifications. Moreover, Barnett *et al.* (1983) brought *Tr. fennicum* into the genus *Candida* as *C. fennica*, although it produces arthrospores.

The last genus, *Candida*, includes all yeast species that cannot be classified in other imperfect genera. As a result it is heterogeneous, including the imperfect state of species of the genera *Citeromyces, Clavispora, Debaryomyces, Hansenula, Issatchenkia, Kluyveromyces, Metschnikowia, Pichia, Saccharomyces, Saccharomycopsis, Stephanoascus, Torulaspora, Wickerhamiella, Filobasidium* and *Leucosporidium*. The perfect state of most *Candida* species is still unknown. Possibly, representatives of new still undescribed genera are among them. Yarrow and Meyer (1978) brought the species of the genus *Torulopsis* into the genus *Candida*. The distinction between the two genera was the formation of pseudo- and true mycelium; *Candida* now includes species with and without hyphae.

Most *Candida* species are ascomycetous, some 25 are basidiomycetous. Three proposals have been made to reclassify the latter. First, Moore (1980) brought them together into a new genus *Vanrija* with *Vanrija humicola* as the type species. This genus is heterogeneous and includes species of different groups, for instance species with their perfect state in the genus *Leucosporidium* (teliospore-forming yeasts), in the genus *Filobasidium* (Filobasidiaceae), and possibly in the genus *Tremella* (Tremellaceae, Tremellales). A second proposal for the removal of basidiomycetous yeasts from *Candida* was made by von Arx and Weijman (1979). They transferred these species to the genus *Apiotrichum*, with the type species *Apiotrichum porosum*, and to *Rhodotorula*, the first with xylose in the hydrolysate of cells, the latter without. *Candida humicola* was renamed *Apiotrichum humicolum*. The genus *Apiotrichum* was classified, together with other xylose-positive species, in the family Filobasidiaceae. The new *Rhodotorula* species, in contrast with the others, had no red or orange cultures. They were classified with other xylose-negative species in the Sporobolomycetaceae. The genus *Rhodotorula*, in the concept of von Arx and Weijman, may include the imperfect state of both *Rhodosporidium* and *Leucosporidium* species. A third proposal for reclassification of basidiomycetous *Candida* species was made by Golubev (1981). He transferred a number of xylose-positive species to the genus *Cryptococcus* (among them *Candida humicola*). These species closely resemble *Cryptococcus* species but produce pseudo- and true mycelium. Barnett *et al.* (1983) accepted Golubev's classifications. Meyer *et al.* (1984) have retained ascomycetous and basidiomycetous species in the genus *Candida* which now includes 196 species.

Apparently, new distinctions are made between existing genera. The availability of methods for distinguishing between imperfect ascomycetous and basidiomycetous yeasts (see p. 9 and p. 16) has first induced such revisions. In the second place, chemical data have played a role, especially the presence of xylose in the cell wall which, so far, seems to correlate with the occurrence of the dolipore in the septum of hyphae, a structure considered to be of great taxonomic value. However, not all proposals for reclassifications of *Candida* spp. include new definitions of the genera concerned, which is confusing.

D. Some other fungi

As mentioned in the Introduction, the yeast species in this chapter are not the only fungi with a yeast phase. Moreover, some fungi without a yeast state may be narrowly related to certain yeast species. A few of these fungi, which are occasionally discussed in relation to the yeasts included in the foregoing part of this chapter, will be mentioned here.

Taphrina species of the ascomycetous Taphrinales (Kramer, 1973) may reproduce by budding as yeasts with pink colonies like *Rhodotorula* and *Cryptococcus* species. Growth is usually slow, and the *Taphrina* species differ from *Cryptococcus* species by the absence of inositol assimilation, and from *Rhodotorula* species by the production of starch-like compounds (Lodder, 1970).

One of the "black yeasts", *Aureobasidium pullulans*, forms hyaline or light-pink slimy colonies of yeast cells surrounded by submerged true hyphae. On suitable media the colonies may turn dark green and black by the production of chlamydospores and thick-walled hyphae. The genus *Aureobasidium* is ascomycetous. Black yeasts have been described by Ellis (1971) and Hermanides-Nijhof (1977).

Haskins and Spencer (1967) have described an organism in a new genus *Trichosporonoides* as *Trichosporonoides oedocephalis*. This species resembles

Trichosporon in producing budding yeast cells and true hyphae with arthrospores, but it forms conidiospores on the swollen tip of aseptate sporophores. The conidia have a dark brown wall. Haskins (1975) showed the presence of a dolipore in the hyphal septa. The species ferments glucose, galactose, sucrose and maltose and assimilates nitrate and erythritol. It had been isolated from a honeycomb.

Species of the ascomycetous genus *Endomyces* and its imperfect state *Geotrichum* are characterized by the formation of true hyphae and arthrospores and the absence of budding yeast cells. The type species, *Endomyces decipiens*, has been described by de Bary (1859), Tulasne and Tulasne (1865) and Brefeldt (1891), and named by Reess (1870). It was observed on the honey fungus, *Armillaria mellea*. Redhead and Malloch (1977) suggested that the original description covered two different organisms: one, isolated in pure culture, forming arthrospores but no ascospores, and the other, not isolated, without arthrospores but forming hat-shaped ascospores. Observations on the latter were made from dried material. N.J.W.Kreger-van Rij (unpublished results) observed asci in the arthrospore-forming fungus, but only on the gills of *Armillaria mellea*, and not yet in culture. The asci were formed after gametangiogamy, and the ascospores had a very thick wall and resembled the spores of other *Endomyces* species. This observation confirms the supposition of the existence of two different species. Malloch and Redhead identified the fungus with hat-shaped spores as *Endomyces decipiens*, and reclassified the other *Endomyces* species in three different genera, *Zendera, Galactozyma* and *Magnusiomyces*. Von Arx (1977) accepted their decision about *Endomyces decipiens* and reclassified the other *Endomyces* species in the genus *Dipodascus*. He also changed the name of certain *Saccharomycopsis* species to *Endomyces* since he considered them to belong to the same genus as *Endomyces decipiens*. Changing of names might be prevented if the arthrospore-forming fungus should be indicated as *Endomyces decipiens*. A full description of the fungus with hat-shaped spores is only possible after isolation.

The type species of the genus *Geotrichum* is *Geotrichum candidum*. Butler and Petersen (1970, 1972) discovered its perfect state which was named *Endomyces geotrichum*. Several other *Endomyces* or *Dipodascus* species may have their imperfect state in *Geotrichum*. Some of the ascosporogenous *Trichosporon* species may be closely related to *Geotrichum* species.

V. ACKNOWLEDGEMENTS

I gratefully acknowledge Dr. C.P. Kurtzman's comments on the manuscript and I thank Mr. Jan Zagers for printing the photographs.

References

Ahearn, D.G. and Schlitzer, R.L. (1981). *Mycologia* **73**, 343.

Ahearn, D.G., Roth, F.J., Fell,J.W. and Meyers, S.P. (1960). *Journal of Bacteriology* **79**, 369.

Ainsworth, G.C. (1973). *In* "The Fungi" (G.C.Ainsworth, F.K.Sparrow and A.S.Sussman, eds), vol. IVA, pp. 1–7. Academic Press, New York and London.

Aulakh, H.S., Straus, S.E. and Kwon-Chung, K.J. (1981). *International Journal of Systematic Bacteriology* **31**, 97.

Babjeva, I.P. and Gorin, S.E. (1975). *Antonie van Leeuwenhoek* **41**, 185.

Bacon, J.S.D., Jones, D., Farmer, V.C. and Webley, D.M. (1968). *Biochimica et Biophysica Acta* **158**, 313.

Baillon, E.H. (1889). Traité de Botanique Médicale Cryptogamique. Octave Doin, Paris.

Bandoni, R.J. (1984). *In* "The Yeasts, a Taxonomic Study" (N.J.W. Kreger-van Rij, ed.), pp. 541–544. Elsevier Biomedical Press, Amsterdam.

Bandoni, R.J., Johri, B.N. and Reid, S.A. (1975). *Canadian Journal of Botany* **53**, 2942.

Banno, I. (1967). *Journal of General and Applied Microbiology* **13**, 167.

Baptist, J.N. and Kurtzman, C.P. (1976). *Mycologia* **68**, 1195.

Barker, B.T.P. (1901). *Philosophical Transactions of the Royal Society of London, Series B*, **194**, 467.

Barnett, J.A. (1968). *In* "The Fungi" (G.C. Ainsworth, F.K. Sparrow and A.S. Sussman, eds), vol. III, pp. 557–595. Academic Press, New York and London.

Barnett, J.A. and Pankhurst, R.J. (1974). "A New Key to the Yeasts". North-Holland Publishing Company, Amsterdam.

Barnett, J.A., Payne, R.W. and Yarrow, D. (1979). "A Guide to Identifying and Classifying Yeasts". Cambridge University Press, Cambridge.

Barnett, J.A., Payne, R.W. and Yarrow, D. (1983). " Yeasts: Characteristics and Identification". Cambridge University Press, Cambridge.

Bastide, J.M., El Hadi Hadibi and Bastide, M. (1979). *Journal of General Microbiology* **113**, 147.

Besson, M. (1966). *Bulletin de la Société Mycologique de France* **82**, 489.

Bicknell, J.N. and Douglas, H.C. (1970). *Journal of Bacteriology* **101**, 505.

Black, S.H. and Gorman, C. (1971). *Archiv für Mikrobiologie* **79**, 231.

Boedijn, K.B. (1960). *Mycopathologia et Mycologia Applicata* **12**, 163.

Brefeld, O. (1891). *In* "Untersuchungen aus dem Gesamtgebiete der Mykologie", vol. 9, pp. 119–156. H. Schöningh, Munster.

Bryant, T.N. (1983). *Journal of Applied Bacteriology* **54**, 101.

Butler, E.E. and Petersen, L.J. (1970). *Science* **169**, 481.

Butler, E.E. and Petersen, L.J. (1972). *Mycologia* **64**, 365.

Christensen, W.B. (1946). *Journal of Bacteriology* **52**, 461.

De Bary, A. (1859). *Botanische Zeitung* **17**, 401.

Derx, H.G. (1930). *Annales Mycologia* **28**, 1.

Di Menna, M.E. (1966). *Antonie van Leeuwenhoek* **32**, 29.

Do Carmo-Sousa, L. (1970a). *In* "The Yeasts, a Taxonomic Study" (J. Lodder, ed.), pp. 1161–1166. North-Holland Publishing Company, Amsterdam.

Do Carmo-Sousa, L. (1970b). *In* "The Yeasts, a Taxonomic Study" (J. Lodder, ed.), pp. 1309–1352. North-Holland Publishing Company, Amsterdam.

Do Carmo-Sousa, L. and van Uden, N. (1967). *Mycologia* **59**, 653.

Ellis, M.B. (1971). Dematiaceous Hyphomycetes. Commonwealth Mycological Institute, Surrey, England.

Erke, K.H. (1976). *Journal of Bacteriology* **128**, 445.

Fell, J.W. (1966). *Antonie van Leeuwenhoek* **32**, 99.

Fell, J.W. (1970a). *In* "The Yeasts, a Taxonomic Study" (J.Lodder, ed.), pp. 1229–1234. North-Holland Publishing Company, Amsterdam.

Fell, J.W. (1970b). *In*"Recent Trends in Yeast Research" (D.G. Ahearn, ed.), pp. 49–66. Georgia State University, Atlanta.

Fell, J.W. (1974). *Transactions of the Mycological Society, Japan* **15**, 316.

Fell, J.W. (1984). *In* "The Yeasts, a Taxonomic Study" (N.J.W. Kreger-van Rij, ed.), pp. 491–494. Elsevier Biomedical Press, Amsterdam.

Fell, J.W. and Statzell, A.C. (1971). *Antonie van Leeuwenhoek* **37**, 359.
Fell, J.W. and Statzell Tallman, A. (1980). *International Journal of Systematic Bacteriology* **30**, 206.
Fell, J.W. and Statzell Tallman, A. (1981). *Current Microbiology* **5**, 77.
Fell, J.W. and Statzell Tallman, A. (1982). *Current Microbiology* **6**, 392.
Fell, J.W. and Statzell Tallman, A. (1984a). *In* "The Yeasts, a Taxonomic Study" (N.J.W. Kreger-van Rij, ed.), pp. 496–508. Elsevier Biomedical Press, Amsterdam.
Fell, J.W. and Statzell Tallman, A (1984b). *In* "The Yeasts, a Taxonomic Study" (N.J.W. Kreger-van Rij, ed.), pp. 509–531. Elsevier Biomedical Press, Amsterdam.
Fell, J.W. and Statzell Tallman, A. (1984c). *In* "The Yeasts, a Taxonomic Study" (N.J.W.Kreger-van Rij, ed.), pp. 532–540. Elsevier Biomedical Press, Amsterdam.
Fell, J.W. and Statzell Tallman, A. (1984d). *In* "The Yeasts, a Taxonomic Study" (N.J.W.Kreger-van Rij, ed.), pp. 911–920. Elsevier Biomedical Press, Amsterdam.
Fell, J.W., Statzell, A., Hunter, I. and Phaff, H.J. (1969). *Antonie van Leeuwenhoek* **35**, 433.
Fell, J.W., Hunter, I.L. and Statzell Tallman, A. (1973). *Canadian Journal of Microbiology* **19**, 643.
Fell, J.W., Statzell Tallman, A. and Ahearn, D.G. (1984). *In* "The Yeasts, a Taxonomic Study" (N.J.W. Kreger-van Rij, ed.), pp. 893–905. Elsevier Biomedical Press, Amsterdam.
Fuson, G.B., Price, C.W. and Phaff, H.J. (1979). *International Journal of Systematic Bacteriology* **29**, 64.
Golubev, W.I. (1973). *Mikrobiologia* **42**, 1058.
Golubev, W.I. (1977). *Antonie van Leeuwenhoek* **43**, 317.
Golubev, W.I. (1981). *Mikologiya i Fitopatologiya* **15**, 467.
Golubev, W.I., Okunev, O.N. and Vdavina, N.W. (1974). *Mikrobiologia* **6**, 1047.
Gordon, M.A. (1979). *Sabouraudia* **17**, 305.
Gorin, P.A.J. (1973). *Canadian Journal of Chemistry* **51**, 2105.
Gorin, P.A. and Spencer, J.F.T. (1970). *Advances in Applied Microbiology* **13**, 25.
Gow, N.A.R., Gooday, G.W., Newsam, R.J. and Gull, K. (1980). *Current Microbiology* **4**, 357.
Guilliermond, A. (1909a). *Revue Générale de Botanique* **21**, 353.
Guilliermond, A. (1909b). *Revue Générale de Botanique* **21**, 401.
Hagler, A.N. and Ahearn, D.G. (1981). *International Journal of Systematic Bacteriology* **31**, 204.
Hansen, E.C. (1904). *Centralblatt für Bakteriologie und Parasitenkunde*, Abteilung II, **12**, 529.
Haskins, R.H. (1975). *Canadian Journal of Botany* **53**, 1139.
Haskins, R.H. and Spencer, J.F.T. (1967). *Canadian Journal of Botany* **45**, 515.
Hermanides-Nijhof, E.J. (1977). *In* "Studies in Mycology", No. 15, pp. 141–177. Centraalbureau voor Schimmelcultures, Baarn.
Holzschu, D.L., Presley, H.L., Miranda, M. and Phaff, H.J. (1979). *Journal of Clinical Microbiology* **10**, 202.
Holzschu, D.L., Tredick, J. and Phaff, H.J. (1981). *Current Microbiology* **5**, 73.
Johannsen, E. (1980). *Antonie van Leeuwenhoek* **46**, 177.
Johannsen, E. and van der Walt, J.P. (1978). *Archives of Microbiology* **118**, 45.
Jülich, W. (1981). "Higher Taxa of Basidiomycetes". Bibliotheca Mycologica 85, J. Cramer.
Kamienski, T. (1899). *Travaux de la Société Impériale des Sciences Naturelles de St. Pétersbourg* **30**, 363.
Kawakami, N. (1960). *Memoirs of the Faculty of Engineering, Hiroshima University* **1**, 207.
Keddie, F.M. (1966). *Sabouraudia* **5**, 134.
King, D.S. and Jong, S.C. (1975). *Mycotaxon* **3**, 89.
King, D.S. and Jong, S.C. (1976). *Mycotaxon* **3**, 401.
Kirsop, B.E., Painting, K.A., Henry, J.E. and Fernandes, M.A. (1985). *Binary* (The News Letter of the Society for General Microbiology Computer Club) Issue 3.
Klöcker, A. (1907). *In* "Handbuch der Technischen Mykologie" (F.Lafar, ed.), vol. 4, pp. 168–191. Fischer, Jena.
Klöcker, A. (1909). *Comptes Rendus des Travaux du Laboratoire Carlsberg* **7**, 273.
Klöcker, A. (1924). "Die Gärungsorganismen". 3th Auflage, Urban und Schwarzenberg, Berlin-Wien.
Kocourek, J. and Ballou, C.E. (1969). *Journal of Bacteriology* **100**, 1175.

Kraepelin, G. and Schulze, U. (1982). *Antonie van Leeuwenhoek* **48**, 471.
Kramer, Ch.L. (1973). *In* "The Fungi" (G.C. Ainsworth, F.K. Sparrow and A.S. Sussman, eds), vol. IVA, pp. 33–41. Academic Press, New York and London.
Kreger, D.R. (1954). *Biochimica et Biophysica Acta* **13**, 1.
Kreger-van Rij, N.J.W. (1969a). *In* "The Yeasts" (A.H.Rose and J.S.Harrison, eds), vol. 1, pp. 5–78. Academic Press, London and New York.
Kreger-van Rij, N.J.W. (1969b). *In* "Proceedings of the 2nd Symposium on Yeasts, Bratislava 1966" (A.Kocková-Kratochvílová, ed.), pp. 169–176. The Slovak Academy of Science, Bratislava.
Kreger-van Rij, N.J.W. (1973). *In* "The Fungi" (G.C.Ainsworth, F.K.Sparrow and A.S. Sussman, eds), vol. IVA, pp. 11–32. Academic Press, New York and London.
Kreger-van Rij, N.J.W. (1977a). *Antonie van Leeuwenhoek* **43**, 55.
Kreger-van Rij, N.J.W. (1977b). *Antonie van Leeuwenhoek* **43**, 225.
Kreger-van Rij, N.J.W. (1978a). *Antonie van Leeuwenheok* **44**, 451.
Kreger-van Rij, N.J.W. (1978b). *Archives of Microbiology* **117**, 73.
Kreger-van Rij, N.J.W. (1979). *Archives of Microbiology* **121**, 53.
Kreger-van Rij, N.J.W. (1980). *In* "Biology and Activities of Yeasts" (F.A. Skinner, S.M. Passmore and R.R. Davenport, eds), Society for Applied Bacteriology, Symposium Series no.9, pp. 29–52. Academic Press, London.
Kreger-van Rij, N.J.W. (ed.) (1984a). "The Yeasts, a Taxonomic Study". Elsevier Biomedical Press, Amsterdam.
Kreger-van Rij, N.J.W. (1984b). *In* "The Yeasts, a Taxonomic Study" (N.J.W. Kreger-van Rij, ed.), pp. 130–145. Elsevier Biomedical Press, Amsterdam.
Kreger-van Rij, N.J.W. (1984c). *In* "The Yeasts, a Taxonomic Study" (N.J.W. Kreger-van Rij, ed.), pp. 399-413. Elsevier Biomedical Press, Amsterdam.
Kreger-van Rij, N.J.W. (1984d). *In* "The Yeasts, a Taxonomic Study" (N.J.W. Kreger-van Rij, ed.), pp. 933–962. Elsevier Biomedical Press, Amsterdam.
Kreger-van Rij, N.J.W. and Ahearn, D.G. (1968). *Mycologia* **60**, 604.
Kreger-van Rij, N.J.W. and Veenhuis, M. (1969a). *Journal of General Microbiology* **57**, 91.
Kreger-van Rij, N.J.W. and Veenhuis. M. (1969b). *Journal of General Microbiology* **58**, 341.
Kreger-van Rij, N.J.W. and Veenhuis, M. (1971). *Journal of General Microbiology* **68**, 87.
Kreger-van Rij, N.J.W. and Veenhuis, M. (1973). *Antonie van Leeuwenhoek* **39**, 481.
Kreger-van Rij, N.J.W. and Veenhuis, M. (1974). *Annali di Microbiologia* **24**, 209.
Kreger-van Rij, N.J.W. and Veenhuis, M. (1975a). *Journal of General Microbiology* **89**, 256.
Kreger-van Rij, N.J.W. and Veenhuis, M. (1975b). *Archives of Microbiology* **104**, 263.
Kreger-van Rij, N.J.W. and Veenhuis, M. (1976a). *Canadian Journal of Microbiology* **22**, 960.
Kreger-van Rij, N.J.W. and Veenhuis, M. (1976b). *Antonie van Leeuwenhoek*, **42**, 445.
Kudriavzev, V.I. (1960). "Die Systematik der Hefen". Akademie Verlag, Berlin.
Kurtzman, C.P. (1973). *Mycologia* **65**, 388.
Kurtzman, C.P. (1984a). *Mycotaxon* **19**, 271.
Kurtzman, C.P. (1984b). *In* "The Yeasts, a Taxonomic Study" (N.J.W. Kreger-van Rij, ed.), pp. 165–213. Elsevier Biomedical Press, Amsterdam.
Kurtzman, C.P. (1984c). *In* "The Yeasts, a Taxonomic Study" (N.J.W. Kreger-van Rij, ed.), pp. 214–223. Elsevier Biomedical Press, Amsterdam.
Kurtzman, C.P. (1984d). *In* "The Yeasts, a Taxonomic Study" (N.J.W. Kreger-van Rij, ed.), pp. 295–378. Elsevier Biomedical Press, Amsterdam.
Kurtzman, C.P. (1984e). *Antonie van Leeuwenhoek* **50**, 209.
Kurtzman, C.P. and Smiley, M.J. (1974). *In* "Proceedings of the IV International Symposium on Yeasts, Vienna, Austria" (H. Klaushofer and U.B. Sleytr, eds), vol. I. Hochschülerschaft an der Hochschule für Bodenkultur, Vienna.
Kurtzman, C.P. and Smiley, M.J. (1976). *Antonie van Leeuwenhoek* **42**, 355.
Kurtzman, C.P. and Wickerham, L.J. (1973). *Antonie van Leeuwenhoek* **39**, 81.
Kurtzman, C.P., Rogers, R. and Hesseltine, C.W. (1971). *Applied Microbiology* **21**, 870.
Kurtzman, C.P., Vesonder, R.F. and Smiley, M.J. (1974). *Mycologia* **66**, 580.
Kurtzman, C.P., Smiley, M.J. and Baker, F.L. (1975). *Mycopathologia et Mycologia Applicata* **55**, 29.

Kurtzman, C.P., Johnson, C.J. and Smiley, M.J. (1979). *Mycologia* **71**, 844.
Kurtzman, C.P., Smiley, M.J. and Johnson, C.J. (1980). *International Journal of Systematic Bacteriology* **30**, 503.
Kurtzman, C.P., Smiley, M.J., Johnson, C.J. and Hoffman, M.J. (1983). *Abstracts, 5th International Symposium on Yeasts*, Y-5.2.5 (L), p. 246.
Kwon-Chung, K.J. (1975). *Mycologia* **67**, 1197.
Kwon-Chung, K.J. (1977). *International Journal of Systematic Bacteriology* **27**, 293.
Kwon-Chung, K.J. (1980). *Mycologia* **72**, 418.
Kwon-Chung, K.J. (1984). *In* "The Yeasts, a Taxonomic Study" (N.J.W. Kreger-van Rij, ed.), pp. 468–469. Elsevier Biomedical Press, Amsterdam.
Kwon-Chung, K.J. and Fell, J.W. (1984a). *In* "The Yeasts, a Taxonomic Study" (N.J.W. Kreger-van Rij, ed.), pp. 472–482. Elsevier Biomedical Press, Amsterdam.
Kwon-Chung, K.J. and Fell, J.W. (1984b). *In* "The Yeasts, a Taxonomic Study" (N.J.W. Kreger-van Rij, ed.), pp. 483–491. Elsevier Biomedical Press, Amsterdam.
Lachance, M.-A. (1982). *Canadian Journal of Microbiology* **28**, 1194.
Lodder, J. (1934). "Die Anaskosporogenen Hefen, I. Hälfte", *Verhandelingen van de Koninklijke Akademie van Wetenschappen, Sectie II* **32**, 1–256.
Lodder, J. (1970). *In* "The Yeasts, a Taxonomic Study" (J. Lodder, ed.), pp. 1–30. North-Holland Publishing Company, Amsterdam.
Lodder, J. and Kreger-van Rij, N.J.W. (1952). "The Yeasts, a Taxonomic Study". North-Holland Publishing Company, Amsterdam.
Lodder, J., Slooff, W.Ch. and Kreger-van Rij, N.J.W. (1958). *In* "The Chemistry and Biology of Yeasts" (A.H.Cook, ed.), pp. 1–62. Academic Press, New York.
Lushbaugh, W.B., Rowton, E.D. and McGhee, R.B. (1976). *Journal of Invertebrate Pathology* **28**, 93.
Martini, A. (1973). *Annali della Facultà di Agraria della Università degli studi di Perugia* **28**, 1.
Martini, A., Phaff, H.J. and Douglass, S.A. (1972). *Journal of Bacteriology* **111**, 481.
McCully, E.K. and Robinow, C.F. (1972a). *Journal of Cell Science* **10**, 857.
McCully, E.K. and Robinow, C.F. (1972b). *Journal of Cell Science* **11**, 1.
Metchnikoff, E. (1884). *Archiv für Pathologie, Anatomie und Physiologie von R. Virchow* **96**, 177.
Meyer, S.A., Smith, M.T. and Simione Jr., F.P. (1978). *Antonie van Leeuwenhoek* **44**, 79.
Meyer, S.A., Ahearn, D.G. and Yarrow, D. (1984). *In* "The Yeasts, a Taxonomic Study" (N.J.W. Kreger-van Rij, ed.), pp. 585–844. Elsevier Biomedical Press, Amsterdam.
Miller, M.W. (1984). *In* "The Yeasts, a Taxonomic Study" (N.J.W. Kreger-van Rij, ed.), pp. 890–892. Elsevier Biomedical Press, Amsterdam.
Miller, M.W. and Phaff, H.J. (1958). *Mycopathologia et Mycologia Applicata* **10**, 113.
Miller, M.W. and Phaff, H.J. (1984). *In* "The Yeasts, a Taxonomic Study" (N.J.W. Kreger-van Rij, ed.), pp. 279–284. Elsevier Biomedical Press, Amsterdam.
Miller, M.W. and van Uden. N. (1970). *In* "The Yeasts, a Taxonomic Study" (J. Lodder, ed.), pp. 408–429. North-Holland Publishing Company, Amsterdam.
Miller, M.W., Yoneyama, M. and Soneda, M. (1976). *International Journal of Systematic Bacteriology* **26**, 286.
Moore, R.T. (1980). *Botanica Marina* **23**, 361.
Nakase, T. and Suzuki, M. (1985a). *Journal of General and Applied Microbiology* **31**, 49.
Nakase, T. and Suzuki, M. (1985b). *Journal of General and Applied Microbiology* **31**, 71.
Nieuwdorp, P.J., Bos, P. and Slooff, W.Ch. (1974). *Antonie van Leeuwenhoek* **40**, 241.
Nyland, G. (1949). *Mycologia* **41**, 686.
Olive, L.S. (1968). *The Journal of the Mitchell Society* **84**, 261.
Phaff, H.J. (1970a). *In* "The Yeasts, a Taxonomic Study" (J. Lodder, ed.), pp. 209–225. North-Holland Publishing Company, Amsterdam.
Phaff, H.J. (1970b). *In* "The Yeasts, a Taxonomic Study" (J. Lodder, ed.), pp. 725–732. North-Holland Publishing Company, Amsterdam.
Phaff, H.J. (1970c). *In* "The Yeasts, a Taxonomic Study" (J. Lodder, ed.), pp. 756–766. North-Holland Publishing Company, Amsterdam.
Phaff, H.J. (1970d). *In* "The Yeasts, a Taxonomic Study" (J. Lodder, ed.), pp. 815–821. North-Holland Publishing Company, Amsterdam.

Phaff, H.J. (1970e). *In* "The Yeasts, a Taxonomic Study" (J. Lodder, ed.), pp. 831–862. North-Holland Publishing Company, Amsterdam.
Phaff, H.J. (1971). *In* "The Yeasts. Physiology and Biochemistry of Yeasts" (A.H. Rose and J.S. Harrison, eds), vol. 2, pp. 135–210. Academic Press, London.
Phaff, H.J. (1977). *In* "Improvement through chemical and enzymatic modification. Advances in Chemistry Services" (R.E. Feeney and J.R. Whitaker, eds.), pp. 244–282. American Chemical Society, Washington D.C.
Phaff, H.J. (1984a). *In* "The Yeasts, a Taxonomic Study" (N.J.W. Kreger-van Rij, ed.), pp. 22–24. Elsevier Biomedical Press, Amsterdam.
Phaff, H.J. (1984b). *In* "The Yeasts, a Taxonomic Study" (N.J.W. Kreger-van Rij, ed.), pp. 125–129. Elsevier Biomedical Press, Amsterdam.
Phaff, H.J. and Kurtzman, C.P. (1984). *In* "The Yeasts, a Taxonomic Study" (N.W.J. Kreger-van Rij, ed.), pp. 252–262. Elsevier Biomedical Press, Amsterdam.
Phaff, H.J., Miller, M.W. and Shifrine, M. (1956). *Antonie van Leeuwenhoek* **22**, 145.
Phaff, H.J., Miller, M.W. and Miranda, M. (1976). *International Journal of Systematic Bacteriology* **26**, 326.
Phaff, H.J., Lachance, M.A. and Presley, H.L. (1978a). Abstracts of the XII International Congress of Microbiology 1978, München, p.38.
Phaff, H.J., Starmer, W.T., Miranda, M. and Miller, M.W. (1978b). *International Journal of Systematic Bacteriology* **28**, 326.
Pitt, J.I. and Miller, M.W. (1968). *Mycologia* **60**, 663.
Price, C.W. and Phaff, H.J. (1979). *Mycologia* **71**, 444.
Price, C.W., Fuson, G.B. and Phaff, H.J. (1978). *Microbiological Reviews* **42**, 161.
Prusso, D.C. and Wells, K. (1967). *Mycologia* **59**, 337.
Redhead, S.A. and Malloch, D.W. (1977). *Canadian Journal of Botany* **55**, 1701.
Reess, M. (1870). "Botanische Untersuchungen über die Alkoholgärungspilze". Verlag von Arthus Felix, Leipzig.
Rhodes, J.C., Kwon-Chung, K.J. and Popkin, T.J. (1981). *Journal of Bacteriology* **145**, 1410.
Richle, R. and Scholer, H.J. (1961). *Pathologia et Microbiologia* **24**, 783.
Roberts, G.D., Horstmeier, C.D., Land, G.A. and Foxworth, J.H. (1978). *Journal of Clinical Microbiology* **7**, 584.
Rodrigues de Miranda, L. (1972). *Antonie van Leeuwenhoek* **38**, 91.
Rodrigues de Miranda, L. (1978). *Antonie van Leeuwenhoek* **44**, 439.
Rodrigues de Miranda, L. (1979). *Antonie van Leeuwenhoek* **45**, 479.
Rodrigues de Miranda, L. (1980). Progress Report Centraalbureau voor Schimmelcultures, pp. 6–7.
Rodrigues de Miranda, L. (1984a). *In* "The Yeasts, a Taxonomic Study" (N.J.W. Kreger-van Rij, ed.), pp. 577–584. Elsevier Biomedical Press, Amsterdam
Rodrigues de Miranda, L. (1984b). *In* "The Yeasts, a Taxonomic Study" (N.J.W. Kreger-van Rij, ed.), pp. 845–872. Elsevier Biomedical Press, Amsterdam.
Rodrigues de Miranda, L. and Bandoni, R.J. (1984). *In* "The Yeasts, a Taxonomic Study" (N.J.W. Kreger-van Rij, ed.), pp. 546–555. Elsevier Biomedical Press, Amsterdam.
Rooney, L. and Moens, P.B. (1973a). *Canadian Journal of Microbiology* **19**, 1383.
Rooney, L. and Moens, P.B. (1973b). *Canadian Journal of Microbiology* **19**, 1389.
Ruinen, J. (1963). *Antonie van Leeuwenhoek* **29**, 425.
Ruinen, J. and Deinema, M.H. (1964). *Antonie van Leeuwenhoek* **30**, 377.
Sabouraud, R. (1904). "Maladie du Cuir Chevelu. II. Les Maladies Desquamatives". Masson et Cie., Paris.
Saccardo, P.A. and Sydow, P. (1902). Sylloge Fungorum omnium hucusque cognitorum," vol. 16, Supplementum Universale Pars V. Patavii.
Scheda, R. and Yarrow, D. (1966). *Archiv für Mikrobiologie* **55**, 209.
Scott, D.B. and van der Walt, J.P. (1971). *Antonie van Leewenhoek* **37**, 177.
Seeliger, H.P.R. (1956). *Journal of Bacteriology* **72**, 127.
Seymour, F.R., Slodki, M.E., Plattner, R.D. and Stodola, R.M. (1976). *Carbohydrate Research* **48**, 225.
Slodki, M.E. and Wickerham, L.J. (1966). *Journal of General Microbiology* **42**, 381.

Slodki, M.E., Wickerham, L.J. and Bandoni, R.J. (1966). *Canadian Journal of Microbiology* **12**, 489.
Slooff, W.Ch. (1970a). *In* "The Yeasts, a Taxonomic Study" (J. Lodder, ed.), pp. 379–402. North-Holland Publishing Company, Amsterdam.
Slooff, W.Ch. (1970b). *In* "The Yeasts, a Taxonomic Study" (J. Lodder, ed.), pp. 733–755. North-Holland Publishing Company, Amsterdam.
Slooff, W.Ch. (1970c). *In* "The Yeasts, a Taxonomic Study" (J. Lodder, ed.), pp. 1167–1186. North-Holland Publishing Company, Amsterdam.
Slooff, W.Ch. (1970d). *In* "The Yeasts, a Taxonomic Study" (J. Lodder, ed.), pp. 1353–1357. North-Holland Publishing Company, Amsterdam.
Smith, M.T. (1974). *Antonie van Leeuwenhoek* **40**, 441.
Smith, M.T. and Batenburg-van der Vegte, W.H. (1984). *International Journal of Systematic Bacteriology* **34**, 80.
Smith, M.T., van der Walt, J.P. and Johannsen, E. (1976). *Antonie van Leeuwenhoek* **42**, 119.
Smith, M.T., Batenburg-van der Vegte, W.H. and Scheffers, W.A. (1981). *International Journal of Systematic Bacteriology* **31**, 196.
Soneda, M. (1956). *Nagaoa* **6**, 1.
Staib, F. (1981). *Zentralblatt für Bakteriologie und Hygiene, I.Abteilung Originale A* **248**, 575.
Starkey, R.L. (1946). *Journal of Bacteriology* **51**, 33.
Starmer, W.T., Phaff, H.J., Miranda, M. and Miller, M.W. (1978a). *International Journal of Systematic Bacteriology* **28**, 318.
Starmer, W.T., Phaff, H.J., Miranda, M. and Miller, M.W. (1978b). *International Journal of Systematic Bacteriology* **28**, 433.
Starmer, W.T., Phaff, H.J., Miranda, M., Miller, M.W. and Barker, J.S.F. (1979). *International Journal of Systematic Bacteriology* **29**, 159.
Stelling-Dekker, N.M. (1931). Die sporogenen Hefen. *Verhandelingen der Koninklijke Akademie van Wetenschappen, Afdeling Natuurkunde (sectie II)* **28**, 1.
Stodola, F.H., Deinema, M.H. and Spencer, J.F.T. (1967). *Bacteriological Reviews* **31**, 213.
Streiblová, E. and Wolf, A. (1972). *Zeitschrift für Allgemeine Mikrobiologie* **12**, 673.
Tsuchiya, T., Fukazawa, Y., Taguchi, M., Nakase, T. and Shinoda, T. (1974). *Mycopathologia et Mycologia Applicata* **53**, 77.
Tulasne, L.R. and Tulasne, C. (1865). *Selecta Fungorum Carpologia* **3**, 61.
Van der Walt, J.P. (1956). *Antonie van Leeuwenhoek* **22**, 190.
Van der Walt, J.P. (1957). *Antonie van Leeuwenhoek* **23**, 23.
Van der Walt, J.P. (1970a). *In* "The Yeasts, a Taxonomic Study" (J. Lodder, ed.), pp. 34–113. North-Holland Publishing Company, Amsterdam.
Van der Walt, J.P. (1970b). *In* "The Yeasts, a Taxonomic Study" (J. Lodder, ed.), pp. 157–165. North-Holland Publishing Company, Amsterdam.
Van der Walt, J.P. (1970c). *In* "The Yeasts, a Taxonomic Study" (J. Lodder, ed.), pp. 316–378. North-Holland Publishing Company, Amsterdam.
Van der Walt, J.P. (1970d). *In* "The Yeasts, a Taxonomic Study" (J. Lodder, ed.), pp. 555–718. North-Holland Publishing Company, Amsterdam.
Van der Walt, J.P. (1970e). *Antonie van Leeuwenhoek* **36**, 49.
Van der Walt, J.P. (1978). *Bothalia* **12**, 563.
Van der Walt, J.P. and Hopsu-Havu, V.K. (1976). *Antonie van Leeuwenhoek* **42**, 157.
Van der Walt, J.P. and Johannsen, E. (1975). *CSIR Research Report* no. 325, p.1, Pretoria, South Africa.
Van der Walt, J.P. and Liebenberg, N.V.D.W. (1973). *Antonie van Leeuwenhoek* **39**, 121.
Van der Walt, J.P. and Scott, D.B. (1971). *Mycopathologia et Mycologia Applicata* **43**, 279.
Van der Walt, J.P. and von Arx, J.A. (1980). *Antonie van Leeuwenhoek* **46**, 517.
Van der Walt, J.P. and Yarrow, D. (1984a). *In* "The Yeasts, a Taxonomic Study" (N.J.W. Kreger-van Rij, ed.), pp. 45–104. Elsevier Biomedical Press, Amsterdam.
Van der Walt, J.P. and Yarrow, D. (1984b). *South African Journal of Botany* **3**, 340.
Van der Walt, J.P., Weijman, A.C.M. and von Arx, J.A. (1981). *Sydowia, Annales Mycological* Serie II, **34**, 191.
Van Uden, N. and Castelo-Branco, R. (1961). *Journal of General Microbiology* **26**, 141.

Vidal-Leiria, M., Buckley, H. van Uden, N. (1979). *Mycologia* **71**, 494.
Von Arx, J.A. (1972). *Antonie van Leeuwenhoek* **38**, 289.
Von Arx, J.A. (1977). *Antonie van Leeuwenhoek* **43**, 333.
Von Arx, J.A. and van der Walt, J.P. (1976). *Antonie van Leeuwenhoek* **42**, 309.
Von Arx, J.A. and Weijman, A.C.M. (1979). *Antonie van Leeuwenhoek* **45**, 547.
Von Arx, J.A., Rodrigues de Miranda, L., Smith, M.T. and Yarrow, D. (1977). *Studies in Mycology*, no. 14, pp. 1–42. Centraalbureau voor Schimmelcultures, Baarn.
Weijman, A.C.M. (1979). *Antonie van Leeuwenhoek* **45**, 119.
Wickerham, L.J. (1951). Taxonomy of Yeasts. Technical Bulletin no. 1029, United States Department of Agriculture, Washington, D.C.
Wickerham, L.J. (1958). *Science* **128**, 1504.
Wickerham, L.J. (1964a). *Mycologia* **56**, 253.
Wickerham, L.J. (1964b). *Mycologia* **56**, 398.
Wickerham, L.J. (1970). *In* "The Yeasts, a Taxonomic Study" (J.Lodder, ed.), pp. 226–315. North-Holland Publishing Company, Amsterdam.
Wickerham, L.J. and Burton, K.A. (1948). *Journal of Bacteriology* **56**, 363.
Wickerham, L.J. and Burton, K.A. (1956). *Journal of Bacteriology* **71**, 290.
Wickerham, L.J. and Burton, K.A. (1962). *Bacteriological Reviews* **26**, 382.
Wickerham, L.J., Kurtzman, C.P. and Herman, A.I. (1970). *Science* **167**, 1141.
Winge, Ö. and Laustsen, O. (1939). *Comptes Rendus des Travaux du Laboratoire Carlsberg, Série Physiologique* **22**, 337.
Yamada, Y. and Banno, I. (1984a). *Transactions of the Mycological Society, Japan* **25**, 455.
Yamada, Y. and Banno, I. (1984b). *Journal of General and Applied Microbiology* **30**, 523.
Yamada, Y. and Kondo, K. (1972). *In* "Fermentation Technology Today, Proceedings of the IV International Fermentation Symposium" (G.Terui, ed.), pp. 63–69. Society of Fermentation Technology, Osaka, Japan.
Yamada, Y., Arimoto, M. and Kondo, K. (1973a). *Journal of General and Applied Microbiology* **19**, 353.
Yamada, Y., Kanematsu, Y., Ohashi, M. and Kondo, K. (1973b). *Agricultural and Biological Chemistry* **37**, 621.
Yamada, Y., Okada, T., Ueschima, O. and Kondo, K. (1973c). *Journal of General and Applied Microbiology* **19**, 189.
Yamada, Y., Arimoto, M. and Kondo, K. (1976a). *Journal of General and Applied Microbiology* **22**, 293.
Yamada, Y., Nojiri, M., Matsuyama, M. and Kondo, K. (1976b). *Journal of General and Applied Microbiology* **22**, 325.
Yamada, Y., Arimoto, M. and Kondo, K. (1977). *Antonie van Leeuwenhoek* **43**, 65.
Yamada, Y., Takinami-Nakamura, H., Tahara, Y. and Smith, M.T. (1980). *Antonie van Leeuwenhoek* **46**, 595.
Yamazaki, M. and Komagata, K. (1981). *International Journal of Systematic Bacteriology* **31**, 361.
Yarrow, D. and Meyer, S.A. (1978). *International Journal of Systematic Bacteriology* **28**, 611.

3 Molecular Taxonomy

C.P. Kurtzman and H.J. Phaff

Northern Regional Research Center, Agricultural Research Service, United States Department of Agriculture, Peoria, Illinois 61604, USA and Department of Food Science and Technology, University of California, Davis, California 95616, USA.

I. INTRODUCTION

The yeasts are a phylogenetically diverse group of fungi whose teleomorphs (sexual states) are distributed between the Ascomycotina and the Basidiomycotina. The grouping of yeasts as compared to the higher fungi is one of convenience, bound together by the characteristics of predominantly unicellular growth resulting from either budding or fission and a sexual stage not enclosed in a complex fruiting body. The criteria accepted for use in speciation have, over time, paralleled many other areas of microbiology, first

The Yeasts Vol. 1, 2nd edition
ISBN 0-12-596411-0

relying upon the morphology of vegetative and sexual stages, followed later by definition of taxa through biochemical and genetic tests, and finally, owing to a partial failure of the foregoing to produce definitive results, a reliance upon nucleic-acid sequence relatedness (Phaff, 1980) and other molecular criteria (Holzschu, 1981).

The first modest use of physiological tests in yeast taxonomy, as an adjunct to the already accepted morphological criteria, may be found in the monumental works of Stelling-Dekker (1931), Lodder (1934) and Diddens and Lodder (1942). Here, reliance was placed on ability to ferment glucose, galactose, sucrose, maltose, lactose and raffinose followed by growth on certain of these sugars as well as ethanol, and the nitrogen sources ammonium sulphate, asparagine, urea, peptone and nitrate. Proteolysis of gelatin (liquefaction) was also used as a criterion. Although helpful at first, the resolution afforded by this small number of compounds soon reached its limits, and further differentiation of taxa could only result from expanding the spectrum of potentially assimilatable compounds. Bedford (1942) and especially Wickerham and Burton (1948) and Wickerham (1951) proposed the use of about 40 carbon-containing compounds for differentiation of taxa, a convention now widely accepted in yeast systematics (Lodder, 1970; Kreger-van Rij, 1984; Barnett *et al.*, 1983).

The realization that a single gene could confer on a yeast the ability to ferment or assimilate a disaccharide (Winge and Roberts, 1949; Lindegren and Lindegren, 1949) raised the possibility that species might be inadvertently and artificially established simply because of the functionality of one or a few genes. Verification of species relationships through sexual hybridization studies is an ideal way in which to define taxa, but lack of fertility does not preclude conspecificity because only a few genes may affect the ability to mate (Fell *et al.*, 1973; Hicks and Herskovitz, 1976). The possibility of whole genome comparisons through DNA–DNA re-association and DNA–rRNA re-association has been realized during the past 15 years, and it is now practical to define species as well as to assess present taxonomic methods using these molecular techniques. Consequently, relationships determined from nucleic acid studies will be used in the ensuing discussion as a reference for evaluating many current taxonomic schemes and methodologies.

II. NUCLEIC ACID ISOLATION AND PURIFICATION

The methodology for isolation and purification of yeast DNA has been dealt with in detail elsewhere, and the following account will serve only as a brief review to direct attention to some of the more pertinent references. For most yeast strains, DNA isolation is relatively straightforward. Cell wall breakage may be effected by mechanical or lytic means. Mechanical methods have the advantage of being relatively rapid but, unless cell breakage is carefully monitored, considerable shearing of the DNA may result. The Braun cell homogenizer is one of the more frequently used mechanical devices, and causes breakage by shaking of sucrose-buffered cell suspension in the presence of 0.45 mm glass beads (Bak and Stenderup, 1969; Price *et al.*, 1978; Kurtzman *et al.*, 1980a). An inexpensive variation of this technique involves placing the buffered cell and bead mixture into a test tube and "Vortexing" the mixture (Lipinski *et al.*, 1976; Van Etten and Freer, 1978). Less frequently reported methods include grinding acetone-dried cells in a ball mill (Martini and Phaff, 1973) and grinding cells in liquid nitrogen with a mortar and pestle (Storck and Alexopoulos, 1970; Mendonca-Hagler *et al.*, 1974). Sonication is seldom used for cell breakage, because it is very inefficient for yeasts and it causes extreme and uncontrolled DNA shearing.

Lytic methods offer the possibility of isolating DNA with a higher molecular weight than might be obtained by mechanical cell disruption. Meyer and Phaff (1969) and Bicknell and Douglas (1970) used a modification of the Smith and Halvorson (1967) technique in which cells undergo autolysis in the presence of saline containing EDTA (ethylene diamine tetra-acetate), 2-mercaptoethanol and sodium dodecyl sulphate. The occasional failure of this method has prompted the widespread use of mechanical disruption. The commercial enzyme preparations Glusulase and Zymolyase may be used to prepare osmotically fragile sphaeroplasts by cell-wall digestion. Following this step, high molecular-weight DNA is isolated from the lysed sphaeroplasts. Cryer *et al.* (1975) have given details of these procedures. Zymolyase generally promotes more rapid cell-wall dissolution than Glusulase, but neither enzyme preparation is particularly effective with basidiomycetes. Brady (1981), however, reported relatively rapid dissolution of walls of *Rhodotorula* spp. by an enzyme preparation from culture filtrates of *Penicillium lilacinum*. Another success in this area was achieved by Rhodes and Kwon-Chung (1982) who produced protoplasts from blastospores of *Cryptococcus neoformans* with the multi-enzyme preparation mutanase.

Following rupture or lysis of the cells, DNA isolation and purification may be carried out by the methods of Marmur (1961) and Bernardi *et al.* (1970) as detailed by Price *et al.* (1978). Basically, this consists of making the cell suspension 1 M in sodium perchlorate and 1% in sarcosate and emulsifying with an equal volume of chloroform/3-methylbutan-1-ol (24:1, v/v). After incubation and phase separation by centrifugation, DNA is precipitated from the aqueous phase with cold ethanol. The DNA is treated with a combination of α-amylase and pancreatic ribonuclease (RNAase), and finally with Pronase. Following enzyme removal, the DNA is again precipitated with ethanol, spooled, redissolved and further treated with α-amylase, pancreatic RNAase and T_1-RNAase. Additional purification can include hydroxyapatite fractionation and banding in a caesium chloride gradient.

Some yeasts have active deoxyribonucleases (DNAases) that escape denaturation during the usual extraction procedures and cause sufficient degradation of the DNA to prevent spooling. Several methods have produced various degrees of success for deactivating nucleases and merit testing if problems are encountered. Diethyl pyrocarbonate, used at 0.5–1.0% at the time of cell breakage, gave encouraging results in studies of *Lipomyces* spp. (D.L. Holzschu and H.J. Phaff, unpublished observations) and *Sporidiobolus* spp. (Holzschu *et al.*, 1981). Ehrenberg *et al.* (1976) discussed the interaction of diethyl pyhrocarbonate with nucleic acids and nucleases and emphasized some of the precautions to be taken. Allergy to the compound may be a problem in some individuals. C.P. Kurtzman and his coworkers (unpublished observations) found that the spermine/spermidine/sucrose buffer of Morris, as cited by Timberlake (1978), was much more effective at inactivating nucleases in *Aspergillus* spp. and *Neurospora* spp. than the sucrose buffer frequently used in yeast studies (Price *et al.*, 1978). This was confirmed by Guého *et al.* (1984, 1985) for species of *Geotrichum, Dipodascus* and *Trichosporon.* Heating cells to 60°C for 20 minutes was also reasonably effective in deactivating nucleases in species of *Aspergillus* and *Neurospora* (unpublished results of C.P. Kurtzman and his colleagues). Higher temperatures, if they do not result in denaturation of the DNA, may be even more effective. Methods for assessing DNA purity include chemical analyses as well as spectrophotometric measurements and have been summarized by Price *et al.* (1978) and Johnson (1981).

Techniques for isolation of RNA species have been presented by Gillespie and Spiegelman (1965), Bicknell and Douglas (1970) and Johnson (1981). These references also include procedures for ribosomal RNA–DNA reassociation experiments. Although cellular RNA concentration is many-fold higher than that of DNA, the nearly ubiquitous occurrence of RNAases and their marked stability make isolation of undegraded RNA more difficult than isolation of high molecular-weight DNA.

III. DEOXYRIBONUCLEIC ACID BASE COMPOSITION

A. Methodology

Deoxyribonucleic acid base composition of the nuclear or mitochondrial DNA, expressed as molar percentages of guanine plus cytosine (G + C), may be determined from thermal denaturation profiles (melts), buoyant density in caesium salt gradients generated by ultracentrifugation, chemical analysis, absorbance ratios, or high-pressure liquid chromatography of nucleotides or free bases (Johnson, 1981). Many investigators use thermal denaturation (Marmur and Doty, 1962), but Bak (1973) showed that results obtained by this method can be greatly affected by sample impurities and/or minor DNA species and must therefore be interpreted with caution. Caesium chloride buoyant-density determinations (Schildkraut *et al.*, 1962) generally show the greatest accuracy since they are unbiased by the presence of contaminating RNA, mitochondrial DNA and other impurities, such as carbohydrates and proteins.

B. Taxonomic uses of guanine plus cytosine (G + C) values

The taxonomic uses of G + C values are mainly exclusionary, because the 600 or so yeast species range in G + C content from approximately 27 to 70 mol%, and overlap between unrelated species is inevitable. For example, *Zygosaccharomyces mrakii* and *Zygosacch. florentinus* have G + C contents of 42.2 and 42.4 mol%, respectively, yet they show only 6.0% base sequence complementarity (Price *et al.*, 1978). Consequently, since similar G + C content does not necessarily mean a similar base sequence in the DNA and thus that strains belong to the same species, we must ask how dissimilar G + C values need be to state categorically that two strains represent different species.

The answer to this question depends to some extent upon the method used for determination of base composition. When buoyant density is used, the span of G + C values among strains of a species is usually less than 1% (Price *et al.*, 1978; Kurtzman *et al.*, 1980 a,b; Phaff *et al.*, 1985), which is approximately coincident with the accuracy of the method, i.e. a standard deviation of 0.00–0.50 mol%. Thermal denaturation shows greater variation, and strains of a species may exhibit a G + C range of as much as 2% (Meyer *et al.*, 1978). Thus, strains showing a difference in G + C contents in excess of 1.0–1.5% by buoyant density or 2.0–2.5% by thermal denaturation may be expected to represent different species.

Representative G + C values for yeasts are given in Table I. These values illustrate the interesting fact that the G + C content of ascomycetous yeasts is about 27–50%, whereas that of basidiomycetous yeasts is approximately 50–70%. Except for the narrow range of 48–52%, where some overlap occurs, the taxonomic affinity of imperfects can be reliably determined from their base composition. For species in the doubtful range, affinity determination may be made by examining cell-wall structure under the transmission electron microscope (Kreger-van Rij and Veenhuis, 1971) or through use of the diazonium blue B test (van der Walt and Hopsu-Havu, 1976; Hagler and Ahearn, 1981).

The range of G + C contents among species within a genus is quite often

Table I. Nuclear deoxyribonucleic acid base composition of selected yeast taxa

Species	G + C content (mol%)	Reference
Ascomycetes		
Arthroascus javanensis (Klöcker) von Arx	32.5[a]	Nakase and Komagata (1968b)
Citeromyces matritensis (Santa Maria) Santa Maria	45.5[a]	Nakase and Komagata (1968a)
Debaryomyces		
Deb. hansenii (Zopf) Lodder et Kreger-van Rij	38.6	Price *et al.* (1978)
Deb. marama di Menna	39.1	Price *et al.* (1978)
Deb. tamarii Ohara et Nonomura	46.4	Price *et al.* (1978)
Deb. vanriji (van der Walt et Tscheuschner) Abadie, Pignal et Jacob	33.2	Price *et al.* (1978)
Dekkera bruxellensis van der Walt	38	Nakase and Komagata (1968a)
Hanseniaspora		
H'spora. valbyensis Klöcker	30.0	Meyer *et al.* (1978)
H'spora. occidentalis Smith	34.9	Meyer *et al.* (1978)
H'spora vineae van der Walt et Tscheuschner	40.2	Meyer *et al.* (1978)
Hansenula[b]		
H. ciferrii Lodder	32.2	Nakase and Komagata (1971a)
H. alni Phaff, Miller et Miranda	41.9	Phaff *et al.* (1979)
H. polymorpha de Morais et Maia	48.0	H.J. Phaff and D.L. Holzschu (unpublished observations)
H. henricii Wickerham	49.6	Fuson *et al.* (1979)
H. populi Phaff, Yamada, Tredick et Miranda	44.2	Phaff *et al.* (1983)
Issatchenkia		
I. scutulata (Phaff, Miller et Miranda) Kurtzman, Smiley et Johnson	34.0	Kurtzman *et al.* (1980b)
I. orientalis Kudriavzev	40.3	Kurtzman *et al.* (1980b)
Kluyveromyces		
K. blattae Henninger et Windisch	33.1	Price *et al.* (1978)
K. polysporus van der Walt	35.3	H.L. Presley and H.J. Phaff (unpublished observations)
K. phaffii (v.d. Walt) v.d. Walt	35.3	H.L. Presley and H.J. Phaff (unpublished observations)
K. lodderi (v.d. Walt et Tscheuschner) v.d. Walt	35.7	H.L. Presley and H.J. Phaff (unpublished observations)
K. africanus v.d. Walt	38.5	H.L. Presley and H.J. Phaff (unpublished observations)
K. aestuarii (Fell) v.d. Walt	39.8	H.L. Presley and H.J. Phaff (unpublished observations)
K. delphensis (v.d. Walt et Tscheuschner) v.d. Walt	40.2	H.L. Presley and H.J. Phaff (unpublished observations)
K. lactis (Dombrowski) v.d. Walt	40.3	H.L. Presley and H.J. Phaff (unpublished observations)
K. fragilis (Jörgensen) van der Walt	41.1	H.L. Presley and H.J. Phaff (unpublished observations)
K. wickerhamii (Phaff, Miller et Shifrine) v.d. Walt	41.9	H.L. Presley and H.J. Phaff (unpublished observations)
K. dobzhanskii (Shehata, Mrak et Phaff) v.d. Walt	42.5	H.L. Presley and H.J. Phaff (unpublished observations)
K. thermotolerans (Philippov) Yarrow	46.2	H.L. Presley and H.J. Phaff (unpublished observations)
Lipomyces		
L. anomalus Babjeva et Gorin	43.8	H.J. Phaff and D.L. Holzschu (unpublished observations)
L. lipofer Lodder et Kreger-van Rij	49.3	H.J. Phaff and D.L. Holzschu (unpublished observations)
Lodderomyces elongisporus van der Walt	39.7	Meyer and Phaff (1969)

Table I. Continued

Species	G + C content (mol%)	Reference
Metschnikowia		
M. reukaufii Pitt et Miller	42.2	Meyer and Phaff (1969)
M. pulcherrima Pitt et Miller	48.3	
Pachysolen tannophilus Boidin et Adzet	43	Nakase and Komagata (1968a)
Pichia		
P. deserticola Phaff, Starmer, Tredick et Miranda	27.8	Phaff *et al.* (1985)
P. mexicana Miranda, Holzschu, Phaff et Starmer	42.6	Miranda *et al.* (1982)
P. kluyveri Bedford ex Kudriavzev	30	Nakase and Komagata (1968a)
P. cactophila Starmer, Phaff, Miranda et Miller	36.3	Starmer *et al.* (1978a)
P. amethionina Starmer, Phaff, Miranda et Miller	33.1	Starmer *et al.* (1978b)
P. scolyti (Phaff et Yoneyama) Kreger-van Rij	40.1	Fuson *et al.* (1980)
P. strasburgensis (Ramirez et Boidin) Phaff	40.6	
P. pijperi v.d. Walt et Tscheuschner	42.5	
P. pinus (Holst) Phaff	43.0	
P. fermentans Lodder	43.1	
P. ohmeri (Etchells et Bell) Kreger-van Rij	44.3	
P. guilliermondii Wickerham	44.4	
P. wickerhamii (v.d. Walt) Kreger-van Rij	45.6	
P. haplophila Shifrine et Phaff	39.5	Nakase and Komagata (1968b)
P. membranaefaciens Hansen	44.3	Fuson *et al.* (1980)
P. rhodanensis (Ramirez et Boidin) Phaff	51.9	Kurtzman *et al.* (1980a)
Saccharomyces		
Sacch. cerevisiae Hansen	39.9	Price *et al.* (1978)
Sacch. Kluyveri Phaff, Miller et Shifrine	40.0	Meyer and Phaff (1969)
Saccharomycopsis		
S. fibuligera (Lindner) Klöcker	40.5	Nakase and Komagata (1968a)
S. capsularis Schionning	43	
Schizosaccharomyces		
Schiz. pombe Lindner	42	Rost and Venner (1964)
Schiz. octosporus Beijerinck	42	Storck (1966)
Schwanniomyces occidentalis Klöcker	35.3	Price *et al.* (1978)
Torulaspora		
T'spora delbrueckii Lindner	43.8	Price *et al.* (1978)
T'spora globosa (Klöcker) v.d. Walt et Johannsen	47.5	
Wickerhamia fluorescens (Soneda) Soneda	38	Nakase and Komagata (1968a)
Wingea robertsii (v.d. Walt) v.d. Walt	42.7	Price *et al.* (1978)
Yarrowia lipolytica (Wickerham, Kurtzman et Herman) van der Walt et von Arx	49.6	Storck (1966)
Zygosaccharomyces bisporus Naganishi	44.0	Price *et al.* (1978)
Zygosaccharomyces florentinus Castelli ex Kudriavzev	42.4	

Table I. Continued

Species	G + C content (mol%)	Reference
Basidiomycetes		
Filobasidiella neoformans Kwon-Chung	51.5	Storck *et al.* (1969)
Rhodosporidium toruloides Banno	60	Nakase and Komagata (1971b)
Sporidiobolus		
Sporid. salmonicolor Fell et Tallman	62.0	Holzschu *et al.* (1981)
Sporid. johnsonii Nyland	63.5	Holzschu *et al.* (1981)
Deuteromycetes		
Candida		
C. albicans (Robin) Berkhout	35.5	Nakase and Komagata (1968a)
C. freyschussii Buckley et van Uden	43.8	Holzschu *et al.* (1979)
C. utilis (Henneberg) Lodder et Kreger-van Rij	45.8	Kurtzman *et al.* (1979)
C. lusitaniae van Uden et do Carmo-Sousa	44.7	Holzschu *et al.* (1979)
C. zeylanoides (Cast.) Langeron et Guerra	57.6	Stenderup and Bak (1968)
C. parapsilosis (Ashford) Langeron et Talice	39.5	Holzschu *et al.* (1979)
Phaffia rhodozyma Miller, Yoneyama et Soneda	48.3	Miller *et al.* (1976)
Rhodotorula		
Rh. minuta (Saito) Harrison	53	Storck *et al.* (1969)
Rh. glutinis (Fres.) Harrison	67	Nakase and Komagata (1968a)
Sterigmatomyces		
St. penicillatus Rodrigues de Miranda	49.5	Kurtzman *et al.* (1980c)
St. halophilus Fell	56.2	Kurtzman *et al.* (1980c)
Trichosporon		
Tr. pullulans (Lindner) Diddens et Lodder	57	Nakase and Komagata (1968a)
Tr. beigelii (Küchenmeister et Robenhorst) Vuillemin	63.5	Guého *et al.* (1984)
Trigonopsis variabilis Schachner	47	Nakase and Komagata (1968a)

[a] The values reported by Nakase and Komagata (1968a,b) have been increased by 3% as suggested by Meyer and Phaff (1970).

[b] These species were transferred to the genus *Pichia* (Kurtzman, 1984b).

10% or less as found in *Debaryomyces, Hanseniaspora, Issatchenkia, Kluyveromyces, Metschnikowia* and several other genera (Table I). By contrast, species in the genus *Pichia* differ by as much as 24% while the disparity in base composition among *Rhodotorula* spp. is 14%. It seems likely that *Pichia* is an amalgam of several genera, but this is not certain of *Rhodotorula*. It is also apparent from G + C values, as well as from other tests, that the genus *Candida* and its synonym *Torulopsis* (Yarrow and Meyer, 1978) is composed of anamorphs of both ascomycetes and basidiomycetes.

IV. DEOXYRIBONUCLEIC ACID RELATEDNESS

A. Methodology

The methods for assessing DNA relatedness vary, but short of sequencing they rely on measuring the extent and stability of renatured DNA strands from test pairs, i.e. the fidelity of complementary base pairing. Depending upon the method, the DNA may or may not need to be labelled with radioisotopes. An interesting variation of the DNA–DNA hybridization technique was employed by Bak and Stenderup (1969). Radioactive complementary RNA was synthesized *in vitro* with each test DNA serving as a template, and the relatedness between taxa was then assessed from hybridization of single-standed DNA with the synthesized RNA.

In vivo labelling of DNA is frequently done using ^{14}C, ^{3}H or ^{32}P. Each isotope has its particular advantages and disadvantages. The DNA may also be labelled *in vitro* with ^{125}I (Commerford, 1971; Mendonca-Hagler and Phaff, 1975; Holzschu *et al.*, 1979) or by nick translation (Kelley *et al.*, 1970; Maniatis *et al.*, 1975). Deoxyribonucleic acid labelled with ^{125}I *in vitro* must be highly purified to remove RNA, protein and carbohydrate completely. This requires an extra purification step of banding the DNA in caesium chloride in a preparative ultracentrifuge. Following labelling, DNA re-association reactions may be carried out using the membrane method which involves immobilizing single strands of one of the DNA species onto nitrocellulose filters and allowing sheared single strands of the second DNA to react with the immobilized DNA (Denhardt, 1966). Alternatively, both DNA species may be allowed to react in free-solution, and the extent of re-association assessed by the extent of binding of the resulting duplexes to hydroxyapatite (Brenner, 1973) or resistance to hydrolysis by S_1 nuclease (Crosa *et al.*, 1973). Price *et al.* (1978) and Johnson (1981) provide references and detailed procedures for these methods.

Besides following DNA re-association with radioactive preparations, the reactions may also be monitored spectrophotometrically by measuring the kinetics of duplex formation. This technique has been refined to the point where results are comparable to data obtained using isotopes (Subirana and Doty, 1966; De Ley *et al.*, 1970; Seidler and Mandel, 1971; Kurtzman *et al.*, 1980a).

A number of factors affect DNA renaturation reactions and so may have a marked effect on the results obtained. Highly purified DNA preparations are essential. Contaminating protein or RNA can result in falsely high or low relatedness values depending on whether the contaminants act as single-stranded or double-stranded DNA in the assay system (Brenner, 1973).

Several classes of DNA occur in eukaryotes, and their presence affects estimates of relatedness between taxa since it is the intent of most studies to measure genomic similarity from complementarity of single-copy nuclear DNA sequences. Christiansen *et al.* (1971) reported up to 15% rapidly renaturing sequences in the nuclear DNA (nDNA) of yeasts, which may be interpreted as analogous to the repeated sequences found in higher eukaryotes (Britten and Kohne, 1968). However, the presence of rapidly renaturing, plasmid-like 2 μm circular DNA in *Sacch. cerevisiae* (Bak *et al.*, 1972) with a buoyant density similar to nDNA suggests the proportion of repeated sequences in nDNA to be smaller than first thought. Similar circular DNA species have been reported from *Hansenula wingei* and *Torulopsis glabrata* (O'Connor *et al.*, 1975, 1976). Linear DNA plasmids have been discovered in *Kluyveromyces lactis* (Gunge *et al.*, 1981). These plasmids, which are lighter than the densities of both nuclear and mitochondrial DNA, appear to be associated with the killer character of this yeast. The work of Timberlake (1978) indicates the genome of the fungus *Aspergillus nidulans* to consist of approximately 97 to 98% unique sequences and 2 to 3% repeated sequences. The repeated sequences appear to code for ribosomal RNA.

Nuclear DNA represents the preponderance of cellular DNA in yeasts, and it has a kinetic complexity of 6.5 to $14 \cdot 10^9$ daltons or about two to five times that found in *Escherichia coli* (Christiansen *et al.*, 1971; Petes *et al.*, 1973). Between 5 and 20% of the total DNA in *Sacch. cerevisiae* is mitochondrial in origin. It is present in a variable number of copies and is arranged in closed circles with a G + C content noticeably lower than the nDNA (Hartwell, 1974). Sedimentation studies (Blamire *et al.*, 1972) and measurements of contour length (Hollenberg *et al.*, 1970; Petes *et al.*, 1973) suggest a molecular mass of $46 \cdot 10^6$ daltons. The mitochondrial DNA (mtDNA) contour length for several other yeasts has been reported to be 25 to 50% of that of *Sacch. cerevisiae* (O'Connor *et al.*, 1975, 1976).

If comparisons of taxa are to be made from single-copy nDNA, it is necessary to remove mtDNA, plasmid-like circular DNA and repeat sequence nDNA. Circular plasmid-like DNA as well as circular mtDNA may be relatively easily separated from nDNA by centrifugation in caesium chloride in the presence of ethidium bromide. Williamson and Fennell (1975) emphasized that resolution is poor should mtDNA break into linear fragments during isolation, a not uncommon happening. However, they point out that linear mtDNA may be effectively resolved in caesium chloride gradients when treated with 4′,6-diamidino-2-phenylindole, a fluorescent dye that preferentially binds with AT-rich DNA. Repeated sequences may be removed from nDNA by partial re-association of sheared DNA fragments to a C_0t value of 0.1–0.2 followed by fractionation with hydroxyapatite as described in detail by Price *et al.* (1978). Since the rate of re-association is affected by concentration, multiple copies duplex most rapidly and can be removed because of their ability to bind to hydroxyapatite (Britten and Kohne, 1968). Price *et al.* (1978) and Fuson *et al.* (1979) used this method to remove repeated sequences from yeast DNA for taxonomic studies. The fractionation undoubtedly also removes mtDNA and plasmid-like DNA because both occur as multiple copies and their small molecular weight would allow rapid re-association when in significant concentration. The same effect may be obtained in spectrophotometric determinations of relatedness if the time between melting and cooling to the re-association temperature ($T_m - 25$) is not too rapid. Under their conditions, Kurtzman *et al.* (1980a) determined a cooling period of 12 minutes to allow duplexing of most repetitive sequences as well as of mitochondrial DNA. Relatedness values of strains common to the work of Price *et al.* (1978) and of Kurtzman *et al.* (1980a) were essentially identical, showing that similar results can be obtained with rather different methodologies.

For most studies, DNA fragment size and the ionic strength of the incubation medium may be maintained at any constant value within a fairly wide range without significantly affecting the results (Brenner, 1973). However, increasing ionic strength increases the reaction rate. Seidler *et al.* (1975) and Kurtzman *et al.* (1980a) used relatively high concentrations of standard saline citrate (150 mM sodium chloride/15 mM sodium citrate, pH 7.0) when assessing DNA relatedness spectrophotometrically in order to complete the determination within a few hours. Thus, when C_0t values (C_0t is initial nucleotide concentration in moles per litre multiplied by the time in seconds; Britten and Kohne, 1968) are used as a measure of the extent of re-association, the ionic strength of the reaction mixture must be known if values obtained in different studies are to be compared. For example, the incubation mixture used by Kurtzman *et al.* (1980a) allowed DNA re-association to proceed nearly five times faster than that which occurred in the buffer system employed by Price *et al.* (1978). As noted in the definition of C_0t values, initial nucleotide concentration also affects the reaction rate; consequently a solution containing 100 μg of DNA per ml would reassociate twice as fast as one containing 50 μg.

Optimal DNA re-association occurs at 25 to 30°C below the mid-point of the melting curve (T_m), and this is dependent on base composition since GC base pairs with their triple hydrogen bonds have greater thermal stability than do AT pairs with their double hydrogen bonds (Marmur and Doty, 1962; Marmur *et al.*, 1963). Assuming that two yeasts have melting temperatures of 85°C and thus the optimum re-association temperature ($T_m - 25$) is 60°C, pairs that are highly related at 60°C also show high relatedness at more stringent temperatures such as 66–67°C or 75°C. Pairs showing only moderate relatedness at 60°C show much less binding at higher temperatures than closely related pairs, reflecting the greater mismatch of base pairs present. These studies point out the need for proper re-annealing temperatures, and also show that higher incubation temperatures allow an estimate of stability that indicates the extent of base pair mismatch. Temperatures below $T_m - 25$ would show even greater re-association for sparingly related pairs. Relatedness may also be estimated from the thermal stability of the renatured heterologous DNA as compared with renatured homologous DNA. This may be done spectrophotometrically by monitoring thermal-melts or with labelled DNA by thermal elution from hydroxyapatite columns (Ullman and McCarthy, 1973), or by the batch method of Brenner *et al.* (1969) as modified by Price *et al.* (1978). Brenner *et al.* (1972) equated one degree of difference in thermal stability with 1% mismatch in base pairs. Lower thermal stability (ΔT_m) usually correlates well with lower DNA homology. In conclusion, we must view DNA relatedness values obtained from whole-genome comparisons as an estimate of overall base sequence similarity under defined conditions rather than as a certain percentage of the DNA that has precisely the same base sequence.

B. Correlation with genetic hybridization

On the basis of extensive comparisons between species of enteric bacteria, Brenner (1973) proposed that strains showing 70% or greater DNA relatedness should be considered members of the same species. Johnson (1973) compared numerous anaerobic bacteria, and arrived at essentially the same guidelines. Price *et al.* (1978) examined yeast species in four genera and proposed that strains having 80% or greater DNA relatedness were conspecific, an opinion expressed earlier by Martini and Phaff (1973) from their studies in *Kluyveromyces* spp. Further work on *Kluyveromyces* spp. by H.L. Presley and H.J. Phaff (Phaff, 1980) made it necessary to extend the lower limit of the above range to approximately 65% on the basis of experiments with strains of *K. drosophilarum*. Two strains of this species isolated in Japan showed approximately 65% DNA homology with that of the type strain isolated in California. Phaff and Holzschu (1981) reported that DNA complementarity between the type strains of *P. amethionina* var. *amethionina* and *P. amethionina* var. *pachycereana* was 64%. Speciation in basidiomycetous yeasts may be different from that in ascomycetous yeasts. Baharaeen *et al.* (1982), who studied evolutionary divergence among biotypes of the *Cryptococcus vishniacii* complex, found DNA–DNA homology values among members of the complex ranging from 21 to 79%. They arbitrarily considered that homology values of 60% and above indicated conspecificity with the type strain of *Cr. vishniacii*. Seven additional species were recognized with homology values ranging from 21 to 52% as compared to the reference species *Cr. vishniacii.*

The preceding guidelines concerning relatedness were based on the observation that strains of species characterized by traditional means may share 65–100% of their DNA base sequences. Species boundaries can be prescribed on the basis of a number of criteria, but for sexually reproducing species ability to hybridize and produce fertile successive generations would seem to be of cardinal importance. Dobzhansky (1976) noted that, among sexually reproducing and outbreeding organisms, species can be defined as Mendelian populations or arrays of populations that are reproductively isolated from other population arrays. He cited a number of examples of higher eukaryotes where interspecific hybridizations have been documented (*Drosophila* spp., mammals, frogs, fishes and plants). Fertility varies among hybrids, being rather rare among animal species but much more common among plants. Since conjugation in bacteria is infrequent and may not involve the entire genome, genetic studies could not form the basis for interpretation of DNA complementarity studies among these prokaryotes. Many yeasts, on the other hand, have known sexual stages and with such species hybridization offers a means for correlating results from DNA studies with actual biological phenomena.

Van der Walt and Johannsen (1979) and Johannsen (1980) approached this problem through hybridization studies in the essentially homothallic genus *Kluyveromyces*, where they were able to compare their results with DNA relatedness data obtained earlier by Martini (1973) and Martini and Phaff (1973). Van der Walt and Johannsen (1979) mated auxotrophic yeast strains, obtained by *N*-methyl-*N'*-nitro-*N*-nitrosoguanidine mutagenesis, on media that supported growth of prototrophic recombinant clones. Utilizing this method they observed from various combinations prototrophic colonies at a frequency of 10^{-2} to 10^{-8}. Taxa with high DNA relatedness showed a higher frequency of recombination, but in a few cases taxa having low DNA complementarity also gave a considerable number of recombinants. Lack of crosses through successive generations of progeny has made interpretation of these data difficult. For example, mixes of *K. thermotolerans* and *K. marxianus*, which have significantly different nuclear DNA-base composition values, namely 46.2 mol% G + C and 41.2 mol% G + C, respectively, and essentially no DNA homology (H.L. Presley and H.J. Phaff, unpublished observations; Martini and Phaff, 1973; Martini *et al.*, 1972), yielded recombinant clones. The F_1 generation from a cross of *K. marxianus* and *K. thermotolerans* appears to be most similar to *K. marxianus* on the basis of DNA composition, i.e. F_1 40.6 mol% G + C (H.L. Presley and H.J. Phaff, unpublished observations). These data suggest that the F_1 generation isolated from mixes of *K. marxianus* and *K. thermotolerans* by Johannsen and van der

Walt (1978) represents an aneuploid yeast strain. Additionally, other laboratories have not been able to confirm the fertility of the F_1 generation isolated from this cross (S.R. Snow, personal communication) or in a similar cross between the relatively unrelated species *K. fragilis* and *K. dobzhanskii* (Douglas *et al.*, 1969). Consequently, interspecific fusion seems to have occurred, but there appears to be insufficient genetic complementarity to permit analysis of meiosis products.

Kurtzman and his coworkers (1980a,b) were able to correlate genetic relatedness and DNA complementarity by working with heterothallic strains, thereby circumventing problems inherent in the study with *Kluyveromyces* spp. where auxotrophic mutants of homothallic strains were employed. In addition, Kurtzman *et al.* (1980a,b) isolated progeny from crosses by micromanipulation. In the first of these studies, the newly described *P. amylophila* and *P. mississippiensis* were shown to mate well, but ascospores were not viable. The two species showed only about 25% DNA relatedness, thus correlating lack of fertility with low DNA complementarity. However, the relationship between *Issatchenkia scutulata* var. *scutulata* and *I. scutulata* var. *exigua* was somewhat different. These two taxa also showed only about 25% DNA relatedness, but genetic crosses gave 3–6% viable ascospores (Table II). Furthermore, crosses between these F_1 progeny gave 17% viability among F_2 progeny. Reciprocal crosses among the F_1 progeny were fertile, as were backcrosses to the parents. From this, it appears that all stocks had essentially homologous chromosomes and that progeny were neither amphidiploids nor aneuploids. Nevertheless, in nature the two varieties of *Issatchenkia* are reproductively isolated, and thus gene flow cannot occur. *Issat-*

Table II. Ascospore viability of intra- and intervarietal crosses of *Issatchenkia scutulata* var. *scutulata* NRRL Y-7663 and *Issatchenkia scutulata* var. *exigua* NRRL Y-10,920, taxa which show low DNA relatedness. From Kurtzman *et al.* (1980b)[a].

NRRL Number of parent[b]	Number of ascospores viable/total of ascospores isolated	(%)[c]
Y-7663	23/23	(100)
Y-7663-2 × Y-7663-3	41/68	(60)
Y-10,920	8/36	(22)
Y-10,920-3 × Y-10,920-4	38/46	(83)
Y-7663-2 × Y-10,920-3	5/85	(6)
Y-7663-3 × Y-10,920-4	3/102	(3)
(Y-7663-2 × Y-10,920-3)-1 × (Y-7663-3 × Y-10,920-4)-1	9/53	(17)

[a] Comparisons of DNA relatedness between two strains of each variety gave a range of 21–26% for intervarietal pairings and 92–100% for intravarietal pairings.

[b] NRRL Y-7663 and NRRL Y-10,920 are sporogenous heterothallic diploids isolated from nature. The other strains are single-ascospore isolates derived from these diploids, as indicated. When grown alone, mating types showed neither conjugations nor ascospores.

[c] The frequency of conjugation and the number of apparently mature ascospores from intervarietal crosses were similar to the frequency and number from intravarietal crosses.

chenkia scutulata var. *scutulata* occurs in exudates of trees (*Myoporum sandwicense*) on the island of Hawaii and *I. scutulata* var. *exigua* has been isolated only from exudates of several species of broad-leaved trees in the Pacific northwest of North America. In terms of Dobzhanski's definition of species already stated, the two varieties of *Issatchenkia* could be considered as separate species or perhaps as species *in statu nascendi*. It appears therefore that the lower limit of DNA–DNA homology values suggesting species delimitation is not well defined, but base sequence divergence, as estimated from whole-genome comparisons, may be as great as 75% before genetic exchange can no longer occur.

Data in Table III summarize the preceding findings and provide further comparisons of other taxa. In all of these studies, progressively less DNA relatedness parallels decreasing mating competence and fertility. Exceptions to this trend would include chromosomal changes that affect fertility between strains such as inversions, translocations, autopolyploidy and amphidiploidy (allopolyploidy). Most of these changes would not be detectable in whole genome DNA comparisons. However, amphidiploidy is recognizable if parental species have been included in the study or should be suspected if genome sizes vary between strains showing substantial relatedness. For example, *Saccharomyces carlsbergensis* shows relatively high DNA homology with both *Sacch. cerevisiae* and *Sacch. bayanus* even though these latter two species demonstrate little relatedness (Fig. 1). A comparison of genome sizes suggests *Sacch. carlsbergensis* to be a partial amphidiploid which arose as a natural hybrid of *Sacch. cerevisiae* and *Sacch. bayanus* (Vaughan-Martini and Kurtzman, 1985). In this case, we would predict *Sacch. carlsbergensis* to be infertile with the proposed parents despite relatively high DNA relatedness.

In addition to providing information on DNA divergence and speciation, the data in Table III demonstrate the limits of taxonomic resolution provided by DNA–DNA re-association. Resolution appears to go no further than the detection of sibling species. Beyond this, all species, regardless of extent of kinship, show essentially less than 10% complementarity. Perhaps this should not be surprising since it has already been pointed out that detectable DNA homology presupposes a relatively high degree of nucleotide complementarity (Wilson *et al.*, 1977; Baumann and Baumann, 1978).

C. Comparisons of taxa

Bak and Stenderup (1969) appear to have been the first to make extensive use of nucleic acid re-association as a means for determining relatedness among yeasts. Their study included several phenotypically similar *Candida* species of

Table III. Correlation of mating reaction and deoxyribonucleic acid complementarity among closely related heterothallic ascomycetous and basidiomycetous yeasts

Species	Mating reaction	DNA relatedness (%)	Reference
Filobasidiella neoformans × *Filobasidiella bacillispora*	Fair conjugation, 0–30% basidiospore viability (F_1 progeny; F_2 not determined)	55–63	Aulakh *et al.* (1981)
Issatchenkia scutulata var. *scutulata* × *Issatchenkia scutulata* var. *exigua*	Good conjugation, ascopores viable (F_1 = 5%; F_2 = 17%)	21–26	Kurtzman *et al.* (1980b)
Pichia amylophila × *Pichia mississippiensis*	Good conjugation, ascopores not viable	20–27	Kurtzman *et al.* (1980a)
Hansenula bimundalis × *Hansenula americana*	Poor conjugation, ascopores not produced	21	Kurtzman (1984)
Pichia opuntiae var. *opuntiae* × *Pichia opuntiae* var. *thermotolerans*	Infrequent conjugation, ascopores produced rarely	28	Phaff and Holzschu (1981), Holzschu (1982)
Pichia opuntiae var. *opuntiae* × *Pichia antillensis*	No conjugation, ascospores not produced	55	Phaff *et al.* (1984)
Hansenula alni × *Hansenula wingei*	Poor conjugation, ascopores not produced	6–10	Phaff *et al.* (1979), Fuson *et al.* (1979)
Issatchenkia orientalis × *Issatchenkia occidentalis*	Infrequent conjugation, ascopores not produced	3–8	Kurtzman *et al.* (1980b)

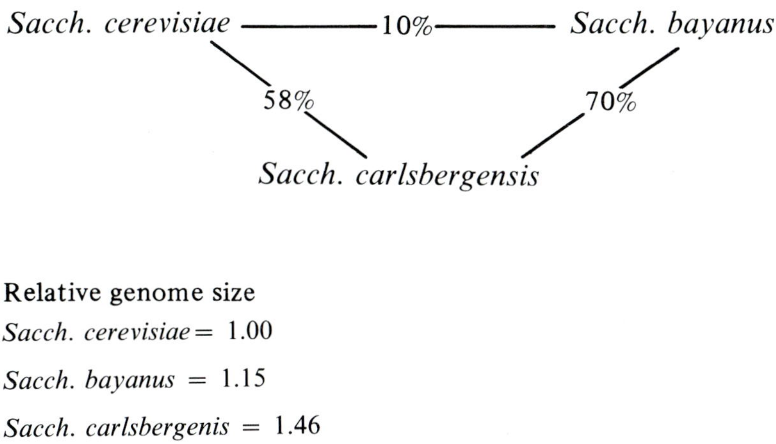

Fig. 1. DNA relatedness between type strains of *Sacch. cerevisiae, Sacch. baynus* and *Sacch. carlbergensis*. Comparison of genome sizes suggests *Sacch. carlsbergensis* to be a partial amphidiploid which may have arisen as a natural hybrid between *Sacch. cerevisiae* and *Sacch. bayanus* (after Vaughan Martini and Kurtzman, 1985).

medical interest. Specifically, their data showed sufficiently high base-sequence complementarity to regard *C. albicans, C. stellatoidea* and *C. claussenii* as conspecific (Table IV). *Candida tropicalis* showed little relatedness to these three taxa.

Additional advances in the clarification of imperfect taxa through DNA studies were made by Meyer and Phaff (1972) who found high degrees of homology between *C. lusitaniae* and *C. obtusa* and between *C. salmonicola* and *C. sake*. Meyer *et al.* (1975) demonstrated *C. cloacae* and *C. subtropicalis* to be synonyms of *C. maltosa*, and Kurtzman *et al.* (1980b) showed *C. krusei* and *C. sorbosa* to represent, respectively, the imperfect states of *Issatchenkia orientalis* and *I. occidentalis*. These authors also showed the synonymy between *Candida krusei* and *C. requinyii*. These findings are described in Table V. Kurtzman *et al.* (1979) and Manachini (1979) demonstrated *C. utilis* to be

Table IV. Deoxyribonucleic acid relatedness among medically important species of *Candida*. From Bak and Stenderup (1969).

Species	Relative binding[a] (%)			
	Candida albicans	*Candida claussenii*	*Candida stellatoidea*	*Candida tropicalis*
Candida albicans	100	69	81	9
Candida claussenii	56	100	79	10
Candida stellatoidea	65	66	100	12
Candida tropicalis	6	4	6	100

[a] Determined from the extent of complementarity between single-stranded nDNA and RNA synthesized from nDNA templates.

Table V. Extent of deoxyribonucleic acid re-association between species of *Issatchenkia* and their imperfect forms in the genus *Candida*. From Kurtzman *et al.* (1980b).

Organisms[a]	DNA re-association (%) with:			
	Issatchenkia orientalis Y-5396[b]	*Issatchenkia occidentalis* Y-7552[b]	*Issatchenkia terricola* YB-4310[b]	*Candida krusei* Y-7179[b]
Issatchenkia orientalis Y-7550	93	8		
Issatchenkia orientalis Y-7724	100	5		
Issatchenkia occidentalis Y-7552[b]	3			
Issatchenkia terricola Y-4310[b]	5	3		
Issatchenkia terricola Y-8218			92	
Candida krusei Y-7179[b]	99	8		
Candida krusei Y-10,930				99
Candida requinyii Y-10,940[b]	97			96
Candida sorbosa Y-7767[b]	4	98		1
Issatchenkia scutulata var. *scutulata* Y-7663[b]	5	2	4	
Issatchenkia scutulata var. *exigua* Y-10,920[b]	4	3	0	

[a] Strain numbers are Northern Region Research Center designations.
[b] Indicates the type strain.

the imperfect form of *Hansenula jadinii*. The work of Mendonca-Hagler and Phaff (1975) provided clear evidence that *Sacch. telluris*, *Torulopsis bovina*, *T. pintolopesii* and *Candida slooffii* are conspecific. This study also convincingly demonstrated that pseudomycelium formation provides no significant distinction between *Torulopsis* and *Candida* species.

The genus *Kloeckera* has long been regarded as the imperfect counterpart of *Hanseniaspora*, but the small differences in sugar assimilation reactions prevented unambiguous assignment of species within the two genera and uncertain correlation between perfect and imperfect species. Meyer *et al.* (1978) established the presence of six species in the homothallic genus *Hanseniaspora* and which of the members of *Kloeckera* corresponded to these perfect species (Fig. 2.). Most of the *Hanseniaspora* species exhibited less than 20% DNA relatedness, but *H'spora. osmophila* and *H'spora. vineae* showed a mean relatedness of 47%. In view of more recent interpretations of DNA relatedness (Kurtzman *et al.*, 1980b), this would suggest these two taxa to be sibling species or possibly varieties of the same species.

Further clarification of speciation in ascosporogenous yeasts resulted from the work of Price *et al.* (1978). Of particular interest were the rough-spored species of *Saccharomyces*, which are currently classified in *Torulaspora* (Kreger-van Rij, 1984). Nine of these species (*Sacch. delbrueckii, Sacch. rosei, Sacch. saitoanus, Sacch. inconspicuus, Sacch. florenzani, Sacch. fermentati, Sacch. microellipsodes* var. *osmophilus, Sacch. vafer, Debaryomyces nilssonii*)

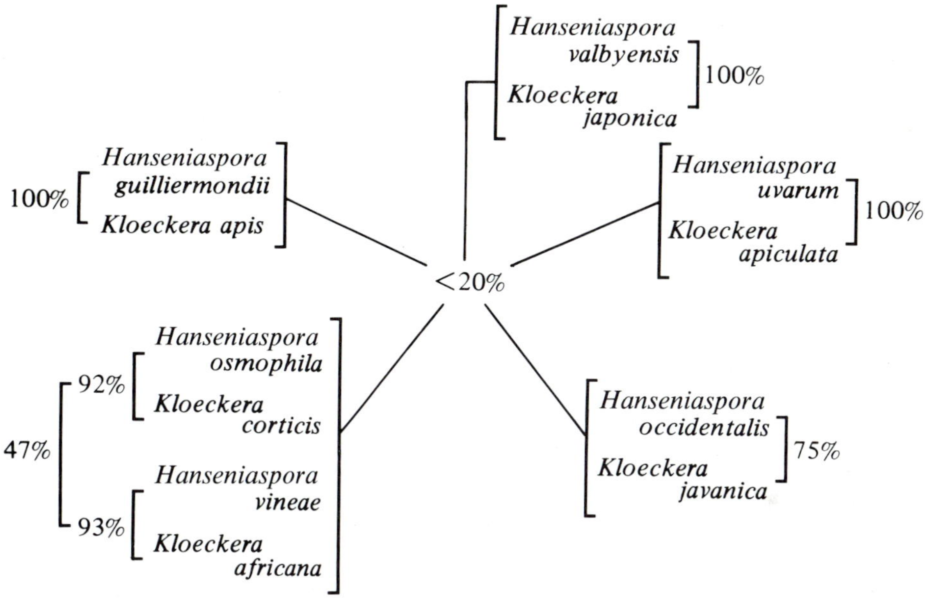

Fig. 2. Schematic representation of DNA relatedness among species of *Hanseniaspora* and their imperfect counterparts in the genus *Kloeckera*. Of the ascosporogenous species, only *Hanseniaspora osmophila* and *Hanseniaspora vineae* show greater than 20% DNA base sequence complementarity. Adapted from Meyer *et al.* (1978).

showed 98% or greater DNA relatedness and must be considered conspecific. These species were formerly separated by various differences in sugar assimilation and fermentation. All are now classified as *Sacch. delbrueckii* Lindner, which has priority (Price *et al.*, 1978). Further evidence for the inadequacy of traditional taxonomic criteria was their finding that *Sacch. microellipsodes* var. *osmophilus* was synonymous with *Sacch. delbrueckii* but *Sacch. microellipsodes* var. *microellipsodes* with low DNA relatedness to these organisms represented a separate species.

Price *et al.* (1978) also compared seven morphologically similar strains of *Debaryomyces hansenii* (Table VI) which originally had been described as separate species but later were placed into synonymy (Kreger-van Rij, 1984). The strains differ in assimilation of L-sorbose, trehalose, lactose, melibiose, soluble starch, D-arabinose, D-ribose, L-rhamnose, glycerol, erythritol, glucono-δ-lactone, resistance to cycloheximide and pellicle-forming ability. As the DNA–DNA complementarity values in Table VI show, there was no

Table VI. Extent of deoxyribonucleic acid relatedness between labelled DNA from *Debaryomyces hansenii* and other strains of *Debaryomyces*. From Price et al. (1978).

Species[a]	Relative binding corrected for self re-association of labelled DNA (%)
Debaryomyces hansenii (type strain)	100
Debaryomyces hansenii (*membranaefaciens*)	96.7
Debaryomyces hansenii (*nicotianae*)	96.5
Debaryomyces hansenii (*guilliermondii*)	93.9
Debaryomyces hansenii (*matruchoti*)	87.2
Debaryomyces hansenii (*kloeckeri*)	85.6
Debaryomyces hansenii (*tyrocola*)	77.0
Debaryomyces coudertii	15.8
Debaryomyces nepalensis	15.4
Debaryomyces marama	8.4
Debaryomyces castellii	4.6
Debaryomyces melissophilus	2.7

[a] Names in parentheses refer to the original designations of these strains.

obvious correlation between these conventional physiological characters and the observed levels of DNA relatedness.

In an ultrastructural comparison of ascospores from species assigned to *Pichia* and *Debaryomyces*, Kurtzman and Smiley (1974) observed spores of *P. polymorpha* and *P. pseudopolymorpha* to be typical of those found in *Debaryomyces*, i.e. spheroidal and roughened. They suggested *P. polymorpha* and *Deb. cantarellii* to be conspecific because of this and other phenotypic similarities. This premise was confirmed by Price *et al.* (1978; Table VII), and their DNA studies also showed *Deb. phaffii* to be a member of the taxon now designated *Deb. polymorphus* (Price and Phaff, 1979). In a subsequent experiment, Price *et al.* (1978) demonstrated the conspecificity between *Deb. vanriji* and *Deb. formicarius* (96% homology).

Table VII. Extent of deoxyribonucleic acid relatedness between *Debaryomyces polymorphus* and phenotypically similar species with roughened ascospores. From Price et al. (1978).

Species	Relative binding corrected for self-re-association of labelled DNA (%)
Debaryomyces polymorphus	100
Debaryomyces cantarellii	96.4
Debaryomyces phaffii	96.1
Debaryomyces vanriji	23.9
Debaryomyces formicarius	23.7
Pichia pseudopolymorpha	21.0
Debaryomyces castellii	19.8
Debaryomyces hansenii	8.0
Pichia terricola	1.9

V. MACROMOLECULES OTHER THAN DEOXYRIBONUCLEIC ACID AS INDICATORS OF RELATEDNESS

As we have already discussed, complementarity studies involving whole-genome DNA seem to resolve taxa no further than to the sibling-species level. Analysis of other macromolecules does, however, provide a means for determining more distant phylogenetic relationships. Some genes diverge quite slowly and that portion (approximately 2%) of genomic DNA coding for ribosomal RNA (rRNA) is highly conserved. Bicknell and Douglas (1970) appear to have been the first to examine distant relationships among the yeasts using the technique of rRNA–DNA homology. Their data indicate, as expected, that intrageneric relationships obtained by this method are in most cases not meaningful because of the high degree of conservation of the DNA sequences coding for rRNA (Table VIII). However, because of this conservation, intergeneric distances can be resolved, and it should be possible to construct an evolutionary hierarchy of the yeasts using rRNA–DNA homology. Baharaeen and Vishniac (1984) examined the 25S rRNA complementarity among basidiomycetous yeasts and provided evidence that species assignments were correct for most teleomorphic genera. However, species placed in *Candida* showed considerable phylogenetic diversity. Additional rRNA–DNA complementarity studies among the yeasts were reported by Segal and Eylan (1974a, b, 1975). Comparative cataloguing of oligonucleotides from T_1 ribonuclease digests of ribosomal RNA has provided information on phylogeny of bacteria and might prove useful in defining relationships among yeasts (Fox *et al.*, 1977). Another approach was that of Walker and Doolittle (1982, 1983) who used 5S rRNA sequences to investigate relationships among the basidiomycetes, including some yeasts. Their work identified two distinct clusters that correlated with the presence or absence of septal dolipores rather than with the traditional separation of these species into the subclasses Heterobasidiomycetidae and Homobasi-

Table VIII. Extent of deoxyribonucleic acid and ribosomal ribonucleic acid relatedness among selected yeasts. From Bicknell and Douglas (1970).

Species	Relatedness with *Saccharomyces cerevisiae* (%)	
	DNA	rRNA
Saccharomyces cerevisiae	100	100
Saccharomyces uvarum	40	100
Saccharomyces kloeckerianus (Syn. *Torulaspora globosa*)	20	97
Saccharomyces rouxii (= *Zygosacch. rouxii*)	0	94
Kluyveromyces wickerhamii	0	87
Hansenula jadinii (*Candida utilis*)	0	77
Pichia pastoris	0	50

diomycetidae. Further, the data suggested that capped dolipores evolved from capless dolipores which in turn may have evolved from single septal pores.

Phylogenetic relationships of species within genera have not, for the most part, been satisfactorily resolved using either DNA–DNA or rRNA–DNA homologies, because the first method is too specific and the second detects relatedness more distant than will usually be found within a genus. However, DNA coding for certain enzymes such as glutamine synthetase and superoxide dismutase, although less conserved than that coding for rRNA, is still less divergent than is apparent from comparisons of whole genome bacterial DNA (Baumann and Baumann, 1978; Baumann *et al.*, 1980). By determining changes in the amino acid sequence of these enzymes through such techniques as quantitative microcomplement fixation, an estimate of intermediate relatedness may be found. Immunological studies of protein similarity are very sparse for yeasts, and the resultant information is highly dependent on the kinds of protein studied. Lachance and Phaff (1979) used exo-β-glucanases from species of *Kluyveromyces* but found that the enzyme was poorly conserved and immunological distances between most species were too great to be reliable for determining evolutionary relationships in that genus.

Electrophoresis of allozymes presents another means for estimating molecular diversity and, as with immunological studies, resolution depends upon the extent of sequence conservation. The proportion of point mutations that are electrophoretically detectable is estimated at approximately 0.27 owing to redundancy in the genetic code and the large proportion of amino acids that are electrically neutral (Baptist *et al.*, 1971; Selander, 1976; Holzschu, 1981). The nearly universal occurrence of extensive protein polymorphisms in natural populations has led some workers to believe that variation would prove physiologically important and therefore to be under selective control; others regard it as without phenotypic effect and thus selectively neutral (Selander, 1976). Regardless, data derived by this technique do provide insight into evolutionary processes and taxonomy. Baptist and Kurtzman (1976) utilized comparative enzyme patterns to separate sexually active strains of *Cryptococcus laurentii* var. *laurentii* from nonreactive strains and from the varieties *magnus* and *flavescens*. Two new species, *Cryptococcus magnus* and *Cr. heveanensis*, were proposed from the strains comprising the variety *magnus*.

Yamazaki and Komagata (1981) made an electrophoretic comparison of seven enzymes in 108 strains of *Rhodotorula* and *Rhodosporidium* and concluded that this technique was useful in establishing taxonomic relationships among these red yeasts. For example, *Rhodotorula pilimanae* and *Rh. rubra* showed identical enzymic patterns. This, together with identical G + C contents in their DNA species and identical coenzyme-Q (Co-Q) system (Q_{10}), suggested that the two species are synonymous. On similar evidence, *Rh. minuta* var. *texensis* was found to be synonymous with *Rh. minuta* var. *minuta*. Some strains of *Rhodosporidium toruloides* that could not assimilate nitrate were found to be identical in enzymic pattern to nitrate-negative strains, supporting our view expressed in Section VI (p. 86) that the ability to assimilate nitrate has limited value in taxonomy. They also clarified the postulated relationships between some species of *Rhodotorula* and species of *Rhodosporidium*. In a further study, Yamazaki and Komagata (1982) made an allozyme study of several asporogenous yeast species and their postulated ascogenous states. Close relationships were observed between 14 teleomorphs and their anamorphic counterparts, but there was no apparent relation between *Candida valida* and *Pichia membranaefaciens, Torulopsis molischiana* and *Hansenula capsulata*, and *Torulopsis candida* and *Deb. hansenii* or *Deb. marama*. Some strains of *Candida parapsilosis* showed the same enzymic pattern as *Lodderomyces elongisporus*, while others (including the type strain) showed no apparent relationship to *L. elongisporus*. Meyer and Phaff (1972) had shown earlier by DNA homology that the DNA of the type strain of *C. parapsilosis* showed little homology with that of *L. elongisporus*. A teleomorph of the type strain of *C. parapsilosis* is not known at present. Further work on the importance and usefulness of electrophoretic comparisons of enzymes has been reported on species of ballistosporogenous yeasts (Yamazaki and Komagata, 1983a), species of *Saccharomyces, Zygosaccharomyces* and *Torulaspora* (Yamazaki and Komagata, 1983b), species of *Hansenula* (Yamazaki and Komagata, 1983c) and in strains of wine yeasts (Subden *et al.*, 1982).

Holzschu (1981) recently studied the evolutionary relationships among some 400 strains of various cactophilic species of *Pichia*. He determined the banding patterns in starch gels of 14 metabolic enzymes, using four buffer systems. The type strain of *P. heedii* was used as the reference species for determining the relative mobilities of corresponding enzymes from other yeasts. Several species could be recognized by unique banding patterns of certain enzymes. For example, Holzschu *et al.* (1983) found that certain strains of *P. cactophila* that are heterothallic and produce four-spored asci represented a separate, although closely related species, which they described as *Pichia pseudocactophila*. Direct comparison of four loci studied (glucose 6-phosphate dehydrogenase, hexokinase-1, hexokinase-2 and phosphoglucomutase) enabled diagnosis of *P. cactophila* and *P. pseudocactophila* with greater than 99.9% probability of being correct. He similarly discovered diagnostic alleles (see Ayala and Powell, 1972 for a definition of diagnostic alleles) for other common cactophilic *Pichia* species. Holzschu *et al.* (1985) showed that *Pichia opuntiae* var. *opuntiae* and *P. opuntiae* var. *thermotolerans* genomes were clearly different in 8 out of 12 allozymes tested. These differences together with low DNA complementarity (see Table III) and different habitats induced these authors to raise *P. opuntiae* var. *thermotolerans* to the rank of species, *P. thermotolerans*. Allozyme variation was also used by Holzschu (1981) to calculate genetic distances among populations (both intraspecific and interspecific) of cactophilic yeasts. The data indicated that enzyme electrophoresis is a much more sensitive probe for revealing relationships among very closely as well as more distantly related yeasts than is DNA– DNA homology. The method is very efficient and can be used for identification of large numbers of yeast strains.

Sidenberg and Lachance (1983) using polyacrylamide-gel electrophoresis of 11 isofunctional enzymes examined the type strains of 20 species of the genus *Kluyveromyces*. Their results supported the division of the genus into 13 species, a conclusion consistent with studies on DNA relatedness (Martini, 1973; Martini and Phaff, 1973; H.L. Presley and H.J. Phaff, unpublished observations).

VI. COMPARISON OF RELATEDNESS FROM NUCLEIC ACID STUDIES WITH THAT DETERMINED BY OTHER METHODOLOGIES

In recent years it has become apparent that any of the criteria traditionally used for speciation of yeasts have their basis in only a very small portion of the genome. As is clear from genetic studies (Winge and Roberts, 1949; Lindegren and Lindegren, 1949; Kurtzman and Smiley, 1976; Starmer *et al.*, 1978a,b), the assimilation and (or) fermentation of certain carbon compounds, especially of di-, tri- and polysaccharides, is frequently under the control of one or two structural genes. Consequently, carbon-utilization tests, for the most part, should be regarded as convenient laboratory means for separating taxa rather than as the foundation of yeast systematics. This is also true for the nitrate and nitrite assimilation reactions. Ability to assimilate nitrate has been considered of cardinal importance in the definition of many species and even genera since Stelling-Dekker (1931) first proposed its use over 50 years ago for separating *Hansenula* (nitrate-positive) from *Pichia* (nitrate-negative). Genetic studies show nitrate assimilation to be under the control of only a few genes (Marzluf, 1981). Recently, Kurtzman (1984b) proposed that *Hansenula* and *Pichia* be considered congeneric because *H. minuta* and *P. lindneri* exhibited 75% DNA relatedness. This pair is not unique as *H. mrakii* and *P. sargentensis* showed 68% DNA relatedness (C.P. Kurtzman, M.J. Smiley, Z.A. Brandt and C.Y. Kelley, unpublished works). It was also found that the nitrate-positive *Sterigmatomyces halophilus* shared 100% of its DNA base sequences with the nitrate-negative

Sterigmatomyces indicus (C.P. Kurtzman, M.J. Smiley, Z.A. Brandt and C.J. Johnson, unpublished works). In the genus *Brettanomyces*, the ability to assimilate nitrate is reported to be positive or negative for *Br. bruxellensis* and *Br. intermedius* (van der Walt, 1970a).

Nucleic acid studies have not proved supportive in all cases of conclusions drawn from numerical analysis, or from proton-magnetic-resonance spectra and serology of cell-wall mannans. For example, using numerical analysis, Campbell (1973) proposed reduction of *H. wingei*, *H. canadensis*, and both varieties of *H. bimundalis*, to a single species. Fuson *et al.* (1979) showed from DNA–DNA complementarity studies that *H. canadensis* and *H. wingei* are conspecific, while the latter species showed little relatedness with either of the two varieties of *H. bimundalis*. Kurtzman (1984a) further demonstrated only 21% complementarity between the DNA species of the two varieties of *H. bimundalis* and he proposed elevating them to the rank of species (*cf* Table III, p. 78). The disparity of the numerical and molecular methods probably results from the relatively limited amount of phenotypic data available for numerical analysis, whereas DNA studies have the whole genome as a data base.

This also appears to be true for separations made using proton magnetic resonance, because Spencer and Gorin (1969) found the spectra of cell-wall mannans from *H. beckii* and *H. canadensis* to be similar, but different from the pattern shared by *H. wingei* and the varieties of *H. bimundalis*. Fuson *et al.* (1979) found little relatedness between *H. beckii* and *H. canadensis*. In further comparisons of this group, Tsuchiya *et al.* (1974) reported the cell-surface antigens of *H. wingei*, *H. canadensis* and *H. beckii* to be indistinguishable. The reason for this discrepancy is not at all clear, but Ballou (1974) has shown, at least for mannans from *Saccharomyces* spp. that single-gene changes can impart significant differences in immunological reaction. The difference between the two main proton-magnetic-resonance spectral types of mannan in *Sacch. cerevisiae* was also found to be controlled by a single gene (Spencer *et al.*, 1971). This indicates that serotypes and proton-magnetic-resonance spectra of yeasts are not likely to be highly stable because of mutational modification of the cell envelope.

Reliance on the presence or absence of pseudohyphae and true hyphae for taxonomic assignment has been questioned. Van Uden and Buckley (1970) as well as Yarrow and Meyer (1978) have pointed out that some taxa do not consistently form pseudohyphae. For primarily this reason, the latter authors transferred species of *Torulopsis* to *Candida*, the genus whose name has priority over *Torulopsis*. In 1979, Fuson *et al.* provided evidence that the presence or absence of true hyphal development is not necessarily a good trait for differentiation. This conclusion was based on the observation that the type strain of *H. canadensis*, which is differentiated from *H. wingei* by its homothallism, ascospore number and lack of true hyphae, has a higher degree of DNA homology (77.6%) with the type strain of *H. wingei* (which forms true hyphae) than do some other strains identified as *H. wingei* (62.2%).

Differentiation of many ascosporogenous genera is based on ascospore shape and surface structure. An opportunity to test the soundness of this convention in the genera *Schwanniomyces*, *Saccharomyces*, *Debaryomyces* and *Pichia* arose through a comparison of the DNA data of Price *et al.* (1978) with the scanning electron-microscope studies by Kurtzman and Kreger-van Rij (1976), Kurtzman and Smiley (1974, 1979) and Kurtzman *et al.* (1972, 1975). All four species of *Schwanniomyces* had high DNA relatedness (Price *et al.*, 1978) and had similar spore architecture (Kurtzman *et al.*, 1972; Kreger-van Rij, 1977). The least related to *Schw. occidentalis*, *Schw. persoonii* at about 80% complementarity, showed fewer and less pronounced spore protuberances than the other three *Schwanniomyces* species, all four of which are now regarded as conspecific. Ascospore topography for many species in *Torulaspora* (van der Walt's, 1970b, group III *Saccharomyces* spp.) was essentially identical and also frequently indistinguishable from the majority of species in *Debaryomyces*. Although many species within their respective genera were found conspecific by DNA homology, others proved distinct despite spore similarity. The few species with recognizably different spores showed little DNA relatedness. Thus, species with unlike spores can be expected to show low DNA complementarity, but no prediction can be made concerning species with similar spores. A few exceptions are known, however. Ascospores of *Pichia ohmeri* may be either spheroidal or hat-shaped, depending upon mating type (Wickerham and Burton, 1954; Fuson *et al.*, 1980), and spore shape in *Yarrowia lipolytica* is mating-type dependent (Wickerham *et al.*, 1969). Spore shapes among *Kluyveromyces* spp., with high degrees of DNA complementarity, may be either spheroidal (e.g. *K. lactis* and *K. vanudenii*) or kidney shaped (such as *K. phaseolosporus* and *K. drosophilarum*). The last three species were found to be synonymous with *K. lactis* (H.L. Presley and H.J. Phaff, unpublished observations).

Kreger-van Rij and Veenhuis (1973) demonstrated, by transmission electron microscopy, a basic difference in hyphal septa among ascomycetous yeasts. Three classes of septum ultrastructure were observed: a single central pore or closure line, plasmodesmata, and a dolipore configuration. All mycelial species of *Hansenula* and *Pichia* have the central pore, whereas plasmodesmata are common to all *Yarrowia* species except *Y. lipolytica*, which has a central pore. Dolipores were found in species of *Ambrosiozyma*. Future comparisons of these species by molecular techniques that detect distant or close relatedness might show ultrastructural differences in septa to be of evolutionary significance.

An analysis of coenzyme Q or ubiquinones (Yamada and Kondo, 1972, 1973; Yamada *et al.*, 1973a, b, c, 1976a, b, 1977, 1980, 1981, 1983; Yamada and Kondo, 1984; Yamada and Banno, 1984) has revealed that the number of isoprene units in the side chains of this molecule varies from 5 to 10 (Q-5 to Q-10) among yeast taxa. In many genera, all species have the same Q number, but heterogeneity exists in such genera as *Pichia*, *Hansenula* and *Saccharomycopsis*. This may indicate that the atypical species are incorrectly assigned, and a comparison of these species by other molecular techniques could have interesting phylogenetic implications.

Nakase and Suzuki (1985a, b) studied 50 strains of *Debaryomyces hansenii* and related yeasts by a combination of DNA base composition, DNA–DNA hybridization, ubiquinone system, isozyme electrophoresis, proton magnetic resonance of the cell-wall mannans and serological methods. They recognized three groups of strains representing three distinct species on the basis of DNA hybridization experiments. Group I was found to consist of two subgroups representing two distinct varieties. All strains investigated showed similar G + C contents of the nuclear DNA species as well as ubiquinone systems. The most useful methods for rapid differentiation between members of the groups and subgroups were ability to utilize propylene glycol for growth, maximum temperature for growth and relative electrophoretic mobilities of glucose 6-phosphate dehydrogenase and malate dehydrogenase.

VII. MOLECULAR TAXONOMY AS AN AID TO GENETIC RESEARCH

The discipline of systematics fulfils our intellectual need to understand the relationships among organisms and has the practical advantage of providing a means for recognizing species of importance to many human activities. Numerous advances in basic and applied yeast genetics have stemmed from exploitation of the diversity found in natural isolates. The source of these isolates is frequently the microbiologist who is looking for additional strains to clarify taxonomic concepts. Once these strains have been properly identified, they are usually placed in culture collections for future reference and may in this way become available to the geneticist looking for new materials. Molecular techniques have served as powerful tools to ensure the most precise identification of strains for these potential genetic studies.

The fermentation industry has been highly dependent on strain improvement through breeding, and molecular taxonomy offers the potential for locating new stocks. An additional recent technique for strain improvement involves protoplast fusion followed by cell-wall regeneration. This approach may be used to add vigour to important strains or to combine useful properties not occurring together in nature (Ferenczy, 1981). Protoplasts fused from closely related species are generally more stable than fusion products from distantly related taxa. Consequently, if the desired characteristics can be obtained from similar or closely related species, a stable fusion product is more likely. To date, molecular techniques have provided much better guides for discerning which species are most closely related than traditional morphological and physiological criteria.

References

Aulakh, H.S., Straus, S.E. and Kwon-Chung, K.J. (1981). *International Journal of Systematic Bacteriology* **31**, 97.

Ayala, F.J. and Powell, J.R. (1972). *Proceedings of the National Academy of Sciences of the United States of America* **69**, 1094.

Baharaeen, S. and Vishniac, H.S. (1984). *Canadian Journal of Microbiology* **30**, 613.

Baharaeen, S., Bantle, J.A. and Vishniac, H.S. (1982). *Canadian Journal of Microbiology* **28**, 406.

Bak, A.L. (1973). *Current Topics in Microbiology and Immunology* **61**, 89.

Bak, A.L. and Stenderup, A. (1969). *Journal of General Microbiology* **59**, 21.

Bak, A.L., Christiansen, C. and Christiansen, G. (1972). *Biochimica et Biophysica Acta* **269**, 527.

Ballou, C.E. (1974) *Advances in Enzymology* **40**, 239.

Baptist, J.N. and Kurtzman, C.P. (1976). *Mycologia* **68**, 1195.

Baptist, J.N., Shaw, C.R. and Mandel, M. (1971). *Journal of Bacteriology* **108**, 799.

Barnett, J.A., Payne, R.W. and Yarrow, D. (1983). "Yeasts: Characteristics and Identification". Cambridge University Press, Cambridge.

Baumann, L. and Baumann, P. (1978). *Archives of Microbiology* **119**, 25.

Baumann, L., Bang, S.S. and Baumann, P. (1980). *Current Microbiology* **4**, 133.

Bedford, C.L. (1942). *Mycologia* **34**, 628.

Bernardi, G., Faures, M., Piperno, G. and Slonimski, P.P. (1970). *Journal of Molecular Biology* **48**, 23.

Bicknell, J.N. and Douglas, H.C. (1970). *Journal of Bacteriology* **101**, 505.

Blamire, J., Cryer, D.R., Finkelstein, D.B. and Marmur, J. (1972). *Journal of Molecular Biology* **67**, 11.

Brady, R.J. (1981). *Abstracts of the Annual Meeting of the American Society for Microbiology* **k65**, 148.

Brenner, D.J. (1973). *International Journal of Systematic Bacteriology* **23**, 298.

Brenner, D.J., Fanning. G.R., Rake, A. and Johnson, K.E. (1969). *Analytical Biochemistry* **28**, 447.

Brenner, D.J., Steigerwalt, A.G. and Fanning, G.R. (1972). *International Journal of Systematic Bacteriology* **22**, 193.

Britten, R.J. and Kohne, D.E. (1968). *Science* **161**, 529.

Campbell, I. (1973). *Journal of General Microbiology* **77**, 427.

Christiansen, C., Bak, A.L., Stenderup, A. and Christiansen, G. (1971). *Nature (London) New Biology* **231**, 176.

Commerford, S.L. (1971). *Biochemistry* **10**, 1993.

Crosa, J.H., Brenner, D.J. and Falkow, S. (1973). *Journal of Bacteriology* **115**, 904.

Cryer, D.R., Eccleshall, R. and Marmur, J. (1975), *In* "Methods in Cell Biology" (D.M. Prescott, ed.), vol. 12, pp. 39–44. Academic Press, New York.

De Ley, J., Cattoir, H. and Reynaerts, A. (1970). *European Journal of Biochemistry* **12**, 133.

Denhardt, D.T. (1966). *Biochemical and Biophysical Research Communications* **23**, 641.

Diddens, H.A. and Lodder, J. (1942). "Die anaskosporogenen Hefen, II Hälfte". North-Holland Publishing Co., Amsterdam.

Dobzhansky, T. (1976). *In* "Molecular Evolution" (F.J. Ayala, ed.), pp. 95–105. Sinauer Associates, Sunderland, Mass.

Douglas, H.C., Grindall, D.P. and Talbott, H. (1969). *Journal of Bacteriology* **99**, 287.

Ehrenberg, L., Fedorcsak, I. and Solymosy, F. (1976). *Progress in Nucleic Acids Research* **16**, 189.

Fell, J.W., Hunter, I.L. and Tallman, A.S. (1973). *Canadian Journal of Microbiology* **19**, 643.

Ferenczy, L. (1981). *In* "Genetics as a Tool in Microbiology" (S.W. Glover and D.A. Hopwood, eds.), pp. 1–34. Society for General Microbiology Symposium 31. Cambridge University Press, Cambridge.

Fox, G.E., Pechman, K.R. and Woese, C.R. (1977). *International Journal of Systematic Bacteriology* **27**, 42.

Fuson, G.B., Price, C.W. and Phaff, H.J. (1979). *International Journal of Systematic Bacteriology* **29**, 64.

Fuson, G.B., Price, C.W. and Phaff, H.J. (1980). *International Journal of Systematic Bacteriology* **30**, 217.

Gillespie, D. and Spiegelman, S. (1965). *Journal of Molecular Biology* **12**, 829.

Guého, E., Tredick, J. and Phaff, H.J. (1984). *Antonie van Leeuwenhoek* **50**, 17.

Guého, E., Tredick, J. and Phaff, H.J. (1985). *Canadian Journal of Botany* **63**, 961.

Gunge, N., Tamuru, A., Ozawa, F. and Sakaguchi, K. (1981). *Journal of Bacteriology* **145**, 382.

Hagler, A.N. and Ahearn, D.G. (1981). *International Journal of Systematic Bacteriology* **31**, 204.

Hartwell, L.H. (1974). *Bacteriological Reviews* **38**, 164.

Hicks, J.B. and Herskowitz, I. (1976). *Genetics* **83**, 245.

Hollenberg, C.P., Borst, P. and van Brugger, E.F.J. (1970). *Biochimica et Biophysica Acta* **209**, 1.

Holzschu, D.L. (1981). Ph.D. Thesis: University of California, Davis.

Holzschu, D.L. and Phaff, H.J. (1982). *In* "Ecological Genetics and Evolution — The Cactus–Yeast–*Drosophila* Model System" (J.S.F.Barker and W.T.Starmer, eds), pp. 127–141. Academic Press, Sydney.

Holzschu, D.L., Presley, H.L., Miranda, M. and Phaff, H.J. (1979). *Journal of Clinical Microbiology* **10**, 202.

Holzschu, D.L., Tredick, J. and Phaff, H.J. (1981). *Current Microbiology* **5**, 73.

Holzschu, D.L., Phaff, H.J., Tredick, J. and Hedgecock, D. (1983). *Canadian Journal of Microbiology* **29**, 1314.

Holzschu, D.L., Phaff, H.J., Tredick, J. and Hedgecock, D. (1985). *International Journal of Systematic Bacteriology* **35**, 457.

Johannsen, E. (1980). *Antonie van Leeuwenhoek* **46**, 177.

Johannsen, E. and van der Walt, J.P. (1978). *Archives of Microbiology* **118**, 45.

Johnson, J.L. (1973). *International Journal of Systematic Bacteriology* **23**, 308.

Johnson, J.L. (1981). *In* "Manual of Methods for General Bacteriology" (P. Gerhardt, ed.), pp. 450–472. American Society for Microbiology, Washington, D.C.

Kreger-van Rij, N.J.W. (1970). *In* "The Yeasts–A Taxonomic Study" (J. Lodder, ed.), pp. 129–156. North Holland Publishing Co., Amsterdam.

Kreger-van Rij, N.J.W. (1977). *Antonie van Leeuwenhoek* **43**, 55.

Kreger-van Rij, N.J.W. (1984). "The Yeasts, a Taxonomic Study", third edn. Elsevier Biomedical Press, Amsterdam.

Kreger-van Rij, N.J.W. and Veenhuis, M. (1971). *Journal of General Microbiology* **68**, 87.

Kreger-van Rij, N.J.W. and Veenhuis, M. (1973). *Antonie van Leeuwenhoek* **39**, 481.

Kurtzman, C.P. (1984a). *Mycotaxon* **19**, 271.

Kurtzman, C.P. (1984b). *Antonie van Leeuwenhoek* **50**, 209.

Kurtzman, C.P. and Kreger-van Rij, N.J.W. (1976). *Mycologia* **68**, 422.

Kurtzman, C.P. and Smiley, M.J. (1974). *In* "Proceedings of the Fourth International Symposium on Yeasts. Vienna, Austria. Part I" (H. Klaushofer and U.B. Sleytr, eds), pp. 231–232. Hochschulerschaft an der Hochschule für Bodenkultur, Vienna.

Kurtzman, C.P. and Smiley, M.J. (1976). *Antonie van Leeuwenhoek* **42**, 355.

Kurtzman, C.P. and Smiley, M.J. (1979). *Mycologia* **71**, 658.

Kurtzman, C.P., Smiley, M.J. and Baker, F.L. (1972). *Journal of Bacteriology* **112**, 1380.

Kurtzman, C.P., Smiley, M.J. and Baker, F.L. (1975). *Mycopathologia et Mycologia Applicata* **55**, 29.

Kurtzman, C.P., Johnson, C.J. and Smiley, M.J. (1979). *Mycologia* **71**, 844.

Kurtzman, C.P., Smiley, M.J., Johnson, C.J., Wickerham, L.J. and Fuson, G.B. (1980a). *International Journal of Systematic Bacteriology* **30**, 208.

Kurtzman, C.P., Smiley, M.J. and Johnson, C.J. (1980b). *International Journal of Systematic Bacteriology* **30**, 503.

Kurtzman, C.P., Smiley, M.J., Johnson, C.J. and Hoffman, M.J. (1980c). *Abstracts 5th International Symposium on Yeasts, Y-5.2.5.*, p. 246.

Lachance, M.A. and Phaff, H.J. (1979). *International Journal of Systematic Bacteriology* **29**, 70.

Lindegren, C.C. and Lindegren, G. (1949). *Proceedings of the National Academy of Sciences of the United States of America* **35**, 23.

Lipinski, C., Ferro, A.J. and Mills, D. (1976). *Molecular and General Genetics* **144**, 301.

Lodder, J. (1934). *Verhandelingen-Koninklijke Akademie van Wetenschappen Amsterdam, Afdeeling Natuurkunde, Sectie II* **32**, 1.

Lodder, J. (1970). *In* "The Yeasts — A Taxonomic Study" (J. Lodder, ed.), pp. 1–33. North-Holland Publishing Co., Amsterdam.

Manachini, P.L. (1979). *Antonie van Leeuwenhoek* **45**, 451.

Maniatus, T., Jeffrey, A. and Kleid, D.G. (1975). *Proceedings of the National Academy of Sciences of United States of America* **72**, 1184.

Marmur, J. (1961). *Journal of Molecular Biology* **3**, 208.

Marmur, J. and Doty, P. (1962). *Journal of Molecular Biology* **5**, 109.

Marmur, J., Rownd, R. and Schildkraut, C.L. (1963). *Progress in Nucleic Acids Research and Molecular Biology* **1**, 231.

Martini, A. (1973). *Annals of the Facilty of Agriculture of the University of Perugia* **28**, 157.

Martini, A. and Phaff, H.J. (1973). *Annals of Microbiology* **23**, 59.

Martini, A., Phaff, H.J. and Douglass, S.A. (1972). *Journal of Bacteriology* **111**, 481.

Marzluf, G.A. (1981). *Microbiological Reviews* **45**, 437.

Mendonca-Hagler, L.C. and Phaff, H.J. (1975). *International Journal of Systematic Bacteriology* **25**, 222.

Mendonca-Hagler, L.C., Travassos, L.R., Lloyd, K.O. and Phaff, H.J. (1974). *Infection and Immunity* **8**, 674.

Meyer, S.A. and Phaff, H.J. (1969). *Journal of Bacteriology* **97**, 52.

Meyer, S.A. and Phaff, H.J. (1970). *In* "Recent Trends in Yeast Research" (D.G. Ahearn, ed.), vol. 1, pp. 1–29. Spectrum (Georgia State University), Atlanta.

Meyer, S.A. and Phaff, H.J. (1972). *In* "Yeasts, Models in Science and Technics" (A. Kockova-Kratochvilova and E. Minarik, eds), pp. 375–386. Proceedings of the First Specialized International Symposium on Yeasts. Slovak Academie of Science, Bratislava.

Meyer, S.A., Anderson, K., Brown, R.E., Smith, M. Th., Yarrow, D., Mitchell, G. and Ahearn, D.G. (1975). *Archives of Microbiology* **104**, 225.

Meyer, S.A., Smith, M.Th. and Simione, F.P., Jr. (1978). *Antonie van Leeuwenhoek* **44**, 79.

Miller, M.W., Yoneyama, M. and Soneda, M. (1976). *International Journal of Systematic Bacteriology* **26**, 286.

Miranda, M., Holzschu, D.L., Phaff, H.J. and Starmer, W.T. (1982). *International Journal of Systematic Bacteriology* **32**, 101.

Nakase, T. and Komagata, K. (1968a). *Journal of General and Applied Microbiology* **14**, 345.

Nakase, T. and Komagata, K. (1968b). *Yeast News Letter* **17**, 4.

Nakase, T. and Komagata, K. (1971a). *Journal of General and Applied Microbiology* **17**, 77.

Nakase, T. and Komagata, K. (1971b). *Journal of General and Applied Microbiology* **17**, 121.

Nakase, T. and Suzuki, M. (1985a). *Journal of General and Applied Microbiology* **31**, 49.

Nakase, T. and Suzuki, M. (1985b). *Journal of General and Applied Microbiology* **31**, 71.

O'Connor, R.M., McArthur, C.R. and Clark-Walker, G.D. (1975). *European Journal of Biochemistry* **53**,137.

O'Connor, R.M., McArthur, C.R. and Clark-Walker, G.D. (1976). *Journal of Bacteriology* **126**, 959.

Petes, T.D., Beyers, B. and Fangman, W.L. (1973). *Proceedings of the National Academy of Sciences of the United States of America* **70**, 3072.

Phaff, H.J. (1980). *In* "Current Developments in Yeast Research" (G.G. Stewart and I. Russell, eds), pp. 635–643. Pergamon Press, Toronto.

Phaff, H.J. and Holzschu, D.L. (1981). "Molecular Genetics in Yeast" (D. von Wettstein, J. Friis, M. Kielland-Brandt and A. Stenderup, eds), pp. 413–427. Alfred Benzon Symposium 16, Munksgaard, Copenhagen.
Phaff, H.J., Miller, M.W. and Miranda, M. (1979). *International Journal of Systematic Bacteriology* **29**, 60.
Phaff, H.J., Yamada, Y., Tredick, J. and Miranda, M. (1983). *International Journal of Systematic Bacteriology* **33**, 375.
Phaff, H.J., Starmer, W.T., Tredick, J. and Miranda, M. (1985). *International Journal of Systematic Bacteriology* **35**, 211.
Price, C.W. and Phaff, H.J. (1979). *Mycologia* **71**, 444.
Price, C.W., Fuson, G.B. and Phaff, H.J. (1978). *Microbiological Reviews* **42**, 161.
Rhodes, J.C. and Kwon-Chung, K.J. (1982). *Abstracts of the Annual Meeting of the American Society for Microbiology* **F61**, 336.
Rost, K. and Venner, H. (1964). *Zeitschrift für Physiologische Chemie* **339**, 230.
Schildkraut, C.L., Marmur, J. and Doty, P. (1962). *Journal of Molecular Biology* **4**, 430.
Segal, E. and Eylan, E. (1974a). *Microbios* **9**, 25.
Segal, E. and Eylan, E. (1974b). *Microbios* **10**, 133.
Segal, E. and Eylan, E. (1975). *Microbios* **12**, 111.
Seidler, R.J. and Mandel, M. (1971). *Journal of Bacteriology* **106**, 608.
Seidler, R.J., Knittel, M.D. and Brown, C. (1975). *Applied Microbiology* **29**, 819.
Selander, R.K. (1976). *In* "Molecular Evolution" (F.J. Ayala, ed.), pp. 21–45. Sinauer Associates, Sunderland, Mass.
Sidenberg, D.G. and Lachance, M.A. (1983). *International Journal of Systematic Bacteriology* **33**, 822.
Smith, D. and Halvorson, H.O. (1967). *In* "Methods in Enzymology" (L.Grossman and K. Moldave, eds), pp. 538–541, vol. 12, part A. Academic Press, New York.
Spencer, J.F.T. and Gorin, P.A.J. (1969). *Canadian Journal of Microbiology* **15**, 375.
Spencer, J.F.T., Gorin, P.A.J. and Rank, G.H. (1971). *Canadian Journal of Microbiology* **17**, 1451.
Starmer, W.T., Phaff, H.J., Miranda, M. and Miller, M.W. (1978a). *International Journal of Systematic Bacteriology* **28**, 318.
Starmer, W.T., Phaff, H.J., Miranda, M. and Miller, M.W. (1978b). *International Journal of Systematic Bacteriology* **28**, 433.
Stelling-Dekker, N.M. (1931). *Verhandelingen-Koninklijke Akademie van Wetenschappen Amsterdam, Afduling Natuurkunde, Sectie II* **28**, 1.
Stenderup, A. and Bak, A.L. (1968). *Journal of General Microbiology* **52**, 231.
Storck, R. (1966). *Journal of Bacteriology* **91**, 227.
Storck, R. and Alexopoulos, C.J. (1970). *Bacteriological Reviews* **34**, 126.
Storck, R., Alexopoulos, C.J. and Phaff, H.J. (1969). *Journal of Bacteriology* **98**, 1069.
Subden, R.E., Irwin, D., Cunningham, J.D. and Meiering, A.G. (1982). *Canadian Journal of Microbiology* **28**, 1047.
Subirana, J.A. and Doty, P. (1966). *Biopolymers* **4**, 171.
Timberlake, W.E. (1978). *Science* **202**, 973.
Tsuchiya, T., Fukazawa, Y., Taguchi, M., Nakase, T. and Shinoda, T. (1974). *Mycopathologia et Mycologia Applicata* **53**, 77.
Ullman, J.S. and McCarthy, B.J. (1973). *Biochimica et Biophysica Acta* **294**, 405.
Van der Walt, J.P. (1970a). *In* "The Yeasts–A Taxonomic Study" (J.Lodder, ed.), pp. 863–892. North Holland Publishing Co., Amsterdam.
Van der Walt, J.P. (1970b). *In* "The Yeasts–A Taxonomic Study" (J.Lodder, ed.), pp. 555–718. North-Holland Publishing Co., Amsterdam.
Van der Walt, J.P. and Hopsu-Havu, V.K. (1976). *Antonie van Leeuwenhoek* **42**, 157.
Van der Walt, J.P. and Johannsen, E. (1979). *Antonie van Leeuwenhoek* **45**, 281.
Van Etten, J.L. and Freer, S.N. (1978). *Applied and Environmental Microbiology* **35**, 622.
Van Uden, N. and Buckley, H. (1970). *In* "The Yeasts–A Taxonomic Study" (J. Lodder, ed.), pp. 893–1087. North-Holland Publishing Co., Amsterdam.
Vaughan-Martini, A. and Kurtzman, C.P. (1985). *International Journal of Systematic Bacteriology* **35**, 508.

Walker, W.F. and Doolittle, W.F. (1982). *Nature, London* **299**, 723.
Walker, W.F. and Doolittle, W.F. (1983). *Nucleic Acids Research* **11**, 7625.
Wickerham, L.J. (1951). *United States Department of Agriculture, Technical Bulletin* **1029**, 1.
Wickerham, L.J. and Burton, K.A. (1948). *Journal of Bacteriology* **56**, 363.
Wickerham, L.J. and Burton, K.A. (1954). *Journal of Bacteriology* **68**, 594.
Wickerham, L.J., Kurtzman, C.P. and Herman, A.I. (1969). *In* "Recent Trends in Yeast Research" (D.G. Ahearn, ed.), vol. 1, pp. 81–92. Spectrum Monograph Series, Georgia State University, Atlanta.
Williamson, D.H. and Fennell, D.J. (1975). *In* "Methods in Cell Biology" (D.M. Prescott, ed.), vol. 12, pp. 335–351. Academic Press, New York.
Wilson, A.C., Carlson, S.S. and White, T.J. (1977). *Annual Review of Biochemistry* **46**, 573.
Winge, O. and Roberts, C. (1949). *Comptes Rendus des Travaux du Laboratoire Carlesberg, Série Physiolique* **24**, 263.
Yamada, Y. and Banno, I. (1984). *Transactions of the Mycological Society of Japan* **25**, 455.
Yamada, Y. and Kondo, T. (1984). *Journal of General and Applied Microbiology* **30**, 313.
Yamada, Y. and Kondo, K. (1972). *In* "Fermentation Technology Today" (G. Terui, ed.), pp. 781–784. Proceedings of the Fourth International Fermentation Symposium. Society of Fermentation Technology, Osaka, Japan.
Yamada, Y. and Kondo, K. (1973). *Journal of General and Applied Microbiology* **19**, 59.
Yamada, Y., Arimoto, M. and Kondo, K. (1973a). *Journal of General and Applied Microbiology* **19**, 353.
Yamada, Y., Kanematsu, Y., Ohashi, M. and Kondo, K. (1973b). *Agricultural and Biological Chemistry* **37**, 621.
Yamada, Y., Okada, T., Ueshima, O. and Kondo, K. (1973c). *Journal of General and Applied Microbiology* **19**, 189.
Yamada, Y., Arimoto, M. and Kondo, K. (1976a). *Journal of General and Applied Microbiology* **22**, 293.
Yamada, Y., Nojiri, M., Matsuyama, M. and Kondo, K. (1976b). *Journal of General and Applied Microbiology* **22**, 325.
Yamada, Y., Arimoto, M. and Kondo, K. (1977). *Antonie van Leeuwenhoek* **43**, 65.
Yamada, Y., Takinami-Nakamura, H., Tahara, Y. and Smith, M.Th. (1980). *Antonie van Leeuwenhoek* **46**, 595.
Yamada, Y., Sasaki, J., Hou, C.F., Tahara, Y., Miranda, M. and Phaff, H.J. (1981). *Agricultural and Biological Chemistry* **45**, 2671.
Yamada, Y., Onishi, T. and Kondo, K. (1983). *Journal of General and Applied Microbiology* **29**, 51.
Yamazaki, M. and Komagata, K. (1981). *International Journal of Systematic Bacteriology* **31**, 361.
Yamazaki, M. and Komagata, K. (1982). *Journal of General and Applied Microbiology* **28**, 119.
Yamazaki, M. and Komagata, K. (1983a). *Journal of General and Applied Microbiology* **29**, 115.
Yamazaki, M. and Komagata, K. (1983b). *Journal of General and Applied Microbiology* **29**, 305.
Yamazaki, M. and Komagata, K. (1983c). *Journal of General and Applied Microbiology* **29**, 365.
Yarrow, D. and Meyer, S.A. (1978). *International Journal of Systematic Bacteriology* **28**, 611.

4 The Typological Yeast Species, and its Delimitation

J.P. van der Walt

Microbiology Research Group, Council for Scientific and Industrial Research, Pretoria, South Africa

I. INTRODUCTION

Because yeasts are fungi, their classification is subject to the requirements of the International Code of Botanical Nomenclature (Stafleu, 1983). The requirements of the Code demand that every yeast strain be treated as belonging to a number of taxonomic groups or taxa of consecutively subordinate rank culminating in the rank of *species* which is basal. Each of the taxonomic groups is based separately on a nomenclatural type which is the decisive element of the group and to which the name of the taxon is permanently attached. A particular strain is chosen as the *type strain* of the species. A permanently preserved specimen of some non-living material of

The Yeasts Vol. 1, 2nd edition
ISBN 0-12-596411-0

the type strain constitutes the *holotype* of the species. The Code also makes provision for ranks below the species (subspecies, varieties, subvarieties, forms and subforms). It does not, however, recognize concepts such as superspecies, ecospecies, semispecies, incipient species or races — terms that apply in evolutionary biology. Taxonomically, the yeast species is *typological* and, because it is typological, its delimitation hinges on differences between individuals or strains.

The provision of a taxonomic framework by which yeast strains can be classified into species and the subsequent differentiation or delimitation of those yeast species may be undertaken for different purposes. Such a taxonomic framework may be required: (i) for the purpose of general reference and communication in the classification of the entire yeast domain in terms of the requirements of the Code, much along the same lines and for the same purposes as those that apply in the case of the higher fungi and plants — such a framework would be comprehensive; (ii) it may be needed merely to provide a contracted reference frame for a limited number of yeasts employed for a special purpose, e.g. the industrial yeasts used in brewing. Such a special purpose classification would need to be much more restricted and utilitarian; it would, by comparison with the former, be artificial in that it would be based on the deliberate selection of a limited number of criteria relevant to the particular purpose for which it is to be used.

Although the International Code of Botanical Nomenclature provides the basis for the nomenclature of species, it gives no definition of the species; nor does it indicate how it should be delimited. As will become evident, there are different ways by which yeast species may be delimited but, because of the heterogeneity of the yeast domain, not all approaches are equally applicable to all yeasts.

As now known, the yeast domain comprises representatives of the Ascomycota, Basidiomycota (van der Walt and von Arx, 1985) and the Deuteromycetes (Fungi Imperfecti). On the basis of sexual stages, two groups are recognized: (i) the *perfect* or *teleomorphic yeasts* which have distinct sexual stages expressed as either the ascus or the basidium, and (ii) the *imperfect* or *anamorphic yeasts* which are those that lack sexual stages whether expressed as the ascus or the basidium — the majority of the yeasts belong in this category. It has been demonstrated, however, that these so-called imperfect yeasts can be related to either the ascomycetous or basidiomycetous yeasts on the basis of such affinitive characteristics as: (i) the ultrastructure of the cell-wall and conidiation (Kreger-van Rij and Veenhuis, 1971; von Arx, 1980); (ii) cell-wall composition (Bartnicki-Garcia, 1970; Weijman, 1979); and (iii) a colour reaction with certain diazonium dyes (van der Walt and Hopsu-Havu, 1976; Hagler and Ahearn, 1981).

Since the introduction of yeast systematics nearly a century ago, few taxonomists have stated explicity what they understood the yeast species to be or how they tackled the problem of delimiting it. What they did was to provide a comprehensive list of characters which could be used for descriptive purposes and which they had used in one way or another for delimiting the species. Consequently, any consideration of these vague, inferred criteria requires, first, an evaluation of the concept of the yeast species which has evolved in practice over the past 90 years.

II. DELIMITATION OF THE SPECIES ON THE BASIS OF PHENOTYPIC DISCONTINUITY

Although Meyen introduced the genus *Saccharomyces* as early as 1837, the first glimmering on the concept of the yeast species only emerged 51 years later when Hansen introduced pure-culture techniques in his studies on brewery yeasts. Hansen (1888) differentiated his isolates on the basis of cellular and ascosporal morphology, characteristics of growth in liquid media, optimal growth temperatures and their ability to ferment different sugars.

On the basis of differences in these characters and in keeping with the prevailing nineteenth Century taxonomic practices, Hansen assigned strains with different but definite phenotypic characters to different species. In so doing, he laid the basis for the use of phenotypic characters for species delimitation, a method which for almost a century has held sway in yeast systematics. As Hansen's work primarily entailed selection of brewing yeasts, it is important to note that yeast classification has decidedly applied roots.

With the discovery of new yeast forms, subsequent taxonomists extended Hansen's list of phenotypic characters to include other features, such as the mode of vegetative reproduction, the ultrastructure of the cell wall and septa, the presence or absence of sexual stages, properties of the ascus and basidium, the morphology of both ascospores and basidiospores, the pattern of the alternation of generations, ploidy of the vegetative phase, pigment formation, oxidative utilization of some 40 carbon sources, utilization of nitrate and other nitrogenous compounds as sole source of nitrogen, vitamin requirements, liquefaction of gelatin, growth at various temperatures, tolerance of high concentrations of glucose, resistance to cycloheximide and the structure of the coenzyme Q involved in the electron transport system. A full list of these *differentiae* is to be found in the monograph by Barnett *et al.* (1983) and a second edited by Kreger-van Rij (1984). Other characters which have been studied, include immunological properties (Richards, 1972; Tsuchiya *et al.*, 1974), proton magnetic resonance of cell wall mannans (Gorin and Spencer, 1970), determination of intracellular glycosidases (Fiol,

1972) and the electrophoretic comparison of enzymes (Baptist and Kurtzman, 1976; Yamazaki and Komagata, 1981). Although the use of these characters has been limited, their wider application can, none-the-less, be foreseen (Nakase and Suzuki, 1985a, b).

The yeast species delimited exclusively on grounds of differences in phenotypic properties, or phenotypic discontinuity among strains, has much in common with the Linnean species and agrees in essence with what Lotsy (1925) regarded as the *linneon* and which Mayr (1976) refers to as the *morphological species*. It must be emphasized that this method of delimiting species is so rigid that its application leads to artificial systems, and to taxonomic entities which come adrift if the phenotypic properties used for delimiting the species have undergone mutational change.

By 1970, the weaknesses inherent in the delimitation of species based exclusively on morphological, physiological and biochemical differences between strains had become obvious. The inadequacies of morphological criteria were apparent from the fact that Ditlevsen (1944) had found a strain of *Saccharomyces italicus* to be heterozygous for short and long cells, and Lindegren and Lindegren (1951), Mundkur (1953) and Townsend and Lindegren (1954) reported that, in the case of *Saccharomyces* species, the size of the cells was proportional to the degree of ploidy. Similarly, the shape of the ascospores, a character long considered to be constant for a species, proved to be subject to variation when Wickerham and Burton (1954b) reported formation of both spherical and hat-shaped ascospores by strains of *Pichia ohmeri*. With this potential for variation, it was clear that the use of simple morphological criteria was no longer practicable. It had by now also come to be recognized that the fermentative and assimilatory characters which had been employed to delimit species in *Saccharomyces* were, in fact, coded by relatively few genes (Winge and Roberts, 1948, 1950, 1952, 1954, 1956, 1957; Gilliland, 1954a, b; Losada, 1957) representing a very small portion of the yeast genome. Moreover, these poorly conserved characters, which previously had been considered more or less stable for given strains, were observed by Kossikov (1957) and Roberts and van der Walt (1960) to be subject to mutation. Scheda (1966) and Scheda and Yarrow (1966, 1968) subsequently confirmed and extended these observations and concluded that "We can assume that every physiological property of a yeast belonging to any genus may be unstable under certain conditions". As it was normal practice for differences in only one or two physiological properties to be used to differentiate species, the potential instability of such characters did away with the belief that any yeast species could be effectively delimited solely on the basis of physiological criteria.

Although it must be conceded that species delimitation on the basis of differences in poorly conserved morphological and physiological characters is taxonomically impractical, it must not be inferred that phenotypic classification serves no purpose or that all morphological characters are unstable. Although the form of vegetative cells may be subject to some variation among strains of a species, conidiation or the mode of formation of new vegetative cells is constant to the extent that this highly conserved character serves to differentiate supraspecific taxa.

III. DELIMITATION OF THE SPECIES ON THE BASIS OF REPRODUCTIVELY ISOLATED POPULATIONS

The second approach to the delimitation of the yeast species stems from the proposal by Winge and Laustsen (1939) that

> "(1) two types generally ought to be reckoned to be the same species when hybridizing and their hybrid being fully fertile whereas (2) two types would be regarded as specifically different when they are unable to hybridize or their hybrid shows reduced fertility".

This point of view was subsequently restated by Wickerham and Burton (1956a).

This approach to delimitation of species clearly reflects the fundamental developments and trends in biology since the turn of the century and the change which the concept of the species has undergone in that period. This development is expressed by a movement away from delimiting species on the basis of phenotypic discontinuity towards defining the species in terms of genetic relatedness and as reproductively isolated populations. For an historical account, Mayr (1976) may be consulted. The Linnean species of the nineteenth century, which had been the common starting point for early naturalists and taxonomists alike, had in biology made way for the *biological species concept*. The following recent definitions of the species indicate the present position in this regard.

1. "Species are accordingly systems of populations; the gene exchange between these systems is limited or prevented in nature by a reproductive isolating mechanism or perhaps by a combination of several such mechanisms. In short, the species is the most inclusive Mendelian population" (Dobzhansky, 1970).
2. "A species consists of a group of populations which replace each other geographically or ecologically and of which the neighboring ones intergrade or hybridize wherever they are in contact or which are potentially capable of doing so (with one or more of the populations) in those cases where contact is prevented by geographical or ecological barriers" (Mayr, 1976).

These broad, biological definitions indicate four separate aspects of the species. It is (i) a reproductive community, (ii) an ecological unit, (iii) a genetic unit and (iv) an evolutionary unit (Mayr, 1976).

To the yeast taxonomist these definitions raise two fundamental issues: (i) the biological species concept is restricted to sexually reproducing organisms where karyogamy implicates the fusion of two independent, free-occurring, haploid cells, and is not applicable to the imperfect yeasts which constitute the greater part of the yeast domain; (ii) that the biologically defined species is based on populations in which no individual shows the precise mean of all of the characters of the population and therefore has no type.

To distinguish between the biologically defined entity and the typologically differentiated entity, Cain (1954) and Grant (1963) proposed the terms *biological species* and *taxonomic species*. As Grant (1971) points out, the biological species implies genetic relationships, but this is not a concept that can be easily applied in practical taxonomy. The taxonomic or typological species, on the other hand, which is much more convenient does not imply biological or genetic relationships. It must, however, be borne in mind that, although the biological species and the taxonomic species are not equivalent, they may on occasion show close agreement.

Because of the weaknesses inherent in species delimitation based on phenotypic discontinuity, taxonomists have consequently come to lean progressively towards delimiting the typological species along the lines of the biological species concept. However, both Dobzhansky (1970) and Mayr (1976) caution that a yardstick such as the concept of the biological species is not automatic: "To apply it effectively, skill and experience are required, particularly in situations where it cannot be applied directly" (Mayr, 1976).

A. Yeast species as populations

Although the validity of the concept that yeast species represent reproductively isolated populations cannot be questioned, its application nevertheless raises practical issues. Yeast taxonomists, unlike plant taxonomists, cannot study yeast populations in a natural habitat directly, but only as axenic cultures under laboratory conditions. When examining specimens of a particular habitat or substrate, taxonomists frequently rely on aerobic cultivation and the use of strongly acidified agar media (pH 3.5–3.7) which permit isolation of most, but not of all, yeasts. It is only after the physiology of the prevalent species is known that the populations of those species may be selectively enumerated (e.g. by the use of special media containing particular carbon and nitrogen sources) in specimens of the substrate. However, many yeast species are only recovered very infrequently and years may elapse before a second strain is encountered. Irrespective of these difficulties, ecological studies along the lines indicated by Starmer and his coworkers (Starmer, 1981a, b, c; Starmer *et al.*, 1976, 1978a, b, 1979, 1980) will undoubtedly provide not only a realistic approach to the recognition and characterization of species as populations or communities, but also a better insight into their evolutionary systematics.

The study of species as populations also raises the question of how many strains must be studied to ensure that the population on which the species is based adequately represents that species. An experienced and skilled taxonomist such as Wickerham (1951) with his extensive knowledge of the genus *Hansenula*, confidently described *Hansenula minuta* on the basis of a single strain. On the other hand, the recent delimitation of *Saccharomyces cerevisiae* by Barnett *et al.* (1983) and Yarrow (1984) hinged on the study of well over 200 strains by numerous authors over a period of almost a century. The answer to this question depends not only on the experience and skill of the taxonomist, and the particular taxonomic area in which he is working, but also on how well the taxonomy of the relevant area has been developed. In principle, as many strains as possible — preferably from divergent localities — should be studied.

B. Habitat isolation

Yeast populations may be prevented from interbreeding because of their ecological or habitat isolation. Whereas some species are recoverable from highly divergent terrestrial as well as marine sources (e.g. *Debaryomyces hansenii*), others (e.g. *Cyniclomyces guttulatus* and certain insect-associated arboreal species) appear to be restricted to very specific habitats or narrow ecological niches. This restriction in habitat does not invariably imply that all individuals are confined exclusively to such ecological niches. *Pichia acaciae* serves as an example. Using non-specific isolation techniques, strains of this species were initially recovered from arboreal insect sources only (van der Walt, 1966, 1972). Subsequent use of more specific techniques confirmed that strains of this species could also be recovered from uncultivated grassland soils (van der Walt and Johannsen, 1975). Although taxonomists affirm that habitat specificity greatly accentuates the distinctness of species, they are unlikely to accept habitat specificity as a decisive criterion when it comes to classification of a single strain.

C. Genetic isolation

Apart from their ecological isolation, yeast species may also be genetically isolated by so-called isolating mechanisms. Mayr (1976) defined these isolating mechanisms as biological properties of individuals that prevent interbreeding of populations that are potentially or actually sympatric, i.e. occurring together. Dobzhansky (1970) classified these isolating mechanisms as either premating, in which case formation of hybrid zygotes is prevented, or postmating whereby the viability or fertility of hybrid zygotes is decreased.

As far as yeasts are concerned, premating mechanisms mainly involve isolation by natural species barriers imposed by normal mating-type systems. Mutual attraction or sexual recognition between strains is absent or weak so that they do not hybridize. Postmating mechanisms involve: (i) hybrid inviability — crosses of different strains form zygotes, but these will be either non-viable or show diminished viability; (ii) hybrid sterility — crosses yield a viable first-filial generation which is sterile so that ascospores (or basidiospores) are either not produced or, if they are produced, are non-viable or show decreased viability; (iii) hybrid breakdown — the first-filial (F_1) generation is fertile but further crosses between first-filial generations or between first-filial generations and parental strains show decreased viability or fertility.

Whereas failure to recover fertile hybrids from crosses of yeast strains may involve an array of genetic factors, interfertility between strains represents the virtual consummation of sexual reproduction. This implies functional homology of their mating-type loci which, it is believed, comprises a cluster of genes, both regulatory and structural, and which govern haploid and diploid functions in yeast cells (Friss and Roman, 1968; Fogel and Mortimer, 1971; Mackay and Manney, 1974; Crandall *et al.*, 1977).

As most sexual yeasts can be hybridized under controlled laboratory conditions, the presence or absence of interbreeding becomes of primary importance when it comes to the more effective delimitation of the perfect species. This is reflected by Wickerham's (1970a) remark that "The usefulness of hybridization studies to taxonomy has barely begun".

The application of hybridization studies for delimitation of species raises practical considerations. The first qualification is the number of strains that need to be studied before a population is adequately represented. To cross n strains in all combinations, $(n^2 - n)/2$ crosses have to be made. This imposes limits to what is practically feasible. Before crosses are made, strains still have to be characterized phenotypically, partly to permit confirmation of combined parental phenotypes of possible hybrids and partly to ensure that taxonomically irrelevant crosses are avoided. Consequently, taxonomists always resort to some degree of phenotypic classification before proceeding with hybridization studies. Taxonomists may tend to avoid extensive hybridization studies when dealing with strains that are phenotypically more or less indistinguishable but, in doing so, they incur the risk of missing sibling or cryptic species that are phenotypically very similar, but which do not hybridize. Taxonomists are much more likely to undertake hybridization studies on phenotypically dissimilar strains which, on the basis of experience, they consider may possibly be related. To assess possible conspecificity of strains, taxonomists must resort to the study of type strains as, in these and all other cases, it must be ensured that the strains used are sexually competent. If, as a result of prolonged maintenance, type strains show regression of their sexual activity (e.g. as evinced by the absence of ascospores), other sexually competent strains, judged to be conspecific with the type, must be selected. In cases where a species is represented by a single strain, this is impossible.

While failure to recover hybrids from an adequate number of crosses would provide reasonable grounds to assume that parental strains are not conspecific, Kurtzman *et al.* (1980b) point out that, because of the numerous factors involving interfertility, the failure of strains to mate or to form viable progeny need not mean complete lack of genetic relatedness. However, if hybrids are recovered, it is still necessary to assess whether they are fertile. Kurtzman *et al.* (1980b) propose that, where hybridization is used to delimit species, not only should the progeny of the initial crosses be viable and fertile but at least one additional generation of crosses should be made, and these progeny should also be tested for sexual competence. It still remains to be seen whether taxonomists involved with the heterothallic basidiomycetous species will follow these guidelines.

However, it is impracticable to delimit species as Mendelian populations by hybridization of large numbers of strains. Mayr (1976) maintains that Dobzhansky's (1970) definition of the species is of limited use in taxonomic work. It appears that the compass of hybridization studies required to delimit a species must, in practice, be left to the judgment and skill of the taxonomist.

1. Hybridization studies

Several techniques are available for detecting hybrid formation between yeast strains. The taxonomist's choice depends on whether strains are heterothallic or homothallic, or haploid or diploid.

(a) Heterothallic species

In the case of the heterothallic yeasts, hybridization studies hinge on the availability of compatible mating types and simple mass-mating techniques. When dealing with ascomycetous strains, recovery of mating types involves selection of sexually compatible strains from crosses of either a collection of anascogenous, presumed-unisexual, haploid strains which, on the basis of phenotypic characters, appear to be related, or single-spore isolates obtained from diploid ascogenous strains presumed to be heterothallic. For details, the publications of Wickerham and Burton (1954a, b) and van der Walt and Yarrow (1984) should be consulted. Once mating types are available, it becomes relatively easy to assign strains to species on the basis of their ability to hybridize with mating types constituting type strains with the recovery of a fertile ascogenous diplophase. Such studies not only serve as the basis for delimitation of new species (e.g. Wickerham and Burton, 1954a; Wickerman, 1965; van der Walt, 1959; Kurtzman and Wickerham, 1973; Smith *et al.*, 1976; Rodrigues de Miranda, 1978; Kurtzman *et al.*, 1980a, b), but also to rationalize classification of the imperfect states of perfect heterothallic species (Wickerham and Burton, 1954a, b; Slooff, 1964).

Application of heterothallism as a taxonomic tool for delimitation of heterothallic basidiomycetous species stems from Banno's (1967) successful crossing of complementary auxotrophic mutants of *Rhodotorula* strains when he delimited the first known perfect heterothallic basidiomycetous species, namely *Rhodosporidium toruloides*. Banno's (1967) studies were soon followed by more extensive investigations by other workers (Fell *et al.*, 1969, 1973; Newell and Fell,1970; Fell and Tallman, 1980a, b, 1981; Rodrigues de Miranda, 1972; Kwon-Chung, 1975, 1976, 1977) which subsequently led to the recovery and successful delimitation of numerous new species. This success hinges on the fact that heterothallic basidiomycetous yeasts are, as a rule, recovered in nature as budding haploid yeast phases, so that hybridization tests inevitably precede the recovery of the dikaryotic clamp-bearing, perfect states. For experimental details, the publications of Banno (1967), Fell *et al.* (1969, 1973), Newell and Fell (1970), Fell and Tallman (1980a, b, 1981), Rodrigues de Miranda (1972) and Kwon-Chung (1975, 1976, 1977) should be consulted.

(b) Homothallic species

By comparison, detection of hybrid formation between homothallic yeast strains is more laborious and hinges on techniques developed for hybridization and genetic studies of *Saccharomyces cerevisiae*. These techniques have been reviewed by Fowell (1969).

The first technique involves micromanipulation and relies on the pairing of individual ascospores or haploid vegetative cells. Pairing of individual spores is, for practical reasons, restricted to strains of high ascosporal viability and which are vegetatively stabilized in the diplophase. The second technique involves application of the prototrophic selection technique of Pomper and Burkholder (1949) as modified by Wickerham and Burton (1956a, b), Johannsen (1980, 1981), Johannsen and van der Walt (1979, 1980) and van der Walt and Johannsen (1980). For experimental details, the publications cited should be consulted. This method has the advantage that: (i) it can be readily applied to haploid strains; (ii) it obviates problems arising from low ascosporal viability of parental strains; (iii) it permits detection of hybrids when they are present in very low numbers; and (iv) it eliminates non-viable or poorly viable hybrids. It must be borne in mind that the prototrophic selection technique has its limitations and disadvantages, and these are fully discussed in the publication by Johannsen (1980).

Because hybridization of homothallic strains is laborious, it is not surprising that, in the past, taxonomists have generally avoided hybridization studies for differentiation and delimitation of homothallic species.

Yarrow (1972), following Winge and Laustsen's (1939) directive, nevertheless became the first to rationalize delimitation of *Saccharomyces* species by successfully hybridizing 17 phenotypically dissimilar representatives of the so-called *Saccharomyces* sensu stricto (van der Walt, 1970a). On the basis of results obtained by use of micromanipulation as well as mass-mating, Yarrow (1972) confirmed the earlier observations of Spiegelman *et al.* (1945), Winge and Roberts (1948), Gilliland (1949) and Lindegren (1949) who, from similar crosses, obtained highly fertile hybrids showing Mendelian segregations of characters. Because of the absence of genetic isolation, Yarrow (1972, 1984), Yarrow and Nakase (1975) and Barnett *et al.* (1983) consequently were able to assign the 17 species formerly accepted in *Saccharomyces* sensu stricto to the status of a single species, namely *Sacch. cerevisiae*.

However, hybridization studies do not invariably result in a clear-cut delimitation of species. Johannsen's (1980) studies of the genus *Kluyveromyces*, which were based on use of the prototrophic selection technique, showed that, unlike the phenotypically dissimilar strains of *Sacch. cerevisiae*, interbreeding groups of strains or *syngameons* (Lotsy, 1931) were not always distinct but tended to overlap so that not all species could be sharply delimited on the basis of hybridization. Johannsen (1980) nevertheless resolved this difficulty taxonomically by retaining the oldest legitimate specific epithets for the two genetically overlapping syngameons.

Moreover, hybridization studies cannot be undertaken on all ascogenous yeasts. Nevertheless, notwithstanding the extreme sensitivity of the prototrophic selection technique, Johannsen and van der Walt (1980), in their study of the soil-borne haploid genus *Schwanniomyces*, were unable to detect hybrid formation in crosses of complementary auxotrophic mutants originating from either single strains or from different strains of the same phenotypically delimited species. The failure to detect hybrid formation even in intrastrain crosses can only be explained on the basis that in these haploid homothallic ascogenous yeasts, zygote formation and karyogamy do not involve fusion of two independent free-occurring haploid cells. In this genus, karyogamy obviously involves autogamous or automictic fusion of two sister nuclei within the confined structure of a mother cell and its attached bud. Because this precludes genetic exchange between strains, each individual strain is, in fact, genetically isolated.

It therefore seems feasible in the case of perfect yeasts with sexual cycles which implicate fusion of two free-occurring independent haploid cells to delimit and define the typological species in terms of the biological species concept, i.e. as the sum total of interbreeding populations of strains. This approach cannot, however, be applied to automictic ascogenous yeasts such as the genus *Schwanniomyces* which may, possibly, have evolved sexually by eliminating genetic exchange between strains and, in this way, retaining a presumably successful genotype.

2. Artificial hybridization by protoplast fusion

Because natural hybridization is excluded in automictic and imperfect yeasts, it is necessary to consider artificial hybridization by fusion of complementary auxotrophic protoplasts in the presence of poly(ethylene glycol) (van Solingen and van der Plaat, 1977; Svoboda, 1976; Stahl, 1978; Máraz *et al.*, 1978; Whittaker and Leach, 1978; Sipiczki, 1979; Russell and Stewart, 1979; Morgan *et al.*, 1980; Sarachek *et al.*, 1981; Freeman and Peberdy, 1983) as an alternative to evaluating the conspecificity of strains.

As zygote formation by two haploid yeast cells may be regarded as analogous to fusion of two protoplasts, reports of hybrid formation by protoplast fusion of sexually compatible yeast strains are not surprising. The analogy between formation of natural hybrids by fusion of two intact cells and fusion of protoplasts is nevertheless superficial. Zygote formation hinges on a sexual recognition between conjugal partners which is dependent on the mating-type locus. Crandall *et al.* (1977) stressed that zygote formation must involve contact between two intact cells and the ensuing dissolution of their cell walls, and that these events are mandatory for triggering subsequent processes of conjugation. In artificially induced fusion of protoplasts, the specificity of the enzyme systems involved in dissolution of cell walls is abolished so that hybridization involves non-specific glycol-induced fusion of aggregated protoplasts into sometimes multinucleate *syncytia* (Svoboda, 1978) from which reconstituted hybrid cells finally emerge. As Sarachek *et al.* (1981) point out, the recovery of such hybrids depends not only on the genetic background of the protoplasts, but also on their surface properties, their regenerative abilities and their respective auxotrophy on the behaviour of the intermediate syncytium.

By avoiding natural species barriers inherent in normal mating systems, some unusual intraspecific, interspecific and even intergeneric hybrids have been produced which, in the present state of our knowledge, detract from the possible use of protoplast fusion for taxonomic purposes. In the case of the heterothallic species *Schizosaccharomyces pombe* and *Saccharomycopsis lipolytica*, Sipiczki and Ferenczy (1977) and Stahl (1978) respectively, obtained intraspecific hybrids by fusion of protoplasts originating from one mating type. Similarly, Sipiczki (1979) reported hybrid formation by fusion of protoplasts of strains of *Schizosaccharomyces pombe* and *Schiz. octosporus* which Johannsen (1981) was unable to confirm by natural hybridization using intact cells. Examples of intergeneric hybrid formation have involved protoplast fusions implicating *Saccharomycopsis fibuligera* and *Candida tropicalis* (Provost *et al.*, 1978), *Sacch. cerevisiae* and *Kluyveromyces lactis* (Stewart *et al.*, 1981), *Candida utilis* and the hyphomycetous fungus *Trichoderma reesei* (Beydon and Heslot, 1980), *Sacch. diastaticus* and *Hansenula capsulata* (Spencer and Spencer, 1981), but the majority of these intergeneric hybrids have either proved genetically unstable or to be only partial recombinants which lack the full genetic complement of both parental strains.

More recently, Johannsen *et al.* (1984) investigated protoplast fusion between hybridizing and non-hybridizing strains of the genus *Kluyveromyces*. This more taxonomically orientated study confirmed that stable fusion products with combined parental phenotypes were only recoverable from crosses involving hybridizing parental strains. The authors consequently concluded that for the recovery of stable recombinants in this genus, "the protoplast fusion technique did not offer a convenient or reliable alternative to hybridization via the prototrophic selection technique". On the basis of their study Johannsen and her associates nevertheless raised the possibility of applying protoplast fusion for the delimitation of anamorphic or imperfect species.

Possible future application of the protoplast-fusion technique for taxonomic purposes (e.g. for delimitation of automictic or imperfect species) will, of necessity, require further calibration studies in which the results obtained by this technique will have to be evaluated in terms of natural hybridization and nuclear deoxyribonucleic acid (nDNA) re-association data.

IV. DELIMITATION OF THE SPECIES ON THE BASIS OF NUCLEIC ACID ANALYSES

Because hybridization cannot be applied as a criterion in the delimitation of species of yeast which seemingly lack sexual stages or where sexuality has evolved to a state of autogamy or automixis, a third approach to delimitation of the yeast species was developed. This approach, which is applicable to both perfect and imperfect yeasts, had its origins in modern developments in bacterial systematics resulting from comparative studies of the prokaryotic genome (Marmur *et al.*, 1963). It hinged on an assessment of relatedness by means of the so-called comparison of genomes by nucleic acid analyses. It was only subsequent to this development in bacterial systematics and chiefly as a result of many exploratory studies (Belozersky and Spirin, 1960; Storck, 1966; Storck *et al.*, 1969; Stenderup and Bak, 1968; Nakase and Komagata, 1968, 1969, 1970a, b, 1971a, b, c, d, e; Meyer and Phaff, 1969, 1972; Bicknell and Douglas, 1970; Ouchi *et al.*, 1970; Martini and Phaff, 1973; Martini, 1973; Mendoca-Hagler and Phaff, 1975; Meyer *et al.*, 1977) that comparison studies of genomes were found to be of taxonomic significance in delimitation of yeast species. For experimental details, the publications of Meyer *et al.* (1978) and Price *et al.* (1978) should be consulted.

Price *et al.* (1978) in an evaluation maintain that the rationale for such genome comparison studies is accepted to be the restatement at molecular level of the evolutionary statement of the principle of common descent. "If two organisms are related, they must retain in their genomes, base sequences that are descendant from a common ancestral base sequence; closely related organisms will have retained a greater portion of base sequences in common that have highly diverged". It must, nevertheless, be emphasized that, although this rationale may be tenable, it evades the issue as to whether or not it is possible, in practice, to distinguish between or determine relative proportions of ancestral and non-ancestral nDNA base sequences.

At present, two particular features of the yeast genome are found to be useful in delimitation of species. The first involves the composition of the nuclear DNA bases. "A significant difference" in accurate determination of the mean molar percentage of guanine-plus-cystosine (mol% G + C) present in two yeast strains is accepted as *prima facie* evidence that, reputedly, " . . . precludes the sharing of similar base sequences between two yeasts and indicates that they are not recently descendent from a common ancestor". It should be noted that, since base composition determinations do not reflect base sequences, strains with virtually identical mol% G + C values may still be representative of completely unrelated species. This second feature hinges on determination of the complementarity of the nDNA base sequence between strains on the basis of re-association of their fragmented denatured single-stranded nDNA. Given the appropriate re-association data, strains having about 80–100% of their base sequences in common may reasonably be considered to comprise the same species. Degrees of re-association between 20% and 80% are, reputedly, rarely encountered in the yeasts and many phenotypically similar yeasts have been observed to show nDNA relatedness values of 10% or less. Price *et al.* (1978) proposed using the term *genospecies* for the taxonomic unit delimited along these lines. This usage, however, is confusing and inappropriate since Ravin (1963) had 15 years previously coined this term to denote a prokaryotic taxonomic unit delimited on the basis of genetic transfer and recombination.

In delimiting species on the basis of a comparison of genomes, taxonomists must again resort to some degree of phenotypic differentiation of strains before undertaking such studies in order to ensure that such comparative studies are taxonomically relevant. A parallel situation obtains in the case of hybridization studies which emphasize the indispensability of phenotypic classification in delimitation of yeast species.

Whether significant differences in accurately determined nDNA base compositions are invariably indicative of species not being either recent descendants from a common ancestor or closely related remains on open question. Price *et al.* (1978) avoided quantifying the significant differences that could distinguish between recent and less recent descendants of an ancestral species. In this connection, it is worth noting that Wickerham (1970b) stressed the fact that strains of the closely related species *Hansenula fabianii* and *Hansenula subpelliculosa*, which he placed adjacently on line two of his phylogenetic diagram of the genus, differ in their base composition by no less than 11% (Nakase and Komagata, 1969). Whether differences of 1–2% in accurately determined base compositions invariably preclude the conspecificity of strains, as Phaff (1977) suggests, also requires further evaluation. This is evident from the fact that, although strains of *K. dobzhanskii* and *K. lactis* differ in their base composition by 2.2% (H.J. Phaff, personal communication), fertile hybrids with combined parental phenotypes have nevertheless been obtained from crosses of these strains (Johannsen, 1980). This is in sharp contrast to the fact that although strains of *K. lactis* and *K. delphensis* differ in their base composition by only 0.1% (H.J. Phaff, personal communication), these strains fail to hybridize (Johannsen, 1980). Wickerham (1970b) consequently seems justified in cautioning yeast taxonomists to be conservative in denying close relationships to occasional taxa that have similar taxonomic characters, but moderately-to-widely different base compositions.

It should clearly be understood that there are no formal or definitive guidelines for construction of bacterial species based on nucleotide homology. Consequently, not only Marmur *et al.* (1963) but also Mandel (1969) and Moore (1974) all stressed the need to calibrate such re-association methods in terms of genetic tests, using transformation and conjugation. When applying genome-comparison studies to the yeasts, such calibration must, of necessity, evaluate nDNA nucleotide homology and DNA base composition in terms of hybridization studies and assessment of interfertility.

A survey of the literature shows that this evaluation has been inadequate. In their otherwise elaborate exposition of species delimitation based on nDNA re-association data, Price *et al.* (1978) cite only the studies of Bicknell and Douglas (1970) and Ouchi *et al.* (1970) in which some interfertile strains of *Sacch. cerevisiae* were reported to show an 80% degree of nDNA complementarity. However, they omit to mention that Bicknell and Douglas (1970), and also Ouchi and his associates, had reported degrees of nDNA complementarity as low as 40% between strains of *Sacch. cerevisiae* and its interfertile melibiose-fermenting biotypes cited as *Sacch. carlsbergensis* and *Sacch. uvarum*. These discrepancies between low degrees of base-sequence complementarity and hybrid formation are not restricted to strains of *Sacch. cerevisiae*. Although Martini and Phaff (1973) reported degrees of re-association between nDNA from strains of *K. fragilis* and *K. dobzhanskii* to be as low as 13–27%, Wickerham and Burton (1956b) nevertheless obtained hybrid progeny with combined parental phenotypes from crosses of *K. fragilis* × *K. dobzhanskii*. The true hybrid nature of the progeny from this cross was, moreover, confirmed by Bicknell and Douglas (1970) who, on the basis of nDNA re-association data and characteristics of the hybrid's β-glucosidase, concluded that the hybrid in fact contained the genomes of both parents. Some further comparison of nDNA base-sequence complementarity and genetic compatibility became possible from van der Walt and Johannsen's (1979) hybridization studies of the genus *Kluyveromyces* using the prototrophic selection technique. These authors noted that, whereas reported degrees of nDNA re-association above 70% invariably tallied with the recovery of recombinants, degrees of re-association of less than 70% and as low as 12% did not, in the case of crosses involving strains of *K. marxianus* and *K. lactis*, preclude recovery of fertile recombinants capable of further recombination. Van der Walt and Johannsen (1980) consequently concluded that, although degrees of nDNA re-association above 70% between strains probably provided reliable grounds for regarding such strains as conspecific, degrees of re-association less than 70% would not, necessarily, preclude interfertility between strains. Kurtzman *et al.* (1980b), on the basis of studies in the genus *Issatchenkia*, reached similar conclusions and stressed that the range of 80–100% nDNA complementarity, rather prematurely proposed by Price *et al.* (1978), probably needed to be extended considerably.

Kurtzman *et al.* (1983) subsequently emphasized the fact that the lower limit of nDNA re-association values as criterion for conspecificity of strains was still not well defined, and that base sequence divergence might be as great as 75% before genetic exchange can no longer occur. The wider application of nDNA re-association data for species delimitation purposes unavoidably hinges on the determination of this lower limit. The need to have this lower limit defined more accurately has become all the more urgent in view of more recent studies on the genus *Hansenula*. Poncet *et al.* (1984) reported strains of the diploid species, *H. anomala* and *H. beckii*, as showing nDNA base complementarity as high as 84.8% — well within the range which Price *et al.* (1978) accept as indicative of conspecificity. However, when sexually competent strains of these species were examined in the author's laboratory for actual genetic recombination by means of the prototrophic selection technique, no recombinants could be recovered.

In trying to account for these discrepancies in relatedness in terms of hybrid formation on the one hand, and nDNA re-association methods on the other, it must be kept in mind that the re-association methods used for estimating similar nucleotide sequences (i) are crude in comparison with the available analytical procedures that permit direct determination of actual base sequences and (ii) were initially developed on the basis of the structure of the simple prokaryotic genome. The eukaryotic yeast genome, however, differs from its prokaryotic counterpart not only in size, localization and complexity but probably in its structure as well, since it has been established that the coding sequences of higher eukaryotes differ fundamentally from those of prokaryotes in being spaced by stretches of "silent DNA" or introns (Gilbert, 1978; Crick, 1979; Doel, 1980).

Despite these qualifications, nDNA re-association and DNA base-composition studies have the advantage that they are applicable to both perfect as well as imperfect yeast strains. The meaningful interpretation of results obtained with these techniques, however, must unavoidably hinge on further correlative studies involving more perfect ascomycetous and basidiomycetous yeasts.

V. DELIMITATION OF THE SPECIES BY NUMERICAL ANALYSES

It is only comparatively recently that yet a fourth approach has been attempted (e.g. Campbell, 1972, 1973, 1974; Kocková-Kratochvílová *et al.*, 1972; Jones, 1975). This approach also stems from developments in modern prokaryotic systematics. Numerical classification is based on the so-called "overall similarities" of strains calculated on the basis of multiple characters or character states of the strains (Sneath, 1962, 1972; Sokal and Sneath, 1963; Sokal, 1965; Sneath and Johnson, 1972). Because of the multiplicity of the chosen characters, the "overall similarities" are calculated by computer analyses. These results are usually presented as matching coefficients and dendrograms which convey statistical estimates of the number of differences and similarities between strains. For a critical analysis of the methods and principles applied in numerical classification, the publications by Mayr (1965, 1969) and Johnson (1970) should be consulted.

As Dobzhansky (1970) pointed out, all critics of numerical taxonomy stress that the neglect to recognize genetic criteria and the treatment of randomly chosen characters as equivalent are liable to result in misestimation of genetic similarities which are basic to any natural classification. The misestimation is particularly evident from Campbell's (1972) numerical studies in the genera *Saccharomyces, Kluyveromyces, Pichia* and *Hansenula*. To cite only two examples; firstly, Campbell (1972) "splits" the phenotypically dissimilar strains of *Sacch. cerevisiae* into 10 species irrespective of the fact that such phenotypically dissimilar strains are capable of interbreeding (Yarrow, 1972); conversely Campbell's (1973) results lead him into "lumping" strains of *Pichia wickerhamii* and *Pichia rhodanensis* irrespective of the fact that these heterothallic strains do not hybridize (van der Walt, 1959). An even more illuminating example of such misestimation is found in Jones' (1975) numerical analysis of 26 yeast strains, which established that a strain of the ascomycetous species *Lipomyces starkeyi* is more closely related to a strain of the imperfect, basidiomycetous species *Bullera alba*, than are two strains of the ascomycetous species *Sacch. cerevisiae* and *Hanseniaspora valbyensis*.

Although computer programs may well serve for storing taxonomic data and facilitate construction of extensive taxonomic keys (Barnett *et al*., 1983), yeast taxonomists, when resorting to numerical analyses for species delimitation, should seriously consider Mayr's (1969) dictum that "Some users of electronic data processing have suggested that thinking and theory become unnecessary if we merely trust our fate to the computer".

VI. DELIMITATION OF THE IMPERFECT SPECIES

Whether it be regarded from the biological or taxonomic viewpoint, nowhere is the intractability of the species concept more apparent than in the case of the delimitation of the so-called imperfect species. Not only is their delimitation on the basis of phenotyic discontinuity inadequate but, because they lack sexual cycles which involve fusion of free-occurring cells, hybridization studies are also precluded. Consequently their delimitation must, for the present, hinge on nucleic acid analyses. While degrees of nDNA re-association greater than 80% may serve as a criterion for the conspecificity of imperfect strains, the lower limit of nDNA re-association values indicative of conspecificity among perfect species are as yet not clearly defined (Kurtzman *et al*., 1983). It follows that degrees of re-association below 80% between imperfect strains need not imply that such strains are invariably representative of different species. In the absence of more reliable criteria and from experience with perfect species, delimitation of imperfect species is precarious — even to the extent that some taxonomists have refrained from describing imperfect species. Nevertheless, as the greater part of the yeast domain comprises imperfect yeasts, their classification cannot be avoided.

Apart from the limited criteria available for delimitation of imperfect species, the imperfect yeasts themselves are inadequately understood. As a rule, these yeasts have been regarded as lacking sexual stages and only capable of vegetative reproduction. Consequently, it was usual to regard haploid strains of a given mating type of a heterothallic species as imperfect, until compatible strains of opposite mating types had been recovered. Ploidy levels and affinitive characters consequently consitute important criteria when it comes to a more effective delimitation and a more natural classification of these species (van der Walt and Johannsen, 1975; van der Walt *et al.*, 1978, 1981).

Nevertheless, although some imperfect yeast strains with ascomycetous affinitive characters appear to be neither haploid nor representative of mating types, they have been found to show ploidy levels higher than haploid, and an alternation of presumed haploid and diploid generations in the absence of an ascosporal state.

Van der Walt and Pitout (1969) and van der Walt (1970b) submitted evidence for a budding, postulated haplophase and diplophase in *Candida albicans* on the basis of DNA content per cell. Further evidence for the diploid condition was subsequently provided by Oliaya and Sogin (1979) on the basis of data obtained by ultraviolet inactivation, mutagenesis and microfluorometry, and by Whelen and Magee (1981) on the basis of the natural heterozygosity of strains. Suzuki *et al.* (1982) subsequently reported a tetraploid phase of this species. Sukroongreung and Rodrigues de Miranda (1973) concluded, on the basis of chromosome counts, that the related species, *Candida tropicalis*, was likewise characterized by two ploidy levels. Van der Walt and Johannsen (1973) on the basis of X-ray inactivation data, showed a budding haploid and diploid state for *Torulopsis magnoliae*.

A comparison of the postulated life cycles for *C. albicans* (van der Walt, 1970b), *C. tropicalis* (Sukroongreung and Rodrigues de Miranda, 1973) and *T. magnoliae* (van der Walt and Johannsen, 1973) draws attention to eight salient features: (i) strains have affinitive characters that relate them to the ascomycetous yeasts; (ii) strains form chlamydospores permitting survival during unfavourable changes in the environment; (iii) vegetative strains are as a rule stabilized in the presumed-diploid, uninucleate condition; (iv) the postulated occurrence of reduction division; (v) the haplophase, irrespective of the ascomycetous affinitive characters of the strains, is delimited externally by budding; (vi) the postulated site of reduction division is not a basidium as was originally supposed (van der Walt, 1970b) but a structure homologous with the ascus, a so-called meioconidiophore (van der Walt and Johannsen, 1973, 1974); (vii) in culture, the sexually reactive haplophase is evanescent and rapidly overgrown by the diplophase, and (viii) diploidization is presumably effected by autodiploidization or autogamous fusion of two sister nuclei of a budded, haploid cell. Such yeasts would therefore be completely inbreeding, as seems to be the case with the genus *Schwanniomyces* (p. 106). To account for the absence of an ascosporal state during alternation of generations, van der Walt and Johannsen (1973, 1974) suggested that, although the strains had the genes for meiosis, the genes for ascospore formation were either lacking or not expressed.

The information available, consequently, points to the possibility that the "imperfect" yeasts such as *C. albicans, C. tropicalis* and *T. magnoliae* may represent a stage in the presumed evolution of the ascomycetous yeasts which is characterized by (i) a stabilized diplophase, (ii) meiosis, (iii) autodiploidization and (iv) the absence or non-expression of genes mediating ascospore formation. In such completely inbreeding yeasts, meiosis would provide a mechanism by which the ageing of the germ-line could be prevented (Bernstein, 1979). Although such postulated life cycles resist easy elucidation, understanding and hence the natural classification of the imperfect yeasts nevertheless hinges on further studies along such lines.

VII. DELIMITATION OF THE YEAST SPECIES IN PRACTICE

Because the yeast domain is heterogeneous and the procedures for species delimitation are all subject to some or other qualification, in general, no automatic or universal approach to delimitation of the typological yeast species is possible. Although application of different criteria employed for species delimitation, to a given set of strains, may or may not result in the same taxonomic entity, these definitive criteria cannot, in practice, be separated one from the other. This becomes obvious from the simple fact that delimitation of the species on the basis of either hybridization or nucleic acid analyses is, invariably, limited to strains that have been selected on the basis of their phenotypic characters and subjected to some prior degree of phenotypic differentiation. It must be emphasized that, once the species has been delimited along either of these lines, its description still hinges on phenotypic characters, which not only serve to distinguish between strains of different species, but provide a basis for the construction of practical taxonomic keys and routine identification of strains.

In summary, practical realistic delimitation of the yeast species hinges on an integrated approach in which as many criteria as possible, based on the morphological, sexual, genetic, physiological, biochemical and ecological features of as many strains as are available, are taken into account.

In practice, yeast taxonomists occasionally find it necessary to introduce infraspecific taxa for classification of certain strains. As a rule, these infraspecific taxa have been limited to *varieties* which have been distinguished on the basis of consistent phenotypic differences among closely related or presumed-related strains. The following serve as examples. In the case of the species *Hansenula bimundalis*, Wickerham (1965) distinguished between two varieties, *H. bimundalis* var. *bimundalis* and *H. bimundalis* var. *americana*, on the basis of differences in utilization of D-arabinose, growth at 37°C and the fact that crosses of these varieties form zygotes which do not produce ascospores. Wickerham (1965) pointed out that allopatry or differences in geographical distribution might account for the genetic diversion of strains. Kurtzman (1984a) subsequently raised these two varieties to the rank species on the basis that strains of these taxa showed only 19% nDNA base sequence relatedness. Taxonomists committed to the use of nDNA complementarity for species delimitation have, however, not quantified differences in nDNA base sequence relatedness that would discriminate between species and varieties. Price *et al.* (1978) separated the two varieties *Schwanniomyces occidentalis* var. *occidentalis* and *Schw. occidentalis* var. *persoonii* on the grounds of 80–83% base sequence relatedness. Kurtzman (1984b), on the other hand, merged the two species described as *Hansenula minuta* and *Pichia lindneri* on the basis of only 75% complementarity. In the case of *K. marxianus*, Johannsen (1980) introduced seven varieties for classification of hybridizing strains which had previously been considered to be representative of phenotypically delimited species. Apart from such cases where there is a genetic or ecological basis for recognition of varieties, strains have also been assigned to infraspecific taxa as a matter of expediency. Taxonomists may, while still uncertain of the precise status of closely related strains, prefer to treat such strains as varieties of known species rather than to publish new species. Although the term "variety" has been applied rather loosely, future developments in yeast classification may be marked by the avoidance of infraspecific taxa.

VIII. CLASSIFICATION OF THE INDUSTRIAL YEASTS

The application of more reliable criteria, such as the interfertility of strains, for delimitation of species often requires that two or more species have to be merged. Consequently, well-known names of species pass into synonomy with earlier published names. A pertinent example is the so-called *Saccharomyces* sensu stricto (cf. van der Walt, 1970a). As discussed on p. 105, these 17 taxa are now accepted by Yarrow (1984) as being representative of the single species, *Sacch. cerevisiae*.

Taxonomists dealing with the yeast domain as a whole view such changes as inevitable as they consider they are the result of a more effective delimitation of the typological species. On the other hand, industrial microbiologists who deal with relatively few yeasts, the names of which have, over the years, come to be associated with a particular fermentation process, become disquieted when they find that well-known names with industrial connotations, fall into taxonomic disuse (Gilliland, 1971; Rose, 1980). This disenchantment with taxonomic developments clearly reflects the smaller and more specialized reference frame which industrial microbiologists, such as brewers, use to characterize their industrially connotated taxonomic units. It also reflects the fact that most authors of the earliest described species in *Saccharomyces* were actively engaged in the brewing industry. With this background, it is inevitable that the brewing industry is biased in favour of the use of industrially relevant descriptive characters.

To the brewer the names *Sacch. cerevisiae, Sacch. bayanus, Sacch. carlsbergensis* and *Sacch. diastaticus* connote strains with definite brewing characters whereas, to the taxonomist, these names currently imply phenotypically different, interfertile strains of a single species, namely *Sacch. cerevisiae*.

Nevertheless, as industrial microbiologists are obliged to differentiate between industrial strains, the fermentation industries should feel free to develop a *special purpose taxonomy* for the utilitarian classification of industrial yeasts. Provided such a special purpose taxonomy has only a limited application and this is fully appreciated by both industrial microbiologists and vocational taxonomists, there is an argument for retaining industrially connotated names for distinguishing between strains used in various industrial applications. For delimitation of such industrially characterized taxonomic units, industrial microbiologists would obviously rely on characters and criteria which do not find any meaningful application in more modern approaches to delimitation of the typological species.

There are also other applied fields, apart from the fermentation industries, which use plants where much the same situation obtains regarding the continued use of species names which have passed into synonomy. In this connection it is interesting to note that the Thirteenth International Botanical Congress finally accepted, in principle, the proposal for conservation of species names. It must, however, be stressed that this will apply only to species of major economical importance, which would include industrial yeasts. Proposals for the conservation of such species names, of course, must conform to formal procedures. These have been set out in Article 14.2 in the Code adopted by the Thirteenth Congress (Staflue, 1983).

References

Banno, I. (1967). *Journal of General and Applied Microbiology* **13**, 167.
Baptist, J.N. and Kurtzman, C.P. (1976). *Mycologia* **68**, 1195.
Barnett, J.A., Payne, R.W. and Yarrow, D. (1983). "Yeasts: Characteristics and Identification". Cambridge University Press, Cambridge, London, New York, New Rochelle, Melbourne and Sydney.
Bartnicki-Garcia, S. (1970). *In* "Phytochemical Phylogeny" (J.B. Harborne, ed.), pp. 81–103. Academic Press, London.
Belozersky, A.N. and Spirin, A.S. (1960). *In* "The Nucleic Acids" (E. Chargraff and J.N. Davidson, eds), vol. 3, pp. 147–185. Academic Press, New York.
Bernstein, C. (1979). *Perspectives in Biology and Medicine* **22**, 539.
Beydon, M.H. and Heslot, H. (1980). *Abstracts of Tenth International Conference of Yeast Genetics and Molecular Biology*, Louvain-la-Neuve, Belgium, Sept. 8–12, 1980, p. 47.
Bicknell, J.N. and Douglas, H.C. (1970). *Journal of Bacteriology* **101**, 505.
Cain, A.J. (1954). "Animal Species and Their Evolution". Hutchinson, London and Harper and Row, New York.
Campbell, I. (1972). *Journal of General Microbiology* **73**, 279.
Campbell, I. (1973). *Journal of General Microbiology* **77**, 427.
Campbell, I. (1974). *Advances in Applied Microbiology* **17**, 135.
Crandall, M., Egel, R. and Mackay, V.L. (1977). *Advances in Microbial Physiology* **15**, 307.
Crick, F.H.C. (1979). *Science* **204**, 264.
Ditlevsen, E. (1944). *Comptes Rendu des Travaux du Laboratoire Carlsberg, Série Physiologique* **24**, 223.
Dobzhansky, Th. (1970). "Genetics of the Evolutionary Process." Columbia University Press, New York and London.
Doel, M.T. (1980). *Cell Biology International Reports* **4**, 433.
Fell, J.W. and Tallman, A.S. (1980a). *International Journal of Systematic Bacteriology* **30**, 206.
Fell, J.W. and Tallman, A.S. (1980b). *International Journal of Systematic Bacteriology* **30**, 658.
Fell, J.W. and Tallman, A.S. (1981). *Current Microbiology* **5**, 77.
Fell, J.W., Statzell, A., Hunter, I.L. and Phaff, H.J. (1969). *Antonie van Leeuwenhoek* **35**, 433.
Fell, J.W., Hunter, I.L. and Tallman, A.S. (1973). *Canadian Journal of Microbiology* **19**, 643.
Fiol, J.B. (1972). *Mycopathologia et Mycologia Applicata* **48**, 303.
Fogel, S. and Mortimer, R.K. (1971). *Annual Review of Genetics* **5**,219.
Fowell, R.R. (1969). *In* "The Yeasts" (A.H. Rose and J.S. Harrison, eds), vol. 1, pp. 301–383. Academic Press, London and New York.
Freeman, R.F. and Peberdy, J.F. (1983). *In* "Yeast Genetics, Fundamental and Applied Aspects" (J.F.T. Spencer, D.M. Spencer and A.R.W. Smith, eds), pp. 243–253. Springer-Verlag, New York, Berlin, Heidelberg and Tokyo.
Friss, J. and Roman, H. (1968). *Genetics* **59**, 33.
Gilbert, W. (1978). *Nature, London* **271**, 501.
Gilliland, R.B. (1949). *Comptes Rendu des Travaux du Laboratoire Carlsberg, Série Physiologique* **24**, 347.
Gilliland, R.B. (1954a). *Nature, London* **173**, 409.
Gilliland, R.B. (1954b). *Wallerstein Laboratory Communications* **17**, 165.
Gilliland, R.B. (1971). *Journal of the Institute of Brewing* **77**, 276.
Gorin, P.A.J. and Spencer, J.F.T. (1970). *Advances in Applied Microbiology* **13**, 25.
Grant, V. (1963). "The Origin of Adaptations." Columbia University Press, New York.
Grant, V. (1971). "Plant Speciation." Columbia University Press, New York and London.
Hagler, A.N. and Ahearn, D.G. (1981). *International Journal of Systematic Bacteriology* **31**, 204.
Hansen, E.C. (1888). *Meddeleser fra Carlsberg Laboriet* **2**, 220.
Johannsen, E. (1980). *Antonie van Leeuwenhoek* **46**, 177.
Johannsen, E. (1981). *Canadian Journal of Microbiology* **27**, 184.
Johannsen, E. and van der Walt, J.P. (1979). *Archives of Microbiology* **118**, 45.
Johannsen, E. and van der Walt, J.P. (1980). *Canadian Journal of Microbiology* **26**, 1199.

Johannsen, E., Halland, L. and Opperman, A. (1984). *Canadian Journal of Microbiology* **30**, 540.
Johnson, L.A.S. (1970). *Systematic Zoology* **19**, 203.
Jones, G.R. (1975). *Journal of General Microbiology* **89**, 151.
Kocková-Kratochvílová, A., Blagodatskaya, V. and Hronská, L. (1972). *In* "Yeasts–Models in Science and Tecnics" (A. Kocková-Kratochvilová and E. Minarik, eds), pp. 375–387. Publishing House of the Slovak Academy of Sciences, Bratislava.
Kossikov, K.V. (1957). *In* "Ciba Foundation Symposium on Drug Resistance in Micro-organisms", pp. 102–135.
Kreger-van Rij, N.J.W. (1984). "The Yeasts, a Taxonomic Study." Elsevier Biomedical Press, Amsterdam.
Kreger-van Rij, N.J.W. and Veenhuis, M. (1971). *Journal of General Microbiology* **68**, 87.
Kurtzman, C.P. (1984a). *Mycotaxon* **19**, 271.
Kurtzman, C.P. (1984b). *Antonie van Leeuwenhoek* **50**, 209.
Kurtzman, C.P. and Wickerham, L.J. (1973). *Antonie van Leeuwenhoek* **39**, 81.
Kurtzman, C.P., Smiley, M.J., Johnson, C.J., Wickerham, L.J. and Fuson, G.B. (1980a). *International Journal of Systematic Bacteriology* **30**, 208.
Kurtzman, C.P., Smiley, M.J. and Johnson, C.J. (1980b). *International Journal of Systematic Bacteriology* **30**, 503.
Kurtzman, C.P., Phaff, H.J. and Meyer, S.A. (1983). *In* "Yeast Genetics, Fundamental and Applied Aspects" (J.F.T. Spencer, D.M. Spencer and A.R.W. Smith, eds), pp. 139–166. Springer-Verlag, New York, Berlin, Heidelberg and Tokyo.
Kwon-Chung, K.J. (1975). *Mycologia* **67**, 1197.
Kwon-Chung, K.J. (1976). *Mycologia* **68**, 942.
Kwon-Chung, K.J. (1977). *International Journal of Systematic Bacteriology* **27**, 293.
Lindegren, C.C. (1949). "The Yeast Cell, its Genetics and Cytology". Educational Publishers Inc., Saint Louis.
Lindegren, C.C. and Lindegren, G. (1951). *Journal of General Microbiology* **5**, 885.
Losada, M. (1957). *Comptes Rendu des Travaux du Laboratoire Carlsberg, Série Physiologique* **25**, 460.
Lotsy, J.P. (1925). *Genetica* **7**, 487.
Lotsy, J.P. (1931). *Genetica* **13**, 1.
Mandel, M. (1969). *Annual Review of Microbiology* **23**, 239.
Mackay, V.L. and Manney, T.R. (1974). *Genetics* **76**, 225.
Maráz, A., Kiss, M. and Ferenczy, L. (1978). *FEMS Microbiology Letters* **3**, 319.
Marmur, J., Falkow, S. and Mandel, M. (1963). *Annual Review of Microbiology* **17**, 329.
Martini, A. (1973). *Annali della Facoltà di Agraria della Università degli studi di Perugia* **28**, 1.
Martini, A. and Phaff, H.J. (1973). *Annali di Microbiologia (Milano)* **23**, 59.
Mayr, E. (1965). *Systematic Zoology* **14**, 73.
Mayr, E. (1969). "Principles of Systematic Zoology". McGraw Hill, New York.
Mayr, E. (1976). "Evolution and the Diversity of Life, Selected Essays". Belknap Press of Harvard University Press, Cambridge, and London.
Mendonca-Hagler, L.C. and Phaff, H.J. (1975). *International Journal of Systematic Bacteriology* **25**, 222.
Meyen, J. (1837). *Wiegmanns Archiv für Naturgeschichte* **4** (2), 100.
Meyer, S.A. and Phaff, H.J. (1969). *In* "Recent Trends in Yeast Research" (D.G. Ahearn, ed.), vol. 1, pp. 1–29. Spectrum Monograph Series in Arts and Sciences, Georgia State University, Atlanta.
Meyer, S.A. and Phaff, H.J. (1972). *In* "Yeasts–Models in Science and Tecnics" (A. Kocková-Kratochvílová and E. Minarik, eds), pp. 375–387. Publishing House of the Slovak Academy of Sciences, Bratislava.
Meyer, S.A., Brow, R.E. and Smith, M. Th. (1977). *International Journal of Systematic Bacteriology* **27**, 162.
Meyer, S.A., Smith, M. Th. and Simione, F.P. (1978). *Antonie van Leeuwenhoek* **44**, 79.
Moore, R.L. (1974). *Currrent Topics in Microbiology and Immunology* **64**, 105.
Morgan, A.J., Heritage, J. and Whittaker, P.A. (1980). *Microbios Letters* **4**, 103.
Mundkur, B.D. (1953). *Experientia* **9**, 373.

Nakase, T. and Komagata, K. (1968). *Journal of General and Applied Microbiology* **14**, 345.

Nakase, T. and Komagata, K. (1969). *Journal of General and Applied Microbiology* **15**, 85.

Nakase, T. and Komagata, K. (1970a). *Journal of General and Applied Microbiology* **16**, 241.

Nakase, T. and Komagata, K. (1970b). *Journal of General and Applied Microbiology* **16**, 511.

Nakase, T. and Komagata, K. (1971a). *Journal of General and Applied Microbiology* **17**, 43.

Nakase, T. and Komagata, K. (1971b). *Journal of General and Applied Microbiology* **17**, 121.

Nakase, T. and Komagata, K. (1971c). *Journal of General and Applied Microbiology* **17**, 161.

Nakase, T. and Komagata, K. (1971d). *Journal of General and Applied Microbiology* **17**, 227.

Nakase, T. and Komagata, K. (1971e). *Journal of General and Applied Microbiology* **17**, 259.

Nakase, T. and Suzuki, M. (1985a). *Journal of General and Applied Microbiology* **31**, 49.

Nakase, T. and Suzuki, M. (1985b). *Journal of General and Applied Microbiology* **31**, 71.

Newell, S.Y. and Fell, J.W. (1970). *Mycologia* **62**, 272.

Oliaya, A.F. and Sogin, S.J. (1979). *Journal of Bacteriology* **140**, 1043.

Ouchi, K., Saito, H. and Ikeda, Y. (1970). *Agricultural and Biological Chemistry, Tokyo* **34**, 95.

Phaff, H.J. (1977). *Abstracts of the Second International Mycological Congress*, 27 August–3 September, 1977, Tampa, vol. M–Z, p. 519.

Pomper, S. and Burkholder, P.R. (1949). *Proceedings of the National Academy of Sciences of the United States of America* **35**, 456.

Poncet, S., Fiol, J.B. and Billon-Grand, G. (1984). *Mycopathologia* **86**, 71.

Price, C.W., Fuson, G.B. and Phaff, H.J. (1978). *Microbiological Reviews* **42**, 161.

Provost, A., Bourguignon, C., Fournier, C., Ribet, A.M. and Heslot, H. (1978). *FEMS Microbiology Letters* **3**, 313.

Ravin, A.W. (1963). *American Naturalist* **97**, 305.

Richards, M. (1972). *Antonie van Leeuwenhoek* **38**, 177.

Roberts, C. and van der Walt, J.P. (1960). *Comptes Rendu des Travaux du Laboratoire Carlsberg, Série Physiologique* **32**, 19.

Rodrigues de Miranda, L. (1972). *Antonie van Leeuwenhoek* **38**, 91.

Rodrigues de Miranda, L. (1978). *Antonie van Leeuwenhoek* **44**, 439.

Rose, A.H. (1980). *In* "The Biology and Activities of Yeasts", Society for Applied Bacteriology, Symposium Series No. 9. (F.A. Skinner, S.M. Passmore and R.R. Davenport, eds), pp. 103–121. Academic Press, London, New York, Toronto, Sydney and San Francisco.

Russell, I. and Stewart, G.G. (1979). *Journal of the Institute of Brewing* **85**, 95.

Sarachek, A., Rhoades, D.D. and Schwarzhoff, R.H. (1981). *Archives of Microbiology* **129**, 1.

Scheda, R. (1966). *Die Branntweinwirtschaft* **106**, 373.

Scheda, R. and Yarrow, D. (1966). *Archiv für Mikrobiologie* **55**, 209.

Scheda, R. and Yarrow, D. (1968). *Archiv für Mikrobiologie* **61**, 310.

Sipiczki, M. (1979). *Current Microbiology* **3**, 37.

Sipiczki, M. and Ferenczy, L. (1977). *Molecular and General Genetics* **151**, 77.

Slooff, W. Ch. (1964). *Antonie van Leeuwenhoek* **30**, 129.

Smith, M. Th., van der Walt, J.P. and Johannsen, E. (1976). *Antonie van Leeuwenhoek* **42**, 119.

Sneath, P.H.A. (1962). *In* "Microbial Classification", Twelth Symposium of the Society for General Microbiology (G.C. Ainsworth and P.H.A. Sneath, eds), pp. 289–332. University Press, Cambridge.

Sneath, P.H.A. (1972). *In* "Methods in Microbiology" (J.R. Norris and D.W. Ribbons, eds), vol. 7A, pp. 29–98. Academic Press, New York.

Sneath, P.H.A. and Johnson, R. (1972). *Journal of General Microbiology* **72**, 377.

Sokal, R.R. (1965). *Biological Reviews* **40**, 337.

Sokal, R. and Sneath, P.H.A. (1963). "Principles of Numerical Taxonomy." Freeman Press, San Francisco.

Spencer, J.F.T. and Spencer, D.M. (1981). *Current Genetics* **4**,177.

Spiegelman, S., Lindegren, C.C. and Lindegren, G. (1945). *Proceedings of the National Academy of Sciences of the United States of America* **31**, 95.

Stafleu, F.A. (ed.) (1983). "International Code of Botanical Nomenclature", adopted by The Thirteenth International Botanical Congress, Sydney, August 1981. Bohn, Scheltema and Holkema, Utrecht/ Antwerpen.

Stahl, U. (1978). *Molecular and General Genetics* **160**, 111.

Starmer, W.T. (1981a). *Evolution* **35**, 38.
Starmer, W.T. (1981b). *In* "Current Developments in Yeast Research" (G.G. Stewart and I. Russell, eds), pp. 493–498. Pergamon Press, Toronto.
Starmer, W.T. (1981c). *In* "The Fungal Community: Its Organization and Role in the Ecosystem" (D.T. Wicklow and O. Carroll, eds), pp. 129–156. Marcel Dekker Inc., New York and Basel.
Starmer, W.T., Heed, W.B., Miranda, M., Miller, M.W. and Phaff, H.J. (1976). *Microbial Ecology* **3**, 11.
Starmer, W.T., Phaff, H.J., Miranda, M. and Miller, M.W. (1978a). *International Journal of Systematic Bacteriology* **28**, 318.
Starmer, W.T., Phaff, H.J., Miranda, M. and Miller, M.W. (1978b). *International Journal of Systematic Bacteriology* **28**, 433.
Starmer, W.T., Phaff, H.J., Miranda, M., Miller, M.W. and Barker, J.S.F. (1979). *International Journal of Systematic Bacteriology* **29**, 159.
Starmer, W.T., Kirchner, H.W. and Phaff, H.J. (1980). *Evolution* **34**, 137.
Stenderup, A. and Bak, A.L. (1968). *Journal of General Microbiology* **52**, 231.
Stewart, G.G., Russell, I. and Panchal, C.J. (1981). *In* "Current Developments in Yeast Research" (G.G. Stewart and I. Russell, eds), pp. 17–24. Pergamon Press, Toronto.
Storck, R. (1966). *Journal of Bacteriology* **91**, 227.
Storck, R., Alexopoulos, C.J. and Phaff, H.J. (1969). *Journal of Bacteriology*, **98**, 1069.
Sukroongreung, S. and Rodrigues de Miranda, L. (1973). *Antonie van Leeuwenhoek* **39**, 65.
Suzuki, T., Nishibayashi, S., Kuroiwa, T., Kanabe, T. and Tanaka, K. (1982). *Journal of Bacteriology* **152**, 893.
Svoboda, A. (1976). *Archives of Microbiology* **110**, 313.
Svoboda, A. (1978). *Journal of General Microbiology* **109**, 169.
Townsend, G.F. and Lindegren, C.C. (1954). *Journal of Bacteriology* **67**, 480.
Tsuchiya, T., Fukazawa, Y., Taguchi, M., Nakase, T. and Shimoda, T. (1974). *Mycopathologia et Mycologia Applicata* **53**, 77.
van der Walt, J.P. (1959). *Antonie van Leeuwenhoek* **25**, 344.
van der Walt, J.P. (1966). *Antonie van Leeuwenhoek* **32**, 159.
van der Walt, J.P. (1970a). *In* "The Yeasts, a Taxonomic Study" (J. Lodder, ed.), pp. 555–718. North-Holland Publishing Co., Amsterdam and London.
van der Walt, J.P. (1970b). *Mycopathologia et Mycologia Applicata* **40**, 231.
van der Walt, J.P. (1972). *In* "Fermentation Technology Today" (G. Terui, ed.), Proceedings of the Fourth International Fermentation Symposium, pp. 797–803. Society of Fermentation Technology, Osaka, Japan.
van der Walt, J.P. and Hopsu-Havu, V.K. (1976). *Antonie van Leeuwenhoek* **42**, 157.
van der Walt, J.P. and Johannsen, E. (1973). *Antonie van Leeuwenhoek* **39**, 635.
van der Walt, J.P. and Johannsen, E. (1974). *Antonie van Leeuwenhoek* **40**, 185.
van der Walt, J.P. and Johannsen, E. (1975). *Antonie van Leeuwenhoek* **41**, 361.
van der Walt, J.P. and Johannsen, E. (1979). *Antonie van Leeuwenhoek* **45**, 281.
van der Walt, J.P. and Johannsen, E. (1980). *Journal of General and Applied Microbiology* **26**, 217.
van der Walt, J.P. and von Arx, J.A. (1985). *South African Journal of Science* **81**, 156.
van der Walt, J.P. and Yarrow, D. (1984). *In* "The Yeasts, a Taxonomic Study" (N.J.W. Kreger-van Rij, ed.), pp. 45–104. Elsevier Biomedical Press, Amsterdam.
van der Walt, J.P. and Pitout, M.J. (1969). *Antonie van Leeuwenhoek* **35**, 227.
van der Walt, J.P., Johannsen, E. and Yarrow, D. (1978). *Antonie van Leeuwenhoek* **44**, 97.
van der Walt, J.P., Weijman, A.C.M. and von Arx, J.A. (1981). *Sydowia, Annales Mycologica. Series II* **34**, 191.
van Solingen, P. and van der Plaat, J. (1977). *Journal of Bacteriology* **130**, 946.
von Arx, J.A. (1980). *In* "Biology and Activities of Yeasts". Society for Applied Bacteriology Symposium, Series No. 9, (F.A. Skinner, S.M. Passmore and R.R. Davenport, eds), pp. 53–61. Academic Press, London, New York, Toronto, Sydney and San Francisco.
Weijman, A.C.M. (1979). *Antonie van Leeuwenhoek* **57**, 119.
Whelan, W.L. and Magee, P.T. (1981). *Journal of Bacteriology* **145**, 896.

Whittaker, P.A. and Leach, S.M. (1978). *FEMS Microbiology Letters* **4**, 31.
Wickerham, L.J. (1951). "Taxonomy of Yeasts". Technical Bulletin No. 1029, United States Department of Agriculture, Washington, D.C.
Wickerham, L.J. (1965). *Mycopathologia et Mycologia Applicata* **26**, 87.
Wickerham, L.J. (1970a). *In* "The Yeasts, a Taxonomic Study" (J. Lodder, ed.), pp. 226–315. North-Holland Publishing Co., Amsterdam and London.
Wickerham, L.J. (1970b). *In* "Recent Trends in Yeast Research" (D.G. Ahearn, ed.), vol. 1, pp. 31–47. Spectrum Monograph Series in Arts and Sciences, Georgia State University, Atlanta.
Wickerham, L.J. and Burton, K.A. (1954a). *Journal of Bacteriology* **67**, 303.
Wickerham, L.J. and Burton, K.A. (1954b). *Journal of Bacteriology* **68**, 594.
Wickerham, L.J. and Burton, K.A. (1956a). *Journal of Bacteriology* **71**, 290.
Wickerham, L.J. and Burton, K.A. (1956b). *Journal of Bacteriology* **71**, 296.
Winge, O. and Laustsen, O. (1939). *Comptes Rendu des Travaux du Laboratoire Carlsberg, Série Physiologique* **22**, 337.
Winge, O. and Roberts, C. (1948). *Comptes Rendu des Travaux du Laboratoire Carlsberg, Série Physiologique* **24**, 263.
Winge, O. and Roberts, C. (1950). *Comptes Rendu des Travaux du Laboratoire Carlsberg, Série Physiologique* **25**, 35.
Winge, O. and Roberts, C. (1952). *Comptes Rendu des Travaux du Laboratoire Carlsberg, Série Physiologique* **25**, 141.
Winge, O. and Roberts, C. (1954). *Comptes Rendu des Travaux du Laboratoire Carlsberg, Série Physiologique* **25**, 285.
Winge, O. and Roberts, C. (1956). *Nature, London* **177**, 383.
Winge, O. and Roberts, C. (1957). *Comptes Rendu des Travaux du Laboratoire Carlsberg, Série Physiologique* **25**, 419.
Yamazaki, M. and Komagata, K. (1981). *International Journal of Systematic Bacteriology* **31**, 361.
Yarrow, D. (1972). *Verhandelingen der Koninklijke Nederlandse Akademie van Wetenschappen, Afdeling Natuurkunde, Sectie II* **61**, 50.
Yarrow, D. (1984). *In* "The Yeasts, a Taxonomic Study" (N.J.W. Kreger-van Rij, ed.), pp. 379–395. Elsevier Biomedical Press, Amsterdam.
Yarrow, D. and Nakase, T. (1975). *Antonie van Leeuwenhoek* **41**, 81.

5 Yeasts Associated with Plants, Insects and Soil

Herman J. Phaff and William T. Starmer

Department of Food Science and Technology, University of California, Davis, California, U.S.A. and Department of Biology, Syracuse University, Syracuse, New York, U.S.A.

I. INTRODUCTION

The field of yeast ecology is concerned with the manner in which yeasts and yeast-like organisms propagate in nature, the kinds of substrates that support specific yeasts, competition among yeast species and between yeasts and other micro-organisms, the effect of the chemical composition of substrates on the microflora and the vectors responsible for introducing yeasts to suitable habitats. Numerous surveys of substrates capable of supporting growth of yeasts have revealed that in Nature nearly all yeasts or groups of species have highly specialized habitats. As a result of this knowledge it is

The Yeasts Vol. 1, 2nd edition
ISBN 0-12-596411-0

possible in many instances to isolate specific yeasts at will from suitable substrates and in geographic areas where they are known to occur. Early surveys of substrates were largely directed to food products undergoing spoilage and to the study of so-called wild yeasts that sometimes invaded industrial processes, such as brewing, wine and saké fermentations, and baker's yeast production, where they caused undesirable changes or spoilage. Other early surveys of the microflora of tree saps flowing from wounds on tree trunks (so-called slime fluxes or exudates) revealed the presence of quite unusual yeasts that were very different from those found associated with spoiled foods. These discoveries stimulated further surveys into other substrates and led, up to the present time, to the finding of numerous new habitats for specific yeasts. Phaff *et al.* (1978a) have given a broad overview of yeasts associated with plants, animals, soils and water. The present review will be limited to plants, insects (as the only groups of animals) and soils. The rationale for these three categories is that photosynthesis and the soil furnish the necessary nutrients for the growth of plants and they in turn can support the growth of yeasts. Insects are by far the most important vectors in the distribution of yeasts in nature, whereas soil is the ultimate repository for organic and inorganic materials and constitutes a suitable medium for storage and even development of certain species of yeast.

Although yeasts are sometimes thought to occur primarily in sugar-containing substrates such as fruits, where they can carry out a fermentative metabolism, most species of yeast are also endowed with the ability to respire sugars and other organic compounds. In fact, the majority of all known yeast species lack the ability to ferment or they ferment hexose sugars at a negligible rate. Thus, the ability of many yeasts to utilize a large spectrum of non-fermentable organic compounds through the respiratory process greatly expands the ability of specific yeasts to occupy ecological niches where compounds are present that can be metabolized only by specific respiratory enzymes. Examples of such compounds are pentose sugars (D-xylose, L- and D-arabinose, D-ribose), methylpentoses (D-rhamnose, D-fucose), primary alcohols (methanol, ethanol, propanol), polyols or sugar alcohols (glycerol, erythritol, ribitol, D-arabitol, galactitol, mannitol, glucitol), amino sugars (D-glucosamine, *N*-acetyl-D-glucosamine, D-galactosamine), organic acids (lactic, succinic, citric) and miscellaneous compounds such as acetone, propan-2-ol, ethyl acetate, *meso*-inositol, glucuronic acid, hydrocarbons and aromatic compounds (phenolics and their derivatives). The utilization of these compounds by yeasts is highly selective and is often related to niche specificity, because most are metabolic products of specific plant species or products of bacteria or fungi that produce them from precursors in plants.

Because yeasts cannot fix nitrogen gas from the air (Babjeva *et al.*, 1977) they depend on oxidized or reduced forms of nitrogen for the synthesis of proteins and nucleic acids. A balanced mixture of amino acids is rarely present in natural substrates, but from glutamic acid other amino acids can be synthesized by transamination. Ammonium and urea nitrogen is suitable for all yeasts associated with plants, soils and insects, but the ability to utilize nitrate and nitrite varies among yeast species and this trait may be related to the availability of these oxidized nitrogen compounds in certain plants and soils. Examples of such yeasts are species of *Hansenula* and *Citeromyces*. Some species of *Hansenula* (e.g. *H. anomala*) utilize both nitrate and nitrite for growth, but others (e.g. *H. polymorpha*) are very sensitive to nitrite at pH 4.5; these yeasts assimilate only nitrate and cannot grow on nitrite supplied in the medium. Still others (e.g. some strains of *Debaryomyces hansenii* and *Cryptococcus laurentii*) are unable to reduce nitrate to nitrite but grow well on nitrite as the sole nitrogen source. Also the ability to assimilate amines is variable among yeast species. These differences among yeast species to assimilate various specific carbon and nitrogen sources undoubtedly play a role in habitat specificity. Mineral and vitamin requirements for yeast growth are normally fulfilled adequately in plants and soils.

Specificity of yeast habitats in plants and soils is determined not only by the composition of nutrients present, but also by insect vectors that utilize certain plants and soils for breeding and feeding, and further by the presence of compounds that are inhibitory to insect vectors and/or compounds to which some yeasts are sensitive, and are therefore eliminated, while others are resistant. Some examples will illustrate this aspect of yeast ecology. Soils are known to contain antibiotic-producing strains of actinomycetes. Cycloheximide is an antibiotic produced by *Streptomyces griseus*. Some yeasts (e.g. *Pichia cactophila*) are extremely sensitive to this antibiotic (1.0 mg per litre of medium completely inhibits growth), whereas others tolerate somewhat higher concentrations (*Prototheca* spp.) and still others are completely resistant (*Candida sonorensis*). These differences in sensitivity to cycloheximide and other antibiotics could have a pronounced effect on the composition of yeast communities in soils (di Menna, 1962). Acetic-acid bacteria of the genus *Gluconobacter* oxidize ethanol produced by yeast fermentation to acetic acid, which in its undissociated form inhibits growth of many yeast species and may even kill the yeasts that produced the ethanol initially. The ability to oxidize ethanol to acetic acid is also a property of yeasts belonging to the genera *Brettanomyces* and *Dekkera*. Such yeasts, although fairly resistant to acetic acid, can ultimately kill themselves in unbuffered media. Species of these two genera are best known as spoilage yeasts of table wines and sometimes of soft drinks, but their natural habitat is not known at present.

Some plants contain unusually high concentrations of the lower fatty acids (C_6–C_{12}) whose fungistatic properties have been established (Wyss *et al.*,

1945). The organpipe cactus (*Stenocereus thurberi*) has a high content of these short-chain fatty acids (Kircher, 1982). The yeast *Candida ingens*, endowed with strong lipolytic activity and ability to oxidize these fatty acids by β-oxidation, is particularly common in necrotic tissue of this cactus species and has a competitive advantage over several other cactophilic yeast species (Starmer, 1982). A second example related to yeasts associated with necrotic cactus tissue concerns the inhibitory effect of triterpene glycosides present in cacti belonging to the subtribe Stenocereinae, and lacking in cacti belonging to the subtribe Pachycereinae; both are giant columnar cacti of the North American Sonoran Desert (Gibson and Horak, 1978; Starmer *et al.*, 1980). A cactophilic yeast species, *Pichia amethionina*, was described with two varieties (Starmer *et al.*, 1978b). *Pichia amethionina* var. *amethionina* occurs only in rots of cacti belonging to the subtribe Stenocereinae and is tolerant to these triterpene glycosides. In contrast, *P. amethionina* var. *pachycereana* occurs in cacti of the subtribe Pachycereinae (Gibson and Horak, 1978) and is sensitive to triterpene glycosides. Starmer *et al.* (1980) have shown that this difference in resistance to triterpene glycosides is controlled by a single temperature-sensitive gene (ttg^+ and ttg^-) and that this genetic difference may be singularly responsible for the divergence of the two varieties.

Species of *Hansenula* and *Pichia* that are associated with bark beetles that attack various species of pine, fir and other coniferous trees, grow vigorously in the sap of these trees underneath the bark, and in the presence of high concentrations of oleoterpenes that are inhibitory to the great majority of other yeast species (Shifrine and Phaff, 1956). One species of pine, *Pinus jeffreyi*, has a high content of n-heptane in its oleoterpenes, and this hydrocarbon inhibits the growth of some bark-beetle yeasts that can grow in other species of pine (e.g. *Pichia pinus*).

Finally, the composition of yeast communities is likely to be affected by the production and excretion of killer toxin by yeasts possessing the two species of encapsidated double-stranded ribonucleic acid (RNA) molecules responsible for their synthesis (for a review of the killer factor in yeast, see Bussey, 1981). Killer strains are widespread among members of various yeast genera. Each killer type appears to produce a different kind of killer protein; each is immune to its own toxin but can kill sensitive strains in its own genus as well as in different genera. For example, Middelbeek *et al.* (1980) have shown that the killer toxin of *Pichia kluyveri* is strongly active against various strains of *Saccharomyces cerevisiae, Candida albicans, C. krusei, C. parapsilosis, C. tropicalis* and *Torulopsis glabrata.*

Other general environmental factors that influence the selective development of yeasts are oxygen concentration and temperature. Among the yeasts associated with cactus soft rots, some are strict aerobes (e.g. *Pichia heedii, P. amethionina* and *C. mucilagina*), whereas others are facultative aerobes that can also carry out an alcoholic fermentation (e.g. *C. sonorensis* and *P. mexicana*). Thus, in the deeper layers of the thick decomposing plant tissue of the giant columnar cacti, the fermenting species have a selective advantage over the strictly aerobic species. The same separation between fermenting and aerobic yeasts is found during the lactic-acid fermentation of cucumbers in brine (Etchells *et al.*, 1953) where respiratory film yeasts grow on the surface and the fermenting yeasts in the subsurface brine. The same principle applies to the fermentation of Spanish-type olives.

A. Pitfalls in correct identification of yeast isolates

It is obvious that accurate identification of yeast isolates during ecological studies is essential for an analysis of the ecosystem under study. In spite of significant improvements in the taxonomic keys and the criteria on which they are based during the past 30 years, our present systems of yeast identification still have many shortcomings (Barnett *et al.* 1983; Kreger-van Rij, 1984). Recent molecular taxonomic studies have revealed that the traditional fermentation and assimilation tests of hexose sugars and di- and oligosaccharides, often determined by a single gene difference coding for a specific hydrolytic enzyme, are usually of little or no value in species differentiation (Price *et al.*, 1978; Meyer *et al.*, 1978). Conversely, nuclear genome comparisons of yeast deoxyribonucleic acid (DNA) species from phenotypically similar isolates have shown that some have only negligible DNA base sequences in common, and therefore represent different species (Price *et al.*, 1978; Kurtzman *et al.*, 1980, 1983; Phaff *et al.*, 1983, 1985). Another example, out of many, concerns our own initial misidentification of a number of cactus rot-associated yeasts as *P. membranaefaciens* (Heed *et al.*, 1976) to which they keyed by the system of Kreger-van Rij (1970a). Determination of the nuclear DNA base composition of representative strains revealed that *P. membranaefaciens* (mol% G + C 44.3) does not occur in necrotic tissue of cacti, but instead, seven new species of *Pichia* were recognized, i.e. *P. cactophila* (mol% G + C 36.9; Starmer *et al.*, 1978a), *P. pseudocactophila* (mol% G + C 36.8; Holzschu *et al.*, 1983), *P. heedii* (mol% G + C 32.5; Phaff *et al.*, 1978b), *P. amethionina* (mol% G + C 33.1; Starmer *et al.*, 1978b), *P. opuntiae* (mol% G + C 33.4; Starmer *et al.*, 1979a), *P. antillensis* (mol% G + C 33.7; Starmer *et al.*, 1984) and *P. deserticola* (mol% G + C 27.4; Phaff *et al.*, 1985). Because in yeasts a difference in G + C content greater than about 1.5% (by the buoyant-density method in CsCl) virtually precludes synonymy of two organisms, these seven species are clearly different from *P. membranaefaciens*. Later studies using DNA hybridization also showed them to be different from each other (Holzschu, 1981; Holzschu and Phaff, 1982;

Holzschu *et al.*, 1983, 1985; Starmer *et al.*, 1984; Phaff *et al.*, 1985). Subsequently, we recognized less commonly used phenotypic properties to differentiate most of these species from *P. membranaefaciens* and from each other as shown in Table I.

Pichia heedii can be differentiated from *P. membranaefaciens* only by its specific habitat in necrotic tissue of some columnar cacti of the subtribe Pachycereinae (Starmer *et al.*, 1982), whereas *P. membranaefaciens* has never been isolated from cactus tissue rots. A further search for unusual nitrogen- or carbon-containing compounds may lead to the discovery of some that can differentiate between phenotypically similar species, that may be the result of evolutionary convergence. For the differentiation of some other phenotypically similar cactus-specific and as yet undescribed new species, we have with ethyl acetate, acetone,propanol-2-ol, *N*-acetylglucosamine and hexadecane as sole carbon sources and with differential sensitivity to various concentrations of cycloheximide and sodium chloride (W.T. Starmer and H.J. Phaff, unpublished results). Examples could also be cited for other (non-cactus specific) yeasts where strains of existing species have been shown by DNA analysis or other molecular techniques to be mixtures of several unrelated taxa (e.g. *C. sake*; Meyer *et al.*, 1975). This situation is especially common among the imperfect yeasts, where a sexual state is lacking.

There is no universal agreement as to how much base sequence complementarity there has to be for two yeasts to belong to the same species. If interfertility (mating competence and spore viability) is included in the definition of species, there do not appear to be absolute values for the delimitation of species between 25 and 65% homology. Values below 25% and above 65% normally refer to different species and conspecificity, respectively. Questions on species delimitation have been discussed by Kurtzman and Phaff (see Chapter 3 of this volume) and by Phaff (1981). It is obvious, however, that in cases of misidentification of yeasts, data on ecological habitats and niche specificity would have little meaning.

II. METHODS OF ISOLATION AND STUDY

The purpose of studying the yeast flora of a natural substrate or representative samples thereof is to obtain a picture of the yeast population that is typical of the habitat under study, and to estimate as closely as possible the distribution of the various yeast species present in the samples taken for analysis. Since very few yeast species can be recognized with a fair degree of certainty by direct microscopic observation of samples of the substrate under study, the only recourse is to isolate representative members of each yeast

Table I. Phenotypic properties used to differentiate various cactophilic species of *Pichia* from *Pichia membranaefaciens* and from each other.

Species	Phenotypic properties						
	Strong growth on glucosamine as carbon source	Number of spores/ascus	Good growth on D-xylose as carbon source	Requires L-methionine for growth. Good growth on gluconate	Good growth on salicin as carbon source	Growth in presence of 1 μg cycloheximide ml^{-1}	Strong growth on ethylacetate as carbon source
Pichia cactophila	+	2 (Ho)[a]	−	−	−	−	−
Pichia pseudocactohphila	+	4 (He)[a]	−	−	−	−	−
Pichia heedii	−	4 (He)	+	−	−	−	−
Pichia opuntiae	−	2-4 (He)	−	−	+	−	−
Pichia antillensis	−	4 (He)	−	−	−	+	−
Pichia deserticola	−	2 (Ho)	−	−	−	−	+
Pichia amethionina	−	4 (He, Ho)	−	+	−	−	−
Pichia membranaefaciens	−	2-4 (He, Ho)	− or +	−	−	−	−

[a] Ho, Homothallic; He, heterothallic

community and to estimate the relative and absolute numbers of members of each species present. Because yeast identification requires pure cultures of each species present in a natural habitat, the preferred procedure is to weigh an aliquot of the substrate and streak suitable dilutions in sterile water or other diluent on an agar medium that supports growth of all yeasts that may be present, and that inhibits or lowers bacterial growth. Inhibition or diminution of bacterial growth is often accomplished by streaking on acidified yeast extract–malt extract agar (AYM:Difco YM agar plus 0.7%, v/v, 1 M HCl, final pH 3.7–3.8) directly in the field (Starmer *et al.*, 1976). An alternative method to repress bacterial growth is to use one or more broad-spectrum antibiotics in the medium. After yeast colonies develop fully they are carefully inspected with a dissection microscope, counts are made of each colony type present and one colony of each morphological type is picked and purified by replating. Another technique, which does not yield quantitative results or absolute numbers of cells of each species in a sample, is by direct streaking of a loopful of sample on agar media as described above. Control of spreading moulds constitutes another problem that is encountered in some substrates surveyed for yeasts. Since yeasts are also fungi and thus related, they frequently respond in the same way to inhibitory conditions. Fortunately, only a limited number of natural substrates, for example soil samples and spoiled fruits, contain spreading fungi in abundance. Some antifungal agents have been used with a fair degree of success. Addition of 0.1–0.15% propionic acid to the AYM medium described above lowers mould growth significantly, but it also slows the growth of some obligately aerobic yeasts (Buhagiar and Barnett, 1971). Addition of 0.003% of the dye rose bengal or 1% ox gall to potato–glucose agar also helps in decreasing rapid growth of fungi (Miller and Webb, 1954). An effective method to separate yeasts from moulds is to place the material to be studied (e.g. a soil sample) in an acidified liquid nutrient medium on a shaker. Under such conditions the fungal conidia germinate and form small mycelial balls that do not form new conidia. After a few days of growth the fungal mycelium can be filtered off over sterilized glass wool and the yeasts in the filtrate can be plated in various dilutions or by direct streaking with a loop. This is a type of enrichment procedure that does not give information on the total yeast population in the original sample or very reliable data on the proportion of the various species present. The reason for the latter is the likelihood that vegetative cells, dormant cells (e.g. chlamydospores) and sexual spores may show a very uneven rate of germination during the period of aeration. Banno and Mikato (1977) used an enrichment method in test tubes without aeration for samples very low in yeast content. A disadvantage of this procedure is that non-fermentative yeasts are likely to be overgrown by fermentative species. They noted a significant effect of the composition of the enrichment medium on the yeast flora recovered.

A. Pretreatment of samples

Some materials may require a pretreatment before plating. Substrates that are in a desiccated condition, such as a tree gum, should be rehydrated in a small amount of sterile water or in an atmosphere saturated with water vapour. In the case of insects or insect larvae that feed on substrates that contain yeasts (e.g. *Drosophila* species, bark beetles, dried fruit beetles) it is useful to differentiate between yeast cells actually consumed, those adhering to various exterior body parts and those present in the substrate. As soon as possible after capturing the insects or larvae, they should be surface sterilized by submerging them for one minute in 70% ethanol, followed by a rinse in 10 ml of sterile water. The abdomen is then dissected open and the crop contents streaked directly on AYM agar plates in the field. If plating cannot be done immediately it is best to store the insects in an ice chest to lower the rate of digestion of the ingested yeasts. In order to obtain a true evaluation of the yeast flora present on fruits, berries, leaves and soil samples, a more accurate picture is obtained after disruptive pre-isolation treatments of the sample (Martini *et al.*, 1980). Scanning electron-microscopy studies of the distribution of microbial cells on natural surfaces, such as soil particles (Gray, 1967; Kilbertus and Proth, 1979), grapes (Belin, 1972), leaves (Beech and Davenport, 1970) and the rhizosphere (Locci *et al.*, 1977), have shown that micro-organisms are present in the form of microcolonies that adhere firmly to the substrate by gummy, waxy or mucous secretions. Martini *et al.* (1980) showed that pretreatment of these substrates by vigorous shaking, percolation with an excess of water, and sonication with an ultrasonic disintegrator under carefully controlled conditions, allowed recovery of a much higher number of species than by traditional isolation procedures. Rosini *et al.* (1982) confirmed the usefulness of these pretreatment procedures in a study on the yeast flora of grape berries during their ripening.

B. Estimating the proportion of each species occurring in a yeast community

Recognizing different yeast species by colony morphology on media that support growth of all species is the only way to estimate the proportion of each type in the sample. However, at least two significant drawbacks are inherent in this procedure. One is that some yeast species form identical or nearly identical colonies and can therefore not be recognized as separate entities. Conversely, species of yeast that spend their vegetative reproduction as haploid cells, often exhibit variation in colony morphology, which are thought to be caused by mutations of genes involved in the control of morphological changes. Mutations in haploid yeasts are directly expressed. Such changes often result in more hyphal development and a tougher texture.

Such colonies could thus be picked as putative separate species. A second problem is related to minority species present in the yeast community under study. If dilution plating is practised in order to have a reasonable number of colonies on a plate (preferably less than 300), population densities of less than about 0.3% of the total would not be expected to occur on such plates. For example, in decaying stems or cladodes of cacti the total yeast population is often well over 10^6 cells for each gram of rotting tissue. When this soft rot material is serially diluted and plated so that about 300 colonies show, it is possible that representatives of as many as 3000 yeast cells of the original sample could go undetected in the final plating. Fortunately, the total number of species in a natural substrate is rather small, usually ranging from one to four in a particular habitat.

C. Use of selective media to isolate specific yeasts

If the yeast communities of a certain habitat have been studied in sufficient detail and the prevalent species are known, it may then become possible to use quantitative methods to enumerate their relative numbers, even including minority populations. This approach is possible by using positive selective synthetic media containing carbon and/or nitrogen compounds that are utilized by only one, or possibly two, of the species known to occupy a particular niche, or to use a negative medium to which an inhibitor such as cycloheximide has been added, or to use a synthetic medium lacking an essential nutrilite, such as a sulphur-containing amino acid required for the growth of *P. amethionina* (Starmer *et al.*, 1978b). After obtaining the necessary experience about dilutions that are suitable in the plating of dilutions of particular substrates, and obtaining the expertise to recognize various yeast species by their morphology on such special media, quantitative counts of populations of the species present can usually be made. In our laboratories the technique of using selective media to enumerate quantitatively the various species of yeast occurring in necrotic tissue of some cactus species has been applied. Yeast Nitrogen Base (YNB, Difco) agar with *meso*-inositol as the sole carbon source enumerates *Cryptococcus cereanus* (Phaff *et al.*, 1974). Occasional isolates of *Cr. albidus* and *Cr. laurentii* also are found in cactus rots but these species are rather easily distinguished from the cactus-specific *Cr. cereanus* because of the slimy consistency of their colonies due to their highly capsulated cells. *Cryptococcus cereanus* is also known since 1978 by its ascosporogenous state as *Sporopachydermia cereana* (Rodrigues de Miranda, 1978). Yeast Nitrogen Base agar with 0.5% methanol as the sole source of carbon enumerates the cactus-specific *Candida sonorensis* (synonym

Torulopsis sonorensis; Miller *et al.*, 1976b). Occasionally strains of *C. boidinii* also occur in necrotic tissue. Colonies of this methanol-positive, but not cactus-specific yeast can be easily distinguished from those of *C. sonorensis*. Similarly, colonies of *Hansenula polymorpha*, another methanol-positive species, are very different because of the pinkish colour of the colonies, caused by the presence of numerous ascospores. This species is found mainly in rotting cladodes of *Opuntia* spp. in the southwestern U.S.A. Yeast Nitrogen Base agar with glucosamine as the carbon source enumerates *P. cactophila* (Starmer *et al.*, 1978a) one of the most common and cosmopolitan cactus-specific yeasts as well as *P. pseudocactophila*, a species found only in rots of closely related columnar cacti belonging to the subtribe Pachycereinae (Holzschu *et al.*, 1983). *Pichia norvegensis* or its anamorph *Candida norvegensis* is non-cactus specific and is occasionally found in rots of columnar cacti; more recently it has been observed fairly commonly in *Opuntia* rots on several Caribbean islands (H.J. Phaff and W.T. Starmer, unpublished observations); it also grows vigorously on glucosamine. It is interesting that the ability to assimilate glucosamine by a yeast does not usually enable it to use *N*-acetylglucosamine as a carbon source and the reverse is true as well. Yeast Nitrogen Base agar with *N*-acetylglucosamine as the carbon source permits vigorous growth of two cactus-specific yeasts, *P. mexicana* (Miranda *et al.*, 1982) and an as yet undescribed species that belongs to the genus *Clavispora*. The former occurs primarily in rots of columnar cacti, belonging to the subtribe Stenocereinae, and the latter in decaying cladodes of *Opuntia* spp. In addition, some non-cactus specific yeasts that are found rather rarely in cacti show strong growth on *N*-acetylglucosamine. These are *Debaryomyces hansenii* (or its anamorph *C. famata*), *Cryptococcus albidus* and *Cr. laurentii*. Colonies of the last three species are easy to differentiate from the cactus-specific species. *Pichia heedii* (Phaff *et al.*, 1978b), which occurs predominantly in two species of columnar cacti in the Sonoran Desert (Starmer *et al.*, 1982), can be enumerated on YNB agar with D-xylose as the carbon source. *Cryptococcus cereanus* also grows on this medium, but the colony and cell morphologies of the two species are distinct.

Single nitrogen sources can also be helpful in the isolation of specific yeasts from natural habitats. In this case the nitrogen source is added to Yeast Carbon Base (YCB, Difco) agar. For example, with nitrate as the sole nitrogen source it is possible to isolate and enumerate species of *Hansenula, Citeromyces* and nitrate-positive members of *Candida* from specific habitats where such species are known to occur. This medium was useful to follow quantitatively the competitive growth of *C. utilis* (nitrate positive) in a commercial baker's yeast fermentation (*Sacch. cerevisiae*, nitrate negative) where the contaminant outgrew the baker's yeast strain; the total cell count was obtained on malt agar (H.J. Phaff, unpublished results). Other differentiating nitrogen compounds are nitrite, ethylamine, lysine, creatinine and creatine.

D. Incubation temperatures and special growth requirements

Incubation temperatures should generally be close to those at which the yeasts propagate in their natural environment, especially those from substrates that are constantly at low temperatures. For example, Vishniac and Hempfling (1979b), who studied yeasts occurring in soils from the Dry Valleys of Antarctica where the mean annual air temperature is −20 to −25°C, reported that *Cryptococcus vishniacii*, which occurs in these soils, was routinely maintained as stock cultures at 4°C and grew best at 10°C (poorly at 20°C). At 10°C colony development took approximately two weeks. Mesophilic species are best incubated at 20–25°C, because many yeasts isolated from sources in temperate zones do not grow at 30°C. Even yeasts that can grow at 45–46°C, a maximum temperature that is common among cactophilic yeasts from hot desert climates, will grow well at 20–25°C. The few psychrophobic yeasts that occur in the intestinal tract of warm-blooded animals and that require incubation temperatures above 30°C (Travassos and Cury, 1971) will not be considered further as this subject falls outside the scope of this review.

Some yeasts have unusual growth requirements and, as a consequence, media and incubation conditions that are suitable for the vast majority of species must be adjusted in these special circumstances. A few illustrations follow.

Schizosaccharomyces octosporus is most commonly found in substrates with high sugar content (e.g. sun-dried raisins or prunes). Strains of this species grow poorly in synthetic media such as YNB agar supplied with a carbon source, unless supplemented with at least 15 mg of adenine l^{-1} (Northam and Norris, 1951).

Pichia amethionina is unable to convert inorganic sulphate into sulphur-containing amino acids (Starmer *et al.*, 1978b) and synthetic media must be supplemented with approximately 10 mg of L-methionine or L--cysteine l^{-1} for growth. Commercial YNB agar contains an adequate concentration.

Although the natural habitat of species of *Brettanomyces* or *Dekkera* (the latter is the ascogenous form of the former) is not yet known, indications suggest that they occur in soils or on grapes around wineries and in this way gain entry as spoilage organisms in bottled or casked wine. They are, therefore, discussed briefly here. Their growth rate is much lower than other yeasts causing spoilage or turbidity in the wine industry. They have been reported in Europe, South Africa and in the United States. Colonies usually do not appear on complex media for 10–12 days and even then are quite small. Thus absence of species of these genera should not be concluded when no growth is observed on standard media after four or five days. A raised concentration of thiamin in the medium to about 10 mg l^{-1} improves the growth rate somewhat. Other additions to the medium helpful in recognizing *Brettanomyces* spp. are 0.5% precipitated calcium carbonate, which reveals the strongly acid-forming *Brettanomyces* colonies, and a supplement of 100 μg of cycloheximide ml^{-1}, which inhibits many other species, including wine yeasts, that may also occur in bottled or other forms of wine.

Finally, a few words should be said on the effect of osmotic pressure of both the subtrate to be analysed and the medium on which the suspected yeasts are to be isolated. Most yeasts that occur in substrates of high salt content, such as species of *Pichia* and *Debaryomyces* in foods preserved by high concentrations of sodium chloride, will grow readily on media devoid of salt. Such species may be regarded as halotolerant rather than as halophilic. Only occasionally are yeasts encountered that have become dependent on salt in natural environments. An example is *Metschnikowia bicuspidata*, strains of which were isolated from diseased brine shrimps grown in seashore ponds with about 10–12% NaCl. These strains did not grow in standard media of low osmotic pressure (e.g. YNB-glucose) but grew well when this medium was supplemented with 2% or more NaCl (Lachance *et al.*, 1976). Malt extract media of 10° Balling or higher also supported growth of this yeast.

In contrast to salt-tolerant yeasts, sugar-tolerant species generally thrive best in media of moderately high sugar content (approximately 40%, w/v) and some grow poorly or not at all when transferred directly from substrates with a high sugar content (e.g. honey) to ordinary isolation media with only a low concentration of sugar (Munitis *et al.*, 1976). Similarly on YNB synthetic media (carbon source concentration usually 0.5%) used for identification tests, growth is often very poor initially, or lacking. To isolate yeasts from substrates of high osmotic pressure it is best to use media with 30–40% sugar and, when pure cultures have been obtained, the isolates can usually be adapted to lower osmotic pressure by transferring them stepwise to lower sugar concentrations. Examples are the haploid *Zygosaccharomyes bisporus* and *Zygosacch. rouxii.* In contrast, the yeast-like organism *Eremascus albus* apparently cannot adapt to low sugar concentrations. It is a spoilage organism of certain food products with high osmotic pressure (Harrold, 1950).

III. SPECIFICITY OF HABITATS

In this section plant materials, insects and soils are described as habitats, hosts and vectors for yeasts that have been investigated in considerable detail by yeast ecologists from many parts of the world. Numerous natural habitats have been discovered where yeasts occur in various population densities and with great specialization of habitat. For most of these specific interactions between yeast and substrate, information on the biochemical and environmental reasons for such high specificities is still very limited. Because of the very extensive literature on the association of yeasts with the substrates covered in this chapter a complete coverage is not possible, and the principal emphasis will be on more recent literature. For more detailed earlier general reviews the reader is referred to Lund (1954), do Carmo-Sousa (1969), Last and Price (1969), Phaff *et al.* (1978a) and Phaff and Starmer (1980). Additional information on yeast habitats and sources of isolation can be found in Lodder (1970) and Kreger-van Rij (1984). Most of these reviews cover the yeast flora by habitat or substrate. However, Phaff and Starmer (1980), Lodder (1970) and Kreger-van Rij (1984) provide information on the habitat or occurrence of specific yeast species. In the present review the approach by substrate is used.

A. Yeasts associated with plants

It is not surprising that plants, with their photosynthetic ability to synthesize not only sugars and polysaccharides, but numerous other carbon compounds as well, offer a great variety of substrates for yeast growth both on living plants and on decaying or necrotic portions. The plant also supplies the necessary mineral salts obtained from the soil as well as micronutrients such as vitamins.

1. Yeasts occurring in tree exudates and tanning liquors

Many tree species on all continents show the phenomenon of fluxing — the flowing of tree sap from a wound. The cause of this condition has been attributed to injuries due to boring insects, frost cracks, or the breaking of limbs, all in conjunction with an initial bacterial infection of the wound. Tree fluxes often persist for several years and do not appear to damage a tree appreciably, if at all, and occasionally heal over. The flowing sap becomes heavily infected with bacteria, yeasts and protozoa, but filamentous fungi are relatively scarce. In addition, many insects find tree exudates an attractive site for oviposition as well as feeding, and larvae develop in them. It is very likely that insects which breed in slime fluxes are responsible for introducing the yeasts and other micro-organisms that are known to occupy these niches to complete their life cycle (Gilbert, 1980). The tree sap usually assumes a thick and slimy consistency, probably due to microbial capsular polysaccharide formation, and for this reason it is usually referred to as a slime flux.

Studies made more than 70 years ago indicated that several species of

Nadsonia as well as *Saccharomycodes ludwigii* were first discovered in exudates of birch or oak in Russia or other European countries (Lodder, 1970). Many subsequent studies have shown that tree fluxes are a rich and often highly specific habitat for yeasts. Tree fluxes often persist for many months, and sometimes for years. It has been shown that the microflora of such fluxes is relatively constant rather than subject to abrupt changes due to visits by different insects or other contaminating influences. Phaff *et al.* (1964a) made a one-year quantitative study of the yeast flora in a single slime flux of *Ulmus carpinifolia* in Davis, California. Samples were taken and analysed at approximately two-week intervals. *Pichia pastoris* was found in all samples and was quantitatively the most common species. Also present in all samples, but in much lower proportion, was *Trichosporon penicillatum* (syn. *Geotrichum penicillatum*). *Prototheca* spp. were abundant in the summer but were virtually absent during the winter months. Occasionally, transient foreign yeasts constituted a small proportion of the total yeast population. They were probably introduced by visiting insects, but did not colonize permanently in the flux. Similarly, *Endomycopsis javanensis* (syn. *Arthroascus javanensis*) was isolated repeatedly from the same slime flux of *Quercus kellogii* in the Sierra Nevada in California (Phaff and Knapp, 1956) over a two-year period.

A very extensive survey of yeasts occurring in slime fluxes of trees on all of the major Japanese islands and of trees in the Pacific north-west area of North America was conducted by Phaff *et al.* (1972) and repeated by Kodama (1974) on the Japanese islands. The first survey revealed interesting geographic separations of some species. For example, *Pichia pastoris* was shown to be common in tree exudates on the west coast of North America (70 strains isolated) but was found only once in an exotic species of *Ulmus* in Japan. This yeast, which was first isolated in 1919 by Guilliermond from the exudate of a chestnut tree in France, appears to be common in tree fluxes in European forests (Kreger-van Rij, 1970a), and was also found in fluxes of elm and oak in forests of eastern Canada and the United States (Lachance *et al.*, 1982; Bowles and Lachance, 1983).

Nadsonia elongata, first isolated by Nadson and Konokotina from slime fluxes of oak and birch in Russia in the period 1911–1913 and later by Golubev *et al.* (1977), also occurs in deciduous tree fluxes in the eastern United States (Lodder, 1970), and was found to be common in Japan, where Soneda (1957) and later Phaff *et al.* (1972) isolated 33 strains from nine genera of trees represented by 11 species occurring in a wide area, with the exception of Hokkaido and northern Honshu. Kodama (1974) also isolated seven strains from several tree species in southern and central Japan. *Nadsonia* species proved to be absent, despite extensive searches, in any species of tree in Alaska or in the western forests of Canada and lower U.S.A. *Nadsonia elongata* is rather easy to recognize on isolation plates, because fresh isolates sporulate heavily and the dark colour of the spores turns the colonies brown. It was somewhat unexpected that this species, with its relatively low maximum temperature for growth ($\sim 27°C$), was isolated only in southern and central Japan but not in the cooler northern portion of Japan or in the Pacific north-west. It can therefore be postulated that its absence from these areas may be related to the absence of a specific insect vector.

Another species that was found mainly in Japan and in eastern European deciduous trees is *Phaffia rhodozyma* (Miller *et al.*, 1976a). It was originally designated as *Rhodozyma montanae* (Phaff *et al.*, 1972) but it was not given a Latin description and therefore constituted a *nomen nudum*. It was later validly described by Miller *et al.* (1976a) as indicated above. This is an unusual fermenting carotenoid-containing yeast, with the salmon-coloured astaxanthin constituting the principal carotenoid pigment. A single strain was isolated from a birch flux in Alaska and nine strains from five species of broad-leafed trees in Japan, mainly in the northern cool areas. The maximum temperature for growth of this species is approximately 26°C. Later another 67 strains were reported by Golubev *et al.* (1977) in fluxes of *Betula verrucosa* in the Moscow region of the U.S.S.R. *Dipodascus aggregatus* was also limited to Japanese forests. Phaff *et al.* (1972) isolated 44 strains from fluxes of 17 genera of trees over a very wide area of the Japanese islands but none were found in fluxes of broad-leafed trees in the Pacific north-west. Isolates of this species are also known from Europe and the eastern U.S.A., but these were associated with bark beetles in coniferous trees (Batra, 1959). Another species with a low maximum temperature for growth is *Trichosporon pullulans*. This species is well known for its specific occurrence in fluxes of deciduous trees in many parts of the world (Lodder, 1970). Phaff *et al.* (1972) reported 56 additional isolates from numerous species of broad-leafed trees in Japan and seven strains from the Pacific north-west. Golubev *et al.* (1977) reported 84 strains of this species from birch fluxes in the Moscow region of the U.S.S.R. It has not been found associated with coniferous trees.

Isolates considered as *Hansenula wingei* and *H. nonfermentans* (Phaff *et al.*, 1972) have since been subjected to molecular taxonomic studies and shown to be two new species of *Hansenula*. Three strains isolated from exudates of *Alnus rubra* in British Columbia and the State of Washington, and representing both mating types, keyed to *H. wingei*. Because this species is normally associated with coniferous trees, its occurrence in exudates of a broad-leafed tree raised doubts about its identity with *H. wingei*. In addition, mating types of the two heterothallic species reacted poorly with each other on mixing and no spores were observed in such mixes. Fuson *et al.* (1979) showed that the strains from alder trees exhibited greater than 98% DNA complementarity among themselves but less than 10% relatedness to the type strain of *H. wingei* and thus they concluded that these yeast strains represented a different species, which was named *Hansenula alni* (Phaff *et al.*, 1979). The only phenotypic difference that we have been able to detect so far between *H.*

wingei and *H. alni*, besides habitat, is the ability of the former to grow at 37°C, whereas the latter has a maximum growth temperature of about 33°C.

Phaff *et al.* (1972) also isolated from *Populus trichocarpa fluxes* (cottonwood) five homothallic strains of *Hansenula* which keyed to *H. nonfermentans*, although they differed in several assimilatory properties and in their maximum temperature for growth. Phaff *et al.* (1983) showed that the DNA base composition of the strains from *Populus* was 1.1–1.5 mol% lower than the G + C content of the only extant strain of *H. nonfermentans* (45.3 mol% G + C). The DNA base composition of another phenotypically similar species, *H. dryadoides* (Scott and van der Walt, 1971), was 31.0 mol% G + C, which precludes conspecificity. DNA–DNA re-association experiments showed that the strains from *Populus* had virtually no base sequences in common with the other two *Hansenula* spp., and the five strains from cottonwood fluxes in British Columbia and Alaska were named *Hansenula populi* (Phaff *et al.*, 1983).

We have also recently re-investigated by molecular techniques 24 strains from fluxes of Populus and Betula in the Pacific north-west which Phaff *et al.* (1972) designated as *Torulopsis* sp. These slimy nitrate-positive isolates showed some similarity to *Torulopsis molischiana* but DNA analysis showed them to represent a new species (Leda Mendonca-Hagler and H.J. Phaff, unpublished results). The new species could be placed either in the genus *Candida* sensu Yarrow and Meyer (1978) or in the new anamorphic genus *Myxozyma* (van der Walt *et al.*, 1981).

Table II shows other more or less commonly occurring yeasts in slime fluxes of broad-leafed trees in Japan, western North America and in eastern forests of the United States and Canada (Phaff *et al.*, 1972; Phaff and Knapp, 1956; Lachance *et al.*, 1982; Bowles and Lachance, 1983). It should be noted that some changes in nomenclature have occurred in a few species since they were reported in the original publications. Several new species have been reported from tree exudates since the appearance of the taxonomic treatise by Lodder (1970). Since most of these represent rather few isolates their true habitat has not been definitely established in all cases. Their taxonomic designations, sources of isolation, and references are given in Table II (see also Kreger-van Rij, 1984).

In connection with yeasts occurring in tree exudates, it seems appropriate to discuss briefly the unusual yeast flora identified in tanning liquors, mainly by Professor J. Boidin and his associates in France. Commercial sources of tannin used for the tanning of skins and hides include the bark of oak and acacia, wood of chestnut and quebracho, sumac leaves and twigs, and others. Besides tannin, extracts of these materials contain mineral salts, nitrogenous compounds and carbohydrates including sugars. The pH value of the tanning solutions varies from about 4.0 to 5.9. Numerous known yeast species have been recovered from tanning liquors, including a number of new species

Table II. Selected yeast species associated with tree exudates or tree bark

Species designation	Source and locality	Reference
Saccharomyces kluyveri	*Populus, Salix, Acer, Alnus;* Pacific north-west of North America; *Ulmus* and *Pterocarya*, Japan	Phaff *et al.* (1972) Miller *et al.* (1962)
Saccharomyces montanus	Fluxes of 17 genera of trees, Japan	Kodama (1974)
Kluyveromyces waltii	Flux of *Ilex integra*, Japan	Kodama (1974)
Schizosaccharomyces japonicus var. *versatilis*	Outer bark of deciduous oaks, Japan	Yoneyama (1957a)
	Exudates from 13 genera of trees, Japan	Kodama (1974)
	Pine honey, Japan	Yoneyama (1957b)
Sporopachydermia quercuum	*Quercus rubra* fluxes Ontario, Canada	Lachance (1982)
Citeromyces matritensis	*Myoporum sandwicense*, Island of Hawaii	H.J. Phaff and W.T. Starmer (unpublished work)
Pichia fluxuum	*Quercus kellogii* and *Abies concolor* fluxes, Siera Nevada, California	Phaff and Knapp (1956); Bowles and Lachance (1983)
Pichia trehalophila	Exudate of *Populus trichocarpa*, Sierra Nevada, California	Phaff *et al.* (1964b)
	Exudate of *Fagus crenata*, Japan	Kodama (1974)
Pichia salictaria	*Salix* sp., riverbank willow trees, California central valley	Phaff *et al.* (1964b)
Pichia angophorae	Australian red gum tree *Angophora costata*	Miller and Barker (1968)
Pichia scutulata var. *scutulata*	*Myoporum sandwicense* Hawaii	Phaff *et al.* (1976)
Pichia scutulata var. *exigua*	*Populus, Salix, Acer,* British Columbia, Canada	Phaff *et al.* (1976)
Pichia naganishii	*Camellia japonica*, Japan	Kodama (1974), Kodama and Kyono (1974a)
Pichia veronae	*Aesculus turbinata*, Japan	Kodama (1974), Kodama and Kyono (1974a)
Pichia nakazawae var. *nakazawae*	*Quercus mersinaeforia*, Japan	Kodama (1975)
Pichia nakazawae var. *akitaensis*	*Salix* sp., Japan	Kodama (1975)

(Boidin and Abadie, 1954). Species unique to this source and listed in Lodder (1970) include *Pichia chambardii, P. rhodanensis, P. strasburgensis, P. toletana, P. pseudopolymorpha* (=*Debaryomyces pseudopolymorphus*), *C. boidinii* and *T. molischiana*. Additional species not included in Lodder (1970) are *P. abadieae* and *P. adzetii* (Jacob, 1969) and *P. pignaliae* (Jacob, 1970). Since many of the species recovered have not been found in other habitats, it is possible that these species are associated in some way with tree bark in small numbers and develop in the tanning liquor. They are very resistant to the high tannin content of these solutions. The identity of yeasts in tanning liquors is generally different from the yeast flora of tree fluxes.

2. Yeasts occurring in necrotic tissue of cacti

Anyone who has explored desert areas where cacti occur naturally will have observed skeletal remains of giant columnar cacti or of the pads of smaller *Opuntia* cacti. These are the last stages of a rotting process that is quite widespread in the desert. The process begins when the succulent tissue is stressed and injured in some way, rendering it vulnerable to infection and subsequent breakdown by pectinolytic soft-rot bacteria (Lightle *et al.*, 1942; Alcorn and May, 1962). The bacterial flora is joined at some later time by saprophytic yeasts. Adult drosophilids, and possibly other insects, are presumed to disperse the yeasts from one rot to another according to the particular host preference of the flies, whereas the larvae spread the yeast within the soft rot. Both adult flies and their larvae feed on the microorganisms present which supply them with vitamins, sterols and concentrated protein. The rotting tissues may be localized and short-lived in some of the smaller cactus species, but often they are long-lived (months) and involve large areas of tissue in larger cactus species. We have discovered a highly specific yeast population in necrotic cactus tissue.

An initial survey, limited both by the number of plants sampled and by the geographic region covered, indicated that one yeast, at that time thought to represent or resemble *Pichia membranaefaciens*, was common and dominant (Heed *et al.*, 1976; Starmer *et al.*, 1976). These studies also revealed that *Candida ingens* and two cactus-specific yeasts, *Torulopsis sonorensis* (Miller *et al.*, 1976b) and *Cryptococcus cereanus* (Phaff *et al.*, 1974), were frequently recovered from rotting cactus tissue and from the associated species of *Drosophila*. *Torulopsis sonorensis* was transferred to the genus *Candida* by Yarrow and Meyer (1978) and will be designated as *C. sonorensis* in this chapter. The ascogenous state of *Cr. cereanus* was discovered by Rodrigues de Miranda (1978) and described as *Sporopachydermia cereana*. However, in our experience not all strains of *Cr. cereanus* appear capable of sporulation. In the spring of 1976 a greatly expanded survey was conducted on the Baja California peninsula and part of the State of Sonora, Mexico, the results of which were published by Starmer *et al.* (1982). The use of molecular taxonomic criteria revealed that those isolates previously identified as *P. membranaefaciens*, and similar isolates obtained in later surveys actually represented six or more cactus-specific species, distinct from *P. membranaefaciens* (*cf.* Table I). These were described as *P. cactophila* (Starmer *et al.*, 1978a), *P.*

pseudocactophila (Holzschu *et al.*, 1983), *P. opuntiae* (Starmer *et al.*, 1979), *P. amethionina* (Starmer *et al.*, 1978b), *P. heedii* (Phaff *et al.*, 1978b), *P. antillensis* (Starmer *et al.*, 1984) and *P. deserticola* (Phaff *et al.*, 1985). Likewise, isolates designated *C. tenuis* and obtained in the early as well as later surveys to which they keyed by the taxonomic system in Lodder (1970), proved to be representatives of a new heterothallic cactus-specific species, which was later named *Pichia mexicana* (Miranda *et al.*, 1982). Another new species recognized during the Mexican survey was named *Candida mucilagina* (Phaff *et al.*, 1980). For the systematics of the columnar species of cacti of Mexico the reader is referred to Gibson and Horak (1978).

Pichia cactophila is probably the most widespread among the species occurring in cactus necroses. It is common in the columnar cacti throughout Mexico as well as in *Opuntia* cacti introduced into Australia (Barker *et al.*, 1983, 1984), Spain, Hawaii, Caribbean Islands, Venezuela and other areas where cacti are native or have been introduced (W.T. Starmer and H.J. Phaff, unpublished results). In addition, a certain degree of host specificity of cactus yeasts has been established (Starmer *et al.*, 1982). For example, *Pichia heedii* is restricted to the subtribe Pachycereinae as a result of the sensitivity of this species to triterpene glycosides found in the cacti of the subtribe Stenocereinae (Starmer *et al.*, 1980; Gibson and Horak, 1978). The two major host plants of *P. heedii* are the senita and saguaro columnar cacti, members of the subtribe Pachycereinae. Strains of *P. heedii* occurring in senita assimilate citrate whereas those of saguaro do not. This difference is due to a change at a single genetic locus and could be in response to the chemistry of the host plants (Starmer *et al.*, 1980). Mating experiments and DNA–DNA reassociation experiments have shown that the two metabolic types are completely compatible with about 95% of the spores from hybrids being viable and the DNA species of the two parents showing greater than 95% complementarity (Phaff and Holzschu, 1981; Holzschu and Phaff, 1982). Presumably the three *Drosophila* species that utilize senita (*Drosophila pachea*) and saguaro (*Drosophila nigrospiracula* and *D. mettleri*) maintain the partition of these metabolic subpopulations because of the exclusive chemical composition of their host cacti (Starmer *et al.*, 1980; Heed, 1978). *Drosophila pachea* is restricted to senita rots because this cactus contains a unique sterol necessary for the development and fertility of *Drosophila pachea* (Heed and Kircher, 1965). In addition, Kircher *et al.* (1967) have demonstrated that other drosophilae (e.g. *Drosophila nigrospiracula*, which breeds in saguaro) cannot use senita for breeding because of the presence of toxic alkaloids in this cactus. The resulting exclusive utilization of senita as a breeding site by *Drosophila pachea* could therefore limit the distribution of the citrate-positive *P. heedii* to this cactus.

The two varieties of *P. amethionina* occupy separate subtribes of columnar cacti. *Pichia amethionina* var. *amethionina* is found in cacti of the subtribe Stenocereinae, whereas *P. amethionina* var. *pachycereana* is found in rots of the subtribe Pachycereinae. The cacti of the Stenocereinae contain triterpene glycosides (Gibson and Horak, 1978) which inhibit the growth of *P. amethionina* var. *pachycereana*, whereas the variety *amethionina* is resistant to these triterpene glycosides. This difference has been shown to be controlled by a single temperature-sensistive gene and this genetic difference is thought to be singularly responsible for the divergence of the two varieties (Starmer *et al.*, 1980). Since the permissive temperature is high it has been postulated that the gene product is a heat-labile protein involved in the transport of the toxic triterpene glycoside into or across the plasmalemma of the yeast cell. Starmer *et al.* (1980) considered the evolutionary divergence of the two varieties as the beginning of the speciation stage. Holzschu and Phaff (1982) conducted DNA hybridization experiments with strains of *P. amethionina*, including some recently isolated strains of *P. amethionina* var. *amethionina* from *Opuntia* cacti in Australia by Barker *et al.* (1983). DNA complementarity values at the intravarietal level approached 100% for strains of *P. amethionina* var. *amethionina, P. amethionina* var. *pachycereana* and the *P. amethionina* strains collected in Australia. Intervarietal DNA re-associations between *P. amethionina* var. *amethionina* and the variety *pachycereana* provided a mean complementarity value of 64.9 $\pm$ 3.2% based on 16 pair-wise comparisons, including reverse experiments. Their study also showed that some atypical strains of *P. amethionina* var. *pachycereana*, identified on the basis of triterpene glycoside sensitivity and positive mannitol utilization, were actually strains of the variety *amethionina* (85% homology) and that true strains of the varieties *amethionina* and *pachycereana* do not produce viable spores when mated, in contrast to an earlier report by Starmer *et al.* (1980). Holzschu and Phaff (1982) suggested that the lack of interfertility between the heterothallic *P. amethionina* var. *amethionina* and *P. amethionina* var. *pachycereana* justifies raising the rank of these two varieties to the species level. DNA relatedness of 65% indicates that they are closely related and have evolved to separate species only relatively recently. Unfortunately, with the finding of the above-mentioned atypical strains a physiological parameter separating all strains of the two varieties is not known at present. The Australian strains of *P. amethionina* are homothallic and are more similar to *P. amethionina* var. *amethionina* (about 87% homology) than to the variety *pachycereana* (about 64% homology). These Australian strains are mannitol positive (latent) and triterpene glycoside sensitive (*Opuntia* cacti do not contain this chemical) indicating that by current criteria they would be assigned to the variety *pachycereana*.

Another example of two closely related yeasts that have their habitat in different cactus species is *P. opuntiae* (Starmer *et al.*, 1979). This species shifts between cacti which are members of two different cactus tribes, the Opuntiaeae (*P. opuntiae* var. *opuntiae*) and the Pachycereeae (*P. opuntiae* var. *thermotolerans*). The two varieties differ by several physiological characteristics and by a slight but significant difference in nuclear DNA base composition. Both varieties are heterothallic, but matings between the mating types of the two varieties rarely produce zygotes or asci, precluding gene flow or genetic analysis. Starmer *et al.* (1980) concluded that the two varieties had entered the final stage of the speciation process. Holzschu and Phaff (1982) determined the DNA–DNA base sequence complementarity among strains of this species. The intravarietal DNA homology values varied between 93 and 100%, but the mean *P. opuntiae* intervarietal relative DNA homology values were 28.0 $\pm$ 3.1% based on five pair-wise comparisons. They concluded that the low degree of intervarietal DNA homology in conjunction with pronounced differences in electrophoretic enzyme mobilities and the lack of interfertility suggested species status for the two varieties of *P. opuntiae*, namely *P. opuntiae* and *P. thermotolerans* (Holzschu *et al.*, 1985). It is not known whether the evolutionary divergence between the two species is the result of selective differences imposed by the host plants of the two species (two different cactus tribes) or divergence due to allopatry (two different continents, i.e. Australia versus North America). A study of yeasts associated with *Opuntia* cacti from areas in the Americas from where the cacti were introduced into Australia may throw light on this question (Murray, 1982). Thus far *P. opuntiae* has not been isolated from *Opuntia* plants in North America (H.J. Phaff and W.T. Starmer, unpublished work).

Following the study of the association of yeasts with necrotic tissue of columnar cacti in the Sonoran Desert (Starmer *et al.*, 1982), studies have been made of the yeasts found in cactus rots of *Opuntia* cacti in Australia (Barker *et al.*, 1983, 1984) and in *Opuntia* rots in Arizona and Texas (Starmer and Phaff, 1983). The cactus species sampled in Australia were *Opuntia stricta, Opuntia tomentosa, Opuntia monacantha* and *Opuntia streptacantha*, in Arizona *Opuntia phaeacantha* and *Opuntia ficus-indica*, and in Texas *Opuntia lindheimeri*. Table III summarizes the findings in the different geographical regions. Only the principal species of yeast are included in the compilation. It should be noted that the 205 rot samples from *Opuntia* cacti in Arizona and Texas were taken in October and November of 1981 respectively, and over a fairly wide geographical area. The rot samples from *Opuntia* cacti in Australia were based on periodic sampling over a 15-month period from October 1976 to December 1977 in the Hunter Valley of New South Wales and in Queensland in November 1977, November 1978 and May 1979. In the first Australian survey (Barker *et al.*, 1983) it was shown that the two most

... geographical areas.

Taxonomic designation of the yeast species[a]	Columnar cacti in the North American Sonoran Desert (number found/ total isolates)	Opuntia cacti in the North American Sonoran Desert (number found/ total isolates)	Australian Opuntia cacti (number found/ total isolates)
Pichia cactophila[b]	247/721	65/592	207/944
Pichia heedii[b]	89/721	0/592	0/944
Pichia amethionina var. *amethionina*[b]	42/721	10/592	21/944
Pichia amethionina var. *pachycereana*[b]	11/721	4/592	0/944
Pichia pseudocactophila[b]	15/721	0/592	0/944
Pichia mexicana[b]	21/721	9/592	0/944
Pichia thermotolerans[b]	20/721	0/592	0/944
Pichia opuntiae[b]	0/721	0/592	22/944
Pichia sp. "B"[b]	0/721	0/592	45/944
Pichia deserticola[b]	0/721	21/592	0/944
Candida sonorensis[b]	102/721	59/592	230/944
Candida mucilagina[b]	11/721	50/592	30/944
Candida deserticola[b]	16/721	0/592	0/944
Candida ingens[c]	63/721	4/592	0/944
Candida boidinii[c]	1/721	9/592	15/944
Candida "caseinolytica"[b]	0/721	7/592	0/944
Cryptococcus cereanus[b] "complex"	44/721	92/592	37/944
Hansenula polymorpha[c]	0/721	30/592	2/944
Clavispora "opuntiae"[b]	6/721	23/592	159/944
Prototheca zopfii[c]	4/721	49/592	N[d]
Geotrichum sp.[c]	N[d]	96/592	N[d]

[a] Specific names in quotes are considered to represent new species, as yet undescribed. Data from Starmer *et al.* (1982), Starmer and Phaff (1983), Barker *et al.* (1983, 1984), and unpublished results by W.T. Starmer and H.J. Phaff.

[b] Designated cactus-specific species.

[c] Designated yeast found both on cactus and other sources.

[d] Accurate count not conducted.

common species (*P. cactophila* and *C. sonorensis*) were present in all months and throughout the area sampled, but other relatively common species (3–9% of all isolates) had much more restricted distributions in both time and space. Temporal effects (lower number in the summer and winter months) were the major components contributing to the heterogeneity of the yeast flora. Comparisons of the species compositions of the yeast communities in Queensland indicated that the differences among communities were greater between *Opuntia* species than between different localities within a single cactus species; also that differences between years were greater than average differences between localities within years (Barker *et al.*, 1984). Other factors influencing the structure of yeast communities were temperature, pH value and age of the rots. It is therefore possible that there could also be significant changes in the yeast flora of *Opuntia* rots in Arizona and Texas throughout the year. However, the *Opuntia* yeast system found in Australia was recently introduced (Murray, 1982) to that continent and may not be representative of the long-standing North American yeast communities occupying *Opuntia* cactus. Table III also shows that the great majority of yeast species recovered from cactus necroses are cactus specific, and thus the majority of yeast community evolution in the cacti has presumably proceeded by evolution *in situ* rather than by additions and replacements from outside the system.

Apart from the presence of inhibitory compounds in certain cactus tissues, such as triterpene glycosides in the Stenocereinae, not a great deal is known about the carbon sources that support yeast growth in rotting cactus tissue. Although some compounds have been found useful in differentiating certain cactus-specific yeasts in the laboratory, such compounds may or may not play significant roles in determining the composition of the yeast communities in cactus rots. Moreover, the cacti undoubtedly contain compounds not tested in the laboratory which are utilizable by the yeasts. Until the decaying cactus tissue is analysed in detail for utilizable carbon compounds, the exact nature of the adaptive physiological attributes of the yeasts remains speculative (Starmer, 1981a). Nevertheless, some observations are of interest (Starmer, 1981b; H.J. Phaff, unpublished results). *Pichia cactophila*, the most cosmopolitan cactus yeast, utilizes very few compounds. However, its very strong growth on the aminosugar D-glucosamine is shared with only a few minor species occasionally found in some cactus species (*P. pseudocactophila* and *P. norvegensis*). D-Glucosamine has not been demonstrated in cactus rots but it could arise from the breakdown of chitosan, a polymer of D-glucosamine, which is a known major constituent of the cell wall of some fungi and yeast-like fungi (Wessels and Sietsma, 1981). Similarly the ability of *P. mexicana* and *Clavispora* "*opuntiae*" to utilize *N*-acetylglucosamine, might be the result of hydrolysis of chitin, a polymer of *N*-acetylglucosamine which is a major component of the exoskeleton of insects and of the cell wall of some fungi and basidiomycetous yeasts (Wessels and Sietsma, 1981). *Candida sonorensis* is another common cactus yeast: it is capable of growing on methanol which is liberated from cactus pectin by the pectin methylesterase of soft-rot bacteria (Fogleman, 1982). Other more rarely isolated yeasts from cacti that also can utilize methanol are *C. boidinii* and *H. polymorpha*. *Candida sonorensis* can also carry out an active alcoholic fermentation. Under aerobic conditions it grows poorly on the ethanol it has produced fermentatively, but for other cactus-specific yeasts (e.g. *P. cactophila* and several others) ethanol is an excellent carbon source. Other volatile compounds of microbial origin include acetone, ethyl acetate and propan-2-ol. Not only do such compounds play an important role in attracting *Drosophila* larvae and adults to areas where preferred yeast species are present (Fogleman, 1982), but they also can serve as selective carbon substrates for a limited number of cactus-specific yeasts. *Candida ingens* is almost unique among the yeasts isolated from cactus for its strong lipolytic activity. *Pichia mexicana* also has lipolytic activity, but less strong than that of *C. ingens*. Several cacti, such as organ pipe, contain rather high concentrations of short-chain fatty acids (C_8, C_{10}, C_{12}) which occur as monoesters of certain triterpenes and sterols. The liberated free fatty acids form an excellent substrate for *C. ingens*, whereas for some other species of yeasts they may actually be inhibitory (Wyss *et al.*, 1945). Another cosmopolitan cactus-specific yeast is *Cr. cereanus* (Phaff *et al.*, 1974) whose teleomorphic state has been described as *Sporopachydermia cereana* (Rodrigues de Miranda, 1978). Among the common cactus-specific yeasts, *Cr. cereanus* is the only species that demonstrates strong growth on YNB with *meso*-inositol as the sole carbon source. This cyclic polyol occurs widespread in nature in plants, micro-organisms and animals. In plants it is generally present as phytin, a calcium–magnesium salt of phytic acid, the hexaphosphate ester of *meso*-inositol. A phosphatase, phytase, is capable of hydrolysing the phosphate groups, yielding ultimately free inositol. *meso*-Inositol is also widespread in phospholipids of higher plants and micro-organisms. Although, as far as we are aware, its occurrence in cacti has not been specifically demonstrated, its presence may be assumed and would give *Cr. cereanus* a relatively non-competitive carbon source for its growth in cactus rots.

3. Yeasts associated with flowers and fruits of higher plants

Flowers, and especially their nectars, have long been considered an ideal habitat for yeasts. Firstly, the nectars contain ample sugar for yeast multiplication and, secondly, in at least those cases where pollination is done by bees or other insects, the yeasts are distributed between different host plants by these pollinating insects. Furthermore, in plants where ovaries develop into fruits with a high sugar content, it was logical to suspect that nectar yeasts were responsible for the beginning of yeast development in overripe and senescent fruit. Decaying fruit, either fallen or still attached to the plant, serves as a breeding site for many species of *Drosophila* (Carson, 1971) as well as for other insects, such as the dried fruit beetle *Carpophilus hemipterus* (Miller and Mrak, 1953). These insects usually introduce additional yeast species which thrive in the decaying fruit. The early history of our knowledge in this area has been reviewed by Last and Price (1969). Although yeast nomenclature has changed somewhat between 1969 and the present, the knowledge of the yeast species occupying this niche has not changed very much. Principal emphasis in the following will be on data reported in the literature since 1969 and on some earlier data not included in the review by Last and Price (1969).

The importance of yeasts in the nutrition of wild species of *Drosophila* (Phaff *et al.*, 1956b) led de Camargo and Phaff (1957) to study the yeasts occurring in domestic species of *Drosophila* flies and in fermenting tomato fruits in a commercial field in northern California. The flies were dissected after being swept into a net above the tomato bushes. Their results showed that all major species of yeast recovered occurred both in the crops of the flies and in the fermenting fruit. Of 159 isolates obtained from flies, 78 represented *Hanseniaspora uvarum*, 12 *Kloeckera apiculata*, 61 *P. kluyveri*, and eight *C. krusei*. Although the number of isolates from fermenting tomatoes was much smaller, the ratio of the three principal yeasts isolated was approximately the same, but *C. krusei* was not obtained from spoiled fruit. Vacek *et al.* (1979) did a similar survey of the yeast flora of necrotic oranges and associated *Drosophila* flies in the Tucson, Arizona area. Their 221 isolates were composed of *Kl. apiculata* (75), *P. fermentans* (75), *P. kluyveri* (50), *T. stellata* (17) and three minor species. Qualitatively, the yeast species from the oranges and adult *Drosophila* were very similar. However, the species of *Drosophila* contained a higher proportion of *P. fermentans* and a lower proportion of *Kl. apiculata* than was found in the rotting oranges. This was ascribed to the fact that *P. fermentans* is a film-forming yeast which tends to grow on the surface of the necrotic tissue where the adult flies feed. *Kloeckera apiculata* lacks a pellicle and is more likely found in the deeper layers of the tissue where it grows fermentatively, and may serve there in the larval diet. However, Vacek and his coworkers did not include an analysis of the larval gut contents in this study. The presence of *P. fermentans* in the rotting oranges and its absence from tomatoes (see above) may be due to the different geographical regions, the different host plants and the kinds of drosophilae breeding in them.

A special case of fruit spoilage by yeasts is the fig. The fig is a syconium, that is a hollow receptacle lined with flowers. It contains an opening or eye, partially covered by scales, through which small insects can pass. Caprifigs (an inedible variety) constitute a niche for the tiny fig wasp, *Blastophaga psenes*. Caprifigs produce three successive crops each year. Only the spring crop (profichi) produces both staminate and pistillate flowers. The *Blastophaga* wasp, which lives inside the fig, carries pollen from the profichi crop to the next crop so that seeds are formed. The insect passes from one crop to the next until the yearly cycle repeats itself. Phaff and Miller (1961) found a unique microflora, quite constantly associated with the fig wasp, consisting of a bacterium, *Serratia plymuthica*, and a single species of yeast, *C. guilliermondii* var. *carpophila*. The bacterial symbiont has recently been restudied and reclassified as *Serratia ficaria*, a new speices (Grimont *et al.*, 1979). It seems probable that these two micro-organisms have a function in the nutrition of some stage in the life cycle of the wasp. Further details on the life cycle of *Blastophaga psenes* can be found in Condit (1947). The important commerical Smyrna type of fig differs from other commerical varieties in that it contains only pistillate flowers in the syconium. Pollination is necessary for seeds to form, to insure proper setting of the fruit and to prevent it from falling off the tree prematurely. Pollination is effected by hanging perforated bags with profichi caprifigs, containing fig wasps and pollen, in each tree of Smyrna figs. The wasp also introduces its associated microflora (*C. guilliermondii* var. *carpophila* and *Serratia ficaria*) into the Smyrna figs, but these two organisms do not cause spoilage in spite of a moderate amount of growth (Miller and Phaff, 1962). However, this flora does appear to increase the attractiveness of the fruit to drosophilae which use the fruit for ovipositing and in the process introduce spoilage yeasts and bacteria, leading to fermentation and souring. Miller and Phaff (1962) found that apiculate yeasts (species of *Hanseniaspora* and *Kloeckera*), *T. stellata* and *C. krusei* were the most common fermentative spoilage yeasts. Acetic-acid bacteria were responsible for converting the ethanol formed into acetic acid. Fallen figs are readily invaded by dried-fruit beetles (*Carpophilus hemipterus*) which can introduce additional yeasts and moulds or even initiate spoilage in sound figs (Miller and Mrak, 1953). Successive microbial florae have also been noted on endemic African figs (Lachaise, 1977). Species of *Lissocephala*, the most primitive genus of Drosophilidae, breed in immature figs (see also Section II.B.5). A specialized yeast, *Torulopsis fructus*, appears to colonize the early young green receptacle. When the fig matures the exocarp of the syconium becomes infected with other fermenting yeasts and bacteria which begin the souring process. *Drosophila* and *Zaprionus* spp. use the rotting exocarp for feeding and oviposition. The spoilage yeasts identified by Lachaise included *P. membranaefaciens, H'spora valbyensis, H'spora uvarum, Kl. apiculata* and

C. sorboxylosa, most of which were also identified by Miller and Phaff (1962) in sour California figs.

Yeasts, even though present as part of the natural microflora of fruits, only rarely become spoilage organisms in unprocessed fruits unless mechanical damage occurs due to handling or because of attack by insects, birds or other animals (Dennis and Buhagiar, 1980). For example, Lowings (1956) showed that in injured strawberries the apiculate yeast *Kl. apiculata* developed abundantly and caused a leaky type of rot. However, in healthy fruit the situation is different as shown in a thorough study over two seasons conducted by Buhagiar and Barnett (1971) on the yeasts present on freshly harvested strawberries. Weighed samples of the strawberries were blended in a laboratory blender with sterile phosphate buffer and the yeast count was determined by dilution plating. There were about 10^5 viable yeast cells per g of strawberry purée, most of which were strains of *Cr. albidus* and *Cr. laurentii*, neither of which can cause fermentation or is known as a spoilage yeast of fruits. The authors attributed the source of these species to soil, where *Cryptococcus* spp. are common (do Carmo-Sousa, 1969). The proximity of the strawberries to soil could well explain this unusual occurrence compared to yeasts associated with tree fruits. It is noteworthy that in this survey much higher yeast counts were obtained when the plates were incubated at 5°C than at 25°C. The yeasts on the strawberries could also have come from the flowers, because *Cr. albidus* and *Cr. laurentii* are frequently recovered from flowers (J.F.T. Spencer and H.J. Phaff, unpublished results). In another study of the surface yeast flora of strawberries, Martini and Frederici (1976) pointed out that yeast cells may be firmly embedded in gummy fruit secretions and that vigorous shaking or sonication of the berries yield species not obtained by the usual less stringent isolation procedures. Another study by the same group (Rosini *et al.*, 1982) dealt with the yeast flora on the surface of sound grape berries during ripening. Mechanical agitation procedures were used to dislodge yeast cells from the grape surface, resulting in improved recoveries. The number of cells per cm^2 of skin increased with ripening and was highest in the area surrounding the stem of a berry and in cluster sectors closest to the peduncle. *Kloeckera apiculata* was the normal resident species of yeast, regardless of the cluster sector or the ripeage. This species usually accounts for the initial natural fermentation of grape juice. During the final stages of ripening the numbers of *Metschnikowia pulcherrima* also increased.

J.F.T. Spencer and H.J. Phaff (unpublished results) studied the yeasts present in various flowers from Saskatoon, Saskatchewan, Fort Smith, Northwest Territory (both in Canada) and from the Davis area, California. The effect of geography on the species recovered was minimal. *Cryptococcus albidus* was by far the most common yeast in flowers (38 strains). Also recovered were *T. apicola* (13 strains), *Sporobolomyces roseus* (11 strains),

Metschnikowia reukaufii (haploid, 10 strains), *M. pulcherrima* (haploid, eight strains) and *Cr. laurentii* (eight strains). Occasional isolates of *Cr. macerans, Cr. infirmo-miniatus, Rhodotorula rubra, Rh. graminis, Rh. pilimanae, Rh. glutinis* and *Rh. aurantiaca* were also encountered. The absence of typically fermentative yeasts in flowers is noteworthy. *Metschnikowia reukaufii* and *M. pulcherrima* have been known for many years as *C. reukaufii* and *C. pulcherrima* respectively, until Pitt and Miller (1968) discovered their ascogenous states. The strains from flowers were all haploid (both h^+ and h^-) and they mated with the appropriate opposite mating types and produced the typical needle-shaped spores. *Metschnikowia lunata*, a yeast with lunate vegetative cells and asci with up to two needle-shaped ascospores, was described by Golubev (1977). He had isolated it from flowers of *Viccia cracca* in the Spasski District, Primorski Territory, U.S.S.R. A polyol-producing yeast from a flower in Japan was described by Onishi (1972) as *C. tsukubaensis*. Besides the true yeasts, such as those listed above, all investigators of yeasts associated with flowers of higher plants have reported the presence of black yeasts, yeast-like fungi belonging to the genus *Aureobasidium*. Five strains of a newly described species of *Candida, C. fusiformata,* were isolated from macerated fresh cauliflower florets (Buhagiar, 1979; see also Section III.A.4). Occasionally, uncommon species have been recovered from fresh soft fruits. These include 35 strains of *T. fragaria* from strawberries, five strains from black currents (Barnett and Buhagiar, 1971), four strains of *T. pustula* from black currents, 20 strains of *T. multis-gemmis* from raspberries, 19 strains of *T. bacarum* from black currents and two strains each from raspberries and strawberries (Buhagiar, 1975).

Limited sampling of various kinds of cactus fruits (e.g. prickly pear fruit) in the process of decay have shown that cactus-specific yeasts (see Section III.A.2) do not occur in cactus fruits, but instead species of the apiculate yeast genera *Kloeckera* and *Hanseniaspora* are prevalent (W.T. Starmer, M.A. Lachance and H.J. Phaff, unpublished results).

4. Yeasts associated with leaves of plants

The external surface of the leaf as an environment for micro-organisms has been termed the phyllosphere or phylloplane. Its use as a habitat for yeast has been recognized more recently than the interior of flowers or the fruits of higher plants. Last and Price (1969) reported that the phylloplane is a good substrate for saprophytic organisms, and cereal leaves in particular were a suitable environment for species of *Sporobolomyces*. Ruinen (1961, 1963) demonstrated that healthy leaves of tropical plants normally carry a rich flora of nitrogen-fixing bacteria. As leaves are also the site of photosynthesis, it is not surprising that yeasts are found in the phylloplane, sometimes in large numbers. Ruinen (1963) described three new species which she had isolated from tropical foliage collected in Indonesia, Surinam and the Ivory Coast: *Candida javancia, C. foliarum* and *C. diffluens*. In this survey 65 strains of yeast representing 22 species of the genera *Hansenula, Cryptococcus, Candida, Sporobolomyces, Rhodotorula* and *Aureobasidium* were identified. Among the strains of *Candida* were two strains of *C. bogoriensis*, a species from leaves in Indonesia that had been described earlier (Deinema, 1961).

Another early investigator who contributed greatly to our knowledge of yeasts occurring on leaves was di Menna (1959). She studied the yeasts occurring on pasture plants in New Zealand. di Menna found that the number of yeasts from surfaces of pasture grasses increased seasonally from about 10^5 to 10^7 per g fresh weight of foliage. The numbers were lowest during winter and early spring and reached a peak in the summer. The proportion of species present varied greatly with the season. For example, the proportion of *Rhodotorula* and *Sporobolomyces* species increased from 6% in winter to 92% in late autumn. She also noted that common soil yeasts constituted only a small fraction of the yeast flora (below 8%) in the phylloplane. In contrast, typical phylloplane yeasts (*Cr. laurentii, T. ingeniosa, Rh. marina, Rh. graminis* and *Sp. roseus*) did invade the upper layers of the soil, especially in summer when the numbers were high, but they did not persist and were never dominant. Their inability to persist was attributed to a number of antibiotic-producing bacteria in the soil (di Menna, 1962). Within wide limits, the phylloplane yeast flora was not significantly affected by the various collection sites. Overall, the results from different investigators show that most of the yeast species isolated from leaf surfaces are non-sporogenous, non-fermentative, and most commonly represent the genera *Cryptococcus, Rhodotorula, Sporobolomyces* and, in some areas, *Torulopsis*. *Candida* species are usually in the minority. For further details of earlier studies on the yeasts associated with the phylloplane the reader is referred to the excellent review by Last and Price (1969). Among the more recent accounts on yeasts isolated from the phylloplane the following isolations are relevant. Buhagiar (1979) isolated 47 strains of *C. fusiformata* from macerated samples of the outer wrapper cabbage leaves in the U.K. This inositol-positive yeast species forms its buds on spicules. The buds are not forcefully discharged, however, and Buhagiar therefore placed it in the genus *Candida* rather than in the genus *Sporobolomyces*.

McBride and Hayes (1977) studied the micro-organisms that are present on the leaves of a coniferous tree, the European larch. They found that the yeast population increased with leaf development. The pink yeasts (mainly *Rh. glutinis* and *Sp. roseus*) usually constituted over 50% of the yeast population. *Torulopsis* spp. were also common in the early season of leaf development, but in the later season up to leaf fall they were replaced by *Crypto-*

coccus spp. There does not appear to be, therefore, a significant difference in the yeast flora of broad-leafed plants and this particular coniferous tree.

Andrews *et al.* (1980) studied the positional variation in phylloplane microbial populations within an apple tree canopy. For yeasts, height and lateral position were the most significant factors on microbial density, with $P < 0.05$ at several seasonal periods. There was a trend for the yeast concentration to decrease with height and increase in peripheral leaves of the canopy. The differences were attributed to variation in accessibility to airborn microbes, microclimate, vectors and physiological condition of leaves.

In our own laboratory fermentative yeasts have repeatedly been isolated from *Ulmus* leaves in California (H.J. Phaff, unpublished work). Yeasts such as *Sacch. rosei* (syn. *Torulaspora delbrueckii*), *Sacch. pretoriensis* (syn. *Torulaspora pretoriensis*) and *Kl. veronae* (syn. *Kl. thermotolerans*) occurred in late summer and early autumn when the leaves began to excrete a sugary syrup. Yeasts may also occur in large numbers in or on decaying leaves in humid tropical regions. W.T. Starmer and H.J. Phaff (unpublished observations), in a study of the breeding sites of Hawaiian *Drosophila*, found yeasts in association with rotting chirodendron leaves in which some species breed.

5. Plant pathogenic yeasts

Some yeasts are recognized as plant pathogens, but their pathogenicity is of a relatively mild form compared with the action of phytopathogenic bacteria and higher fungi. Yeasts responsible for plant diseases are primarily members of the genera *Nematospora* and *Ashbya*, the latter a yeast-like organism (see do Carmo-Sousa (1970) for the historical aspects of the taxonomy of these genera). *Nematospora coryli* and *Ashbya gossypii* (both are monotypic genera) produce long needle-shaped ascospores with a whip-like appendage. Yeast cells or spores are introduced into various host plants during the feeding action of piercing–sucking insects and cause a variety of symptoms during their development in the host plant tissue. Diseases transmitted in this way are termed stigmatomycoses (singular stigmatomycosis). They have been found in many countries located in tropical or subtropical areas of the world and in a large number of cultivated or wild host plants. Of considerable economic importance is the internal boll disease of cotton. The infection with spores or cells of *Nematospora* and *Ashbya* takes place by hemipterous insects, especially *Dysdercus* spp. or cotton-stainer bugs. Frazer (1944) demonstrated that transmission of spores and cells of the pathogens is purely mechanical, and that they are carried as external contaminants in stylet pouches. This term refers to invaginations in the insect's head which house the stylets. The yeast is protected in these pouches and spore germination is possible. As the stylets enter the host plant, the contaminating yeast is transferred to the plant tissue and starts the infection. In cotton bolls only the lint and seeds are affected and there is no external evidence of decay. The yellow or brown staining of the lint in bolls infected with *Nematospora* or *Ashbya* is very conspicuous; young bolls, up to four weeks old, are most susceptible.

Stigmatomycosis is not limited to cotton bolls but may affect the seeds and fruits of a large variety of plants. They may show various symptoms, and transmission is caused by species of a number of genera of insects with piercing-sucking mouthparts. Fawcett (1929a,b) compiled findings of *Nematospora* in the following crops: beans in Virginia; tomatoes in southern California; citrus, cotton and 15 other hosts in the Lesser Antilles; citrus in the Philippines, Japan and China; cotton in Burma and Nigeria; pomegranates and citrus fruit in Florida. Wallace (1939) compiled a later list of stigmatomycoses, giving host plants, insect vectors and localities. Fawcett (1929a) found that the Western leaf-footed plant bug *Leptoglossus zonatus* acted as a vector carrying *Nematospora* from pomegranate to citrus. The symptoms in citrus develop in two to three weeks and involve collapse and drying out of the juice sacs just inside the rind, a condition referred to as inspissosis. The symptoms in pomegranates are also evident only inside the fruit. At first there are depressed light-coloured spots in the fleshy covering surrounding the seeds. Later there is a general browning and collapsing of infected regions of the pulp. It was also shown that the yeast was responsible for the symptoms rather than the puncture wound *per se*. *Nematospora* infection of tomatoes is associated with "cloudy spot" caused by insects of the pumpkin and leaf-footed plant bug group. Weber (1933a) inoculated tomato fruits with the yeast which subsequently produced colonies within the seed cavities and a discolouration of the seeds. He also isolated *Nem. coryli* from brown areas on the kernels of pecan nuts in Florida (Weber, 1933b). A condition known as yeast-spot disease is caused by *Nematospora* spp. and occurs in various kinds of beans. Wingard (1922) described this condition in lima beans where the yeast growth occurs on the seed within the pod and causes dark sunken areas on the cotyledons. *Nematospora coryli* also causes yeast-spot disease in soybeans (Daugherty, 1967). The yeast is transmitted by certain species of the Pentatomidae, such as the green stink bug (*Acrosternum hilare*), the brown stink bug (*Euschistus servus*) and other species of stink bugs. Symptoms of the disease include failure of the pods to fill and early drop off, an increase of small seeds, and a large increase in discoloured and mouldy seeds during storage. Also the germination of soybean seeds damaged by stink bugs was decreased. Damage by stink bugs to rice is very similar to that in soybeans. Rice so damaged is often referred to as "pecky rice". Daugherty and Foster (1966) found that all samples of pecky rice damaged by the rice stink bug, *Oebalus pugnax*, were infected with *Nem.*

coryli. Pecky rice is discoloured and thus yields an inferior product with a high proportion of broken damaged grains.

In conclusion, when one takes into consideration the wide range of the distribution of stigmatomycoses and the large number of hosts and vectors, *Nem. coryli* must be recognized as a parasite of significant importance.

B. Yeasts associated with insects

The association of yeasts with insects can be broadly divided into two aspects. The first is a purely mechanical association where insects serve as vectors with the yeasts attached to some exterior parts of the insect's body, and no nutritive benefit is derived from the association. Some examples have been given in Section III.A.5. (p. 153), particularly under plant pathogenic yeasts carried by insects with piercing-sucking mouthparts. The second is a beneficial association where the insect may act not only as a vector but can also derive important nutritional benefits (either as adults or larvae) from the consumption of yeast cells growing in the feeding or breeding substrate, or in some cases growing in the insect's crop itself. The separation of the two categories is not always easy because the occurrence of yeasts in the substrates where adults feed or larvae develop is not necessarily proof that they are essential nutrients in the insect's life cycle.

1. Yeasts associated with bark beetles (Scolitidae)

The regular association of yeasts and blue-staining fungi with the wood and borings (frass) of beetle-infested coniferous trees was first noted by Grosmann (1930) and soon afterwards by Rumbold (1931). Holst (1936) showed that a new species of yeast could be isolated regularly from the bark beetles *Dendroctonus* and *Ips*, and from their galleries in coniferous trees in various parts of the U.S.A. The yeast was originally described as *Zygosacch. pini* and later reclassified as *Pichia pinus* by Phaff (1956). In a later study by Rumbold (1941), the same yeast, together with an asporogenous mycelium-forming species, was isolated repeatedly from the bark beetle *Dendroctonus monticolae* infesting mountain pines growing in a large area of the Rocky Mountains. A much more extensive survey of yeasts associated with pine bark beetles was conducted by Shifrine and Phaff (1956) in northern California. All isolations were made following dissection from the interior parts of adults, larvae and callow adults after sterilization of their outer surfaces by submerging them in a solution of mercuric chloride followed by rinsing. Isolations were made from the beetles *Dendroctonus brevicomis, Dendroctonus monticolae, Dendroctonus jeffreyi, Dendroctonus valens, Ips oregoni, Ips confusus* and *Ips emarginatus* occurring in the phloem of *Pinus ponderosa, Pinus jeffreyi, Pinus monticola, Pinus contorta* and *Pinus lambertiana*. The principal species among the 169 strains isolated were *P. pinus* (32 strains), *H. capsulata* (36) and *C. silvicola* (70). A sexual state of the last species was discovered later by Wickerham (1960), who named it *H. holstii*. This common species occurs in nature as haploid heterothallic cells. Shifrine and Phaff (1956) described several other new species (*P. haplophila, T. nitratophila, T. melibiosum* and *Rh. crocea*) associated with pine bark beetles, but these species occurred in low numbers so that their significance in this habitat is uncertain. Calaham and Shifrine (1960) reviewed the literature on yeasts associated with bark beetles in coniferous trees. They pointed out the rather fragmentary knowledge of the yeast–beetle relationship, especially the role of yeasts in bark beetle attraction. There is general agreement that adult beetles introduce the yeast into the phloem of pine trees and that, after ovipositing, the larvae distribute the yeast in the galleries where they multiply actively. Trees weakened by drought or damaged by lightning are particularly susceptible to attack by pine beetles and build up of yeasts in the affected phloem (Howe *et al.*, 1971). Most investigators feel that the yeasts play at least a supplementary role in the nutrition of larvae in the phloem galleries. Yeasts have been found internally associated not only with every species of bark beetle, but through all stages of beetle metamorphosis from the smallest larvae to the adult beetles flying to the next host tree. In addition, there is little doubt that the adult beetles are also externally carrying yeasts to their next host tree. In general, most yeasts associated with bark beetles occur in most species of beetles and in many species of *Pinus*. An exception is the absence of *P. pinus* and *C. curvata* from *Pinus jeffreyi* phloem. It has been postulated that these two species might be especially sensitive to the oleoterpenes of the jeffrey pine, particularly its unique n-heptane component.

Other species of coniferous trees are infested by different bark beetles and these are usually associated with different yeasts. Some beetles of the genus *Scolytus* are parasites of true firs (species of *Abies*) and Douglas firs (*Pseudotsuga*). Phaff and Yoneyama (1961) found that the fir engraver *Scolytus ventralis* and the Douglas fir engraver *Scolytus unispinosus* contained a new haploid heterothallic species of yeast which they named *Endomycopsis scolyti*. Kreger-van Rij (1964) reclassified this species as *P. scolyti*. The yeast occurred in large numbers in the beetle galleries of both tree species in widely separated areas of northern California. Later studies with pine bark beetles revealed that *P. scolyti* sometimes occurs in the yeast communities associated with species of *Dendroctonus* and *Ips*, as discussed above (H.J. Phaff, unpublished results).

Other species that have their habitat exclusively in bark beetle galleries of coniferous trees are *H. wingei* and *C. tenuis*. Most strains of *H. wingei* have been isolated from species of spruce (*Picea*) although it may occur occasionally in other host trees (*Abies* or *Pinus*) (Wickerham, 1970). Fuson *et al.* (1979) showed by DNA–DNA re-association experiments that *H. wingei* is a synonym of *H. canadensis* of which Wickerham (1970) isolated one strain from frass of a red pine (*Pinus resinosa*) in Ontario, Canada, and another strain from *Pinus strobus* in the state of New Hampshire. *Hansenula canadensis* (syn. *H. wingei*) appears to be confined to coniferous trees in the extreme northern U.S.A. or Canada. Three strains of *C. tenuis* were originally isolated by Holst from bark beetles in the U.S.A. and described in 1942 as a new species by Diddens and Lodder (see van Uden and Buckley, 1970). Phaff *et al.* (1972) reported two additional strains from *Picea* beetle galleries collected in the Yukon Territory, Canada. Additional strains of *C. tenuis* from a variety of beetles and other sources are listed by van Uden and Buckley (1970). DNA hybridization experiments would be desirable to determine if these other strains are truly representative of *C. tenuis* from coniferous trees. A precedent in this connection are 15 strains from cactus rots originally designated as *C. tenuis* by the key of van Uden and Buckley (Starmer *et al.*, 1976), and which were later shown to be haploid heterothallic strains of *P. mexicana* (Miranda *et al.*, 1982). Other, more recently described species of bark beetle-associated yeasts are *H. americana* (Kurtzman, 1984) from *Pinus jeffreyi* frass in California and from *Pinus ponderosa* frass in new Mexico; *H. bimundalis* (Wickerham, 1965), a species closely related to *H. americana* has been recovered from *Pinus sylvestris* in Spain and from *Abies firma* frass in Illinois, U.S.A. and *P. amylophila* and *P. mississippiensis* (Kurtzman *et al.*, 1980) isolated from frass of loblolly pines, *Pinus taeda*, Mississippi, U.S.A. An annotated bibliography on micro-organisms associated with bark and ambrosia beetles was published by Barras and Perry (1975).

2. Yeasts associated with ambrosia beetles and other wood-boring insects (*Coleoptera*)

Ambrosia fungi are mutualistic symbionts of Scolytidae, Platypodidae and Lymexylidae (Coleoptera). Beetles belonging to these families, although related to bark beetles (see Section III.B.1), have different habitats. Whereas the bark beetles complete their life cycle in the phloem or cambium layer between bark and wood, the ambrosia beetles bore tunnels directly into the sapwood or heartwood of weakened or freshly felled coniferous or broad-leafed trees. The beetles carry the spores of certain fungi and, if the moisture conditions are favourable, these begin to grow along the walls of the galleries. Most species of ambrosia beetles have specialized structures, termed mycangia, that contain the symbiotic ambrosia fungi and yeast-like organisms. Both larvae and adults feed on one or more species of ambrosia fungi (Batra, 1966) and this includes yeasts and ambrosial (= yeast) phase of dimorphic ambrosia fungi. The association between ambrosia beetles and fungi has been reviewed by Francke-Grosmann (1963) and Batra (1966). Batra (1967) has also reviewed the taxonomy of the ambrosia fungi, which will not be considered further in this review. Besides higher fungi (Batra, 1967) numerous yeast species have been found in association with ambrosia beetles as well (van der Walt, 1972a,b). The yeasts isolated by van der Walt and colleagues in South Africa from ambrosia beetle-infested indigenous trees (21 genera) yielded numerous new species. This survey probably represents the most thorough recent study of yeasts associated with ambrosia beetles in the world. The 64 yeast species isolated represented not only already known taxa associated with insects not belonging to the ambrosia beetles, but also many new species that appear to be specific in their association with members of the Scolytidae and the Platipodidae in South Africa (Table IV). Additional isolates from xylophagous scolytid beetles in central Africa were reported by

Table IV. Yeasts specifically associated with ambrosia beetles belonging to the families Scolytidae and Platipodidae.

Taxonomic designation	Number of strains	Reference
Ambrosiozyma ambrosiae (= *Pichia ambrosiae*)	23	van der Walt and Scott (1971a)
Ambrosiozyma cicatricosa (syn. *Pichia cicatricosa*)	5	van der Walt (1972b)
Ambrosiozyma monospora (syn. *Endomycopsis monospora; Endomycopsis fasciculata*)	10	van der Walt (1972b) Pignal (1971a)
Ambrosiozyma philentoma	5	van der Walt (1972b)
Ambrosiozyma platypodis (syn. *Endomycopsis platypodis*)	16	van der Walt (1972b)
Saccharomycopsis synnaedendra	12	van der Walt and Scott (1971b)
Hansenula dryadoides	4	Scott and van der Walt (1971)
Hansenula philodendra	7	Scott and van der Walt (1970a)
Pichia acasiae	17	van der Walt (1966)
Candida hylophila	1	van der Walt *et al.* (1971a)
Candida dendronema	4	
Candida silvanorum	3	
Candida entomophila	11	
Candida insectorum	3	van der Walt *et al.* (1972)
Candida berthetii	23	Boidin *et al.* (1963), Pignal (1971a)
Candida nitrativorans	11	van der Walt *et al.* (1972)
Torulopsis dendrica	7	van der Walt *et al.* (1971b)
Torulopsis insectalens	11	
Torulopsis nemodendra	4	
Torulopsis philyla	1	
Torulopsis silvatica	2	
Sterigmatomyces polyborus	2	Scott and van der Walt (1970a)

See van der Walt (1972a) for beetle genera and host trees.

Pignal (1971a). An annotated bibliography on micro-organisms associated with bark and ambrosia beetles was published by Barras and Perry (1975).

The literature abounds with reports of yeasts that were isolated from various wood-boring insects that cannot be considered strictly as ambrosia beetles. Many of these insects belong to the families of the Bostrichidae, Cerambicidae and Bupresidae. Some of the yeasts found in association with beetles of these families are listed in Table V. Many of these yeasts were reported from South African trees. In addition, Pignal (1971b) made a study of Coleoptera larvae belonging to the Buprestidae which had infested forest trees in the province of Quebec, Canada. The principal yeasts isolated by her were four strains of *H. capsulata* from the eastern hemlock (*Tsuga canadensis*) and from the eastern white pine (*Pinus strobus*) and *Debaryomyces vanriji, C. oregonensis* and *C. shehatae* from *Pinus strobus.* A single strain of *T. nitratophila* was recovered from *Pinus strobus*; this yeast was originally reported from the bark beetle *Dendroctonus monticolae* in *Pinus ponderosa* in the western U.S.A. (Shifrine and Phaff, 1956).

3. Yeasts associated with lacewing insects

An interesting association between certain species of Chrysopidae, honeydews and yeast has been described by Hagan *et al.* (1970) and Hagan and Tassan (1972). *Chrysopa carnea* adults (green lacewings) depend primarily on homopterous honeydews (e.g. those produced by aphids) as food for sustenance and production of eggs, but its larval stage is predaceous on a variety of small soft-bodied arthropods, primarily aphids. The green lacewings require a number of essential amino acids in their diet which are often missing or in very low concentration in honeydews. Hagan and his associates have shown that in adult *Chrysopa carnea* an extracellular yeast symbiote, residing in the crop and tracheal trunks, is able to provide such essential nutrients that may be lacking in the honeydew. The predaceous larvae, on the other hand, do not contain the symbiotic yeasts. Adults apparently acquire the yeast in the exterior environment, possibly honeydew or nectar, but it is further spread among adults by trophallaxis, such as the sharing of a drop of regurgitated liquid prior to mating. The yeast actively grows in the crop and can become so abundant that compact spheres of yeast are formed; these may be regurgitated and are sometimes visible to the naked eye in areas of ovipositing. Hagan and his coworkers refer to this yeast as a *Torulopsis* species. A number of strains of *Chrysopa* yeasts from different areas have been studied in the Davis laboratory and all showed identical morphology and physiology. Although the strains show some similarity to *T. multisgemmis*, their DNA base compositions (G + C contents) were significantly different, suggesting that the lacewing yeast is a new species.

Table V. Yeast species found occasionally in association with insect borings in trees or in phloem frass, but excluding ambrosia beetles.

Organism	Source	Reference
Pichia castillae	Frass of insects infesting *Gymnocladus canadiensis*	Santa Maria and Garcia Aser (1970)
Pichia heimii	Decaying insect-invaded wood in Equatorial Africa	Pignal (1970)
Pichia kodamae	Insect infestations in *Euphorbia ingens* in South Africa	van der Walt *et al.* (1982)
Pichia meyerae	Insect infestations in *Euphorbia ingens* in South Africa	van der Walt (1982a)
Pichia segobiensis	Larva of *Calcophora* sp. in *Pinus sylvestris*	Santa Maria and Garcia Aser (1977)
Pichia acasiae	Bostrychid, buprestid and cerembicid beetles	van der Walt (1972a)
Pichia farinosa[a]	Bostrychid beetles	van der Walt (1972a)
Pichia xylopsoci	Bostrychid beetles	van der Walt and Scott (1971c)
Hansenula americana	Frass from coniferous trees, south-west U.S.A.	Kurtzman (1984)
Hansenula bimundalis	Frass from coniferous trees, Europe and Asia	Kurtzman (1984)
Hansenula philodendra	Frass from tunnels of the bostrichid, *Xylion adustus* in *Ficus sycomorus*, South Africa	Scott and van der Walt (1970a)
Hansenula sydowiorum	Frass from tunnels of the bostrichid, *Sinoxylon ruficorne* infesting *Combretum apiculatum*, South Africa	Scott and van der Walt (1970b)
Hansenula euphorbiaphila	Insect borings in *Euphorbia ingens*, South africa	van der Walt (1982b)
Hansenula anomala[a]	Bostrichid beetles	van der Walt (1972a)
Hansenula ciferrii[a]	Bostrichid beetles	van der Walt (1972a)
Hansenula holstii	*Ruguloscolytus rugulosus* beetles in various rosaceous fruit trees	Pignal (1973)
Saccharomycopsis fibuligera[a]	Beetles belonging to the bostrichidae and cerambicidae	van der Walt (1972a)
Trichosporon melibiosaceum	Bostrichid sp. in *Acacia karroo*, South Africa	Scott and van der Walt (1970a)
Candida silvanorum	Bostrichid beetles	van der Walt *et al.* (1972)
Candida entomophila	Bostrichid beetles	van der Walt *et al.* (1972)
Candida guilliermondii var. *guilliermondii*[a]	Bostrichid and buprestid beetles	van der Walt *et al.* (1972)
Candida insectamans	Buprestid beetle larvae	van der Walt *et al.* (1972)
Candida amylolenta	Bostrichid beetles	van der Walt *et al.* (1972)
Candida nitrativorans	Cerambicid beetles	van der Walt *et al.* (1972)
Candida rhagii	Bostrichid and buprestid beetles, Cerambicidae spp.	van der Walt (1972a), Jurzitza *et al.* (1960)
Candida silvicola	Cerambicid beetle	van der Walt (1972a)
Candida silvicultrix	Buprestid larvae	van der Walt *et al.* (1972)
Candida tenuis[a]	Buprestid, bostrichid, cerambicid beetles, cerambicid symbionts	van der Walt (1972a), Jurzitza *et al.* (1960)

Table V. Continued

Organism	Source	Reference
Torulopsis dendrica	Cerambicid larvae	van der Walt (1972a)
Torulopsis insectalens	Cerambicid beetle	
Torulopsis torresii[a]	Bostrichid beetle	
Trichosporon behrendii[a]	Bostrichid and buprestid beetles	
Trichosporon melibiosaceum	Bostrichid beetles	

[a] Designated yeast recovered also from non-insect sources. See Table IV for yeasts associated with ambrosia beetles.

4. Yeasts associated with bees

It is well known that all bees (Hymenoptera:Apoidea), including some 20,000 wild species, are nurtured almost exclusively by the pollen and nectar that they collect from plants. In the process they are responsible for the pollination of many plant species. This is particularly important in the production of commerical crops in which pollination is done by domesticated honey bees (*Apis mellifera*). Because of the ample sugar content for the growth of yeasts in nectar and honey, many investigators have studied the microflora of bees. Batra *et al*. (1973) have published a review of the fungi and yeasts found in floral nectar, in honey stomachs of diverse bees, in the honey and pollen provisions of bee larvae and in dead or moribund adult bees or brood. Some fungi are the cause of significant declines in populations of certain bees, while others appear to be saprophytic or mutualistic. In larval cells of the economically valuable alkali bee (*Nomia melanderi*), yeasts (mainly species of *Saccharomyces, Pichia, Hansenula, Candida* and *Kloeckera*) may initiate fermentation of the provision, causing eggs and larvae to die. This occurs, for example, when rain water enters the subterranean nests of this bee and dilutes the sugar of the provision. Batra *et al.* (1973) observed a much higher yeast concentration in the crops of bees than in the floral nectar on which they foraged. This was attributed to a combination of concentration, caused by the evaporative honey-making behaviour of the bee, and actual multiplication of sugar-tolerant yeasts. They listed all fungi and yeasts that have been found associated with Apoidea, with literature citations. The yeasts directly associated with adult bees and larvae included *Aureobasidium pullulans, C. rugosa*, which may be pathogenic, *C. vini, Eremascus fertilis, H. anomala, Kl. apiculata*, which may be pathogenic, *M. pulcherrima, M. reukaufii, Rh. glutinis, Sacch. cerevisiae, Sacch. mellis* (syn. *Zygosacch. rouxii*), *Sacch. rosei* (syn. *Torulaspora delbrucckii*), *Zygosacch. rouxii, Saccharomycodes ludwigii* and *Schizosaccharomyces* spp. Another listing of yeasts isolated from honey bees was published by Gilliam *et al.* (1974). Their paper included a number of references omitted by Batra *et al.* (1973). Altogether 50 yeast species were listed. Gilliam *et al.* (1974) examined the yeast flora of 388 honey bees in Arizona and found that the most common yeasts in bee guts were *T. magnoliae, C. parapsilosis* and *T. glabrata*. Grilione *et al.* (1981) made an extensive survey of yeasts associated with honey bees in the Davis area of California. In the year-long study foraging bees were captured on their return to the hive and yeasts were isolated from the crop and/or gut contents of 194 pollen-collecting (PC) bees (181 yeast strains) and of 209 nectar-collecting (NC) bees (381 yeasts). The highest number of yeasts in PC bees was obtained in March–April, September–October and in December. The maximum numbers of yeast in NC bees occurred in July to September, which coincided with the lowest numbers in PC bees. Most of the yeast species were found only during a limited period of the year which may be related to the kinds of flowers or decaying fruit supplying nectar, pollen or fruit sugar. Some yeasts were isolated only once or twice and may be considered incidental isolations. The most common species were as follows: *Hanseniaspora guilliermondii*, 5 PC and 21 NC isolates; *Hanseniaspora uvarum*, 7 PC and 3 NC isolates; *Kluyveromyces fragilis*, 7 PC and 5 NC isolates; *Metschnikowia pulcherrima*, 6 PC and 19 NC isolates; *Metschnikowia reukaufii*, 8 PC and 46 NC isolates; *Pichia terricola*, 4 PC and 10 NC isolates; *Candida guilliermondii* var. *guilliermondii*, 6 PC and 15 NC isolates; *Candida krusei*, 1 PC and 27 NC isolates; *Candida pulcherrima*, 9 PC and 16 NC isolates; *Candida* sp. A, 15 PC and 39 NC isolates; *Candida* sp. B, 0 PC and 15 NC isolates; *Candida* sp. C, 0 PC and 9 NC isolates; *Cryptococcus albidus* var. *albidus*, 12 PC and 14 NC isolates; *Cryptococcus macerans*, 6 PC and 7 NC isolates; *Cryptococcus laurentii*, 8 PC and 4 NC isolates; *Kloeckera apiculata*, 1 PC and 15 NC isolates; *Kloeckera apis*, 7 PC and 23 NC isolates; *Rhodotorula glutinis*, 15 PC and 23 NC isolates; *Torulopsis apicola*, 17 PC and 9 NC isolates; *Torulopsis magnoliae*, 17 PC and 18 NC isolates; *Torulopsis stellata*, 6 PC and 7 NC isolates.

Of the species listed here, *T. apicola* was first isolated from the intestinal tract of bees by Hajsig (1958) in Yugoslavia. This species may be synonymous with *T. apis* isolated from the trachea of a bee in France (van Uden and Buckley, 1970). The study by Grilione *et al.* (1981), listed above, shows that *T. apicola* is rather common in association with bees in northern California. Another species shown to be common in northern California by Grilione *et al.* (1981) is *Kl. apis* (Smith *et al.*, 1977), the imperfect form (anamorph) of *H'spora guilliermondii*, another common bee-associated yeast in northern California (see above). Three additional species associated with members of the Apoidea appear to be either rare or insufficiently investigated. Two of these are *C. bombi* (Montrochet, 1967) and *T. bombicola* (Spencer *et al.*, 1970). Both were isolated from bumble bees (*Bombus*) or honey thereof, the first one in the Savoie, France and the second species in Canada. They do not appear to be associated with the domestic honeybee, *Apis mellifera*, according to the literature. The third species is *P. melissophila* recovered by van der Walt and van der Klift (1972) from the digestive tract of African honeybees foraging on *Aloe castanea* and *Echium fastuosum* in Pretoria, South Africa. It is perhaps not surprising, when one compares the common yeast species that have been isolated from flowers (Section III.A.3) with those obtained from bees, that a significant number are similar. Also significant is the correlation between the yeast flora of substrates that serve as food for bees, the season of the year in which they occur, and the yeasts associated with foraging bees (Grilione *et al.*, 1981).

5. Yeasts found in the alimentary canal of *Drosophila* spp.

The following discussion pertains mainly to studies of the yeast flora occurring in the crops of wild species of *Drosophila* captured in forests and other natural areas where the feeding habits of the adult flies are not well understood. Begon (1982) has recently reviewed the literature on yeasts associated with *Drosophila*. Domestic species (e.g. *Drosophila simulans* and *Drosophila melanogaster*) that feed on fermenting fruit have been discussed already in Section III.A.3. Wild *Drosophila* species that feed on yeast are usually collected by placing gallon cans containing mashed bananas and baker's yeast at various locations in a forest. The cans are tightly covered with cheesecloth or silk screen to prevent the flies from feeding on baker's yeast. Fermented banana bait sterilized with ethylene oxide is also a good attractant and the covering can then be omitted (Phaff, 1955). Even though *Drosophila* species have distinct preferences for certain yeasts (Dobzhansky *et al.*, 1956) baker's yeast is quite satisfactory. Flies that are attracted to bait cans are collected by sweeping with a net or by aspiration in collection vials. In an early study in southern California mountains, Shehata and Mrak (1952) studied the intestinal yeast flora of *Drosophilia pseudoobscura* at different times of the year. The yeast flora varied from month to month, but was homogeneous at any one time and usually only one species of yeast, or occasionally two or three, were present in a single fly. The unsatisfactory state of yeast taxonomy at that time precluded accurate identifications of many of the isolates from flies and suspected feeding substrates. However, it was clear that the yeasts isolated from various plant materials were different from those obtained from the crops of flies. Shehata *et al.* (1955) later restudied the yeasts isolated from flies and recognized three new speices, *Kluyveromyces drosophilarum, K. dobzhanskii* and *K. phaseolosporus*, which were originally placed in the genus *Saccharomyces*. Excluding strains of *Sacch. cerevisiae*, which originated from improperly screened bait cans, the most prevalent yeast species were *K.*

veronae (syn. *K. thermotolerans*), *K. drosophilarum, K. fragilis, Kl. apiculata* and *H. polymorpha* (syn. *H. angusta*). These species were either absent from or extremely rare in a large number of natural substrates suspected to be possible feeding places for the flies.

A further study of the yeast flora present in the crops of *obscura* group *Drosophila* and *Aulacigaster* occurring at altitudes of 1300 to 3000 m in the Yosemite region of the Sierra Nevada in California was conducted by Phaff *et al.* (1956a,b). Among the 241 strains isolated the most frequently isolated species were: *Sacch. montanus* (36), *Sacch. cerevisiae* var. *tetrasporus* (22), *K. veronae* (30), *K. drosophilarum* (13), *H. polymorpha* (syn. *H. angusta*) (19), *Kl. apiculata* (15), *Kl. magna* (13), *T. stellata* (10) and *C. krusei* (10). Comparing these results with those of the southern California mountains (about 400 miles apart) it can be seen that several yeast species were common to both locations, but in addition the central California survey included a number of common species not present in the south. The only species present in the south and not in central California was *K. fragilis*. Additional new species recognized during this study included *K. wickerhamii, Sacch. kluyveri*, and *Trichosporon aculeatum*, but their presence was rare (Phaff *et al.*, 1956a). A study of known breeding sites of the flies studied in this area, such as oak and fir fluxes (Phaff and Knapp, 1956), various fungi and other natural breeding sites (Carson *et al.*, 1956) revealed that the yeasts found in the breeding sites were strikingly different from those found in the crops of adult flies of the same species in the same region. This indicates that, at least in these regions, larval and adult yeast foods are different and that adult flies do not regularly feed from their breeding sites. The source of the yeasts found in adult *Drosophila* remains to be elucidated. However, at the 1300 m level at which most of the flies were collected, *Drosophila persimilis, Drosophila azteca, Drosophila pseudoobscura, Drosophila pinicola* and *Aulacigaster* all contained the same common species of yeast (Phaff *et al.*, 1956b). Because Kodama (1974) isolated many strains of *Sacch. montanus* and *Sacch. cerevisiae* var. *tetrasporus* from tree exudates in Japan (yeasts that were common in the crops of California *Drosophila* spp.) it is possible that *Drosophila* species in Japan actually feed on slime exudates.

All known African species of the primitive Drosophilidae genus *Lissocephala* are strictly specialized on fruit of *Ficus* spp.. Its larvae breed inside immature figs. Eggs are laid near the fig ostiole and first-instar larvae enter the fig receptacle through the scale-covered ostiole where they feed on yeast (*T. fructus*) and bacteria, or both, which are thought to have been introduced earlier into the sterile receptacle by the pollinator females of *Ceratosolen* fig wasps (Lachaise, 1977). Third-instar larvae leave the fig by crawling out through the ostiole and drop to the soil where they pupate. Since *T. fructus* was the only yeast species found in adult *Lissocephala* crops it is also possible that during ovipositing the eggs become inoculated with *T. fructus* and that the larvae carry the yeast into the fig receptacle where they can multiply.

Kearney (1982) determined the dominant yeast flora of the major breeding sites of *Drosophila subobscura* and *Drosophila phalerata* in woodlands around Leeds, England, namely rowanberries (*Sorbus aucuparia*) and the stinkhorn fungus *Phallus impudicus* respectively (Kearney and Shorrocks, 1981). Their aims were to investigate the relative importance of yeast species and substrates to larval development and ovipositional behaviour of adults. It was found that both *Drosophila* species required yeasts on their respective substrates for survival. *Drosophila phallerata* (a fungal specialist) used cues for oviposition derived primarily from the natural substrate and not its microflora. *Drosophila subobscura* showed little preference for either, in agreement with its generalist nature; it also uses rotting vegetation and tree sap flows, especially in spring and summer. When in competition for oviposition sites, however, *Drosophila subobscura* is forced onto *Sorbus* tissue. Kearney concluded that the behaviour of *Drosophila subobscura* demonstrated one facet of the superiority of a specialist, such as *Drosophila phallerata*, on its own substrate. He found little discernible pattern concerning the effect of yeast species on larval development time, probably reflecting the spatial and temporal variation in the distribution and abundance of yeast species preventing specialization of the two *Drosophila* species.

Cactophilic drosophilae, on the other hand, rely on yeast-rich cactus rots for both feeding and breeding (Starmer *et al.*, 1982; Barker *et al.*, 1983, 1984). The composition of the cactus yeast community has been discussed in Section III.A.2. Adult flies do not feed randomly on any cactus species with necrotic tissue, but feed mainly on their respective larval host plant. For example, the major host plant for *Drosophila pachea* is the senita cactus, in which *P. heedii, C. ingens* and *Cr. cereanus* occur as the principal species. The same three species constituted 89% of the yeasts isolated from adult *Drosophila pachea* flies. Similarly, the major host plant for *Drosophila mojavensis* in Baja California is the agria cactus in whose rots *P. cactophila, P. amethionina* var. *amethionina* and *C. sonorensis* are the most common yeast species. The same three species constituted 76% of the yeasts isolated from adult *Drosophila mojavensis*. Similar correlations were found for *Drosophila nigrospiracula* breeding in saguaro and cardon cacti, and for *Drosophila mettleri* breeding in rot-wetted soil beneath these two cactus species.

The utilization of yeasts present in rotting cactus tissue by *Drosophila* larvae does not appear to be random in all cases. Fogleman *et al.* (1981) have shown that larvae of *Drosophila mojavensis* did not contain a random sample of the yeasts present in the substrate. *Pichia cactophila* was typically in greater frequency in the larvae than in the substrate, whereas *C. sonorensis* exhibited the opposite relationship. Laboratory tests also showed that larvae are capable of distinguishing between patches of different yeast species and spend more time feeding in patches of preferred yeasts. In Sonora, Mexico and southern Arizona, U.S.A., *Drosophila mojavensis* breeds in organpipe cactus, a species rich in medium-chain fatty acids (Kircher, 1982). Starmer (1982) showed that these fatty acids inhibit the growth of *Drosophila mojavensis* larvae, but *C. ingens*, a common yeast in organpipe rots, can utilize these fatty acids efficiently and thus detoxifies them so that *Drosophila mojavensis* develops faster. *Drosophila mojavensis* appears to be an ecological (host plant) generalist and a physiological (yeast) specialist. Other cactophilic drosophila larvae are more or less restricted to a single host plant and cannot afford to specialize. They (e.g. *Drosophila pachea*) therefore are host plant specialists, but physiological yeast generalists, eating whatever yeasts are available. In a further study by Fogleman *et al.* (1982), the yeast flora in the natural substrates and larval guts of two additional desert species and three non-desert species breeding in necrotic oranges were compared. Larvae of all of the species utilizing oranges (*Drosophila melanogaster, Drosophila pseudoobscura* and *Drosophila arizonensis*) contained non-random samples of the yeasts available in the substrate. The other cactophilic *Drosophila* species (*Drosophila nigrospiracula* and *Drosophila mettleri*) did not exhibit significant differences in substrate and gut flora. Fogleman *et al.* (1982) concluded that selective feeding behaviour in *Drosophila* could be related to the degree of polyphagy of the adults. Larvae of the polyphagous species feed selectively and those of the essentially monophagous species feed more or less randomly. They suggested that the relationship between polyphagy and selective feeding is a special case of the generalism versus specialism paradigm.

Vacek *et al.* (1984) studied the feeding preferences of mature, mated female *Drosophila buzzatii* flies (an Australian species breeding in *Opuntia* spp.) in population cages. Choices consisted of cactus homogenate inoculated with cactus rot bacteria only and bacteria plus one of eight cactophilic yeasts. The ability of the flies to discriminate between yeasts developed with time over a three-day period. Although initially feeding and oviposition were random, after three days there was a significant correlation ($r = 0.96$) between the yeasts preferred for feeding and those preferred for oviposition, with *P. cactophila* being the most preferred yeast.

6. Intracellular symbiotes

Many investigators of insects have recognized the presence of intracellular symbiotes, thought to be yeasts or yeast-like organisms, in specialized cells or structures termed mycetocytes and mycetomes. The anatomic relationships with their hosts and the early research in this area have been covered in detail by Steinhaus (1946). The difficulty of culturing the yeasts observed microscopically in the insect's cells has delayed a thorough knowledge of this unique group of microfungi, so that at present only a few species have been isolated and described in adequate detail. The first yeast species isolated and studied was obtained from larvae and imagoes of Anobiidae beetles. The yeasts occur in cells (mycetocytes) of mycetomes (evaginations of the mid gut), in epithelial cells lining the alimentary tract or in cells of other parts of the insect's body. Muller (1934) isolated the symbiotes of *Ernobius mollis* L. and described the yeast cells as globose, occurring in clusters, and the strain as weakly fermentative. This yeast was later named *T. ernobii* by Lodder and Kreger-van Rij (1952). Muller was unsuccessful in culturing the symbiotes of several other Anobiidae beetles. Following observations by previous workers (see, e.g. Pant and Fraenkel, 1950) that symbiote-free larvae would not grow in substrates that supported growth of larvae with symbiotes (*cf.* Jurzitza, 1962), the concept developed that the main contribution of the symbiote appeared to be the release of various vitamins of the B-complex to the host in exchange for other nutrients from the host. This was also found for another Anobiid beetle, *Stegobium paniceum* L., by Foeckler (1961), who tried unsuccessfully to re-infect sterile larvae with foreign yeast species, although they readily incorporated their normal symbiotes in the mycetocytes. Here also the beneficial action of the intracellular yeasts was ascribed to B vitamins. However, Pant and Fraenkel (1950) also obtained some evidence that the symbiotic yeasts contributed sterols to the developing larvae, as well as vitamins. Since the symbiote of *Stegobium paniceum* requires addition of asparagine or glutamic acid to a medium with an inorganic nitrogen source (Kuhlwein and Jurzitza, 1961), it perhaps obtains these amino acids from its host. Further work by Jurzitza (1970) resulted in the isolation of additional symbiotes of Anobiid beetles that previously resisted isolation: *T. karawaiewi* from *Ernobius abietis* Fabr. and *T. xestobii* from *Xestobium plumbeum* Illig. He also isolated additional strains of *T. ernobii* from *Ernobius mollis* (see above). Although the new isolates were not validated with a Latin description, this will undoubtedly be done by yeast taxonomists because good evidence was given that all three Anobiid yeasts had different physiological properties and could be readily differentiated. Of considerable significance was the discovery by Jurzitza (1970) that strains of *T. karawaiewi* either required RNA in a malt extract medium, or growth was stimulated; this may explain why previous efforts to isolate the symbiotes of *Ernobius abietis* were unsuccessful.

An entirely different symbiote of *Drosophila funebris* was described by Chatton (1913) as *Coccidiascus legeri*, but he was unable to culture the yeast that occurred in intestinal epithelia cells of this fly. The yeast reproduced *in situ* by budding, and produced asci containing up to eight spindle-shaped ascospores. The yeast was rediscovered by Lushbaugh *et al.* (1976) as a parasite within vacuoles of the intestinal epithelium of *Drosophila melanogaster* in the state of Georgia, U.S.A. They provided an expanded description of *Coccidiascus legeri*, based on electron micrographs and light-microscope photographs of material *in situ*. Enlarged vegetative cells developed into asci containing two spindle-shaped bilaterally flattened ascospores without apical appendages. Attempts to cultivate the yeast on artificial media were unsuccessful.

In conclusion, judging on the basis of the large number of observed intracellular symbiotes that have resisted cultivation on artificial media, this area of yeast ecology is likely to be highly rewarding in future research efforts.

C. Yeasts occurring in soils

As do Carmo-Sousa (1969) pointed out in her review on the distribution of yeasts in Nature "yeasts have been found in soils of widely different texture, chemical composition, humidity and pH value at various geographic locations and diverse climatic conditions, in bare soils as well as in soils that support a natural vegetation (forests, bush, grasslands) or are cultivated by man (orchards, vineyards, cropfields, gardens)". This view still holds and has undergone little change in the past 15 years. Her listing of soil-originated yeasts provides a fairly up-to-date picture of yeast species commonly found in soil samples all over the world. Because the soil is the ultimate repository of all organic matter that can support yeast growth, one might expect a large variety of yeasts to occur in soils, more or less representative of the yeast flora associated with plant, fungal and animal life above the soil surface. Although this is true in some instances, for example the very common occurrence of apiculate yeasts of the genera *Hanseniaspora* and *Kloeckera* in soils of fruit orchards from residues of spoiled and fermenting fruit dropping onto the soil, there are many other species of yeast that are typical inhabitants of soil and for which no obvious surface sources are known. Phaff *et al.* (1978a) pointed out that in most cases, especially agricultural soils, the soil should be regarded more as a reservoir for yeasts from sources above it than as a specific habitat. Nevertheless, occasional counts as high as 10^5–10^6 g^{-1}, as reported by several investigators, suggest that yeast multiplication can occur in some soils. This is especially true in certain forest soils, as shown in a study by Danielson and Jurgensen (1973) in north Carolina and Virgina. Most of the yeast species occurring in soils possess a wide spectrum of metabolic abilities, enabling them to utilize many of the hydrolytic products of plant materials generated by fungal and bacterial activities. For example, nearly all soil yeasts can grow on cellobiose, produced during microbial breakdown of cellulose. Others use aromatic compounds formed from lignin breakdown (Henderson, 1961). The production of yeast nutrients by other soil microbes may be the explanation why, in an experiment by Lund (1954), soil yeasts inoculated into a sterilized soil had essentially died after six months, but in a control experiment with the same untreated soil the natural yeast flora had increased significantly during the same period. The reader is referred to the thorough review of the early literature in this field by do Carmo-Sousa (1969). The present review will cover those aspects of soil yeasts not covered by do Carmo-Sousa, as well as more recent surveys and publications describing new species of soil yeasts from various geographic areas.

One of the earliest yeast species isolated from soil was *Schwanniomyces occidentalis*, which Kloecker collected in 1909 on the island of St. Thomas in the West Indies. Subsequently, three additional species were described; *Schw. castellii* from vineyard soil in Spain, *Schw. alluvius* from various soils in the middle west of the U.S.A., and *Schw. persoonii* from South African soils (*cf.* Phaff, 1970). The four species were separated on the basis of sugar fermentation and assimilation reactions. However, Price *et al.* (1978) demonstrated later by DNA–DNA re-association experiments that all belonged to a single species, *Schw. occidentalis*. Hence, all strains of this species have been isolated exclusively from soils in widely separated geographical areas, but it does not appear to be a common species.

Another yeast genus whose species have been isolated solely from soils is *Lipomyces*. The first strains representing this genus were isolated in the period 1935–1944 (Starkey, 1946) during a study of nitrogen-fixing bacteria; they were later placed in the new genus *Lipomyces* by Lodder and Kreger-van Rij (1952) who named the type species of the genus *L. starkeyi*. This species has been isolated from soils in many parts of the world (Slooff, 1970). Lodder and Kreger-van Rij (1952) recognized a second species, *L. lipofera*, previously known in the asexual state as *T. lipofera*. It has been isolated from soils in the Netherlands and in Wales. Three additional species have been described since 1970. Nieuwdorp *et al.* (1974) added *L. kononenkoae*, isolated from soils on the island of Trinidad and from Nigeria, and *L. tetrasporus*, isolated by Krassilnikov and coworkers from various soils in the U.S.S.R. Additional strains of this species have been isolated from soils in the U.S.A., the Netherlands and Africa. The latest species assigned to *Lipomyces* is *L. anomalus* (Babjeva and Gorin, 1975) isolated from podzolic soils of the north taiga subzone, U.S.S.R. It is morphologically and physiologically very different from the other species of the genus.

The monotypic genus *Schizoblastosporion* is also soil specific, and the species *Schizobl. starkeyi-henricii* has been isolated from soils ranging from New Zealand to Norway (Kreger-van Rij, 1970b). Babjeva and Blagodatskaya (1972) later reported that this species was predominant in some marsh soils in the U.S.S.R. Several species of *Cryptococcus* also have been isolated solely from soils, or the majority of strains are from that source. These include *Cr. albidus* and *Cr. terreus* (Phaff and Fell, 1970). Emmons

(1951) presented the first direct evidence that the pathogenic yeast *Cr. neoformans* is present in some soils. This was demonstrated by inoculating mice intraperitoneally with soil suspensions and recovering the yeast after four to six weeks from the spleen and liver. More recently, Helen Vishniac and coworkers have isolated and identified a number of new psychrophilic species of *Cryptococcus* from soil samples in the Dry Valleys of South Victoria Land, Antarctica. The first indigenous species described and dedicated to Wolf Vishniac was *Cr. vishniacii* (Vishniac and Hempfling, 1979a), whose habitat was described in detail by the same authors (Vishniac and Hempfling, 1979b). *Cryptococcus vishniacii* was later regarded as the *Cr. vishniacii* complex as a result of molecular taxonomic analysis using DNA–DNA hybridizations (Baharaeen *et al.*, 1982). It was subsequently subdivided into a number of individual species arbitrarily defined as those biotypes having less than 52% DNA–DNA complementarity with the type strain of *Cr. vishniacii* and among themselves. These new species are *Cr. lupi* (Baharaeen and Vishniac, 1982), and *Cr. asgardensis, Cr. baldrensis, Cr. hempflingii, Cr. tyrolensis* and *Cr. wrightensis* (Vishniac and Baharaeen, 1982). In addition, the last authors also described several varieties of *Cr. vishniacii* whose DNA–DNA complementarity to *Cr. vishniacii* was greater than 52%. All species and varieties recovered from antarctic soils could be separated by various carbon assimilation reactions. Another extensive survey of yeasts in antarctic soils was made by Atlas *et al.* (1978): they found that in some antarctic soils yeasts were the dominant microbial populations, varying in concentration from a few to $5 \cdot 10^5$ (g of soil)$^{-1}$. The most common yeasts were *Cr. albidus* and *Aureobasidium pullulans*. Two soil samples yielded *Rh. minuta*, two others *Sporobolomyces salmonicolor*. *Candida scottii* was found in only two soil samples although di Menna (1960) in an earlier antarctic survey found *C. scottii* to be the most commonly isolated yeast in the dry Wright Valley. The sexual state of *C. scottii* is *Leucosporidium scottii* of which some strains are psychrophilic, growing best from 4 to 15°C and not at 20°C (Fell and Phaff, 1970). Other psychrophilic species of *Leucosporidium* that were obtained from terrestrial Antarctic and Greenland soils are *Leu. frigidum, Leu. nivalis* and *Leu. stokesii*. The similarity of the life cycles of *Leucosporidium* and the smut fungi suggests that *Leucosporidium* species occurring in these antarctic environments may function as plant parasites on the vegetation of these areas (Fell and Phaff, 1970). Many of the yeast species described in Lodder (1970) have been isolated only from soils, but their occurrence in most cases appears to be sporadic and limited to only one location and often a single strain. A partial listing is given below; for further details and references to the original publications the reader is referred to Lodder (1970) and Kreger-van Rij (1984): *Sacch. transvaalensis* and *Sacch.* (syn. *Torulaspora*) *pretoriensis* (South Africa), *K. africanus, K. lodderi, K. phaffii*, and *K. polysporus* (South

Africa), *P. terricola* (South Africa), *H. beijerinckii* (South Africa), *H. californica* (wide geographical distribution), *H. dimennae* (India, New Zealand), *H. saturnus* (widespread in North America, Africa, New Zealand and U.S.S.R.), *Debaryomyces polymorphus* (syn. *Pichia polymorpha*, Denmark), *Deb. vanriji* (syn. *Pichia vanriji*, South Africa; also from slime fluxes of deciduous trees in England and U.S.A.), *C. humicola* (Switzerland, also from mushrooms), *C. muscorum* (rotting sphagnum in a peat bog, New Zealand), *C. vartiovaarai* (Finland) and *T. castellii* (Finland). Two of the species listed above were recently reported in an extensive survey of the major soil-climatic regions of the U.S.S.R. (Vustin and Babjeva, 1981). *Hansenula saturnus* (syn. *Williopsis saturnus*) was found in alluvial marsh and meadow types of soil, whereas *H. californica* (syn. *Zygowilliopsis californicus*) was associated with the root systems of root crop plants.

Additional new species have been discovered from various soils that are not included in Lodder (1970). The principal ones are listed in Table VI.

In conclusion it may be stated, on the basis of an extensive literature on this subject, that yeasts are distributed quite unevenly in soil both from a quantitative and qualitative point of view. The presence or growth of certain species may be related to the deposition of plant or animal residues on a particular soil, its structure, porosity, organic content, rainfall, pH value and other factors. Yeasts can multiply in some soils under favourable conditions, but the rise in numbers is usually followed by a sharp decrease (Lund, 1954). Survival is probably favoured in the case of sporulating yeasts (e.g. species of *Lipomyces* and *Hansenula*). Also, the literature presents evidence that capsulated yeasts (*Lipomyces, Cryptococcus* and *Rhodotorula*) survive better in habitats poor in available nutrients and during periods of desiccation than non-capsulated species (Golubev *et al.*, 1984). At any rate, survival of yeasts in soils, even in small numbers, is a well established fact and a renewal of multiplication is probably conditioned by essential nutrients. Total yeast counts of but a few to up to 10^5 or 10^6 (g of soil)$^{-1}$ suggests that yeasts occur in very small numbers compared to bacteria, actinomycetes and moulds, and thus do not seem to play a major quantitative role in the decomposition of organic matter in the soil. Nevertheless, *C. humicola*, a typical soil yeast, has been observed to produce volatile trimethyl arsine from four arsenic pesticides and dimethylselenide from sodium selenite and sodium selenate that were present in soil (Cox and Alexander, 1974). Since the great majority of yeasts isolated from soil lack fermentative ability, and thus depend on aerobic metabolism for their growth, it is not surprising that yeasts are most common in the top 5–15 cm of soil and become exceedingly rare at depths of 25–30 cm. Surface samples are highly variable in yeast content because of extreme changes in moisture content and the effect of sun rays and high temperatures.

Table VI. Selected species of soil yeasts described after the publication of Lodder (1970). See also Kreger-van Rij (1984).

Organism	Source	Reference
Nadsonia commutata	East Folkland field soil	Golubev (1973)
Pichia lindneri	Forest soil, Germany	Henninger and Windisch (1975)
Pichia methanolica	Soil, Japan	Kato *et al.* (1974)
Pichia mucosa	Forest soil, Illinois; South Africa	Wickerham and Kurtzman (1971)
Pichia philogaea	Grassland soil, South Africa	van der Walt and Johannsen (1975a)
Pichia inositovora	High moor peat, Moscow region, U.S.S.R.	Golubev *et al.* (1981)
Hansenula lynferdii	Grassland soil, South Africa	van der Walt and Johannsen (1975b)
Candida hydrocarbofumarica	Soil, Japan	Yamada *et al.* (1970), Ramirez (1974)
Candida ishiwadae	Core soil sample, 718 mm deep, Japan	Sugiyama and Goto (1969)
Candida podzolica	Soil, U.S.S.R.	Bajeva and Reshetova (1975)
Candida terebra	Core soil sample, 819 mm deep, Japan	Sugiyama and Goto (1969)
Candida paludigena	High moor peat, Moscow region, U.S.S.R.	Golubev *et al.* (1981)
Cryptococcus himalayensis	Soil from Chawa Gassar, Bhutan	Goto and Sugiyama (1970)
Debaryomyces nepalensis	Soil from Nepal	Goto and Sugiyama (1968), Price *et al.* (1978)
Saccharomyces servazzii	Soil in Finland	Capriotti (1967)
Torulopsis navarrensis	Acid-washed brown soil, Spain	Ramirez and Moriyon (1978)
Torulopsis pampelonensis	Beech forest soil, Spain	Ramirez and Martinez (1978)
Torulopsis geochares	Grassland soil near Groblersdal, South Africa	van der Walt *et al.* (1978)
Torulopsis azyma	Grassland soil near Nelspruit, South Africa and from a rupicolous lichen, near Groblersdal, South Africa	van der Walt *et al.* (1978)
Trichosporon terrestre	Grassland soil; Pretoria, South Africa	van der Walt and Johannsen (1975c)

IV. EVOLUTIONARY ECOLOGY OF YEASTS

There have been several recent studies which utilize yeasts in investigations of factors important in the evolutionary ecology of yeasts and yeast communities. Most of these studies have focused on yeasts associated with the habitats of drosophilae and thus include tree flux yeast communities as well as the yeasts found in necroses of cactus stems or cladodes. Common to most analyses of ecological and evolutionary questions in these investigations is the utilization of combined physiological profiles of yeast isolates to characterize the yeast community occupying a particular habitat. The physiological profile of a yeast community results from a compilation of the physiological abilities of the individual yeast species which are found living together. Because yeasts are identified and characterized by studying their physiological abilities and tolerances, the characterization of yeast communities is an extension of this primary information to characterize a higher level of organization. The rationale for using this information is based on the postulate that natural selection has favoured particular physiological traits in the natural habitat where the yeasts were recovered and that those traits can thus be used to relate to and make inferences about that specific environment. In a broad sense, the physiological attributes of the yeasts and yeast communities are useful indicators of habitat characteristics. This conclusion is based on the observation that yeast communities with similar physiological abilities are usually associated with habitats known to be phylogenetically similar (Starmer 1981a). In addition, the dominant carbon sources of the habitats (where known) are often reflected in the physiological profiles of yeasts (Lachance *et al.*, 1982; Starmer, 1982).

These studies have also used the physiological profiles of yeast communities in multivariate comparisons directed at unfolding relevant ecological factors important in the determination of natural yeast distributions. Lachance *et al.* (1982) attempted to discern effects of host plant and geography in their study of yeasts reported from tree fluxes in North America. They found that geographical distance, although a meaningful ecological factor, is largely overshadowed by host tree specificity, provided that yeast community physiological profiles, and not yeast taxa, are used as ecological descriptors. Some yeast communities can have extensive physiological abilities, whereas others appear to be limited to the utilization of only a few carbon compounds and these are often restricted to one class, for example bacterial fermentation products. It appears that evolutionary specialization of yeast communities occupying tree fluxes most likely operates by selection of physiologically limited yeasts, rather than by decreasing the number of yeast species in the different yeast communities (Lachance and Starmer, 1982). This implies that most yeast habitats, tree fluxes in this example, are saturated in terms of the numbers of species occupying them and that this number is relatively constant. Furthermore, as the environment changes, for instance if the habitat itself evolves, the yeasts adapt by changes in physiological organization. In the case of the cactus yeast communities, their evolutionary invasion and subsequent evolutionary changes in the cactus habitat were possibly facilitated by a loss of metabolic activity as an adaptive response to secondary chemicals in the host cactus tissue (Starmer, 1981b).

As already pointed out in this review, these chemicals (triterpene glycosides and unusual lipids) can inhibit yeast growth and thus function to limit the types of yeasts which can live in the decaying tissue. It is interesting to note that the most widespread cactus yeast, *P. cactophila*, has limited physiological abilities, whereas some of the most restricted cactus-specific yeasts (e.g. *P. mexicana*) have extensive metabolic activity. In these cases, it appears that physiological specialization has led to ecological generalization for *P. cactophila.*, and physiological generalization has resulted in ecological specialization in *P. mexciana*. This type of relationship between physiological organization and ecological breadth has been pointed out earlier by McNaughton and Wolf (1970) as a general principle.

In a theoretical study of yeasts associated with columnar cacti, Starmer (1981b) concluded that it is not necessary to invoke competitive interactions between and among yeasts to explain their distributions in the different species of cactus. Instead, the distributions could be adequately explained on the basis of differential physiological abilities and differences in host specificities of the insects which vector the yeasts from host to host. These conclusions are speculative but provide an interesting hypothetical framework for experimentation in the evolutionary ecology of yeasts. It is also possible that in-depth studies on the evolutionary ecology of yeasts and their communities may lead to a better understanding of how and why yeasts become different species.

References

Alcorn, S.M. and May, C. (1962). *Plant Disease Reporter* **46**, 156.

Andrews, J.H., Kenerley, C.M. and Nordheim, E.V. (1980). *Microbial Ecology* **6**, 71.

Atlas, R.M., diMenna, M.E. and Cameron, R.E. (1978). *In* "Terrestrial Biology III, Antarctic Research Series Volume 30" (B.C. Parker, ed.), pp. 27–34. The William Byrd Press, Richmond, Virginia.

Babjeva, I.P. and Blagodatskaya, V.M. (1972). *Mikrobiologya* **41**, 99.

Babjeva, I.P. and Gorin, S.E. (1975). *Antonie van Leeuwenhoek* **41**, 185.

Babjeva, I.P. and Reshetova, I.S. (1975). *Mikrobiologya* **44**, 333.

Babjeva, I.P., Moavad, Kh. and Marchenko, A.I. (1977). *Mikrobiologya* **46**, 270.

Baharaeen, S. and Vishniac, H.S. (1982). *International Journal of Systematic Bacteriology* **32**, 229.

Baharaeen, S., Bantle, J.A. and Vishniac, H.S. (1982). *Canadian Journal of Microbiology* **28**, 406.

Banno, I. and Mikata, K. (1977) *Research Communications of the Institute for Fermentation, Osaka* **8**, 7.

Barker, J.S.F., Toll, G.L., East, P.D., Miranda, M. and Phaff, H.J. (1983). *Canadian Journal of Microbiology* **29**, 6.

Barker, J.S.F., East, P.D., Phaff, H.J. and Miranda, M. (1984). *Microbial Ecology* **10**, 379.

Barnett, J.A. and Buhagiar, R.W.M. (1971). *Journal of General Microbiology* **67**, 233.

Barnett, J.A., Payne, R.W. and Yarrow, D. (1983). "Yeasts: Characteristics and Identification." Cambridge University Press, Cambridge.

Barras, S.J. and Perry, T.J. (1975). "Inter-relationships among Microorganisms, Bark or Ambrosia Beetles, and Woody Host Tissue: an Annotated Bibliography, 1965–1974." United States Department of Agriculture Forest Service General Technical Report SO-10, pp. 1–34.

Batra, L.R. (1959). *Mycologia* **51**, 329.
Batra, L.R. (1966). *Science* **153**, 193.
Batra, L.R. (1967). *Mycologia* **59**, 976.
Batra, L.R., Batra, S.W.T. and Bohart, G.E. (1973). *Mycopathologia et Mycologia Applicata* **49**, 13.
Beech, F.W. and Davenport, R.R. (1970). *In* "Ecology of Leaf Surface Microorganisms" (T.F. Preece and C.H. Dickinson, eds), pp. 139–157. Academic Press, London, New York.
Begon, M. (1981). *In* "The Genetics and Biology of Drosophila" (M. Ashburner, H.L. Carson and J. Thompson, eds), vol. 3B. pp. 345–384. Academic Press, London.
Bélin, J.M. (1972). *Vitis* **11**, 135.
Boidin, J. and Abadie, F. (1954). *Bulletin de la Société Mycologique de France* **70**, 353.
Boidin, J., Pignal, M.C., Mermier, F. and Arpin, M. (1963). *Cahiers de la Maboké* **1**, 86–101.
Bowles, J.M. and Lachance, M.A. (1983). *Canadian Journal of Botany* **61**, 2984.
Buhagiar, R.W.M. (1975). *Journal of General Microbiology* **86**, 1.
Buhagiar, R.W.M. (1979). *Journal of General Microbiology* **110**, 91.
Buhagiar, R.W.M. and Barnett, J.A. (1971). *Journal of Applied Bacteriology* **34**, 727.
Bussey, H. (1981). *Advances in Microbial Physiology* **22**, 93.
Calaham, R.Z. and Shifrine, M. (1960). *Forest Science* **6**, 146.
Capriotti, A. (1967). *Annali di Mirobiologia ed Enzimologia* **17**, 79.
Carson, H.L. (1971) *In* "H.L. Lyon Arboretum Lecture, No. 2", pp. 1–27. University of Hawaii, Honolulu.
Carson, H.L., Knapp, E.P. and Phaff, H.J. (1956). *Ecology* **37**, 538.
Chatton, E.R. (1913). *Comptes Rendues de la Société Biologique* **75**, 117.
Condit, I.J. (1947). "The Fig". Chronica Botanica, Waltham, Massachusetts.
Cox, D.P. and Alexander, M. (1974). *Microbial Ecology* **1**, 136.
Danielson, R.M. and Jurgensen, M.F. (1973). *Mycopathologia et Mycologia Applicata* **51**, 191.
Daugherty, D.M. (1967). *Journal of Economic Entomology* **60** 147.
Daugherty, D.M. and Foster, J.E. (1966). *Journal of Economic Entomology* **59**, 1282.
de Camargo, R. and Phaff, H.J. (1957). *Food Research* **22**, 367.
di Menna, M.E. (1959). *New Zealand Journal of Agricultural Research* **2**, 394.
di Menna, M.E. (1960). *Journal of General Microbiology* **23**, 295.
di Menna, M.E. (1962). *Journal of General Microbiology* **27**, 249.
Deinema, J. (1961). *Mededelingen Landbouwhogeschool, Wageningen* **61** (2), 1.
Dennis, C. and Buhagiar, R.W.M. (1980). *In* "Biology and Activities of Yeasts" (F.A. Skinner, S.M. Passmore and R.R. Davenport, eds), pp. 123–133. Society for Applied Bacteriology Symposium Series No. 9. Academic Press, London.
Dobzhansky, Th., Cooper, D.M., Phaff, H.J., Knapp, E.P. and Carson, H.L. (1956). *Ecology* **37**, 544.
do Carmo-Sousa, L. (1969). *In* "The Yeasts" (A.H. Rose and J.S. Harrison, eds), vol. 1, pp. 79–105. Academic Press, London, New York.
do Carmo-Sousa, L. (1970). *In* "The Yeasts–a Taxonomic Study" (J. Lodder, ed.), 2nd edn, pp. 440–447. North Holland Publishing Co., Amsterdam.
Emmons, C.W. (1951). *Journal of Bacteriology* **62**, 685.
Etchells, J.L., Bell, T.A. and Jones, I.D. (1953). *Farlowia* **4**, 266.
Fawcett, H.S. (1929a). *Phytophathology* **19**, 479.
Fawcett, H.S. (1929b). *Phytophathology* **19**, 1138.
Fell, J.W. and Phaff, H.J. (1970). *In* "The Yeasts–A Taxonomic Study" (J. Lodder, ed.), 2nd edn, pp. 776–802. North Holland Publishing Co., Amsterdam.
Foeckler, F. (1961). *Zeitschrift zur Morphologie und Ökologie der Tiere* **50**, 119.
Fogleman, J.C. (1982). *In* "Ecological Genetics and Evolution — The Cactus–Yeast-Drosophila Model System" (J.S.F. Barker and W.T. Starmer, eds), pp. 191–206. Academic Press Australia, Sydney.
Fogleman, J.C., Starmer, W.T. and Head, W.B. (1981). *Proceedings of the National Academy of Sciences of the United States of America* **78**, 4435.
Fogleman, J.C., Starmer, W.T. and Heed, W.B. (1982). *Oecologia* (*Berlin*) **52**, 187.
Francke-Grosmann, H. (1963). *Annual Review of Entomology* **8**, 415.
Frazer, H.L. (1944). *Annals of Applied Biology* **31**, 271.

Fuson, G.B., Price, C.W. and Phaff, H.J. (1979). *International Journal of Systematic Bacteriology* **29**, 64–69.
Gibson, A.C. and Horak, K.E. (1978). *Annals Missouri Botanical Garden* **65**, 999.
Gilbert, D.G. (1980). *Oecologia (Berlin)* **46**, 135.
Gilliam, M., Wickerham, L.J., Morton, H.L. and Martin, R.D. (1974). *Journal of Invertebrate Pathology* **24**, 349.
Golubev, W.I. (1973). *Mikrobiologiya* **42**, 1058.
Golubev, W.I. (1977). *Antonie van Leeuwenhoek* **43**, 317.
Golubev, W.I., Babjeva, I.P., Blagodatskaya, V.M. and Reshetova, I.S. (1977). *Mikrobiologiya* **46**, 564.
Golubev, W.I., Blagodatskaya, V.M., Suetin, S.O. and Trotsenko, R.S. (1981). *International Journal of Systematic Bacteriology* **31**, 91.
Golubev, W.I., Manukyan, A.R. and Lazarev, P.I. (1984). *Zhurnal Obshchei Biologii* (Journal of General Biology, in Russian) **45**, 507.
Goto, S. and Sugiyama, J. (1968). *Journal of Japanese Botany* **43**, 102.
Goto, S. and Sugiyama, J. (1970). *Canadian Journal of Botany* **48**, 2097.
Gray, T.R.G. (1967). *Science* **155**, 1668.
Grilione, P., Frederici, F. and Miller, M.W. (1981). *In* "Current Developments in Yeast Research" (G.G. Stewart and I. Russell, eds), pp. 599–605. Pergamon Press, Toronto.
Grimont, P.A.D., Grimont, F. and Starr, M.P. (1979). *Current Microbiology* **2**, 277.
Grosmann, H. (1930). *Zeitschrift für Parasitenkunde* **3**, 56.
Hagan, K.S. and Tassan, R.L. (1972). *In* "Insect and Mite Nutrition" (J.G. Rodriguez, ed.), pp. 323–351. North Holland Publishing Co., Amsterdam.
Hagan, K.S., Tassan, R.L. and Sawall, E.F., Jr. (1970). *Bollettino del Laboratorio di Entomolgia Agraria "Filippo Silvestri" di Portici* **28**, 113.
Hajsig, M. (1958). *Antonie van Leeuwenhoek* **24**, 18.
Harrold, C.E. (1950). *Annals of Botany* **54**, 127.
Heed, W.B. (1978). *In* "Ecological Genetics: The Interface" (P.F. Bussard, ed.), pp. 109–126. Springer Verlag, New York.
Heed, W.B. and Kircher, H.W. (1965). *Science* **149**, 758.
Heed, W.B., Starmer, W.T., Miranda, M., Miller, M.W. and Phaff, H.J. (1976). *Ecology* **57**, 151.
Henderson, M.E.K. (1961). *Journal of General Microbiology* **26**, 149
Henninger, W. and Windisch, S. (1975). *Archives of Microbiology* **105**, 47.
Holst, E.C. (1936). *Journal of Agricultural Research* **53**, 513.
Holzschu, D.L. (1981). Ph.D. Thesis: University of California, Davis.
Holzschu, D.L. and Phaff, H.J. (1982). *In* "Ecological Genetics and Evolution–The Cactus-Yeast–Drosophila Model System" (J.S.F. Barker and W.T. Starmer, eds), pp. 127–141. Academic Press Australia, Sydney.
Holzschu, D.L., Phaff, H.J., Tredick, J. and Hedgecock, D. (1983). *Canadian Journal of Microbiology* **29**, 1314.
Holzschu, D.L., Phaff, H.J., Tredick, J. and Hedgecock, D. (1985). *International Journal of Systematic Bacteriology* **35** (3), 457.
Howe, V.K., Oberle, A.D., Keeth, T.G. and Gordon, W.J. (1971). Western Illinois University Bulletin, Series in the Biological Sciences No. 9, vol. L, no. 3, pp. 1–37.
Jacob, F. (1969). *Bulletin Trimestriel de la Société Mycologique de France* **85**, 117.
Jacob, F. (1970). *Annales de l'Institut Pasteur, Paris* **118**, 207.
Jurzitza, G. (1962). *Archiv für Mikrobiologie* **43**, 412.
Jurzitza, G. (1970). *Archiv für Mikrobiologie* **72**, 203.
Jurzitza, G., Kühlwein, H. and Kreger-van Rij, N.J.W. (1960). *Archiv für Mikrobiologie* **36**, 229.
Kato, K., Kurimura, Y., Makiguchi, N. and Asai, Y. (1974). *Journal of General and Applied Microbiology* **20** , 123.
Kearney, J.N. (1982). *Oikos* **39**, 103.
Kearney, J.N. and Shorrocks, B. (1981). *Biological Journal of the Linnean Society* **15**, 39.
Kilbertus, G. and Proth, J. (1979). *Canadian Journal of Microbiology* **25**, 943.
Kircher, H.W. (1982). *In* "Ecological Genetics and Evolution — The Cactus–Yeast–Drosophila Model System" (J.S.F. Barker and W.T. Starmer, eds), pp. 143–158. Academic Press, Australia, Sydney.

Kircher, H.W., Heed, W.B., Russell, J.S. and Grove, J. (1967). *Journal of Insect Physiology* **13** 1869.
Kodama, K. (1974). *Annali di Microbiologia ed Enzymologia* **24**, 215.
Kodama, K. (1975). *Journal of Fermentation Technology* **53**, 626.
Kodama, K. & Kyono, T. (1974a). *Journal of Fermentation Technology* **52**, 1.
Kodama, K. & Kyono, T. (1974b). *Journal of Fermentation Technology* **52**, 605.
Kreger-van Rij, N.J.W. (1964). Ph.D. dissertation, University of Leiden, The Netherlands.
Kreger-van Rij, N.J.W. (1970a). *In* "The Yeasts–a Taxonomic Study" (J. Lodder, ed.), 2nd edn. pp. 455–554. North Holland Publishing Co., Amsterdam.
Kreger-van Rij, N.J.W. (1970b). *In* "The Yeasts–a Taxonomic Study" (J. Lodder, ed.), 2nd edn., pp. 1224–1228. North Holland Publishing Co., Amsterdam.
Kreger-van Rij, N.J.W. (ed.) (1984). "The Yeasts–a Taxonomic Study", 3rd edn. Elsevier Science Publishers, Amsterdam.
Kühlwein, H. and Jurzitza, G. (1961). *Archiv für Mikrobiologie* **40**, 247.
Kurtzman, C.P. (1984). *Mycotaxon* **19**, 271.
Kurtzman, C.P., Smiley, M.J., Johnson, C.J., Wickerham, L.J. and Fuson, G.B. (1980). *International Journal of Systematic Bacteriology* **30**, 208.
Kurtzman, C.P., Phaff, H.J. and Meyer, S.A. (1983). *In* "Yeast Genetics–Fundamental and Applied Aspects" (J.F.T. Spencer, D.M. Spencer and A.R.W. Smith, eds), pp. 139–166. Springer Verlag, New York.
Lachaise, D. (1977). *Oecologia (Berlin)* **31**, 201.
Lachance, M.A. (1982). *Canadian Journal of Microbiology* **28**, 567.
Lachance, M.A., Miranda, M., Miller, M.W. and Phaff, H.J. (1976). *Canadian Journal of Microbiology* **22**, 1756.
Lachance, M.A., Metcalf, B.J. and Starmer, W.T. (1982). *Microbial Ecology* **8**, 191.
Last, F.T. and Price, D. (1969). *In* "The Yeasts" (A.H. Rose and J.S. Harrison, eds), vol. 1, pp. 183–218. Academic Press London, New York.
Lightle, P.C., Standring, E.T. and Brown, J.G. (1942). *Phytopathology* **32**, 303.
Locci, R., Petrolini-Baldan, E.B., Quaroni, S. and Sardi, P. (1977). *Rivista di Patologia Vegetale* **13**, 49.
Lodder, J. (ed.) (1970). "The Yeasts–a Taxonomic Study" 2nd edn. North-Holland Publishing Co., Amsterdam.
Lodder, J. and Kreger-van Rij, N.J.W. (1952). "The Yeasts–a Taxonomic Study". North Holland Publishing Co., Amsterdam.
Lowings, P.H. (1956). *Applied Microbiology* **4**, 84.
Lund, A. (1954). "Studies on the Ecology of Yeasts". Munksgaard, Copenhagen.
Lushbaugh, W.B., Rowton, E.D. and McGhee, R.B. (1976). *Journal of Invertebrate Pathology* **28**, 93.
Martini, A. and Frederici, F. (1976). *Giornale Botanico Italiano* **110**, 297.
Martini, A., Frederici, F. and Rosini, G. (1980). *Canadian Journal of Microbiology* **26**, 856.
McBride, R.P. and Hayes, A.J. (1977). *Transactions of the British Mycological Society* **69**, 39.
McNaughton, S.J. and Wolf, L.L. (1970). *Science* **167**, 131.
Meyer, S.A., Anderson, K., Brown, R.E., Smith, M.Th., Yarrow, D., Mitchell, G. and Ahearn, D. (1975). *Archives of Microbiology* **104**, 225.
Meyer, S.A., Smith, M.T. and Simione, F.P., Jr. (1978). *Antonie van Leeuwenhoek* **44**, 79.
Middelbeek, E.J., Stumm, C. and Vogels, G.D. (1980). *Antonie van Leeuwenhoek* **46**, 205.
Miller, J.J. and Webb, N.S. (1954). *Soil Science* **77**, 197.
Miller, M.W. and Barker, E.R. (1968). *Antonie van Leeuwenhoek* **34**, 183.
Miller, M.W. and Mrak, E.M. (1953). *Applied Microbiology* **1**, 174.
Miller, M.W. and Phaff, H.J. (1962). *Applied Microbiology* **10**, 394.
Miller, M.W., Phaff, H.J. and Snyder, H.E. (1962). *Mycopathologia et Mycologia Applicata* **16**, 1.
Miller, M.W., Yoneyama, M. and Soneda, M. (1976a). *International Journal of Systematic Bacteriology* **26**, 286.
Miller, M.W., Phaff, H.J., Miranda, M., Heed, W.B. and Starmer, W.T. (1976b). *International Journal of Systematic Bacteriology* **26**, 88.

Miranda, M., Holzschu, D.L., Phaff, H.J. and Starmer, W.T. (1982). *International Journal of Systematic Bacteriology* **323**, 101.
Montrocher, R. (1967). *Revue de Mycologie* **32**, 69.
Müller, W. (1934). *Archiv für Mirobiologie* **5**, 84.
Munitis, M.T., Cabrera, E. and Rodriguez-Navarro, A. (1976). *Applied and Environmental Microbiology* **32**, 320.
Murray, N.D. (1982), *In* "Ecological Genetics and Evolution — The Cactus–Yeast–Drosophila Model System" (J.S.F. Barker and W.T. Starmer, eds), pp. 17–29. Academic Press, Australia, Sydney.
Nieuwdorp, P.J., Bos, P. and Slooff, W.C. (1974), *Antonie van Leeuwenhoek* **40**, 241.
Northam, B.E. and Norris, F.W. (1951). *Journal of General Microbiology* **5**, 502.
Onishi, H. (1972). *Antonie van Leeuwenhoek* **38**, 365.
Pant, N.C. and Fraenkel, G. (1950). *Science* **112**, 498.
Phaff, H.J. (1955). *The American Naturalist* **89**, 53.
Phaff, H.J. (1956). *Antonie van Leeuwenhoek* **22**, 113.
Phaff, H.J. (1970). *In* "The Yeasts–a Taxonomic Study" (J. Lodder, ed.), pp. 756–766. North Holland Publishing Co., Amsterdam.
Phaff, H.J. (1981). *In* "Current Developments in Yeast Research" (G.G. Stewart and I. Russell, eds), pp. 635–643. Pergamon Press, Toronto.
Phaff, H.J. and Fell, J.W. (1970). *In* "The Yeasts–a Taxonomic Study" (J. Lodder, ed.), 2nd edn., pp. 1088–1145. North Holland Publishing Co., Amsterdam.
Phaff, H.J. and Holzschu, D.L. (1981). *In* "Molecular Genetics in Yeast" (D. von Wettstein, J. Friis, M. Kielland-Brandt and A. Stenderup, eds), Alfred Benson Symposium No. 16, pp. 413–427. Munksgaard, Copenhagen.
Phaff, H.J. and Knapp, E.P. (1956). *Antonie van Leeuwenhoek* **22**, 117.
Phaff, H.J. and Miller, M.W. (1961). *Journal of Insect Pathology* **3**, 233.
Phaff, H.J. and Starmer, W.T. (1980). *In* "Biology and Activities of Yeasts" (F.A. Skinner, S.M. Passmore and R.R. Davenport, eds), pp. 79–102. Society for Applied Bacteriology Symposium Series No. 9. Academic Press, London.
Phaff, H.J. and Yoneyama, M. (1961). *Antonie van Leeuwenhoek* **27**, 196.
Phaff, H.J., Miller, M.W. and Shifrine, M. (1956a). *Antonie van Leeuwenhoek* **22**, 145.
Phaff, H.J., Miller, M.W., Recca, J.A., Shifrine, M. and Mrak, E.M. (1956b). *Ecology* **37**, 533.
Phaff, H.J., Yoneyama, M. and do Carmo-Sousa, L. (1964a). *Revista di Patologia Vegetale* **4**, 485.
Phaff, H.J., Miller, M.W. and Spencer, J.F.T. (1964b). *Antonie van Leeuwenhoek* **30**, 132.
Phaff, H.J., Miller, M.W., Yoneyama, M. and Soneda, M. (1972). *In* "Fermentation Technology Today: Proceedings IV IFS" (G. Terui, ed.), pp. 759–774. Society of Fermentation Technology, Osaka, Japan.
Phaff, H.J., Miller, M.W., Miranda, M., Heed, W.B. and Starmer, W.T. (1974). *International Journal of Systematic Bacteriology* **24**, 486.
Phaff, H.J., Miller, M.W. and Miranda, M. (1976). *International Journal of Systematic Bacteriology* **26**, 326.
Phaff, H.J., Miller, M.W. and Mrak, E.M. (1978a). "The Life of Yeasts", 2nd edn. Harvard University Press, Cambridge, Massachusetts.
Phaff, H.J., Starmer, W.T., Miranda, M. and Miller, M.W. (1978b). *International Journal of Systematic Bacteriology* **28**, 326.
Phaff, H.J., Miller, M.W. and Miranda, M. (1979). *International Journal of Systematic Bacteriology* **29**, 60.
Phaff, H.J., Starmer, W.T., Miranda, M. and Miller, M.W. (1980). *International Journal of Systematic Bacteriology* **30**, 596.
Phaff, H.J., Yamada, Y., Tredick, J. and Miranda, M. (1983). *International Journal of Systematic Bacteriology* **33**, 375.
Phaff, H.J., Starmer, W.T., Tredick, J. and Miranda, M. (1985). *International Journal of Systematic Bacteriology* **35**, 211.
Pignal, M.-C. (1970). *Antonie van Leeuwenhoek* **36**, 525.

Pignal, M.-C. (1971a). *Cahiers de la Maboké* **9**, 95.
Pignal, M.-C. (1971b). *Naturaliste Canadien* **98**, 69.
Pignal, M.-C. (1973). *Annales de Zoologie — Ecologie Animale* **5**, 325.
Pitt, J.I. and Miller, M.W. (1968). *Mycologia* **60**, 663.
Price, C.W., Fuson, G.B. and Phaff, H.J. (1978). *Microbiological Reviews* **42**, 161.
Ramirez, C. (1974). *Microbiologia Espagñola* **27**, 15.
Ramirez, C. and Martinez, A.T. (1978). *Canadian Journal of Microbiology* **24**, 433.
Ramirez, C. and Moriyon, I. (1978). *Mycopathologia* **63**, 61.
Rodrigues de Miranda, L. (1978). *Antonie van Leeuwenhoek* **44**, 439.
Rosini, G., Frederici, F. and Martini, A. (1982). *Microbial Ecology* **8**, 83.
Ruinen, J. (1961). *Plant and Soil* **15**, 81.
Ruinen, J. (1963). *Antonie van Leeuwenhoek* **29**, 425.
Rumbold, C.T. (1931). *Journal of Agricultural Research* **43**, 847.
Rumbold, C.T. (1941). *Journal of Agricultural Research* **62**, 589.
Santa Maria, J. and Garcia Aser, C. (1970). *Boletin del Instituto Nacional de Investigaciones Agronomicas (Madrid)* **62**, 51.
Santa Maria, J. and Garcia Aser, C. (1977). *Anales del Instituto Nacional de Investigaciones Agrarias. Serie: General* **5**, 45.
Scott, D.B. and van der Walt, J.P. (1970a). *Antonie van Leeuwenhoek* **36**, 389.
Scott, D.B. and van der Walt, J.P. (1970b). *Antonie van Leeuwenhoek* **36**, 45.
Scott, D.B. and van der Walt, J.P. (1971). *Antonie van Leeuwenhoek* **37**, 171.
Shehata, A.M. E-T. and Mrak, E.M. (1952). *Evolution* **6**, 325.
Shehata, A.M. E-T., Mrak, E.M. and Phaff, H.J. (1955). *Mycologia* **47**, 799.
Shifrine, M. and Phaff, H.J. (1956). *Mycologia* **48**, 41.
Slooff, W.C. (1970). *In* "The Yeasts–a Taxonomic Study" (J. Lodder, ed.), 2nd edn., pp. 379–402. North Holland Publishing Co., Amsterdam.
Smith, M.T., Simione, F.P. and Meyer, S.A. (1977). *Antonie van Leeuwenhoek* **43**, 219.
Soneda, M. (1957). *The Journal of Japanese Botany* **32**, 347.
Spencer, J.F.T., Gorin, P.A.J. and Tulloch, A.P. (1970). *Antonie van Leeuwenhoek* **36**, 129.
Starkey, R.L. (1946). *Journal of Bacteriology* **51**, 33.
Starmer, W.T. (1981a). *In* "Current Developments in Yeast Research" (G.G. Stewart and I. Russell, eds), pp. 493–498. Pergamon Press, Toronto.
Starmer, W.T. (1981b). *In* "The Fungal Community–its Organization and Role in the Ecosystem" (D.T. Wicklow and G.C. Carroll, eds), pp. 129–156. Marcel Dekker, New York, Basel.
Starmer, W.T. (1982). *In* "Ecological Genetics and Evolution — The Cactus–Yeast–Drosophila Model System" (J.S.F. Barker and W.T. Starmer, eds), pp. 159–174. Academic Press Australia, Sydney.
Starmer, W.T. and Phaff, H.J. (1983). *Microbial Ecology* **9**, 247.
Starmer, W.T., Heed, W.B., Miranda, M., Miller, M.W. and Phaff, H.J. (1976). *Microbial Ecology* **3**, 11.
Starmer, W.T., Phaff, H.J., Miranda, M. and Miller, M.W. (1978a). *International Journal of Systematic Bacteriology* **28**, 318.
Starmer, W.T., Phaff, H.J., Miranda, M. and Miller, M.W. (1978b). *International Journal of Systematic Bacteriology* **28**, 433.
Starmer, W.T., Phaff, H.J., Miranda, M., Miller, M.W. and Barker, J.S.F. (1979). *International Journal of Systematic Bacteriology* **29**, 159.
Starmer, W.T., Kircher, H.W. and Phaff, H.J. (1980). *Evolution* **34**, 137.
Starmer, W.T., Phaff, H.J., Miranda, M., Miller, M.W. and Heed, W.B. (1982). *In* "Evolutionary Biology" (M.K. Hecht, B. Wallace and C.T. Prance, eds), vol. 14, pp. 269–295. Plenum Press, London.
Starmer, W.T., Phaff, H.J., Tredick, J., Miranda, M. and Aberdeen, V. (1984). *International Journal of Systematic Bacteriology* **34**, 350.
Steinhaus, E.A. (1946). "Insect Microbiology". Comstock Publishing Co., Ithaca, New York.
Sugiyama, J. and Goto, S. (1969). *Journal of the Faculty of Science, Tokyo University, Section 3, Botany* **10**, 97.

Travassos, L.R. and Cury, A. (1971). *Annual Review of Microbiology* **25**, 49.
Vacek, D.C., Starmer, W.T. and Heed, W.B. (1979). *Microbial Ecology* **5**, 43.
Vacek, D.C., East, P.D., Barker, J.S.F. and Soliman, M.H. (1984). *Biological Journal of the Linnean Society of London* **24**, 175.
van der Walt, J.P. (1966). *Antonie van Leeuwenhoek* **32**, 159.
van der Walt, J.P. (1972a). *In* "Proceedings IVth Inernational Fermentation Symposium: "Fermentation Technology Today" (G. Terui, ed.), pp. 797–803. Society of Fermentation Technology, Osaka, Japan.
van der Walt, J.P. (1972b). *Mycopathologia et Mycologia Applicata* **46**, 305.
van der Walt, J.P. (1982a). *Antonie van Leeuwenhoek* **48**, 383.
van der Walt, J.P. ()1982v). *Antonie van Leeuwenhoek* **48**, 465.
van der Walt, J.P. and Johannsen, E. (1975a). *Antonie van Leeuwenhoek* **41**, 173.
van der Walt, J.P. and Johannsen, E. (1975b). *Antonie van Leeuwenhoek* **41**, 13.
van der Walt, J.P. and Johannsen, E. (1975c). *Antonie van Leeuwenhoek* **41**, 361.
van der Walt, J.P. and Scott, D.B. (1971a). *Antonie van Leewenhoek* **37**, 15.
van der Walt, J.P. and Scott, D. B. (1971b). *Mycopathologia et Mycologia Applicata* **44**, 101.
van der Walt, J.P. and Scott, D.B. (1971c). *Mycopathologia et Mycologia Applicata* **44**, 321.
van der Walt, J.P. and van der Klift, W.C. (1972). *Antonie van Leeuwenhoek* **38**, 361.
van der Walt, J.P., Scott, D.B. and van der Klift, W.C. (1971a). *Antonie van Leeuwenhoek* **37**, 449.
van der Walt, J.P., Scott, D.B. and van der Klift, W.C. (1971b). *Antonie van Leeuwenhoek* **37**, 461.
van der Walt, J.P., Scott, D.B. and van der Klift, W.C. (1972). *Mycopathologia et Mycologia Applicata* **47**, 221.
van der Walt, J.P., Johannsen, E. and Yarrow, D. (1978). *Antonie van Leeuwenhoek* **44**, 97.
van der Walt, J.P., Weijman, A.C.M. and von Arx, J.A. (1981). *Sydowia, Annales Mycologici Series II* **34**, 191.
van der Walt, J.P., Yarrow, D., Opperman, A. and Halland, L. (1982). *Journal of General and Applied Microbiology*, **28**,155.
van Uden, N. and Buckley, H. (1970). *In* "The Yeasts–a Taxonomic Study" (J. Lodder, ed.), 2nd edn., pp. 893–1087. North Holland Publishing Co., Amsterdam.
Vishniac, H.S. and Baharaeen, S. (1982). *International Joural of Systematic Bacteriology* **32**, 437.
Vishniac, H.S. and Hempfling, W.P. (1979a). *International Journal of Systematic Bacteriology* **29**, 153.
Vishniac, H.S. and Hempfling, W.P. (1979b). *Journal of General Microbiology* **112**, 301.
Vustin, M.M. and Babjeva, I.P. (1981). *Mikrobiologiya* **50**, 1088.
Wallace, G.B. (1939). *East African Agricultural Journal* **4**, 268.
Weber, G.F. (1933a). *Phytopathology* **23**, 384.
Weber, G.F. (1933b). *Phytopathology* **23**, 1000.
Wessels, J.G.H. and Sietsma, J.H. (1981). *In* "Plant Carbohydrates II, Extracellular Carbonhydrates" (W. Tanner and F.A. Loewus, eds), Encyclopedia of Plant Physiology, New Series 13B, pp. 352–394. Springer-Verlag, Berlin.
Wickerham, L.J. (1960). *Mycologia* **52**, 171.
Wickerham, L.J. (1965). *Mycopathologia et Mycologia Applicata* **26**, 87.
Wickerham, L.J. (1970). *In* "The Yeasts–a Taxonomic Study" (J. Lodder, ed.), 2nd ed., pp. 226–315. North Holland Publishing Co., Amsterdam.
Wickerham, L.J. and Kurtzman, C.P. (1971). *Mycologia* **63**, 1013.
Wingard, S.A. (1922). *Phytopathology* **12**, 525.
Wyss, O., Ludwig, B.J. and Joiner, R.R. (1945). *Archives of Biochemistry* **7**, 415.
Yamada, K., Furukawa, T. and Nakahara, T. (1970). *Agricultural and Biological Chemistry* **34**, 670.
Yarrow, D. and Meyer, S.A. (1978). *International Journal of Systematic Bacteriology* **28**, 318.
Yoneyama, M. (1957a). *Journal of Science of Hiroshima University, Series B, Division 2* (*Botany*) **8**, 19.
Yoneyama, M. (1957b). *Botanical Magazine* (*Tokyo*) **70**, 92.

6 Ecology of Aquatic Yeasts

Allen N. Hagler and D.G. Ahearn

Departmento de Microbiologia Geral, Instituto de Microbiologia, Universidade Federal do Rio de Janeiro, Rio de Janeiro, Brazil, and Department of Biology, Georgia State University, Atlanta, Georgia, U.S.A.

I. INTRODUCTION

Prior to extensive reports by Russian microbiologists on the wide-spread distribution of yeasts in the seas (Kriss *et al.*, 1952, 1967; Kriss and Novozhilova, 1954; Kriss, 1959), the literature contained only occasional reference to yeasts in aquatic habitats (ZoBell, 1946; Johnson and Sparrow, 1961). Subsequently yeasts were shown to be of common occurrence in rivers

The Yeasts Vol. 1, 2nd edition
ISBN 0-12-596411-0

(Spencer *et al.*, 1969, 1970, 1974a; Simard and Blackwood, 1971a,b; Viviani and Tortorano, 1976, 1976; Hinzelin and Lectard, 1978), lakes (van Uden and Ahearn, 1963; Hedrick *et al.*, 1964, 1968; Hedrick and Soyugenc, 1967; Meyers *et al.*, 1970a; Wollett and Hedrick, 1970; Spencer *et al.*, 1974a; Hagler and Mendoca-Hagler, 1978), estuaries (van Uden, 1967) and oceans (Fell and van Uden, 1963; van Uden and Fell, 1968; Morris, 1968, 1975; Fell, 1974, 1976; Kohlmeyer and Kohlmeyer, 1979).

A. Sample collection

The methods of sample collection and yeast isolation and identification have varied somewhat with environment, chronology and the investigator. Surface samples from shallow waters often have been collected in sterile bottles either opened in the atmosphere and submerged, or submerged and opened just below the surface. For the isolation of yeasts from surface slicks, Crow *et al.* (1975, 1976) adsorbed the films to sterile polycarbonate membranes floated on the water. Deep waters were collected initially with non-sterile samplers such as the van Dorn sampler or the Nansen bottle. Early reports do not always state if the non-sterile samplers were disinfected before use. Later studies employed sterile samplers such as the Niskin biosampler (see van Uden and Fell, 1968 and Fell, 1976 for further comments on deep-water samplers). Generally, shallow-water sediments have been scooped up with a dredge or by hand with a sterile container, whereas deep-water sediments have been collected mostly with non-sterile core samplers. There have been relatively few studies on yeasts in aquatic sediments.

Since yeasts are usually in concentrations of only a few cells per litre in open and deep waters, most water samples have been filtered and the yeast cells have been collected on membranes for isolation and enumeration. Kriss (1959) reported that direct observations of yeasts on membrane filters indicated higher frequencies of yeasts in the samples than could be shown with culture procedures. Enrichment, most probable number or indicated number methods have been used in studies of polluted water and sewage (Cooke, 1965; Rand *et al.*, 1976; Hagler *et al.*, 1981). Surface films usually have required dilution for isolation of yeasts because of their dense microbial populations. Sediments, and plant and animal materials have been rinsed and the rinse waters filtered or introduced into enrichment culture. Regardless of their source, samples should be processed immediately or transported to the laboratory under refrigeration and processed within a few hours. Short periods of refrigeration are not known to affect populations of yeasts in waters, but sediments stored for periods of six weeks or more may demonstrate yeasts not observed in routine initial sampling (D.G. Ahearn, unpublished observations).

B. Isolation media

No standardized medium exists for the isolation of yeasts from aquatic habitats. All media are selective, and a given medium that yields maximal yeast counts for one sample is not necessarily optimal for samples from other sites. Furthermore, yeasts are subject to competition and crowding on isolation media by more rapidly reproducing bacterial populations, and to inhibition by products of bacterial metabolism; moulds, usually less numerous than yeasts in waters remote from land, may still obscure slowly developing yeast colonies with their spreading growth.

For the isolation of yeasts, most investigators have employed complex organic media containing ingredients such as malt extract, yeast extract and peptones in combination with glucose (Kriss and Novozhilova, 1954; Rozanova and Novozhilova, 1958; Capriotti, 1962a,b; Fell and van Uden, 1963; van Uden and Castelo-Branco, 1963; Hedrick *et al.*, 1964; Ahearn *et al.*, 1968; Cook, 1970; Spencer *et al.*, 1970; Goto *et al.*, 1972; Buck, 1976; Niewolak, 1976; Hagler and Mendonca-Hagler, 1981). In contrast to the situation with bacteria, there is no evidence that nutritionally limited media such as water- and soil-extract agar are effective for isolation of yeasts. The high nutrient concentration and inherently low pH value of the complex yeast isolation media are not favourable to the growth of most bacteria.

The optimum pH values for the growth of most yeasts are near neutrality or slightly in the acid range; in general, yeasts are more tolerant of acid conditions than are bacteria. Phaff *et al.* (1978) have noted that most known yeasts grow well near pH 3 as long as an inorganic acid is used for acidification, because organic acids may inhibit yeast growth. Some researchers feel that the use of antibiotics eliminates the necessity for acidification to inhibit bacterial growth (Seshadri and Sieburth, 1971). However, chemically unstable antibiotics such as penicillin may not remain active in yeast isolation media for sufficient time to be effective, especially in acid conditions. Antibiotics such as streptomycin or oxytetracycline also may be toxic for certain yeasts (Richards and Elliott, 1966; Engelbrecht *et al.*, 1974). Consequently, chloramphenicol has been the antibiotic of choice in most studies. It is stable in acid conditions and can be autoclaved in acid media without loss in activity. Different concentrations of chloramphenicol (50–500 $mg\,l^{-1}$) have been recommended by different workers, but high concentrations may inhibit yeasts (Seshadri and Sieburth, 1971; Buck, 1975). The concentration of antibacterial agent required depends on the nature and density of the bacterial population; therefore new environments, particularly polluted sites, must be screened with media containing various concentrations and types of antibacterial agents.

In general, mould growth, compared with that of yeasts, is more sensitive to rose bengal, organic acids and low temperature (Cooke, 1976; Fell, 1976), but the selective isolation of yeasts from samples containing dense mould populations is difficult. Simard and Blackwood (1971a) passed water samples through sterile filter paper before concentrating yeasts on a membrane filter. This procedure may remove mould growth, but also may remove filamentous forms of yeasts and yeasts attached to larger particles without retarding the passage of some common mould conidia. When grown in a shaken enrichment culture, moulds often form dense mycelial balls which can be filtered out or allowed to settle, whereas yeasts remain suspended longer in the broth. Densities of yeasts can be calculated under such circumstances by an "indicated number" or a "most probable number" method (Cooke, 1965).

Addition of sea water does not appear necessary for growth of marine yeasts, although Fell *et al.* (1960) obtained higher yeast counts from marine samples when sea water was added to the media. However, addition of sea water, or sodium chloride as substitute, has the favourable effect of repressing growth of many terrestrial bacteria without notable influence on yeast counts. Also sea water may be diluted with distilled water without appreciable effect on indigenous yeasts, whereas the lowered salinity suppresses the growth of marine bacteria. Schrank *et al.* (1979a) noted an apparent synergistic inhibition of marine bacteria by sodium chloride and chloramphenicol.

C. Identification

In spite of the use of media with multiple selective parameters, certain bacteria from aquatic habitats may develop on yeast isolation media (Schrank *et al.*, 1979a,b). When growing on solid media, most bacterial colonies are more translucent than yeast colonies. This difference, however, is not so obvious when colonies are growing on membrane filters, on opaque media or are embedded in solid media. The use of dark-green or black membrane filters does not decrease yeast counts, and may aid in distinguishing the more transparent bacterial colonies from yeast colonies (Buck, 1978). In addition, a dissecting microscope for counting colonies on filters may help in distinguishing bacteria from yeasts. Subculturing of colonies from membranes onto the surrounding agar surface, where yeasts and bacteria may be distinguished more easily, also has been employed (van Uden and Castelo-Branco, 1963).

There has been a constant evolution of yeast taxonomy during the last three decades, when most of the work on aquatic yeasts has been done; recent advances in methodology have indicated the necessity of further modifications in classification. Although many of the more common ascomycetous species are reasonably well defined, often relationships are poorly defined among the fungi imperfecti where anamorphs of both basidiomycetous and ascomycetous species have been included in the same genus. Many isolates from aquatic habitats, particularly of the less common species, likely have been misidentified on occasion. However, this was probably more of a problem when classification was based on relatively few tests, as was true before the general use of Lodder's (1970) second edition of "The Yeasts". Yeasts classified in species, according to this treatise, generally have a high degree of phenotypic similarity, regardless of phylogenetic validity. The nomenclature used in this review has been updated to binomials in current use (Lodder, 1970; Barnett *et al.*, 1979) and, where possible, has been made consistent with the third edition of "The Yeasts" (Kreger van Rij, 1984).

II. DISTRIBUTION OF YEASTS IN AQUATIC ENVIRONMENTS

A. Water

Yeast populations usually are highest in fresh waters, and decrease in marine waters with increased depth and increased distance from land; however, open or deep waters associated with plankton blooms, current boundaries, surface slicks, thermoclines, or with pollution, may show densities of yeasts above 10^3 cells l^{-1} (van Uden and Fell, 1968; Ahearn, 1973). The lowest yeast populations typically are those in open ocean water where yeast counts below 10 l^{-1} are common (Kriss *et al.*, 1952, 1967; Kriss and Novozhilova, 1954; Capriotti, 1962b; Shinano, 1962; Siepmann and Höhnk, 1962; Roth *et al.*, 1962; Fell and van Uden, 1963; Fell, 1967, 1974, 1976; Meyers *et al.*, 1967a,b; Ahearn *et al.*, 1968; Goto *et al.*, 1972, 1974a,b; Yamasato *et al.*, 1974; Ahearn and Crow, 1980). Open waters of clean lakes usually contain below 100 yeasts l^{-1} (Rozanova and Novozhilova, 1958; van Uden and Ahearn, 1963; Hedrick *et al.*, 1964; Hedrick and Soyugenc, 1965, 1967; Ahearn *et al.*, 1968, 1969; Cooke, 1970; Meyers *et al.*, 1970a; Ahearn, 1973; Spencer *et al.*, 1974a). Similar values are found for waters from unpolluted coastal marine areas (Bhat and Kachwalla, 1955; Fell *et al.*, 1960; Roth *et al.*, 1962; van Uden and Castelo-Branco, 1963;Taysi and van Uden, 1964; Norkrans, 1966b; Combs *et al.*, 1971; Hoppe, 1972a,b; De Queiroz and Macedo, 1972; Ahearn, 1973; Colwell *et al.*, 1976; Hagler and Mendonca-Hagler, 1981). Mesotrophic lakes, rivers and coastal marine waters typically have total yeast counts in the range of 100–500 l^{-1} (see Table I). Water which is polluted but is still considered safe for recreational use often has total yeast counts of 500–1000 l^{-1}, and eutrophic waters characteristically have total yeast counts of more than 10^3 l^{-1} (Artagaveytia-Allende, 1960; Capriotti, 1962a; Roth *et al.*, 1962; Fell and van Uden, 1963; Taysi and van Uden, 1964; Artagaveytia-Allende and Teira, 1965; van Uden, 1967; Spencer *et al.*, 1969, 1970; Meyers *et al.*, 1970a; Wollett and Hedrick, 1970; Wollett *et al.*, 1970; Combs *et al.*, 1971; Simard and Blackwood, 1971a,b; Hoppe, 1972a,b; Luehrs, 1972; Ahearn, 1973; Ahearn and Meyers, 1976; Hinzelin and Lectard, 1978; Hagler *et al.*, 1979, 1981; Vaatanmen, 1980a; Hagler and Mendonca-Hagler, 1981) (see Table 2). Aquatic microcosms (e.g. surface films and plankton blooms) other than polluted areas also have been associated with yeast counts above 10^3 l^{-1} (Meyers *et al.*, 1967a; Crow *et al.*, 1976). Yeast counts in sewage are highly variable but are much higher than those found in waters: counts up to $2 \cdot 10^8$ yeasts l^{-1} have been reported, and 10^5 l^{-1} are common for domestic sewage (Cooke *et al.*, 1960; Cooke and Matsuura, 1963, 1969; Cooke, 1965, 1970; Hinzelin and Lectard, 1979). Spencer *et al.* (1974a) found that sewage entering a pulp-mill basin contained an average yeast content of $2.6 \cdot 10^5$ l^{-1}. Yeast numbers have been observed to decline during the progressive stages of sewage treatment (Cooke, 1965; Spencer *et al.*, 1974b).

Table I. Yeast counts in waters of the south-east U.S.A. and the State of Rio de Janeiro, Brazil. From Ahearn (1973) and Hagler and Mendonca-Hagler (1981)

	Fresh water		Estuarine		Marine	
	Rivers	Lakes	Urban	Rural	Littoral	Off-shore
Number of samples	73	106	236	101	131	230
Proporation containing yeast (%)	93	78	95	90	90	73
Yeast counts (cells l^{-1})[a]	448	113	2773	102	53	10

[a] Geometric mean

Table II. Total coliform and yeast populations in Brazilian subtropical estuarine and littoral marine waters with different pollution levels. From Santos *et al.* (1975), Hagler *et al.* (1981a) and Hagler and Mendonca-Hagler (1981)

Pollution level	Coliforms		Yeasts	
	Cells l^{-1}[a]	Number of samples	Cells l^{-1}[a]	Number of samples
Low	$5.2 \cdot 10^3$	59	$2.5 \cdot 10^2$	74
Medium	$2.2 \cdot 10^5$	53	$1.7 \cdot 10^3$	121
High	$3.1 \cdot 10^7$	57	$4.8 \cdot 10^4$	136

[a] Geometric mean

An extensive spectrum of yeasts has been isolated from water, but only a limited number of species is consistently prevalent (Kohlmeyer and Kohlmeyer, 1979). The most frequently reported species occurring in high densities in water are listed in Table 3.

Red yeasts, common in most aquatic yeast studies, often comprise more than 50% of the yeast population, especially in oligotrophic marine or fresh water (Cooke *et al.*, 1960; Ahearn *et al.*, 1962, Roth *et al.*, 1962; Cooke and Matsuura, 1963; Fell and van Uden, 1963; van Uden and Ahearn, 1963; Taysi and van Uden, 1964; Hedrick and Soyugenc, 1965, 1967; Ahearn and Roth, 1966; Fell, 1967; van Uden and Fell, 1968; Spencer *et al.*, 1970, 1974a; Wollett and Hedrick, 1970; Simard and Blackwood, 1971a; Hoppe, 1972b; Ahearn, 1973; Hinzelin and Lectard, 1976, 1978; Schneider, 1977; Hagler and Mendonca-Hagler, 1979a, 1981). Most red yeast isolates represent the genus *Rhodotorula*, but species of *Rhodosporidium* and *Sporobolomyces* also occur. *Sporobolomyces* is a genus of oxidative basidiomycetous red yeasts which form ballistospores and is most often associated with the phyllosphere of terrestrial plants. Yeasts of this genus probably enter aquatic environments with run-off from foliage (Ahearn *et al.*, 1962, 1968; Ahearn and Roth, 1966; Fell, 1967; Hedrick and Soyugenc, 1967; Ahearn, 1973; Hinzelin *et al.*, 1979). Pugh and Lindsay (1975) found *Sporobolomyces* spp. to be washed from marsh grasses by tidal action. The red yeasts are primarily non-fermentative fungi of basidiomycetous affinity. They utilize a wide range of compounds as carbon sources, are mostly mesophilic, but grow relatively well at temperatures below 15°C. *Rhodotorula rubra* and *Rh. glutinis* are the most frequently isolated species. Species of the genus *Rhodosporidium* are teleomorphs of some *Rhodotorula* species, and often may be identified as rhodotorulas in ecological studies (Fell *et al.*, 1973; Fell and Tallman, 1980). The genus *Rhodosporidium* has a life cycle similar to that of smuts, which suggests an association with marine vegetation.

The black yeasts are isolated frequently, but often at low densities, from all types of water, and usually are represented by the ill-defined species *Aureobasidium pullulans* (Fell and van Uden, 1963; van Uden and Ahearn, 1963; van Uden and Castelo-Branco, 1963; Roth *et al.*, 1964; Fell, 1967; Meyers *et al.*, 1967b, 1970a; Wollett and Hedrick, 1970; Combs *et al.*, 1971; Simard and Blackwood, 1971a). On yeast-isolation media colonies are initially yeast-like, cream- to red- to green-coloured with an entire border; later, black pigments and often rhizoid borders may develop. *Aureobasidium pullulans* normally is associated with foliage and soil and probably, like *Sporobolomyces*, enters water with run-off.

The cosmopolitan yeast *Debaryomyces hansenii* and its anamorph *Candida famata* (*Torulopsis candida*) are ubiquitous in fresh and marine waters.

Table III. Prevalent yeasts in water: numbers of references in the literature cited that give a high incidence or density for the species. (Number of selected citations in parenthesis)

Species	Fresh water			Salt water		
	Clean (11)	Polluted (13)	Sewage (5)	Polluted estuarine (10)	Littoral (9)	Offshore (14)
Aureobasidium pullulans	2	3	–	–	2	4
Candida diddensii	–	–	–	–	–	5
Candida glabrata	–	4	–	–	–	–
Candida guilliermondii	4	4	3	5	7	4
Candida krusei	3	7	3	7	–	–
Candida intermedia	–	2	2	–	–	–
Candida lambica	–	2	–	–	–	–
Candida oregonensis	2	–	–	–	–	–
Candida parapsilosis	4	6	3	7	6	6
Candida polymorpha	–	–	–	–	–	2
Candida reukaufii	2	2	–	–	–	–
Candida tropicalis	–	4	3	6	4	2
Candida valida	–	5	–	2	–	–
Cryptococcus albidus	11	8	2	3	4	6
Cryptococcus laurentii	7	3	–	3	2	5
Cryptococcus infirmo-miniata	–	–	–	–	–	2
Debaryomyces hansenii (*Candida famata, Torulopsis candida*)	3	6	3	5	3	11
Geotrichum capitatum	–	3	–	–	–	–
Hansenula anomala	–	2	–	2	2	–
Hanseniaspora uvarum (*Kloerckera apiculata*)	–	4	–	–	–	–
Metschnikowia pulcherrima (*Candida pulcherrima*)	2	–	–	–	–	–
Prototheca spp.	–	3	–	–	–	–
Rhodotorula aurantiaca	3	–	–	–	–	–
Rhodotorula glutinis	8	10	4	6	3	11
Rhodotorula graminis	–	2	–	–	–	–
Rhodotorula minuta	–	–	–	2	4	
Rhodotorula rubra	9	9	5	7	5	13
Saccharomyces cerevisiae	–	3	2	3	–	–
Saccharomyces exiguus	–	2	–	–	–	–
Sporobolomyces salmonicolor	2	–	–	–	–	3
Trichosporon beigelii (*Trichosporon cutaneum*)	–	3	3	4	–	–

Debaryomyces hansenii is the most common ascomycete in marine waters (Cooke *et al.*, 1960; Capriotti, 1962a,b; Cooke and Matsuura, 1963; Taysi and van Uden, 1964; Norkrans, 1966a; Fell, 1967, 1974; Meyers *et al.*, 1967a,b; Ahearn *et al.*, 1968; van Uden and Fell, 1968; Wollett and Hedrick, 1970; Simard and Blackwood, 1971a; Hoppe, 1972b; Ahearn, 1973). Its homothallic character and mode of ascospore formation (mother–daughter cell conjugation), broad salinity tolerance and ability to utilize a wide range of carbon sources probably are factors related to its widespread distribution. Fermentative strains which neither form ascospores nor assimilate lactose may be confused with other species, including non-filamentous strains of *C. guilliermondii*.

Yeasts of the *C. parapsilosis* group are isolated commonly from fresh and marine waters (Fell and Meyer, 1967). *Candida parapsilosis*, a species associated with man, occurs in sewage and persists in fresh and marine waters, whereas *C. diddensii*, a phenotypically similar yeast distinguished by assimilation of cellobiose and erythritol, is more often obtained from open ocean waters. *Candida intermedia* is common in in-shore marine waters as well as in sewage and polluted fresh waters (Cooke *et al.*, 1960; Roth *et al.*, 1962; Capriotti, 1962a; Cooke and Matsuura, 1963; Fell and van Uden, 1963; van Uden and Ahearn, 1963; van Uden and Castelo-Branco, 1963; Taysi and van Uden, 1964; Hedrick and Soyugenc, 1965; Norkrans, 1966b; Meyers *et al.*, 1967a,b; Ahearn *et al.*, 1968; Hedrick *et al.*, 1968; Wollett and Hedrick, 1970; Spencer *et al.*, 1970, 1974b; Combs *et al.*, 1971; Goto *et al.*, 1972; Ahearn, 1973; Fell, 1976).

Species of the genus *Cryptococcus*, especially *Cr. albidus* and *Cr. laurentii*, typically are associated with clean, fresh and marine-coastal waters, but they occur also, usually as a relatively low percentage of the total yeasts, in heavily polluted water and sewage (van Uden and Fell, 1968; Wollett and Hedrick, 1970; Spencer *et al.*, 1970, 1974b; Hagler and Mendonca-Hagler, 1981) This non-fermentative genus is common on plant surfaces and in the soil, and could be expected to enter aquatic environments with run-off.

Psychrophilic yeasts of the genus *Leucosporidium* frequently have been isolated from marine waters of the polar regions. These non-fermentative basidiomycetous yeasts utilize a broad range of carbon compounds but generally marine isolates grow poorly, if at all, at temperatures above 20°C. Other psychrophilic yeasts such as *C. norvegica* and *C. austromarina* are isolated from marine waters of the polar regions and marine fishes (Bruce and Morris, 1973; Fell, 1974, 1976), and isolation of psychrophilic yeasts from antarctic lakes has been reported by Stanley and Rose (1967) and Goto *et al.* (1969).

B. Plants

Yeasts are in higher number in the planktonsphere than in other zones of the open sea (Kriss, 1959; Novozhilova and Popova, 1969). Meyers *et al.* (1967b) found increased numbers of yeasts following intensive algal growth, and, in general, the number of yeast cells increased concurrently with decline and/or death of *Noctiluca* cells concentrated at the water surface. Suehiro (1962) found increased yeast numbers from less than 10 (g plankton)$^{-1}$ at the beginning to 50–100 g^{-1} at the end of a bloom of the diatom *Thalassioira subtilis*. The yeast population of algae decaying *in vitro* was dominated by *Rhodotorula* and *Cryptococcus* in the early stage of storage, and after 17 days was dominated by *Candida* (Suehiro and Tomiyasu, 1962; Suehiro *et al.*, 1962). Seven of nine samples of decomposing giant kelp (*Macrocystitis pyrifera*) from the coast of California had populations of *Metschnikowia zobellii*, a species often associated with zooplankton, up to $3.9 \cdot 10^4$ cells g^{-1}, and one sample contained *C. pulcherrima*, the anamorph of *M. pulcherrima* (van Uden and Castelo-Branco, 1963), at a concentration of 800 cells g^{-1}. The prevalent yeasts from fresh or decaying algae in Japan (Suehiro, 1960, 1962; Suehiro and Tomiyasu, 1962, 1964; Suehiro *et al.*, 1962), Recife, Brazil (De Queiroz, 1972), India (Patel, 1979) and Florida (Capriotti, 1962b) included yeasts prevalent in littoral waters (e.g. *C. famata, C. guilliermondii, C. parapsilosis, C. tropicalis, C. albidus, Cr. laurentii, Rhodotorula glutinis* and *Rh. rubra*). In contrast, yeasts on marine algae from the coast of Rhode Island generally were not typical of littoral water (Seshadri and Sieburth, 1975), and in Florida, Roth *et al.* (1962) found yeasts less frequently on marine plants than in surrounding marine waters. Growth of yeasts in association with living algae is probably at the expense of organic carbon and nitrogen compounds excreted by the algae, especially when the algae are injured by dessication at low tide (Stewart, 1963; Suehiro and Tomiyasu, 1964; Fogg, 1966; Anderson and Zeutschel, 1969; Gunkel *et al.*, 1983). Morris (1955a,b) found enzyme-treated sea-weeds to be excellent media for yeast propagation. In seawater microcosms yeast populations have been noted to increase markedly, apparently at the expense of the alga *Desmarestia viridis* (Gunkel *et al.*, 1983). In addition to excreting compounds which can and do support yeast growth, some algae can produce anti-yeast substances (Burkholder *et al.*, 1960; Welch, 1962; Sieburth, 1968). Some marine bacteria also can inhibit yeast growth (Buck *et al.*, 1962, 1963; Buck and Meyers, 1965).

Existing methods for enumeration of yeasts on certan sea-weeds were reported by Seshadri and Sieburth (1971) to be inadequate because of the release of toxic compounds by the algae, especially brown algae which yield polyphenolic compounds. In their study, yeasts were isolated from the nine species of sea-weed tested, and populations were found to be elevated when the water was warmest. *Rhodotorula* spp. occurred only in late spring; otherwise the yeast population was dominated by the genus *Candida*. Highest yeast populations were found on rhodophyceae and chlorophyceae, up to $6.8 \cdot 10^4$ (g dry weight)$^{-1}$ whereas the lower yeast population on phaeophyceae (rarely above 100 g^{-1}) was probably the result of inhibition by polyphenolic compounds. All divisions of algae showed similar seasonal variaties of yeast population, and the yeast population in the water surrounding the sea-weeds paralleled that on the plants. The surfaces of pebbles from the collection sites had a barely detectable but constant yeast population. Yeast-like forms were observed by scanning electron microscopy to occur on the surfaces of the algae throughout the year (Seshadri and Sieburth, 1975).

The rhizosphere of the marine oyster-grass *Spartina alterniflora* has been shown to yield yeast densities comparable to those of terrestrial grasslands. *Pichia spartiniae* was the prevalent species associated with the live plant culm (10^4–10^8 cells g^{-1}), and *Kluyveromyces drosophilarum* was the most common species associated with rhizosphere sediments where yeast populations of up to $9 \cdot 10^4$ cm^{-3} were found. Species of *Trichosporon, Rhodotorula* and *Rhodosporidium* occasionally attained large population densities in the plant rhizosphere (Ahearn *et al.*, 1970; Meyers *et al.*, 1970b, 1971, 1975; Meyers and Ahearn, 1974; Kurtzman and Ahearn, 1976).

C. Animals

Yeasts have been isolated from a variety of aquatic animals including oysters (Hunter, 1920; Buck *et al.*, 1977; Sawyer and Meyer, 1977), clams (Kobayashi *et al.*, 1953; Dabrowa *et al.*, 1964; Buck *et al.*, 1977), mussels (Buck *et al.*, 1977), shrimps (Phaff *et al.*, 1952; Roth *et al.*, 1962; Koburger *et al.*, 1975), isopods (Manier and Loubes, 1978), amphipods (Roth *et al.*, 1962), crabs (Capriotti, 1962b; Debrowa *et al.*, 1964; Seki and Hardon, 1970; Pitts and Cowley, 1974; Cowley and Chrzanowski, 1980), sponges, sea urchins, polychaete worms and Cnidaria (Roth *et al.*, 1962), fish (Capriotti, 1962b; Roth *et al.*, 1962; Fell and van Uden, 1963; van Uden and Castelo-Branco, 1963; Ross and Morris, 1965; Bruce and Morris, 1973; Hatai and Egusa, 1975), dolphins (Sweeney *et al.*, 1976; Nakeeb *et al.*, 1977; Buck, 1980), whales (Mori, 1973; Ridgway, 1979), penguins (Goto *et al.*, 1969) and gulls and terns (Kawakita and van Uden, 1965). Yeasts may find higher nutrient concentrations when associated with animals than when they are free-living in water, and association of yeasts with motile animals could be important for distribution of yeasts (Potter and Baker, 1961).

The more commonly isolated yeasts from animals typically have represented the species prevalent in water, for example *Deb. hansenii, Au. pullulans, Cr. albidus, Cr. diffluens, C. parapsilosis, C. famata* (syn. *T. candida*, anomorph of *Deb. hansenii*), *C. guilliermondii, C. tropicalis, M. zobellii, Rh. rubra, Rh. glutinis, Rh. minuta, Hansenula anomala* and *Trichosporon cutaneum* (Capriotti, 1962; Roth *et al.*, 1962; van Uden and Castelo-Branco, 1963; Fell and van Uden, 1963; Ross and Morris, 1965; Bruce and Morris, 1973; Buck, 1980). A few species infrequently isolated from sea water have been obtained from animals. *Candida glabrata* has been isolated from shrimps (Phaff *et al.*, 1952), dolphins (Buck, 1980), oysters, northern quahogs and mussels (Buck *et al.*, 1977). *Candida albicans* has been isolated frequently from shellfish (Buck *et al.*, 1977), marine birds and dolphins (Kawakita and van Uden, 1965; Buck, 1980). *Trichosporon beigelii* (syn. *Tr. cutaneum*) has been isolated from shrimps (Phaff *et al.*, 1952) and fish (Roth *et al.*, 1962; van Uden and Fell, 1968). Sawyer and Meyer (1977) found *Sterigmatomyces halophilus* associated with the Japanese oyster.

Pathogenicity toward planktonic crustaceans has been attributed to yeasts of the genus *Metschnikowia* (Metschnikoff, 1884; van Uden and Castelo-Branco, 1961, 1963; Spencer *et al.*, 1964; Seki and Fulton, 1969; Fize *et al.*, 1970). Infection and virulence may be related to the forceful ejection of the needle-shaped ascospores that pierce the intestinal wall of the host (Lachance *et al.*, 1976). Seki and Hardon (1970) found that *M. krissii* may cause disease of *Cancer productus*, a large crab. Possible pathogenic effects have been suggested for several other associations. *Candida albicans* and *C. tropicalis* were observed by Combs *et al.* (1981) to cause deformation of the shells of oyster larvae, and *Rh. minuta* was found to infect the gravid amphipod *Podocercus brasiliensis* (Roth *et al.*, 1962).

Marine animals also lower aquatic yeast populations. The red beard sponge (*Microcionia prolifera*) is able to concentrate yeasts from the surrounding water. The sponge and some of its microbial flora produce anti-yeast substances, and the sponge appears to be able to ingest the yeasts as food (Madri, 1968; Kunen *et al.*, 1971; Madri and Claus, 1975). Volz *et al.* (1974) found fewer yeasts in the conch (*Strombus gigas*) and spiny lobster (*Panularus argus*) than in marine sand and sediment. Van Uden and Castelo-Branco (1963) were unable to isolate yeasts from the gut of sea urchins. Yeasts have been used in feeding experiments for molluscs. Shrimps and crabs and gastropods are the source of enzymes used to make yeast sphaeroplasts; thus yeasts probably serve as a food source for these organisms in nature (Chrzanowski and Cowley, 1977; Epifanio, 1979; Kasymov and Likhodeyeva, 1979). Zooplankton, including protozoa (Suehiro and Tomiyasu, 1962), rotifers and *Daphnia*, appear to use yeasts as food (Shimaya *et al.*, 1967; Fukusho, 1980).

D. Sediments

Relatively high yeast densities, frequently above 100 g^{-1}, have been reported for river and intertidal marine sediments influenced by urban pollution (Suehiro, 1963; Matson *et al.*, 1978; Hagler *et al.*, 1982). Yeasts isolated from fresh water and marine sediments, including deep-ocean samples, with little or no pollution, generally are confined to the upper few centimeters (Fell *et al.*, 1960; Hedrick and Soyugenc, 1965, 1967; Hedrick *et al.*, 1966; Steele, 1967; Westheide, 1968; Kishimoto and Baker, 1969; Novozhilova and Popova, 1969; Ahearn *et al.*, 1970; Goto *et al.*, 1974a; Lazarus and Koburger, 1974; Volz *et al.*, 1974). Westheide (1968) found two population maxima on a North Sea sandy beach, one in sandy mud flats and the other in the supralittoral zone. In the Great Lakes, the type of sediment was more important for distribution of yeasts than the depth from which the sample originated (Hedrick and Soyugenc, 1965, 1967; Hedrick *et al.*, 1966). It has been suggested that yeasts participate in the formation of calcium carbonate crystals and humic seams in marine sediments (Buck and Greenfield, 1964; Deelman, 1976). Prevalent yeast species in the sediments generally were those found in water (Table IV), but some species typical of soil were noted in Great Lakes sediments (Fell *et al.*, 1960; Hedrick and Soyugenc, 1965, 1967; Hedrick *et al.*, 1966; Goto *et al.*, 1974a; Lazarus and Koburger, 1974; Volz *et al.*, 1974; Hagler *et al.*, 1981a). In a heavily polluted subtropical estuary, fermentative yeast species, especially of the *C. krusei* complex, were more prevalent, and strictly oxidative yeasts of basidiomycetous affinity, like *Rhodotorula* and *Cryptococcus*, were at a lower incidence than in the water (Hagler *et al.*, 1981a).

Table IV. Prevalent yeasts in eight studies of sediments

Species	Number of studies where prevalent
Candida guilliermondii	3
Candida holmii	2
Candida krusei	3
Candida parapsilosis	4
Cryptococcus albidus	3
Cryptococcus laurentii	4
Debaryomyces hansenii (*Candida famata, Torulopsis candida*)	3
Hansenula saturnus	3
Rhodotorula glutinis	2
Rhodotorula graminis	2
Rhodotorula marina	2
Rhodotorula minuta	2
Rhodotorula rubra	7
Trichosporon beigelii (*Trichosporon autaneum*)	2

III. POLLUTION

A. General effects

An indicated number method for total yeast counts was included, as a tentative method for detecting pollution, in the 14th edition of the American Public Health Association's Standard Methods for Examination of Water and Wastewater (Rand *et al.*, 1976). Yeast cell densities in polluted water and sewage frequently exceed 100 cells 100 ml^{-1}, whereas this volume of non-enriched water contains only a few cells (Ahearn *et al.*, 1968; Spencer *et al.*, 1970, 1974b; Wollett and Hedrick, 1970; Simard and Blackwood, 1971a; Ahearn, 1973; Buck, 1976; Viviani and Tortorano, 1976; Hagler and Mendonca-Hagler). Correlation between yeast counts and chemical, physical and microbiological parameters indicative of sites with various levels of pollution has been observed (Wollett *et al.*, 1970; Simard and Blackwood, 1971b; Noel and Simard, 1973; Viviani and Tortorano, 1976; Vaatanen, 1980a,b; Hagler *et al.*, 1981b).

Simard (1971) and Simard and Blackwood (1971a,b) have proposed that the total red yeast count could be used as an indicator of pollution in river waters. Red yeasts, however, normally comprise a higher proportion of the total yeast population in clean water than in polluted water. Therefore, there is a smaller difference between the densities of red yeasts in polluted and unpolluted water than between the total yeast counts of such waters (Hinzelin and Lectard, 1978; Hagler and Mendonca-Hagler, 1981).

The yeasts associated with faeces of warm-blooded animals include numerous fermentative species, and there is a higher proportion of fermentative yeasts in polluted water than in clean water (Cooke *et al.*, 1960; Wollett and Hedrick, 1970). Using a most probable number method in which fermentation was determined by formation of gas in Durham tubes, Hagler *et al.* (1981a) found that high densities of fermentative yeasts and the total numbers of yeasts correlated equally well with pollution. *Saccharomyces exiguus* and its anamorph *C. holmii* have been found as prevalent yeasts in water polluted by domestic wastes (Hagler and Mendonca-Hagler, 1981; Hagler *et al.*, 1981b), whereas heavy populations of this and other *Saccharomyces* spp. were found by Ahearn *et al.* (1968) in fresh water receiving sugar-cane wastes. The nitrate-assimilating fermentative yeast *H. anomala* has been found as a common species in polluted waters (Capriotti, 1962a; Ahearn *et al.*, 1968; Hagler and Mendonca-Hagler, 1981). The apiculate yeast *Hanseniaspora uvarum* and its anamorph *Kloeckera apiculata* also have been reported from polluted water (Taysi and van Uden, 1964; Ahearn *et al.*, 1968; Meyers *et al.*, 1970a; Hagler *et al.*, 1979; Hagler and Mendonca-Hagler, 1981).

Yeasts of the *C. krusei* complex (e.g. *C. sorbosa, C. valida, C. lambica, P. membranaefaciens, P. kluyveri* and *P. terricola*) ferment glucose only or are non-fermentative, and they assimilate only a few carbon compounds. These species have low salt tolerance, have high phosphate requirements and do not grow in alkaline conditions (Hagler and Mendonca-Hagler, 1979b). These physiological characteristics do not favour growth in the marine environment, and this group is restricted largely to fresh and estuarine waters, especially those with high levels of organic enrichment and sewage (Cooke *et al.*, 1960; Cooke and Matsuura, 1963; Fell and van Uden, 1963; Taysi and van Uden, 1964; Ahearn *et al.*, 1968; Meyers *et al.*, 1970a; Wollett and Hedrick, 1970; Ahearn, 1973; Spencer *et al.*, 1974b). The relatively high tolerance to alkylbenzenesulphonate detergents by isolates of the *C. krusei* complex may be a factor contributing to its prevalence in sewage and polluted water (Standard and Ahearn, 1970; Ahearn, 1973).

Incubation at elevated temperature also may select for yeasts that are associated with pollution. Buck and coworkers found that incubation at 37°C selected for human-associated yeasts (Buck, 1975; Buck *et al.*, 1977). Incubation at 40°C retarded the growth of some human-associated yeasts such as *C. guilliermondii* and *C. parapsilosis* which are more widespread in waters, but apparently selected for species like *C. tropicalis, C. krusei* and *Sacch. cerevisiae* which have been more closely associated with pollution (Hagler *et al.*, 1981b). The adventitious pathogen *C. albicans* has been proposed as an indicator of contamination of water by human or animal excrement (Ahearn *et al.*, 1968; Cook, 1970; Ahearn, 1973; Jamieson *et al.*, 1976; Buck, 1977). *Candida albicans* has been demonstrated to occur commonly, but in relatively low densities, in sewage effluents, receiving streams and inshore marine habitats (Buck and Bubucis, 1978; Sherry *et al.*, 1979; Cook and Schlitzer, 1981). Atypical *C. albicans* has been isolated from the North Sea in the vicinity of oil platforms (Crow *et al.*, 1977).

Trichosporon beigelii (syn. *Tr. cutaneum*), a filamentous arthrospore-forming yeast of basidiomycetous affinity, has been suggested as a pollution indicator because of its presence in polluted river water and its absence from the zone of recovery (Hinzelin and Lectard, 1976, 1978). *Geotrichum penicillatum* and *Geotrichum capitatum* have been primarily, but not exclusively, associated with pollution and sewage (Cooke *et al.*, 1960; van Uden and Ahearn, 1963; Taysi and van Uden, 1964; Hedrick and Soyugenc, 1967; Ahearn *et al.*, 1969; Meyers *et al.*, 1970a; Spencer *et al.*, 1970, 1974b; Wollett and Hedrick, 1970; Hoppe, 1972a; Hinzelin and Lectard, 1978; Hagler and Mendonca-Hagler, 1981). *Prototheca*, a genus of achlorophylous algae that produces yeast-like colonies on isolation media, occasionally has been associated with polluted water (Spencer *et al.*, 1970; Wollett and Hedrick, 1970; De Camargo and Fischman, 1979; Hagler and Mendonca-Hagler, 1981).

B. Oil pollution

Only a few researchers have examined the response of yeasts to oil pollution. Turner and Ahearn (1970) found that yeast populations in a freshwater stream periodically subjected to crude-oil spillage from an asphalt refinery increased from 30–200 colony-forming units ml^{-1} to over 10^4 colony-forming units ml^{-1} in only a few days following a spill. Increased densities of yeasts were observed in surface waters of the Gulf of Mexico following a well blow-out (Ahearn *et al.*, 1971), and with the development of oil fields the incidence of hydrocarbonoclastic yeasts in water samples from the North Sea appeared to increase (Ahearn and Crow, 1980). Some evidence suggests that catastrophic oil pollution may inhibit the indigenous yeast flora and select for oil-utilizing species (Ahearn and Crow, 1980; Ahearn *et al.*, 1976).

It is well established that yeasts, including species occurring commonly in aquatic habitats, utilize single hydrocarbons, particularly straight-chain alkanes and alkenes, as sole sources of carbon for growth (see Watkinson, 1978). A few yeasts, notably *C. lipolytica*, utilize the branched-chain hydrocarbon pristane for growth (Hagihara *et al.*, 1977). More complex hydrocarbons (e.g. thiophenes and polycyclic aromatics) are most probably utilized by yeast by co-oxidation processes. Cerniglia and Crow (1981) demonstrated that isolates of *C. lipolytica, C. guilliermondii, C. tropicalis, C. maltosa* and *Deb. hansenii* were capable of oxidizing naphthalene, biphenyl and benzo(a)pyrene. Nitro-aromatics have been found to be metabolized by reductive pathways to ring-hydroxylated intermediates by a marine-occurring *Rhodosporidium* sp. (Corbett and Corbett, 1981). The actual contribution of yeasts in the biodegradation of hydrocarbons in the environment is unknown, but the metabolic diversity recently demonstrated for yeasts suggests a more important role than that previously expected.

IV. PHYSICOCHEMICAL PARAMETERS AFFECTING AQUATIC YEASTS

A. Effects on growth

1. Sodium chloride

Regardless of the aquatic habitat from which they were isolated, most yeasts have been shown to retain viability in distilled water and to be capable of growth at sodium chloride concentrations exceeding those normally found in sea water (Norkrans, 1966a; Lodder, 1970; Goto *et al.*, 1972; Hoppe, 1972b). Phaff *et al.* (1952) noted an increase in the lag phase of yeasts with increasing sodium chloride concentrations. Ross and Morris (1962) noted a decrease in cell yield of *Deb. hansenii* and *P. membranaefaciens* with increased sodium chloride concentration, but the growth rate during the logarithmic phase was not affected. The optimum salt concentration for growth of marine *Deb. hansenii* isolates was found to be about that of sea water, but other species (*P. membranaefaciens, Rh. glutinis, C. parapsilosis, C. inconspicua, Sacch. cerevisiae* and *Au. pullulans*) grew best with no salt added to the medium. Hagler and Mendonca-Hagler (1979b) indicated that the more common marine-occurring yeasts had sodium chloride tolerances of at least 9%.

2. Temperature

Temperatures for growth of yeasts range from less than 0°C to about 47°C depending on the species and conditions for growth. Most yeasts are mesophilic with maximum growth temperatures between 28°C and 38°C; no thermophilic yeasts have been reported (Vidal-Leiria *et al.*, 1979). Psychrophilic yeasts, those with optimal growth temperatures below 20°C and maximal growth temperatures below 30°C, have been isolated from antarctic water. Most of these yeasts have been classified as *Leucosporidium* or *Candida*. Psychrophobic yeasts, which have minimum growth temperatures ranging from the low twenties to about 35°C, have been isolated from subtropical but not temperate estuaries (van Uden, 1967; van Uden and Fell, 1968). These yeasts are associated with warm-blooded animals and are commensals in the digestive tracts of gulls and terns.

3. Hydrogen ion concentration

Yeasts grow over a range in pH value from below 2.0 to about 9.0. Goto *et al.* (1972) reported growth up to pH 7.8 for marine isolates of *C. diddensii*, up to pH 8.1 for *Rh. marina* and up to pH 8.6 for *Deb. hansenii, Cr. albidus, C. krusei, Rh. glutinis, Rh. rubra, T. candida, Rh. marina* and *Rhodosporidium infirmo-miniata*. Norkrans (1966a) tested 19 isolates belonging to 11 species for growth in the pH range 3.0–8.7 and found two main groups. The first group, composed of *Rh. glutinis, Deb. hansenii, Cr. laurentii, Cr. luteolus, Cr. albidus* and *T. candida*, gave maximal cell yields at about pH 8.1 or higher, although decreased growth rates and increased lag phases were observed. The second group gave maximal cell yields near pH 7.4 and negligible growth at pH 8.1. In order of decreasing tolerance to alkalinity, this group included *Cr. laurentii, T. candida, Cr. luteolus, C. parapsilosis, Rhodosp. infirmo-miniata, Rhodotorula* sp. and *Sacch. cerevisiae*. Hoppe (1972b) reported pH value growth maxima for marine and estuarine isolates of 9.5 for *C. parapsilosis*, 9.3 for *Rh. rubra*, 9.0 for *Rh. glutinis* and *Sacch. cerevisiae*, 8.6 for *Debaryomyces* sp. and *Cr. neoformans*, and 8.5 to 9.3 for *Deb. hansenii*. Isolates of *C. krusei, P. membranaefaciens* and *Kloeckera apiculata*, species common in fresh but not marine waters, did not grow readily at pH values above 7.0. The inclusion of 3% sodium chloride in the medium reduced the maximum pH for growth of some species, including *C. sorbosa, Au. pullulans, Rh. rubra, Sacch. exiguus, Tr. cutaneum* and *Geotrichum penicillatum* (Santos *et al.*, 1975; Hagler *et al.*, 1977).

4. Phosphate

Low concentrations of available phosphate may affect the persistence of some yeasts in marine waters. *Candida krusei, C. sorboxylosa* and *Prototheca zopfii* were found to require relatively high concentrations of phosphate for growth compared with *Rh. rubra, C. guilliermondii* and *Au. pullulans* which are more prevalent in sea water (Santos *et al.*, 1975).

B. Survival

Most studies on survival of yeasts in aquatic habitats have dealt with species that are adventitious pathogens of man, most notably *C. albicans. Candida albicans* and *Cr. neoformans* have been found to retain both their viability and pathogenicity after prolonged exposure to sea water (Dzawachiszwili *et al.*, 1964; Madri, 1968; Madri and Claus, 1975). Jamieson *et al.* (1976) found that *C. albicans* populations in 0.5% (v/v), 2.0% (v/v) and 3.5% (v/v) sea water were decreased during the first six days incubation at 4°C, 25°C and 37°C, but then remained stable up to 11 days. None of the typical water-borne pathogenic bacteria tested survived for 11 days. *Candida albicans* and *Tr. beigelii* were able to survive for six months in dry sand (sterile or non-sterile) and up to six months in wet sand at 25°C. Salinity, up to 10% (v/v) sea water, was not important for survival in sand, but temperatures of 37°C and 45°C substantially lowered survival (Anderson, 1979). Ahearn (1973) reported that the densities of both *C. albicans* and *Rh. rubra* remained relatively stable during 12 days incubation as pure cultures in sea water, but when inoculated together the population of *C. albicans* decreased while populations of *Rh. rubra* increased slightly. Both strains appeared to excrete nutrients when suspended in water. Evidently *C. albicans* was unable to compete effectively in a host-free habitat for uptake of essential compounds. Culture filtrates of an isolate of *Bacillus subtilis* from sewage inhibited strains of *C. albicans* and *Saccharomyces* spp.; microbial antagonisms were suggested as factors in the limited host-free survival of *C. albicans* in sewage (Coleman *et al.*, 1975). Ahearn *et al.* (1968) reported that in laboratory tests *C. albicans* and *C. tropicalis* populations increased during the first eight days in sea water but had decreased by two to three orders of magnitude at six weeks. *Debaryomyces hansenii* and *C. parapsilosis* populations were stable throughout the experiment, but *Rh. rubra* and *Cr. laurentii* decreased by about 90% in eight days and remained at that level for six weeks. Similar results were obtained for these species in fresh water. Buck (1978) noted that recent field isolates of *C. albicans* had better survival in sea water than a laboratory strain. Unpublished observations by C.N. Hagler showed that *Rh. rubra, C. tropicalis, Deb. hansenii, Tr. cutaneum* and *Geotrichum penicillatum* maintained population densities of $10^3\,ml^{-1}$ at 25°C for 16 weeks as pure cultures in both polluted estuarine water and clean sea water. However, *Kl. apiculata* survived less than one day in either water, *C. krusei* only three weeks in sea water but more than 16 weeks in estuarine water, and *Sacch. exiguus* only two weeks in sea water and four weeks in estuarine water.

V. SUMMARY

Current information on aquatic yeasts is based primarily on studies concerned with incidence, population densities and taxonomy. The data support the premise that most yeasts enter lakes and oceans with terrestrial run-off. Some species, such as *Deb. hansenii* and *Rhodotorula* spp., are physiologically adapted for survival and increase in population on their fortuitous encounter with zones of increased nutrients. A commensal and adventitious pathogen role for planktonic crustaceans is indicated for species of *Metschnikowia*. The widespread occurrence of smut-like basidiomycetous yeasts, including many psychrophilic forms, in marine habitats suggests an association with marine vegetation. This latter area has received scant study and appears particularly promising for the determination of definitive environmental roles for some of the aquatic yeasts.

References

Ahearn, D.G. (1973). *In* "Estuarine Microbial Ecology: The Belle W. Baruch Library in Marine Science, Number 1" (L. Harold Stevenson and R.R. Colwell, eds), pp. 433–440. University of South Carolina Press, Columbia, South Carolina.

Ahearn, D.G. and Crow, S.A. (1980). *Botanica Marina* **23**, 125.

Ahearn, D.G. and Meyers, S.P. (1976). *In* "Recent Advances in Aquatic Mycology" (E.B. Gareth Jones, ed.), pp. 125–133. Elek Science, London.

Ahearn, D.G. and Roth, F.J., Jr. (1966). *Developments in Industrial Microbiology* **7**, 301.

Ahearn, D.G., Roth, F.J., Jr. and Meyers, S.P. (1962). *Canadian Journal of Microbiology* **8**, 121.

Ahearn, D.G., Roth, F.J., Jr. and Meyers, S.P. (1968). *Marine Biology* **1**, 291.

Ahearn, D.G., Meyers, S.P., Cook, W.L. and Hansen, G. (1969). *Antonie van Leeuwenhoek, Supplement* **35**, D 19.

Ahearn, D.G., Yarrow, D. and Meyers, S.P. (1970). *Antonie van Leeuwenhoek* **36**, 503.

Ahearn, D.G., Meyers, S.P. and Standard, P.G. (1971). *Developments in Industrial Microbiology* **12**, 126.

Ahearn, D.G., Crow, S.A., Berner, N.H. and Meyers, S.P. (1976). *In* "Estuarine Processes I" (M. Wiley, ed.), pp. 483–493. Academic Press, New York.
Anderson, G.G. and Zeutschel, R.P. (1969). *Limnology and Oceanography* **15**, 402.
Anderson, J.H. (1979). *Sabouraudia* **17**(1), 1.
Artagaveytia-Allende, R.C. (1960). *Atti dell'Istituto Botanico e Laboratorio Crittomagico dell'-Universita di Pavia.* Serie 5 **18**, 1.
Artagaveytia-Allende, R.C. and Lucy E. Teira. (1965). *Atti dell'Istituto Botanico e Laboratorio Crittomagico dell'Universita di Pavia.* Serie 6 **1**, 117.
Barnett, J.A., Payne, R.W. and Yarrow, D. (1979). "A Guide to Identifying and Classifying Yeasts." Cambridge University Press, Cambridge, London, New York, Melbourne.
Bhat, J.V. and Kachwalla, N. (1955). *The Proceedings of the Indian Academy of Sciences* **41** Section B, 9.
Bruce, J. and Morris, E.O. (1973). *Antonie van Leeuwenhoek* **39**, 331.
Buck, J.D. (1975). *Mycopathologia* **56**, 73.
Buck, J.D. (1976). *Florida Scientist* **39**(2), 111.
Buck, J.D. (1977). *In* "Bacterial Indicators/Health Hazards Associated with Water" (A.W. Hoadley and B.J. Dutka, eds), pp. 139–147. ASTM, Philadelphia.
Buck, J.D. (1978). *Microbial Ecology* **4**, 291.
Buck, J.D. (1980). *Journal of Wildlife Diseases* **16**, 141.
Buck, J.D. and Bubucis, P.M. (1978). *Applied and Environmental Microbiology* **35**(2), 237.
Buck, J.D. and Greenfield, L.J. (1964). *Bulletin of Marine Science of the Caribbean* **14**(2), 239.
Buck, J.D. and Meyers, S.P. (1965). *Limnology and Oceanography* **10**, 385.
Buck, J.D., Meyers, S.P. and Kamp, K.M. (1962). *Science* **138**, 1339.
Buck, J.D., Ahearn, D.G., Roth, F.J. and Meyers, S.P. (1963). *Journal of Bacteriology* **85**, 1132.
Buck, J.D., Bubucis, P.M. and Combs, T.J. (1977). *Applied and Environmental Microbiology* **33**, 370.
Burkholder, P.R., Burkholder, L.M. and Almodovar, L.R. (1960). *Botanica Marina* **2**, 149.
Capriotti, A. (1962a). *Archiv für Mikrobiologie* **41**, 147.
Capriotti, A. (1962b). *Archiv für Mikrobiologie* **42**, 407.
Cerniglia, C.E. and Crow, S.A. (1981). *Archives of Microbiology* **129**, 9.
Chrzanowski, T.H. and Cowley, G.T. (1977). *Mycologia* **69**, 1062.
Coleman, A., Cook, W.L. and Ahearn, D.G. (1975). *Abstracts of the Annual Meeting of the American Society of Microbiology* **N20**, 187.
Colwell, R.R., Walker, J.D., Conrad, B.F. and Seesman, P.A. (1976). *Developments in Industrial Microbiology* **17**, 269.
Combs, T.J., Murchelano, R.A. and Jurgen, F. (1971). *Mycologia* **63**, 178.
Combs, T.J., Braun, P.C. and Blogoslawski, W.M. (1981). *Abstracts of the Annual Meeting of the American Society of Microbiology* **81**, 174.
Cook, W.L. (1970). *In* "Recent Trends in Yeast Research" (D.G. Ahearn, ed.), Spectrum vol. 1, pp. 107–112. Georgia State University, Atlanta, Georgia.
Cook, W.L. and Schlitzer, R.L. (1981). *Applied and Environmental Microbiology* **41**, 840.
Cooke, W.B. (1965). *Mycologia* **57**, 696.
Cooke, W.B. (1970). United States Department of the Interior, Federal Water Pollution Control Administration, Robert A. Taft Water Research Center, Cincinnati, Ohio.
Cooke, W.B. (1976). *In* "Recent Advances in Aquatic Mycology" (E.B. Gareth Jones, ed.), pp. 389–434. Elek Science, London.
Cooke, W.B. and Matsuura, G.S. (1963). *Protoplasma* **57**, 163.
Cooke, W.B. and Matsuura, G.S. (1969). *Ecology* **50**, 659.
Cooke, W.B., Phaff, H.J., Miller, M.W., Shifrine, M. and Knapp, E. (1960). *Mycologia* **52**, 210.
Corbett, M.D. and Corbett, B.R. (1981). *Applied and Environmental Microbiology* **41**, 942.
Cowley, G.T. and Chrzanowski, T.H. (1980). *Botanica Marina* **23**(6), 397.
Crow, S.A., Ahearn, D.G., Cook, W.L. and Bourquin, A.W. (1975). *Limnology and Oceanography* **20**, 644.
Crow, S.A., Cook, W.L., Ahearn, D.G. and Bourquin, A.W. (1976). *In* "Proceedings of the Third International Biodegradation Symposium" (J.M. Sharpley and A.M. Kaplan, eds), pp. 93–99. Applied Science Publishers, London.

Crow, S.A., Bowman, P.I. and Ahearn, D.G. (1977). *Applied and Environmental Microbiology* **33**, 738.

Dabrowa, N., Landau, J.W., Newcomer, V.D. and Olunkett, O.A. (1964). *Mycopathologia et Mycologia Applicata* **24**, 137.

De Camargo, Z. and Fischman, O. (1979). *Sabouraudia* **17**, 275.

Deelman, J.C. (1976). *Soil Science* **121**, 184.

De Queiroz, L.A. (1972). *Publicação do Instituto de Micologia Recife* **677**, 1.

De Queiroz, L.A. and Macedo, S.J. (1972). *Publicação do Instituto de Micologia Recife* **676**, 1.

Dzawachiszwili, N., Landau, J.W., Newcomer, V.D. and Plunkett, O.A. (1964). *Journal of Investigative Dermatology* **43**, 103.

Engelbrecht, R.S., Foster, D.H., Greening, E.O. and Lee, S.H. (1974). Environmental Protection Technology Series. EPA 670/2-73-082.

Epifanio, C.E. (1979). *Aquaculture* **16**, 187.

Fell, J.E., Hunter, I.L. and Tallman, A.S. (1973). *Canadian Journal of Microbiology* **19**, 643.

Fell, J.E. and Tallman, A.S. (1980). *International Journal of Systematic Bacteriology* **30**, 658.

Fell, J.W. (1967). *Bulletin of Marine Science of the Gulf and Caribbean* **17**, 454.

Fell, J.W. (1974). *In* "Effects of the Ocean Environment on Microbial Activities" (R.R. Colwell and R.Y. Morita, eds), pp. 510–523. University Park Press, Baltimore.

Fell, J.W. (1976). *In* "Recent Advances in Aquatic Mycology" (E.B. Gareth Jones, ed.), pp. 93–124. Elek Science, London.

Fell, J.W. and Meyer, S.A. (1967). *Mycopathologia et Mycologia Applicata* **32**, 177.

Fell, J.W. and van Uden, N. (1963). *In* "Symposium on Marine Microbiology" (C.H. Oppenheim, ed.), pp. 329–340. Charles Thomas, Springfield.

Fell, J.W., Ahearn, D.G., Meyers, S.P. and Roth, F.J., Jr. (1960). *Limnology and Oceanography* **5**, 366.

Fize, A., Manier, J.F. and Maurand, J. (1970). *Annales de Parasitologie, Humaine et Comparée (Paris)* **45**(3), 357.

Fogg, G.E. (1966). *Oceanography and Marine Biology Annual Review* **4**, 195.

Fukusho, K. (1980). *Bulletin of the Japanese Society for Scientific Fisheries* **46**, 625.

Goto, S., Sugiyama, J. and Iizuka, H. (1969). *Mycologia* **61**, 748.

Goto, S., Yamasato, K., Okuno, D. and Iizuka, H. (1972). *Proceedings of the Fourth IFS: Fermentation Technology Today*, p. 805.

Goto, S., Yamasato, K. and Iizuka, H. (1974a). *Journal of General and Applied Microbiology* **20**, 309.

Goto, S., Ohwada, K. and Yamasato, K. (1974b). *Journal of General and Applied Microbiology* **20**, 317.

Gunkel, W., Crow, S. and Klings, K.W. (1983). *Marine Biology* **75**, 327.

Hagihara, T., Mishma, M., Tanaka, A. and Fuukui, S. (1977). *Agricultural and Biological Chemistry* **41**, 1745.

Hagler, A.N. and Mendonca-Hagler, L.C. (1978). *Anais de Microbiologia* (Rio de Janeiro) **23**, 79.

Hagler, A.N. and Mendonca-Hagler, L.C. (1979a). *Revista Brasileira do Pesquisa Medica e Biologica* **12**, 63.

Hagler, A.N. and Mendonca-Hagler, L.C. (1979b). *Revista do Microbiologia* **10**, 30.

Hagler, A.N. and Mendonca-Hagler, L.C. (1981). *Applied and Environmental Microbiology* **41**, 173.

Hagler, A.N., Cataão, H.V., Barroso, W.J. and Mendonca-Hagler, L.C. (1977). VII *Congresso Brasiliero de Microbiologia*, p. 14.

Hagler, A.N., Santos, S.S. and Mendonca-Hagler, L.C. (1979). *Revista do Microbiologia* **10**, 36.

Hagler, A.N., Mendonca-Hagler, L.C., Schrank, A. and Santos, E.A. (1981). *Abstracts of the Annual Meeting of the American Society for Microbiology* **81**, 175.

Hagler, A.N., De Oliveira, R.B. and Mendonca-Hagler, L.C. (1982). *Antonie van Leeuwenhoek* **48**, 53.

Hatai, K. and Egusa, S. (1975). *Bulletin of the Japanese Society of Scientific Fisheries* **41**, 993.

Hedrick, L.R. and Soyugenc, M. (1965). *Proceedings of the Eighth Conference on Great Lakes Research*, p. 69.

Hedrick, L.R. and Soyugenc, M. (1967). *Proceedings of the Tenth Conference on Great Lakes Research*, p. 20.
Hedrick, L.R., Soyugenc, M., Dupont, P. and Ambrosini, R. (1964). *Proceedings of the Seventh Conference on Great Lakes Research*, p. 77.
Hedrick, L.R., Soyugenc, M. and Larsen, L. (1966). *Proceedings of the Ninth Conference on Great Lakes Research*, p. 27.
Hedrick, L.R., Cook, W. and Wollett, L. (1968). *Proceedings of the Eleventh Conference on Great Lakes Research*, p. 538.
Hinzelin, F. and Lectard, P. (1976). *Hydrobiologia* **49**, 27.
Hinzelin, F. and Lectard, P. (1978). *Hydrobiologia* **61**, 209.
Hinzelin, F. and Lectard, P. (1979). *Mycopathologia* **69**, 121.
Hinzelin, F., Lectard, P. and Pelt, J.M. (1979). *Revue de Mycologie* **43**, 149.
Hoppe, H-G. (1972a). *Kiel Meeresforschung* **28**, 54.
Hoppe, H-G. (1972b). *Kiel Meeresforschung* **28**, 219.
Hunter, A.C. (1920). *United States Department of Agriculture Bulletin*, No. 819, p. 1.
Jamieson, W., Madri, P. and Claus, G. (1976). *Hydrobiologica* **50**, 117.
Johnson, T.W., Jr. and Sparrow, F.K., Jr. (1961). "Fungi in Oceans and Estuaries". Cramer, Weinhein, Germany.
Kasymov, A.G. and Likhodeyova, N.F. (1979). *Water Air Soil Pollution* **11**, 279.
Kawakita, S. and van Uden, N. (1965). *Journal of General Microbiology* **39**, 125.
Kishimoto, R.A. and Baker, G.E. (1969). *Mycologia* **61**, 537.
Kobayashi, Y., Tsubaki, K. and Soneda, M. (1953). *Bulletin of the Natural Science Museum of Tokyo* **33**, 47.
Koburger, J.A., Norden, A.R. and Kempler, G.M. (1975). *Journal of Milk and Food Technology* **38**, 747.
Kohlmeyer, J. and Kohlmeyer, E. (1979). "Marine Mycology: The Higher Fungi." Academic Press, New York.
Kreger van Rij (1984). "The Yeasts, a Taxonomic Study." Elsevier Biomedical Press, Amsterdam.
Kriss, A.E. (1959). "Morskaja Mikrobiologija." Academy of Sciences, United Soviet Socialist Republic, Moscow. [English translation (1963). "Marine Microbiology." Oliver and Boyd, Edinburgh.]
Kriss, A.E. and Novozhilova, M.I. (1954). *Mikrobiologia* **23**, 669.
Kriss, A.E., Runkina, E.A. and Tikhonenko, A.S.A. (1952). *Zhurnal Obshchei Biologii* **13**, 232.
Kriss, A.E., Mishustina, I.E., Mitskevich, N. and Zemtsova, E.V. (1967). "Microbial Population of Oceans and Seas." [Russian publication, 1964, translated by K. Syers (G.E. Fogg, ed.)]. Edward Arnold, London.
Kunen, S., Claus, G., Madri, P. and Peyser, L. (1971). *Microcionia prolifera. Hydrobiologia* **38**, 565.
Kurtzman, C.P. and Ahearn, D.G. (1976). *Mycologia* **68**, 682.
Lachance, M.A., Miranda, M., Miller, M.W. and Phaff, H.J. (1976). *Canadian Journal of Microbiology* **22**, 1756.
Lazarus, C.R. and Koburger, J.A. (1974). *Applied Microbiology* **27**, 1108.
Lodder, J. (1970). "The Yeasts." North Holland Publishing Company, Amsterdam.
Luehrs, K. (1972). *Mykosen* **15**, 219.
Madri, P. (1968). *Botanica Marina* **11**, 31.
Madri, P. and Claus, G. (1975). *Advancing Frontiers of Plant Science* **30**, 67.
Manier, J.F. and Loubes, C. (1978). *Bulletin de la Société Zoologique de France* **103**, 483.
Matson, E.A., Hornor, S.G. and Buck, J.D. (1978). *Journal of Water Pollution Control Federation* **50**, 13.
Metschnikoff, E. (1884). *Virchow's Archiv für Pathologische Anatomie und Physiologie und Klinische Medizin* **96**, 177.
Meyers, S.P. and Ahearn, D.G. (1974). *Veröffentlichungen des Instituts für Meeresforschung in Bremerhaven*, Supplement **5**, 321.
Meyers, S.P., Ahearn, D.G., Gunkel, W. and Roth, F.J. Jr. (1967a). *Marine Biology* **1**, 118.

Meyers, S.P., Ahearn, D.G. and Roth, F.J. Jr. (1967b). *Bulletin of Marine Science* **17**, 576.
Meyers, S.P., Ahearn, D.G. and Cook, W.L. (1970a). *Mycologia* **62**, 505.
Meyers, S.P., Nicholson, M.L., Rhee, J., Miles, P. and Ahearn, D.G. (1970b). *Louisiana State University Coastal Studies Bulletin* **5**, 111.
Meyers, S.P., Ahearn, D.G. and Miles, P.C. (1971). *Louisiana State University Coastal Studies Bulletin* **6**, 7.
Meyers, S.P., Ahearn, D.G., Alexander, S.K. and Cook, W.L. (1975). *Developments in Industrial Microbiology* **16**, 262.
Mori, H. (1973). *Bulletin of the Japanese Society for Scientific Fisheries* **39**, 333.
Morris, E.O. (1955a). *Journal of Science Food and Agriculture* **6**, 611.
Morris, E.O. (1955b). *Journal of Science Food and Agriculture* **6**, 618.
Morris, E.O. (1968). *Oceanography and Marine Biology Annual Review* **6**, 201.
Morris, E.O. (1975). *Journal of Applied Bacteriology* **38**, 211.
Nakeeb, S., Targowski, S.P. and Spottle, S. (1977). *Journal of the American Veterinary Medicine Association* **171**, 961.
Niewolak, S. (1976). *Acta Mycologica* **12**, 241.
Noel, J.G. and Simard, R.E. (1973). *Abstracts of the Annual Meeting of the American Society for Microbiology* **73**, 33.
Norkrans, B. (1966a). *Archiv für Mikrobiologie* **54**, 374.
Norkrans, B. (1966b). *Svensk Botanisk Tidskrift* **60**, 463.
Novozhilova, M.I. and Popova, L.E. (1969). *Antonie van Leeuwenhoek* **35**, Supplement D23.
Patel, K.S. (1979). *Geobios (Jodhpur)* **6**, 114.
Phaff, H.J., Mrak, E.M. and Williams, D.B. (1952). *Mycologia* **44**, 431.
Phaff, H.J., Miller, M.W. and Mrak, E.M. (1978). "The Life of Yeasts." Harvard University Press, Cambridge, Massachusetts.
Pitts, G.Y. and Cowley, G.T. (1974). *Mycologia* **66**, 669.
Potter, L.F. and Baker, G.E. (1961). *Canadian Journal of Microbiology* **7**, 595.
Pugh, G.J.F. and Lindsay, B.I. (1975). *Transactions of the British Mycological Society* **65**, 201.
Rand, M.C., Greenberg, A.E. and Taras, M.J. (eds) (1976). "Standard Methods for the Examination of Water and Wastewater", 14th edn. American Public Health Association, American Water Works Association and Water Pollution Control Federation, Washington, D.C.
Richards, M. and Elliott, R.R. (1966). *Nature, London* **209**, 536.
Ridgway, S.H. (1979). *Journal of Wildlife Diseases* **15**, 99.
Ross, S.S. and Morris, E.O. (1962). *Journal of Science Food and Agriculture* **9**, 467.
Ross, S.S. and Morris, E.O. (1965). *Journal of Applied Bacteriology* **28**, 224.
Roth, F.J. Jr., Ahearn, D.G., Fell, J.W., Meyers, S.P. and Meyer, S.A. (1962). *Limnology and Oceanography* **7**, 178.
Roth, F.J. Jr., Orput, P.A. and Ahearn, D.G. (1964) *Canadian Journal of Botany* **42**, 375.
Rozanova, E.P. and Novozhilova, M.I. (1958). *Microbiology* **27**(3), 364.
Santos, S.S., Hagler, A.N. and Mendonca-Hagler, L.C. (1975). *VII Congresso Brasileira Microbiologia*, p. 1108.
Sawyer, T.K. and Meyer, S.A. (1977). *Journal of Invertebrate Pathology* **29**, 395.
Schneider, J. (1977). *In* "Ecological Studies Analysis and Synthesis, Microbial Ecology of a Brackish Water Environment" (R. Gerhard, ed.), vol. 25, pp. 90–102. Springer-Verlag, Berlin, New York.
Schrank, A., Hagler, A.N. and Mendonca-Hagler, L.C. (1979a). *X Congresso Brasileiro Microbiologia*, p. 60.
Schrank, A., Hagler, A.N. and Mendonca-Hagler, L.C. (1979b). *X. Congresso Brasileiro Microbiologia*, p. 173.
Seki, H. and Fulton, J. (1969). *Mycopathologia et Mycologia Applicata* **38**, 61.
Seki, H. and Hardon, M. (1970). *Oceanographic Society of Japan* **26**, 38.
Seshadri, R. and Sieburth, J.M. (1971). *Applied Microbiology* **22**, 507.
Seshadri, R. and Sieburth, J.M. (1975). *Marine Biology* **30**, 105.
Sherry, J.P., Kuchma, S.R. and Dutka, B.J. (1979). *Canadian Journal of Microbiology* **25**, 1036.
Shimaya, M., Kanazawa, A. and Kashiwada, K. (1967). *Memoirs of the Faculty of Fisheries Kagoshima University* **16**, 34.

Shinano, H. (1962). *Bulletin of the Japanese Society of Scientific Fisheries* **28**, 1113.
Sieburth, J.M. (1968). *In* "Advances in Microbiology of the Sea, I" (M.R. Droop and E.J.F. Wood, eds), pp. 63–94. Academic Press, New York.
Siepmann, R. and Höhnk, W. (1962). *Veröffentlichungen des Instituts für Meeresforschung in Bremerhaven* **8**, 79.
Simard, R.E. (1971). *Marine Pollution Bulletin* **2**, 123.
Simard, R.E. and Blackwood, A.C. (1971a). *Canadian Journal of Microbiology* **17**, 197.
Simard, R.E. and Blackwood, A.C. (1971b). *Canadian Journal of Microbiology* **17**, 353.
Spencer, J.F.T., Phaff, H.J. and Gardner, N.R. (1964). *Journal of Bacteriology* **88**, 758.
Spencer, J.F.T., Gorin, P.A.J. and Gardner, N.R. (1969). *Antonie van Leeuwenhoek* **35** Supplement, Yeast Symposium, p. 17.
Spencer, J.F.T., Gorin, P.A.J. and Gardner, N.R. (1970). *Canadian Journal of Microbiology* **16**, 1051.
Spencer, J.F.T., Gorin, P.A.J. and Gardner, N.R. (1974a). *Canadian Journal of Microbiology* **20**, 949.
Spencer, J.F.T., Gorin, P.A.J. and Gardner, N.R. (1974b). *Canadian Journal of Microbiology* **20**, 993.
Standard, P.G. and Ahearn, D.G. (1970). *Applied Microbiology* **20**, 646.
Stanley, S.O. and Rose, A.H. (1967). *Philosophical Transactions of the Royal Society, London, Series B, Biological Sciences* **252**, 199.
Steele, C.W. (1967). *Pacific Science* **21**, 317.
Stewart, W.O.P. (1963). *Nature, London* **200**, 1020.
Suehiro, S. (1960). *Science Bulletin of the Faculty of Agriculture Kyushu University* **17**, 443.
Suehiro, S. (1962). *Science Bulletin of the Faculty of Agriculture Kyushu University* **20**, 101.
Suehiro, S. (1963). *Science Bulletin of the Faculty of Agriculture Kyushu University* **20**, 223.
Suehiro, S. and Tomiyasu, Y. (1962). *Science Bulletin of the Faculty of Agriculture Kyushu University* **12**, 163.
Suehiro, S. and Tomiyasu, Y. (1964). *Science Bulletin of the Faculty of Agriculture Kyushu University* **13**, 227.
Suehiro, S., Tomiyasu, Y. and Tanaka, O. (1962). *Science Bulletin of the Faculty of Agriculture Kyushu University* **12**, 155.
Sweeney, J.C., Migaki, G., Vainik, P.M. and Conklin, R.H. (1976). *Journal of the American Veterinary Medicine Association* **169**, 946.
Taysi, I. and van Uden, N. (1964). *Limnology and Oceanography* **9**, 42.
Turner, W.E. and Ahearn, D.G. (1970). *In* "Recent Trends in Yeast Research" (D.G. Ahearn, ed.), pp. 113–123. Spectrum, vol. 1, Georgia State University.
Vaatanen, P. (1980a). *Applied and Environmental Microbiology* **40**, 48.
Vaatanen, P. (1980b). *Applied and Environmental Microbiology* **40**, 55.
Van Uden, N. (1967). *In* "Estuaries" (G.H. Lauf, ed.), pp. 306–310. American Association for the Advancement of Science, Washington, D.C.
Van Uden, N. and Ahearn, D.G. (1963). *Antonie van Leeuwenhoek* **29**, 308.
Van Uden, N. and Castelo-Branco, R. (1961). *Journal of General Microbiology* **26**, 141.
Van Uden, N. and Castelo-Branco, R. (1963). *Limnology and Oceanography* **8**, 323.
Van Uden, N. and Fell, J.W. (1968). *In* "Advances in Microbiology of the Sea" (M.R. Droop and E.J.F. Wood, eds), vol. 1, pp. 167–201. Academic Press, New York.
Vidal-Leiria, M., Buckley, H. and van Uden, N. (1979). *Mycologia* **71**, 493.
Viviani, M.A and Tortorano, A.M. (1976). *Bulletin de la Société Française de Mycologie Medicale* **5**, 109.
Volz, P.A., Jerger, D.E., Worzburg, A.J. and Hiser, J.F. (1974). *Mycopathologia et Mycologia Applicata* **54**(3), 313.
Watkinson, R.J. (1978). "Developments in Biodegradation of Hydrocarbons, 1." Applied Science, London.
Welch, A.M. (1962). *Journal of Bacteriology* **83**, 97.
Westheide, W. (1968). *Marine Biology* **1**, 336.
Wollett, L.L. and Hedrick, L.R. (1970). *Antonie van Leeuwenhoek* **36**, 427.

Wollett, L.L., Hedrick, L.R. and Tarves, M.G. (1970). *Antonie van Leeuwenhoek* **36**, 437.
Yamasato, K., Goto, S., Ohwada, K., Okuno, D., Araki, H. and Iizuka, H. (1974). *Journal of General and Applied Microbiology* **20**, 289.
ZoBell, C.E. (1946). *In* "Marine Microbiology", p. 129. Chronica Botanica Company, Waltham, Massachusetts.

7 Yeasts as Human and Animal Pathogens

Rosalinde Hurley, J. de Louvois and Anne Mulhall

Institute of Obstetrics and Gynaecology and Queen Charlotte's Maternity Hospital, Goldhawk Road, London W6 0XG, U.K.

The Yeasts Vol. 1, 2nd edition
ISBN 0-12-596411-0

I. INTRODUCTION

A. Taxonomy and general considerations

The most recent and comprehensive classification of the yeast-like fungi is that of Kreger-van Rij (1984). According to this, the yeasts, which are taxonomically diverse but had been described by Skinner (1947) as "true fungi, whose usual or dominant form is unicellular" can be grouped as ascomycetes, basidiomycetes and imperfect yeasts with both ascomycetous affinities. Their mode of reproduction is by budding or fission, and budding yeast cells may be a stage in the life cycle of multicellular fungi.

Previous systems of yeast classification were discussed by Lodder and Kreger van Rij (1952) and by Lodder (1970). Their present classification is shown in Table I (Kreger-van Rij, 1984). The important yeast pathogens of Man and animals are classified in the imperfect fungi. Of the fifteen genera of *Cryptococcaceae*, five (*Candida* [*Torulopsis*], *Cryptococcus, Malassezia* [*Pityrosporum*], *Rhodotorula* and *Trichosporon* [*Geotrichum*] include species pathogenic to Man and animals.

The cross connections between *Candida* and *Torulopsis* are so numerous that these genera have usually been discussed jointly in taxonomical treatises (van Uden and Vidal-Leiria, 1970). van Uden and Buckley (1970) pointed out the arbitrary nature and artificiality of the separation of the two genera on the criterion of the ability to form pseudohyphae, concluding that it would be logical to unite them. Yarrow and Meyer (1978) proposed an amendment of the diagnosis of the genus *Candida berkhout* to allow for non-hyphal species: in accordance with the amended description, the species currently classified in *Torulopsis* were transferred to the genus *Candida*. Kreger-van Rij (1984) merges the genera. *Pityrosporum*, too, seems no longer to be a valid generic name, since many observers are convinced of its identify with *Malassezia*. Thus, Yarrow and Ahearn (1984) regard *Pit. furfur, Pit. ovale, Pit. malassezi* and *Pit. orbiculare* as synonyms of *Mal. furfur* (Robin) Baillon (1899), and *Pit. pachydermatis* and *Pit. canis* as synonyms of *Mal. pachyder-*

Table I. Classification of the yeasts in the Eumycota

Ascomycotina		
Hemiascomycetes		
	Endomycetales	Spermophthoraceae
		Saccharomycetaceae
Basidiomycotina		
	Ustilaginales	Filobasidiaceae
		Teliospore-forming yeasts
	Tremellales	Sirobasidiaceae
		Tremellaceae
Deuteromycotina		
	Blastomycetes	Cryptococcaceae
		Sporobolomycetaceae

matis (Weidman) Dodge (1935). Kreger-van Rij (1984) notes that *Geotrichum* is distinguished by the formation of hyphae breaking up into arthrospores and by the absence of loose budding cells and blastospores. In her monograph, species that some would assign to *Geotrichum* have been provisionally retained in the genus *Trichosporon.*

All genera of yeasts form separated cells, and most yeasts reproduce by budding, the exceptions being the marine genus *Sterigmatomyces* and the ascosporogenous fission yeasts of the genus *Schizosaccharomyces*, neither of which contains pathogenic species. Although the perfect forms of some of the pathogenic yeasts are known, in practice perfect fungi are not isolated from pathological lesions, other than rarely (Murphy *et al.*, 1985) although they are occasionally cultivated from biological specimens. Minimal cytopathic effects may be exerted in tissue culture by some perfect yeasts (Hurley and Stanley, 1969) but, nevertheless, the pathogens of Man and animals are imperfect fungi, reproducing asexually by budding. Kreger-van Rij (1984) tabulates the relationship between perfect species and their imperfect states. The perfect states of yeasts of medical importance are shown in Table II.

Certain of the important pathogenic fungi of Man exhibit dimorphism with distinct phases of yeast-like and of mycelial growth (Table II).

Table II. Manifestation of dimorphism in pathogenic fungi

Species	Tissue phase	Culture phase	Classified as:
Malassezia furfur (*Pityrosporum orbiculare*)	M[a]	Y	Yeast
Candida spp.	M and Y	M and Y	Yeast
Histoplasma capsulatum	Y	M	Mould
Sporotrichum schenckii	Y	M	Mould
Paracoccidioides brasiliensis	Y	M	Mould
Blastomyces dermatitidis	Y	M	Mould
Coccidioides immitis	Sporangia (Y)	M	Mould

[a] M indicates mycelial phase and Y single cells or yeast phase.

Malassezia furfur is a pleomorphic fungus, the hyphal phase of which may develop in human skin, whereas in artificial culture the yeast phase (*Pityrosporum*) proliferates. *Candida* species, too, exhibit dimorphism. In infected tissues both yeast (Y) and mycelial (M) forms of growth may be seen, and both appear in culture, the predominant form depending on the nature of the growth medium (Audrey, 1887). Many have sought to equate the phases of fungal growth with virulence, and with the pathogenicity of the various species. Odds (1979) reviewed the subject of dimorphism in *C. albicans* exhaustively, concluding that filamentation alone does not account for its pathogenicity, although production of hyphae, indicating active growth *in vivo*, confers on the species an enhancement of its virulence for mammalian hosts. Thus, the experiments of Stanley and Hurley (1967) and Hurley and Stanley (1969) demonstrated a relationship between cytopathic effect and the ability of species to produce filaments in tissue cultures, which was related to growth rates of the fungi. Recent work on dimorphism of *C. albicans in vitro* suggests that hyphal formation is a transitory response under environmental conditions of high temperature and neutral pH values. In other important mycoses, the causative fungus grows *in vitro* and its natural saprophytic state is a mould, appearing in the tissue phase in yeast-like form with discernible separated cells. Such are *Sporotrichum schenckii, Coccidioides immitis, Paracoccidioides brasiliensis, Histoplasma capsulatum* and *Blastomyces dermatitidis*. These fungi show a true mycelial growth when cultivated on mycological media at room temperature and are, therefore, distinguished from yeasts and regarded as filamentous fungi or moulds (Ainsworth, 1958). Although not yeasts, these fungi, and the mycoses that they cause, will be considered briefly in this account, since cells resembling yeasts are seen in histopathological preparations, thus requiring, in some cases, distinction from the true yeasts before the results of culture are available.

Diseases caused by fungi are known collectively as mycoses, and range from very common mild chronic infections affecting the skin, through deep cutaneous or subcutaneous infections, to acute and chronic infections of the viscera and deep tissues. Other than the dermatophytoses, infections by yeasts outnumber all other mycoses. Amongst the yeasts, the most versatile pathogens are those of the genus *Candida* which cause both superficial and deep-seated disease which is worldwide in distribution. Of relatively low virulence when compared with the bacteria and viruses associated with epidemics and pandemics, candidas are generally classified as opportunistic pathogens, causing disease in hosts that are compromised by some underlying pathological process or deficiency state, either determined genetically or occasioned by extremes of age or youth. They are prone to attack the newborn or debilitated sick (oral thrush), the pregnant (vulvovaginitis), the immunosuppressed or the victims of certain malignant diseases prominent amongst which is leukaemia (candida septicaemia). They cause, also, endocarditis, renal disease and other visceral mycoses. Indeed, no organ in the mammalian body has been spared the depredations of candida. Members of the genus *Rhodotorula*, and yeasts formerly classified in the genus *Torulopsis*, though far less frequently encountered in clinical practice, may cause disease resembling that caused by candidas. Infection and colonization of the newborn by the ascosporogenous yeast *Hansenula anomala* has been reported (Murphy *et al.*, 1985). Following Odds (1979) and, as he states, for the sake of linguistic ease, the term "candidosis" will be used in this chapter as a generalization for infections due to *Candida* species, to yeasts formerly assigned to the genus *Torulopsis* and to *Rhodotorula* species. The term "candidosis" is preferred to "candidiasis" and to the older "moniliasis" since it is that used in both authoritative monographs on diseases caused by candidas (Winner and Hurley, 1964; Odds, 1979).

Next to the candidas, in pathogenic importance, is *Cryptococcus neoformans*, causing acute, subacute or chronic disease of the meninges, lung or bones, and behaving, also, as an opportunistic pathogen. *Pityrosporum* (*Malassezia*) species (*Pit. orbiculare, Pit. ovale*) can be isolated from the lesions of pityriasis (Tinea) versicolor, and from patients with dandruff. *Pityrosporum pachydermatis* has been isolated from lesions of otitis in dogs, and the original isolate (Weidman, 1925) came from the inflamed skin of a rhinoceros. Its pathogenic role is uncertain. *Pityriasis versicolor* is an extremely common skin infection, seen in all parts of the world, and affecting a high proportion of the populations of Africa, South-East Asia, India and Central and South America. There is evidence of genetic susceptibility (Burke, 1961) and onset of the disease, which occurs characteristically in the second, third and fourth decades of life, often coincides with the manifestations of serious illness or during pregnancy. *Trichosporon cutaneum* is the causative agent of white piedra, an infestation of the hairs chiefly of the beard or moustache of Man, and of the hairs of the mane and tail in horses. It is characterized by the presence of soft white- to light-brown-coloured nodules along the hair shaft. The affectation occurs in temperate climates, and the fungus is hardly pathogenic in the ordinary sense although, experimentally, *Tr. capitatum*, a similar yeast sometimes isolated from sputum, can cause fatal disease in rabbits inoculated intravenously (Gilbert and Fetter, 1962).

Most of the yeasts that cause disease in Man and in animals are commensals, exerting pathogenic effects only when the balance between the host and its indigenous flora is deranged. Opportunity for the fungi to invade tissues is afforded by many of the procedures of modern medicine, such as organ transplantation, open-heart surgery, prolonged intravenous hyperalimentation, and the use of cytotoxic and immunosuppressive drugs, as well as of antibiotics that suppress the normal bacterial flora. Deep-seated infections caused by these fungi are serious and life-threatening, for all have an extremely grave prognosis if not diagnosed or predicted promptly. Specific antifungal therapy is available for treatment.

II. PATHOGENS CAUSING CANDIDOSIS

Winner and Hurley (1964) reviewed candidosis, with especial reference to disease caused by *C. albicans*, collating the descriptions of systemic (acute disseminated septicaemic) candidosis, of candida endocarditis, and of renal candidosis from case reports published sporadically since the 1880s augmented by their own experimental and clinical findings and those of others. This, the first monograph, was followed by a series of essays by experts on particular aspects of diseases caused by candidas (Winner and Hurley, 1966) and by a review of the pathogenic *Candida* species (Hurley, 1967). A comprehensive review of English-language contributions to knowledge of the entire field relating to candida and candidosis has been compliled by Odds (1979).

The principal pathogen of the genus *Candida* is the thrush fungus, *C. albicans*, discovered in 1839 by Langenbeck in the aphthae of a patient suffering from typhoid (typhus) fever and believed by him to be causally related to the latter. Berg (1846) appreciated the relationship between the thrush fungus and oral lesions, and Wilkinson (1849) described mycotic vulvovaginitis. In 1875, Haussmann demonstrated that the causative agent was the same both in oral thrush and in mycotic vulvovaginitis. Zenker (1862) and Schmorl (1890) described dissemination of the fungus to deep tissues.

The taxonomic position of the thrush fungus was greatly confused until, in 1923, Christine Berkhout suggested the generic name "Candida", thus distinguishing the so-called "medical monilias" from the fruit- and leaf-rotting moulds more properly assigned to the genus *Monilia*. According to the rules of botanical nomenclature, the generic name *Monilia* was retained for the organism it first described, and *Candida* has been adopted internationally as the botanical *nomen conservandum* for the genus containing the thrush fungus. Many synonyms for *C. albicans* had been used, van Uden and Buckley (1970) and Meyer *et al.* (1984) listing over a 100, with fewer synonyms for other members of the genus. The taxonomic confusion surrounding the yeast was reflected in the name used to describe the disease(s) caused by it. Although "moniliasis" has been virtually expunged, it is still used occasionally in medical parlance.

Castellani, in a protracted series of writings (Castellani, 1912, 1925, 1963; Castellani and Chalmers, 1919), first suggested that yeasts other than *C. albicans*, might be aetiological agents of candidosis, and described the yeasts,

Table III. Ecological sources of 81 *Candida* species

Man	27
Warm-blooded animals	20
Products of fermentation or decayed vegetation	21
Marine	9
Insects	10 + 1 variety
Plants or inanimate sources	66

Note that groups are not mutually exclusive

presently known as *C. guilliermondii, C. krusei, C. pseudotropicalis* and *C. tropicalis*, although he assigned them to the genus *Monilia*.

A. Ecological considerations

The genus *Candida* comprised 81 species (van Uden and Buckely, 1970) to which were added a further 69 species, many formerly classified as species of *Torulopsis* (van Uden and Vidal-Leiria, 1970) but now named as candidas (Yarrow and Meyer, 1978), together with other species such as *C. milleri* (Yarrow, 1978) which have been described since 1970. Meyer *et al.* (1984) now name 196 species. There were nine species of *Rhodotorula* (Phaff and Ahearn, 1970); now, Fell *et al.* (1984) name eight. Many of these yeasts have been isolated from Man, and from a number of mammals and non-mammals; still others have been isolated from inanimate sources (Table III). Many of these yeasts are members, or mating types, of perfect species. Excluding species that are closely related to, or more properly classified with, perfect yeasts, the genus split readily in 1970 into three fairly distinct ecological groups. These comprise those species associated predominantly with Man or

Table IV. *Candida* species isolated principally from Man or warm-blooded animals[a]: Group 1

Man		Warm-blooded animals
aaseri	*melibiosica*	*blankii*
albicans[b]	*membranaefaciens*	*ciferrii*
brumptii	*norvegensis*	*lusitaniae*
catenulata	*parapsilosis*[b]	*silvae*
conglobata	*pseudotropicalis*[b]	*slooffii*
curvata	*stellatoidea*[b]	*zeylanoides*
guilliermondii[b]	*tropicalis*[b]	
krusei[b]	*viswanathii*[b]	
langeronii		

[a]Excluding species formerly classified as *Torulopsis* spp.
[b]Undoubted pathogens also occur in other species of animal.

Table V. *Candida* species[a] isolated occasionally from Man or warm-blooded animals[b]: Group 2

Man	Warm-blooded animals
intermedia *lipolytica* *obtusa* *ravautii* *rugosa* *silvae* *tenuis* *utilis*	*lambica*

[a]Excluding those classified as species of *Hansenula, Kluyveromyces, Leucosporidium* or *Metschnikowia*.
[b]Excluding species formerly classified as *Torulopsis* spp.

mammals, those that are isolated occasionally from Man or mammals, but not predominantly so, and those that are isolated from soil, sea water, products of fermentation, vegetation, cold-blooded animals and various inanimate sources, but not from Man or warm-blooded animals (Tables IV, V and VI). The last group is fairly distinct physiologically from the other two groups. While many species have been added to those recognized, these broad groupings still obtain. Do Carmo Sousa (1969) regarded *C. albicans, C. stellatoidea* and *C. glabrata* as obligatory animal parasites, classifying *C. guilliermondii, C. krusei, C. parapsilosis* and *C. tropicalis* as facultative saprophytes since they can be isolated from sources other than animals.

Yeasts of the genus *Candida* are abundantly distributed in nature on land and sea, associated with animals, plants and inanimate objects (Skinner and Fletcher, 1960). The habitat of pathogenic yeasts is more circumscribed, and has been reviewed by Winner and Hurley (1964), Austwick *et al.* (1966), Hurley (1967), Gentles and La Touche (1969) and Odds (1979). Scientific literature abounds in reports of the distribution of yeasts in Man and both

Table VI. *Candida* species[a] not reported as being isolated from Man or warm-blooded animals[b]: Group 3

aquatica *beechii* *berthetti* *bogoriensis* *boidinii* *buffonii* *cacaoi* *capsuligena* *claussenii* *curiosa*	*diddensii* *diffluens* *diversa* *foliarum* *freyschussii* *friedrichii* *frigida* *gelida* *glaebosa* *ingens* *javenica*	*kefyr* *marina* *maritima* *melinii* *mesenterica* *mogii* *muscorum* *oregonesis* *rhagii* *sake* *salmanticensis*	*salmonicola* *santamariae* *scottii* *shehetae* *silvicola* *solani* *valida* *vartiovaarai* *veronae* *vini*

[a]Excluding those classified as species of *Hansenula, Kluyveromyces, Leucosporidium* or *Metschnikowia*
[b]Excluding species formerly classified as *Torulopsis* spp.

wild and domestic animals. Indeed, these are the most frequent of all communications relating to candidas. Some are little better than lists of fungi isolated from material sent to diagnostic laboratories for routine examination, no effort being made to assess the true incidence and distribution by prospective study, or to correlate the findings with clinical, experimental or immunological data. However, even the more pedestrian reports show which species are likely to be encountered in diagnostic work and, where the series are large enough, they reflect the geographical distribution of the fungi.

The gastrointestinal tract is the most frequent source in the majority of animals. Most surveys attest the pre-eminent position of *C. albicans* which has been recovered from a far wider range of animals than any other species and which, with few exceptions, is also the yeast with the highest relative isolation rate. The principal data accumulated since 1960 have been extensively tabulated by Odds (1979; Table VII). Few generalizations can be made from these data other than to remark that pathogenic yeasts are more frequently isolated from populations of patients than from healthy individuals, that there is a tendency for yeast colonization to increase with age of the patient, particularly in the case of the oral cavity, and that the isolation rate from the vagina increases both during pregnancy and in the presence of local morbidity. Yeasts are not denizens of healthy skin, although they may be isolated, presumably in the course of transient carriage, from the skin of those who attend patients with skin disorders, and in some other individuals (Aly *et al.*, 1979). Some species, such as *C. parapsilosis*, have a tendency to colonize devitalized skin, and can be isolated from the vicinity of chronic lesions, such as varicose ulcers (English *et al.*, 1971).

There are few reports of isolation of pathogenic commensal saprophytic

Table VII. Recovery of yeasts and/or *Candida albicans* from various sites in Man. From Odds (1979)

Site	Status of subjects	Yeasts recovered (% range)	*Candida albicans* recovered (% range)
Oral cavity	Healthy	36.9–2.0 (14.3)	23.1–1.9 (10.3)
	Sick	76.2–12.7 (45.2)	69.6–6.0 (42.9)
Anus and rectum	Healthy	39.1–19.4 (23.7)	23.2–12.2 (14.6)
	Sick	62.2–5.7 (25.5)	48.0–8.5 (22.0)
Vagina	Healthy	13.3–8.1 (9.5)	11.3–0.0 (7.8)
	Sick or pregnant	67.9–7.8 (20.7)	52.2–4.5 (14.9)
	Local disease	76.5–25.0 (29.1)	84.4–14.2 (25.7)
Skin	Healthy	17.4–0.0	17.4–0.0
	Sick	16.5–4.7	12.4–0.4
Urine	Sick	17.6– > 3	4.4–1.6
Outer surface eye	Healthy or Sick	8.7–0.0	0.7–0.0

Values in parentheses are calculated mean frequencies, weighted according to the number of subjects.

Table VIII. Isolation of *Candida albicans* from some exogenous sources

Source	Reference
Containers of drugs of addiction	Joachim and Polayes (1940)
Medicinal cream	Harmon *et al.* (1962)
Blankets, pillows, air	Clayton and Noble (1966)
Disinfectant hand cream	France (1968)
Intravenous fluids	Ashcroft and Leape (1970)
	Deeb and Natsios (1972)
Intravenous cannulas	Collin *et al.* (1975)
Freshwater bathing beaches	Sherry *et al.* (1979)
Arterial catheters	Band and Maki (1979)

species from plants or inanimate objects, although *C. albicans* (Rogers and Beneke, 1964) and *C. tropicalis* (Shome, 1963) have been isolated from soil, and *C. guilliermondii* and *C. parapsilosis*, with other candidas, from fruits (Batista *et al.*, 1961c). It was suggested that fruit may play an epidemiological role in the pathogenesis of some superficial and deep mycoses. Hypersensitivity to *C. tropicalis* has been precipitated by the ingestion of fungus contaminated pineapples (Černík *et al.*, 1960).

Some exogenous sources of candida may play a salient role in the aetiology of superficial and of deep-seated infection. The more important of these are shown in Table VIII.

The relative incidence of yeasts isolated from Man merits consideration. Hurley *et al.* (1973) compared their relative incidence in vaginal swabs studied prospectively (that is, in those who may or may not have had local candidosis) and in those sent for diagnosis of vulvovaginitis, finding that *C. albicans* predominated in both series but that the frequency of its isolation, relative to *C. glabrata* and other less frequently encountered yeasts was lower in the unselected population. In this study more recondite species were represented in greater variety and in higher proportion (Table IX). Odds

Table IX. Recovery of yeasts from the human vagina

Species	Diagnostic specimens		Unselected specimens	
Total	131		182	
Candida albicans	118 (90.5%)		139 (76%)	
Candida albicans and *Candida glabrata*	2		1	
Candida glabrata	8 (6.0%)		27 (15%)	
Candida stellatoidea	2	(2.5%)	3	(8.5%)
Candida tropicalis	0		2	
Candida parapsilosis	0		2	
Candida holmii	0		1	
Candida inconspicua	0		1	
Saccharomyces cerevisiae	1		6	

Table X. Yeast flora of the lower genital tract in 280 unselected pregnant women listing 70 isolates from 68 patients. From Hurley *et al.* (1974)

Organism	Number	Organism	Number
Candida albicans	47	*Saccharomycopsis lipolytica*	1
Candida glabrata	11	*Candida castellii*	1
Candida albicans and *Candida glabrata*	2	*Saccharomyces cerevisiae*	1
Rhodotorula rubra	2	*Saccharomyces uvarum*	1
Candida krusei	1	*Pichia norvegensis*	1

(1979) collated a number of studies on yeast flora, reaching the same conclusion. All of the pathogenic candidas, other than *C. viswanathii*, have been isolated from the vagina, as have *Rhodotorula* species and other yeasts. Some typical findings in unselected women are shown in Table X. The data of Odds (1979) show that all, except for *C. stellatoidea* and *C. viswanathii*, have been isolated from the oral cavity, faeces or anorectal swabs, and most have been isolated from urine. With respect to the latter, unless the specimens are collected with meticulous care, yeasts from the vagina, perineum or gut may contaminate them. *Rhodotorula* species account for less than 0.5% of yeasts in the mouth, but for more than 12% of isolates from faeces or anorectal swabs. *Candida albicans* is recovered most frequently, irrespective of the site sampled. *Candida glabrata* is recovered from about 5% of those with mycotic disease of the vagina, and *C. parapsilosis* is isolated fairly frequently from skin lesions of paronychia, and, with *C. tropicalis*, is second only to *C. albicans* as a cause of deep-seated mycotic disease.

These observations, culled from experience and from analysis of the multitudinous data on the yeast flora of Man, have important consequences in clinical diagnositic practice. Most laboratory personnel have neither the skill nor the resources to engage on the labour-intensive exercise of identifying to species status every yeast isolated from the gastrointestinal tract or the integument. In the majority of cases, it suffices to state whether the yeast isolated is or is not *C. albicans*, for this is the principal pathogen of Man and animals. Its identity is established readily and rapidly from colonial and morphological features, from the presence of germ-tubes (a test of rarely matched specificity) and of chlamydospores (Mackenzie, 1966). The exception to this general rule is cultures, taken in the course of surveillance or otherwise, from any site in those patients who are at risk of systemic fungal infection. Here, the presence of yeast pathogens other than *C. albicans* may predicate serious fungal infection, and the isolate must be identified. The value of surveillance cultures has been established (Sandford *et al.*, 1980). Yeasts isolated from the bloodstream, other body fluids or the deep tissues, should be identified to species level. Although this is usually beyond the expertise of non-specialist laboratories, kits are available commercially and their performance has been compared (de Louvois *et al.*, 1979). The API 20C (Auxanogramme) kit is satisfactory for this purpose.

Many studies of the incidence and distribution of yeasts in wild and domestic animals have been made, but particularly large and comprehensive surveys were made in the 1960s. Bisping (1961) studied over 4000 samples from horses, cattle, pigs, dogs, hens and pigeons, finding the highest incidence of *C. albicans* in hens and pigs. Amongst the fungi isolated were *C. tropicalis, C. krusei, C. pseudotropicalis, C. guilliermondii* and *C. parapsilosis*. Batista *et al.* (1961b) studied fresh dung samples from sheep, goats, pigs, poultry and other animals, reporting an incidence of *C. tropicalis* higher than the average in sheep, goats and pigs. *Candida albicans* was isolated only once from 60 head of poultry. Tewari (1963) studied cattle, buffaloes, dogs, horses, goats, sheep, pigs and poultry and found *Candida* sp. including *C. rugosa* in 56.6% of adult birds and 14.6% of chicks. He isolated *Candida* spp. from the nasal passages of 54.9% of cattle with nasal granuloma, but from only 6.6% of apparently healthy animals. *Candida albicans* predominated in poultry and *C. tropicalis* in cattle.

Clarke and di Menna (1961) commented that, amongst the yeasts isolated from the bovine rumen, were *C. albicans, C. tropicalis* and *C. krusei*, and candidas were isolated more frequently than other yeasts from the cervix and uterus of cows and heifers (Hajsig *et al.*, 1962). Van Uden and do Carmo Sousa (1962) isolated *C. slooffii* from the intestinal tracts of 47% of swine, and Mehnert and Koch (1963) isolated this yeast from 75% of pigs, also isolating *C. krusei* from 60% of the population examined. Batista *et al.* (1961a) grew *C. tropicalis, C. krusei, C. guilliermondii, C. parapsilosis, C. intermedia* and other fungi from faeces of horses, also isolating the first three species from faeces of mules. Yeasts, predominantly *C. albicans*, have been isolated from crops of chickens and turkeys, apparently unassociated with disease in the majority of cases (Bacle and Bigland, 1962). *Candida bovina*, as well as pathogenic candidas, were isolated from the crop as well as the deep tissues of unhealthy fowl (Prophet, 1963). The digestive tracts of gulls and terns yielded *C. tropicalis* and *C. albicans* (Kawakita and van Uden, 1965) as did those of many wild animals, including baboons, hippopotami, zebras, wart hogs and various monkeys (Hurley, 1967). Studies have been made on the faeces and chlorinated salt water from pools containing bottle-nosed dolphins (Buck, 1980) which yield *C. tropicalis, C. parapsilosis, C. glabrata* and *C. albicans*. The yeast flora of a female giant panda (Saez, 1979) was studied over a period of weeks, with some evidence that *C. albicans* did not adapt well to the environment of this host. Yeasts, including species of *Torulopsis, Candida* and *Rhodotorula* were isolated from the cervix of breeding and trotting mares, but from observation on specimens from stallions it seemed that spread of yeasts, including candidas, during mating was uncommon

(Sonnenschein *et al.*, 1978). These studies show that the distribution of candidas is wide in the animal kingdom, and that the prevalence of one or other species is probably determined by the host species, depending, perhaps, on dietary habit. *Candida albicans* predominates in Man and birds, whereas *C. tropicalis* and *C. krusei* (Austwick *et al.*, 1966) are frequently isolated from cattle, and *C. slooffii* is commonly found in pigs and also in wart hogs. Some workers have suggested that domestic animals may be reservoirs for human infection, but this postulate is unnecessary in view of the widespread infection of Man himself.

B. Pathogenic species

The thrush fungus was successfully transmitted in Man in 1846 by Berg, who was able to produce lesions experimentally in four previously healthy babies 3–26 days old. He grew the yeast on media containing blood, a finding he disregarded assuming that his cultures (for he was studying a mycelial fungus) had been contaminated. Delafond (1858) was able to produce thrush in the mouths of lambs but only if the animals were weak or debilitated. Haussmann (1875) elucidated the aetiology of vaginal candidosis by introducing scrapings from the mouth of an infant with thrush into the vagina of a pregnant woman, thus inducing vaginitis. There followed an era of animal experimentation and, by 1898, de Stoechlin was able to write that all experimentalists were agreed that the thrush fungus was pathogenic for laboratory animals. The experiments of Redaelli (1924) and of Rhoda Benham (1931) established *C. albicans* as the only member of the genus of proven pathogenicity. They found that *C. albicans*, injected intravenously into rabbits, gave rise to a fatal disease, whereas other species did not. Unhappily, they failed to vary the dose, the route of infection, the species of host inoculated, to wait sufficiently long before terminating the experiments and to perform adequate histological studies, thereby failing to observe the pathogenic effects of other species, with fatal consequence for many. The criterion of pathogenicity for the rabbit became accepted as part of the species identification of *C. albicans*, but the work of these experimentalists led to the dangerous and erroneous concept that other species of *Candida*, even when isolated from the bloodstream in the demonstrable presence of disease, were of no consequence. The experiments of Hasenclever (Hasenclever, 1959; Hasenclever and Mitchell, 1961a, b) in the U.S.A., and of Hurley (Hurley, 1962, 1965, 1966; Hurley and Winner, 1962) in the U.K., working in the 1950s and the early 1960s refuted these findings, established the pathogenicity of other species and placed the concept of deep-seated candidosis caused by *C. albicans* and other species firmly in medical practice. Discounting earlier reports where mycological documentation was unsatisfactory, by the 1960s disseminated infection with *C. albicans, C. tropicalis* and *C. pseudotropicalis* had been reported (Hurley, 1962, 1964; Winner and Hurley, 1964) and infective endocarditis ascribable to *C. albicans, C. parapsilosis, C. guilliermondii, C. krusei* and *C. tropicalis* had been described (Winner and Hurley, 1964).

Because of the prevalence of the false doctrine that species of *Candida* other than *C. albicans* were likely to be contaminants if isolated from the bloodstream, experimentalists took steps to establish that the putatively pathogenic species obeyed Koch's postulates. In the main, they succeeded. *Candida albicans, C. tropicalis, C. stellatoidea, C. pseudotropicalis* and *C. viswanathii* (Sandhu *et al.*, 1965) can all be shown to produce the typical lesions of systemic candidosis on intravenous injection into mice or other laboratory animals. *Candida parapsilosis* produces visceral lesions in cortisone-treated mice (Goldstein *et al.*, 1965) and typical lesions have been produced by *C. glabrata* in pregnant or cortisone-treated mice (Hasenclever and Mitchell, 1963; Knudtson *et al.*, 1973). No lesions ascribable to *C. guilliermondii* have been demonstrated in intact animals. Partridge *et al.* (1971) showed that *C. albicans, C. tropicalis, C. pseudotropicalis* and *C. parapsilosis* caused lesions on chick chorioallantoic membrane, whereas *C. glabrata* did not. Stanley and Hurley (1967) showed that cultured renal epithelial cells of five-day-old mice were destroyed by *C. albicans, C. tropicalis, C. stellatoidea* (Group 1), *C. krusei, C. parapsilosis, C. pseudotropicalis* (Group 2) and by *C. guilliermondii* (Group 3), in all instances more rapidly and more completely than by the control yeast *Sacch. cerevisiae*. *Candida lusitaniae* seems to be a cause of bovine mastitis (Austwick *et al.*, 1966) and has been identified as an agent of deep-seated candidosis in man (Holzschu *et al.*, 1979). Although species of *Rhodotorula*, notably *Rh. rubra* (Stenderup *et al.*, 1978) and *Rh. glutinis* are often isolated from Man, and rhodotorulas are associated with systemic candidosis, the yeasts are seldom, it seems, speciated in clinical pratice, and there have been few experiments on animal hosts. Gentles and La Touche (1969) regarded *Rh. mucilaginosa* (*rubra*) as an opportunist pathogen, and the criterion of animal pathogenicity has been satisfied for *Rh. rubra* in the case of conjunctivitis (Segal *et al.*, 1975). Phaff and Ahearn (1970) regard *Rh. mucilaginosa* as a synonym for *Rh. rubra*. *Rhodotorula glutinis* has been isolated from the sputum of patients with bronchopneumonia. In Table XI, the yeasts are arranged in their probable order of virulence for Man, based partly on the results of cytopathic effect in tissue-culture systems and partly on clinical experience. Although *C. glabrata* is only rarely associated with deep-seated mycotic disease, it is placed between the Group 1 and 2 yeasts of Stanley and Hurley (1967), since it is the only species other than *C. albicans* to be associated frequently with vaginitis and is, thus, numerically, an important pathogen. The effects of strain

Table XI. Pathogenic yeasts of candidosis in probable descending order of virulence for Man

Candida albicans (Robin) Berkhout
Candida tropicalis (Cast) Berkhout
Candida stellatoidea (Jones et Martin) Langeron et Guerra
Candida glabrata (Anderson) Lodder et de Vries amended Yarrow and Meyer
Candida krusei (Cast) Berkhout
Candida parapsilosis (Ashf) Langeron et Talice
Candida pseudotropicalis (Cast) Basgal
Candida guilliermondii (Cast) Langeron et Guerra
Candida viswanathii Sandhu et Randhawa
Candida lusitaniae van Uden et do Carmo-Sousa
Rhodotorula rubra (mucilaginosa) (Demme) Lodder

virulence are unlikely to be of consequence for they are small, indeed, in these rather stable yeasts. The perfect states known or postulated for some species do not seem to cause disease. Virulence depends, in large part, on the opportunity for invasion offered to commensal flora, and is a property perhaps inherent in the genome, finding phenotypic expression after adaptation to commensalism. This seems to occur seldom in nature (Saez, 1979). Six non-pathogenic yeasts, none isolable from Man, were inoculated into cultured mouse epithelial cells by Hurley and Stanley (1969). Five of them showed cytopathic effects of the same order as *C. guilliermondii.* Further work on four of these yeasts, after passage in intact hosts, showed that passaged strains of *C. blankii*, *C. kefyr*, *C. diddensii* and *C. cacaoi* became moderatley virulent, suggesting that candidas from disparate ecological sources can adapt to parasitism (Hurley and de Louvois, 1980).

C. Identification of yeast species causing candidosis

The species associated with candidosis are shown in Table XI. Their principal characteristics are shown in Tables XII and XIII.

Table XII. Morphological characteristics of major genera of yeasts isolated from biological specimens[a]

Morphological characteristics	Genera of yeasts
Pseudomycelium present	*Candida* spp.
Pseudomycelium absent or rudimentary	*Candida* (*Torulopsis*) spp. *Saccharomyces* spp. *Cryptococcus* spp. *Rhodotorula* spp.
Sporulation confirmed	*Saccharomyces* spp. *Hansenula* spp. *Kluyveromyces* spp. *Pichia* spp.

[a]Most of these yeasts grow well on blood agar, but *Candida glabrata* and *Saccharomyces* spp. do not.

Table XIII. Assimilation and fermentation properties used in identification of *Candida* species

Species	Assimilation of:							
	Glucose	Galactose	Maltose	Sucrose	Lactose	Raffinose	Germ tubes	Chlamydospores
Candida albicans	+	+	+	+	−	NT	+	+
Candida stellatoidea	+	+	+	−	−	NT	+	+
Candida tropicalis	+	+	+	+	−	NT	−	−
Candida krusei	+	−	−	−	−	NT	−	−
Candida pseudotropicalis	+	+	−	+	+	NT	−	−
Candida parapsilosis	+	+	+	+	−	NT	−	−
Candida guilliermondii	+	+	+	+	−	+	−	−
Candida glabrata	+	−	−	−	−	NT	−	−
	Fermentation of:							
Candida albicans	+	+(W)	+	−	−	−		
Candida stellatoidea	+	−	+	−	−	−		
Candida tropicalis	+	+	+	+	−	−		
Candida krusei	+[a]	−	−	−	−	−		
Candida pseudotropicalis	+	+	−	+	+	+/(W)		
Candida parapsilosis	+	±W/−	−	−	−	−		
Candida guilliermondii	+	+/(W)	−	+/(W)	−	+/(W)		
Candida glabrata	+	−	−	−	−	−		

[a] indicates that a surface pellicle is formed, NT that the test was not performed, and (W) that the reaction was weak.

In clinical practice, most yeasts of pathogenic importance are species of *Candida*, with *C. albicans* predominating. Many media are available for their isolation, but most diagnostic laboratories favour formulations of the peptone/glucose or peptone/maltose agars (Sabouraud, 1894), and their low pH value (5.5) suppresses growth of most bacteria. Addition of chloramphenicol (50 mg l^{-1}) enhances this inhibitory effect. Cycloheximide should not be added, as it is itself inhibitory to some pathogenic yeast species. The inoculated media should be incubated for not less than three days at 37°C, and this is certainly essential in the case of putative deep-seated mycotic infections where there is difficulty in isolating a fungus. Following isolation, the yeast is streaked out on purity plates and may be further subcultured and maintained if full species identification by biochemical and physiological tests is required.

Two rapid and reliable tests are available for confirmation of identity of *C. albicans*. In the "germ-tube test" at 37°C only *C. albicans* regularly produces hyphae from blastospores within three to four hours (Taschdjian *et al.*, 1960; Fig. 1). Occasional isolates of *C. stellatoidea* may also produce germ tubes. Pure cultures are not a prerequisite of the test (Mackenzie, 1966) but, clearly, the unknown isolate must be predominantly blastosporic (yeast-like) in primary culture and a light inoculum is essential (Mackenzie, 1962). Chlamy-

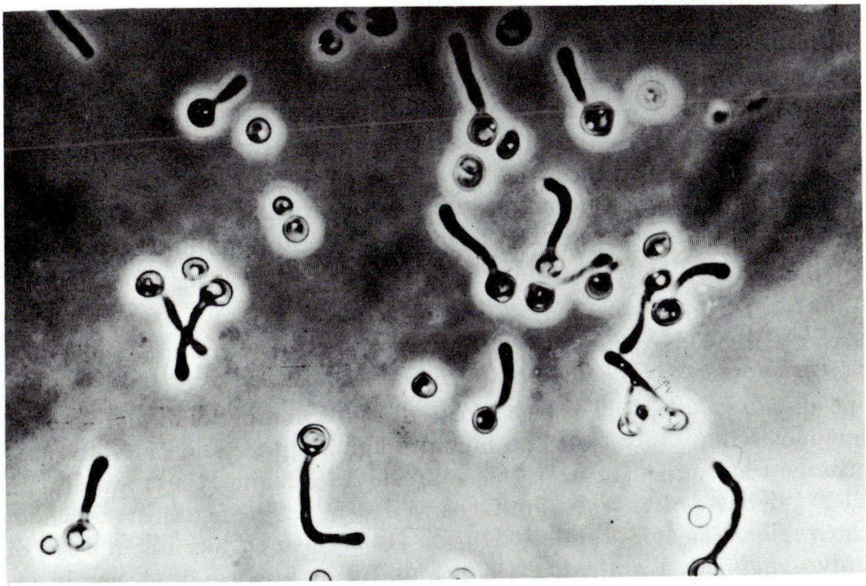

Fig. 1. Phase-contrast micrograph of germ tubes formed by *Candida albicans* grown in egg albumin. Magnification × 1000. From Gentles and La Touche (1969).

dospores are large (8–12 μm diameter), thick-walled, refractile spherical structures, usually borne terminally on hyphae, but occasionally intercalary, or in clusters (Fig. 2). They arise on protochlamydospores, which are round, elongate or pear-shaped. For optimal production, "partial anaerobiosis" is required. Under satisfactory test conditions, chlamydospores are formed abundantly by *C. albicans*, but only rarely or sparsely by *C. stellatoidea, C. tropicalis* and *C. clausenii*, or by other candidas (Pollack and Benham, 1957; Fahlberg *et al.*, 1957). Corn-meal agar (Benham, 1931) prepared from

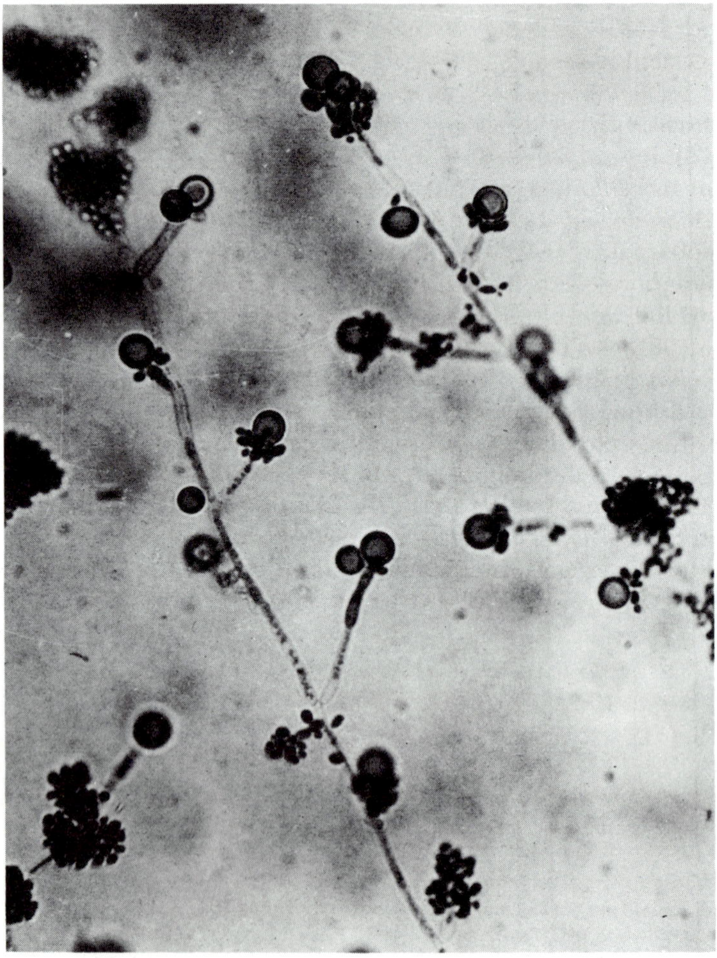

Fig. 2. Micrograph of chlamydospores of *Candida albicans* grown on corn-meal agar. The preparation was stained with lactophenol cotton blue. Magnification × 300. From Gentles and La Touche (1969).

ground maize meal has been widely used, its ability to stimulate the production of pseudohyphae also being valuable in the identification of yeasts. Czapek-Dox (Dawson, 1962) with added Tween 80, or potato/carrot/bile medium (Drouhet and Vieu, 1957) have usually been found to be superior (Mackenzie, 1966). Production of chlamydospores is certainly enhanced by addition of Tween 80 to the various media that have been used. The Dalmau inoculation technique, in which the inoculum is scratched into the medium with a loop, and the inoculated area covered with a glass coverslip, gives good results, and incubation at 25°C is recommended. Combination methods, in which the agar medium is examined after three hours at 37°C for germ-tube production, and then after a further 24 hours for chlamydospore production, are rapid (Beheshti *et al.*, 1975; Schaar *et al.*, 1974).

The ability to assimilate or to ferment individual carbon or nitrogen sources, together with other physiological properties, is the final step in identification of pathogenic candidas. Assimilation tests are performed as auxanograms (Beijerinck, 1889) usually by the paper-disc method, or growth in liquid media (Wickerham and Burton, 1948). The latter is more sensitive and more specific. Many modifications of both types of test have been described. Yeast fermentation of sugars, in solid and liquid media, is also used. Some of these reactions are shown in Table XIII, and a full account of all of the properties of the yeasts is given by van Uden and Buckley (1970). Presumptive identification is afforded by various commercial kits (de Louvois *et al.*, 1979). In practice, if a rare yeast is isolated from biological material from a deep source, it should be referred to an expert mycologist for confirmation of identity, and those who engage in research on clinical yeast infections should ensure that both internal and external quality-assessment control is exercised in the laboratory in the first instance.

Serological methods of yeast identification have not proved useful in practice and, although semi-selective media (Nickerson, 1963; Pagano *et al.*, 1958) have enjoyed some vogue, they are seldom used by experts. Odds (1979) discusses more novel approaches to identification of pathogenic candidas. Biotype, or strain identification of the candidas, which would have great value as an epidemiological tool, is in its infancy, although some schemes seem promising (Warnock *et al.*, 1979; Odds and Abbott, 1980). Yeasts are usually preserved by immersion in sterile distilled water (Castellani, 1963).

D. Diagnosis of infection by *Candida* species

Since members of the genus *Candida* are so widely distributed as commensals, particularly of the gastrointestinal tract, their visualization in smears from mouth, throat, sputum or faeces cannot be regarded as diagnostic of candidosis, except when combined with clinical evidence of mycotic infection (Scherr, 1954). A possible important exception to this rule is discussed below, and relates to prediction of systemic disease. Some writers have attempted to distinguish the pathological significance of the two phases of fungal growth, suggesting that only the demonstration of the mycelial form signifies candida sepsis, particularly in scrapings from skin or from mucous membranes. The balance of evidence suggests that this is not so; the yeast phase is equally significant. Pure cultures are of value when there are signs of infection, but even so are not necessarily diagnostic unless obtained directly from visible lesions of thrush. In all cases, clinical and microbiological data must be considered together.

Since *Candida* spp. are relatively hardy microbes, dry sterile swabs are suitable for transport to the examining laboratory although swabs in Amies or Stuart's transport medium may also be used. Examination of a wet preparation, or Gram-staining, will show budding yeasts, with or without mycelium. Other stains may be used. Some degree of specific staining can be achieved using fluorochrome-labelled antisera, but as yet in practice this is not widely used. Peptone/maltose/agar is a satisfactory medium for isolation, and the "germ-tube" test and demonstration of chlamydospores distinguishes *C. albicans* from other species. Species identification can be accomplished using the methods already outlined.

Culture of *Candida* spp. from sites that should be sterile, such as the bloodstream, body cavities, biopsy material or cerebrospinal fluid, is significant and probably always betokens systemic disease. Culture from catheter or midstream specimens of urine is more problematical in terms of clinical importance. Goldman *et al.* (1960) regarded counts of 10^3 fungus colonies ml^{-1} as signifying active infection, while Schönebeck (1972) concluded that any candida cultured from the sediment of a properly collected specimen of urine should be regarded as significant, commenting that clinically important infection may not develop unless the patient's general condition deteriorates. Schönebeck and Ånsehn (1972) concluded that there was no absolute correlation between the degree of funguria and the clinical picture. Speller (1975) discussed the difficulties of interpretation, suggesting that urine should be re-examined after removal of a catheter or treatment of predisposing factors. Diagnosis of candidosis of the urinary tract is properly established from the persistence of funguria, demonstration of excess leucocytes in the urine, and the presence of signs or symptoms referable to the urinary tract in the absence of bacterial pathogens. Fungi most commonly associated are *C. albicans, C. glabrata* and *C. tropicalis*. It is most unwise to regard *C. albicans* or other candidas as a contaminant from any site in the body. Culture from sites where the organism may play a commensal role must be interpreted in conjunction with the clinical state. Carroll *et al.* (1973) challenged the view that *C. albicans* was commonly a commensal of the female genital tract, though they did not deny the possibility of occasional contamination from the perineum. In their view, isolation of the fungus from this site is good presumptive evidence of active infection or of recent unsatisfactory treatment of mycotic vulvovaginitis. Clinical opinion of the value of serodiagnosis in candidosis varies widely, from the enthusiastic through the cautious, to the downright sceptical (Gaines and Remington, 1972; Taschdjian *et al.* 1973; Stone *et al.*, 1974). Reagents are not standardized, neither are the types of test nor the details of their performance, and the presence of detectable antibody may simply reflect carriage of the fungi. Problems of reproducibility and specificity are reviewed by Taschdijian *et al.* (1973). Candida antibodies belong to the IgA, IgG and IgM immunoglobulin sub-classes, and many types of test have been used for their demonstration. Immunoglobulin G antibody is the major reactant in precipitin and indirect fluorescent antibody tests; IgM is the principal reactant in haemagglutination tests; while all three are involved in direct agglutination. Secretory antibody, principally secretory IgA, has been detected in human vaginal tissues and secretions and in saliva, and IgA, IgM and IgG can all react with candida cells *in vivo*. Skin tests with candida antigens are not used in the diagnosis of candidosis as such, although they are of value in the detection of deficiencies of cellular immunity. Odds (1979) concludes, from his tabulated data, that qualitative tests for candida do not discriminate adequately between those with and those without candidosis, that this lack of discrimination is most evident in those who are predisposed towards candidosis, and that the precipitin test, using cytoplasmic antigens, appears to be most specific. Kozinn *et al.* (1976), analysing their own data and those of others, showed that its sensitivity ranged from 27 to 100%, its specificity from 57 to 96% and its predictive value from 20 to 90%. If serodiagnosis is to be used, it is clear from the foregoing that the tests must be performed serially and quantitatively; false positive reactions can thus be identified by their usually static or slowly falling titres (Odds, 1979). Tests for candida antigen are also available, but have been less thoroughly studied and evaluated (Munoz *et al.*, 1980).

Prediction of systemic candidosis is extremely important in clinical medicine, for the disease has a poor prognosis if not diagnosed and treated promptly. Some recent observations have been particularly helpful. Jarowski *et al.* (1978) reported a distinctive triad of signs — fever, rash and muscle tenderness — which is sufficiently unique to justify a presumptive diagnosis of disseminated candidosis. Goulon *et al.* (1973) and Griffin *et al.* (1973), discussing candida endophthalmitis, suggested that the ocular fundus should always be examined when candida septicaemia is suspected or feared. The discovery of a focal inflammatory intraocular lesion (Fig. 3) is strong evidence that metastatic infection with deep organ involvement has occurred.

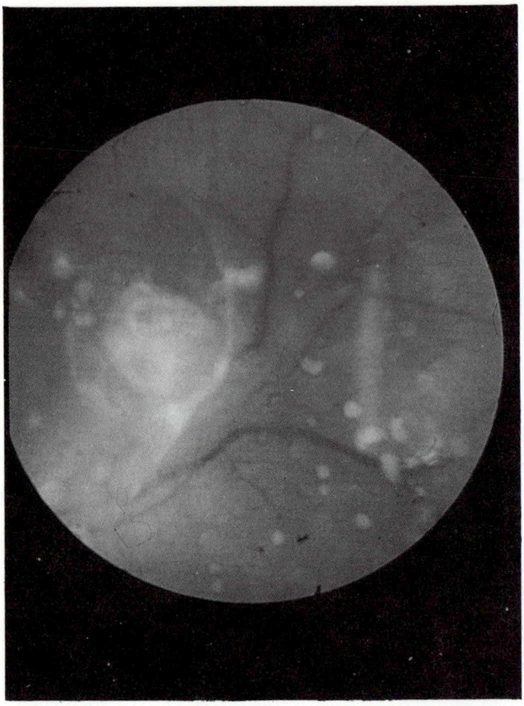

Fig. 3. Paramacular mycotic (candida) retinal lesions with multiple snowball lesions in the vitreum, secondary to candida gingivitis, in a woman with scleroderma being treated with prednisolone (200 mg daily). This eye made a full recovery to normal vision after prolonged treatment with 5-fluorocytosine, despite continuation of massive steroid therapy. Photograph reproduced by permission of Barrie R. Jones.

Sandford *et al.* (1980), although not the first to use fungal surveillance cultures as predictors of systemic fungal infection, gave some important leads to the way in which fungal surveillance might be used in a simple non-statistical manner. The complicated, labour-intensive and costly surveillance schemes used by those authors could be replaced in our opinion by simple culture of selected sites, at weekly or daily intervals, in those at risk of disseminated infection. Since candidas other than *C. albicans* are very rarely isolated as commensals, their isolation from those at risk, particularly if multiple sites were involved, would, in our view, presage deep-seated disseminated infection. Table XIV shows a scheme of clinical and microbiological surveillance for prediction and early diagnosis of septicaemic candidosis. If cutaneous lesions, otherwise unexplained, are observed, particularly if there is a maculo-papular rash involving the trunk, aspiration biopsy may reveal candida. The presence of the triad of fever, rash and muscle tenderness, or of retinal lesions, or the isolation of species other than *C. albicans* indicate

Table XIV. Tests for prediction or early diagnosis of septicaemic candidosis

A. Afebrile patients
Examine once weekly:
1. Stool and throat cultures for *Candida* species
2. Fundus oculi
3. Blood for *Candida* antibody by precipitin test using cytoplasmic antigen

B. Febrile patients
Examine daily
1. Blood and urine for *Candida* species
2. Stool and throat cultures for *Candida* species
3. Fundus oculi
4. Skin of trunk for maculopapular rash
5. Blood for *Candida* antibody, as in A

a need to withdraw antibacterial antibiotics, and to consider therapy with a specific antifungal drug. Serological tests should be evaluated with the clinical findings.

E. Types of infection caused by *Candida* species

1. General considerations

Probably the most important point to emphasize about infections caused by species of *Candida* is that they do not arise unless the host is debilitated or damaged in some way, or lies within the extreme of a physiological norm, such as the state of being newly born or pregnant. Thus, in their biological interactions, candidas exemplify the whole array of states that exist between the host and its indigenous pathogens, ranging from symptomless carriage, through self-limiting disease of the integument, to deeper infections and fulminant mycoses involving many body systems. Many factors are known to predispose to candidosis (Table XV) and the concept of previous morbid state (the "état morbide anterieure" of Parrot, 1877), as prerequisite to sepsis, is well attested. Factors relevant to pathogenesis are the size of the infecting inoculum, the route of entry of the fungus and general or local resistance of the host. Certain diseases are especially associated with candidosis. Diabetes mellitus has long been recognized as predisposing to candida vulvovaginitis (Hesseltine, 1933) and, more recently, to balanoposthitis (Skoglund, 1971), although the mechanism by which this is brought about is unknown. There appear to be no specific immunological deficiencies either in cell-mediated immunity (Ragab *et al.*, 1972) or in phagocytosis (Davies and Denning, 1972), although high glucose concentrations in blood or tissue (Knight and Fletcher, 1971) or low lactate concentrations in the skin (Kandhari *et al.*,

Table XV. Principal factors predisposing to candidosis in Man

Pathological or physiological	Endocrine disorders Defects in cell-mediated immunity Severe microbial infections Malignant disease Genetic defect Infancy Pregnancy
Mechanical	Trauma Local occlusion or maceration
Iatrogenic	Drugs altering or suppressing body flora or local immunity Intravenous therapy Transplant or prosthetic surgery

1969) seem to favour growth of candida. Multiple endocrinopathies, hypothyroidism, hypoparathyroidism and hypo-adrenocorticism are all associated with intractable candidosis of the integument (chronic mucocutaneous candidosis), often described as the candida endocrinopathy syndrome (Lehner, 1966). Amongst the malignant diseases, two, namely leukaemia (Bodey, 1966) and transitional cell carcinoma of the kidney (Robbins *et al.*, 1974), are associated with a particularly high incidence of deep-seated candidosis. Oral thrush occurs frequently in infancy (Parrot, 1877) and vulvovaginitis is far more common during pregnancy. The relevance of trauma to the aetiology of candidosis is less certain, although deep-seated infection certainly follows burns, and one of us has seen intractable thrush established in the scars of hare-lip. Deep-seated candidosis has followed surgical trauma (Winner and Hurley, 1964). The effect of maceration is evident in the proclivity of miners and barmaids to develop cutaneous candidosis, while the role of occlusion in the pathogenesis of skin candidosis was demonstrated in human volunteers by Maibach and Kligman (1962). These local factors operate in Man in the causation of angular cheilitis and denture stomatitis.

Septicaemic candidosis is an important hazard of transplant surgery, of intravenous hyperalimentation, and of immunosuppressive therapy, and is the most frequent fungal complication occurring in compromised hosts, even in parts of the World, such as the U.S.A., where other serious fungal diseases are endemic. The role of antibiotics in contributing to septicaemic candidosis is unproven (Winner and Hurley, 1964; Odds, 1979) but, overall, there is evidence that corticosteroid therapy does enhance susceptibility to candidosis.

Thus, all the clinical evidence and much of the experimental evidence indicates that in most individuals harbouring pathogenic candidas, no harm results. The fungi are, however, able to invade and damage tissues if host defences are impaired, either locally or generally. It is thus unsurprising that candidosis is one of a number of opportunistic infections occurring in the course of HTLV-III disease, in the acquired immunodeficiency syndrome (AIDS). Only in the most artificial of circumstances (Krause *et al.*, 1969) does potentially serious disease arise in those subjects within the physiological norm.

2. Spectrum of candidosis

(a) *Candidosis in Man*

Candidosis in Man is seen in two forms, superficial candidosis and the more severe deep-seated infection.

(i) Superficial candidosis. *Candida* species are associated with infections of the mouth, vagina, external ear, skin or nails. Some of the names given to these infections are listed in Table XVI. Oral thrush, denture stomatitis (Cawson, 1966) and candida vulvovaginitis are very common infections, with a worldwide distribution, that prove costly in terms of medical expertise and of money spent on treatment that, in the case of the last, is often ineffective. The clinical evidence of oral thrush is distinctive, with white patches or aphthae, appearing as discrete lesions on the surfaces of the buccal mucosa or tongue (Fig. 4). *Candida albicans* is the aetiological agent, and rates of 40 to 60% still occur amongst the newborn in some European centres. The disease also occurs in elderly or debilitated patients. Denture stomatitis is regarded as the most frequent form of oral candidosis, and has been reported in 60% of denture wearers aged 65 or more (Budtz-Jorgensen *et al.*, 1975). Candida vulvovaginitis, characterized by pruritus, discharge and the presence of plaques on vaginal, vulvar or cervical epithelium, may afflict up to 16% of pregnant women, and pregnancy thrush may be an initiating event in longstanding vaginal candidosis that proves refractory to treatment (Hurley, 1975). *Candida albicans* and *C. glabrata* are the aetiological agents, although disease caused by the latter does not seem to become inveterate.

Shelmire (1925) reviewed candida infections of the skin, and it is they that best exemplify the historic concept of the great physician and teacher,

Table XVI. Superficial candidoses

Oral thrush	Vulvovaginitis	Allergy
Denture stomatitis	Balanitis	Paronychia
Perlèche	Otitis externa	Onychia
"Leukoplakia"	Intertrigo	Congenital cutaneus candidosis
Glossitis	Napkin rash	Chronic mucocutaneous candidosis

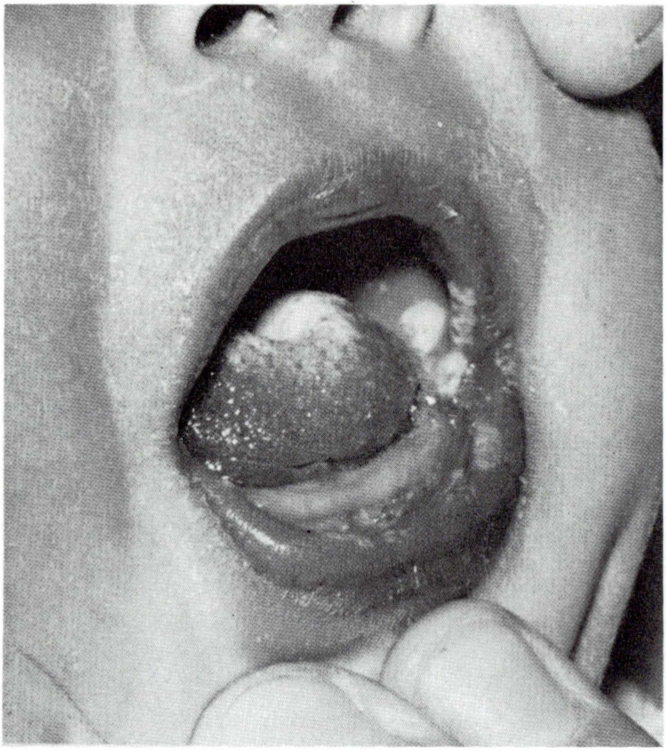

Fig. 4. Photograph showing oral thrush in a child aged eight days. From Winner and Hurley (1966).

Trousseau (1869), who was the first explicitly to expound the view that disease caused by candida is always secondary to some host-determined defect. Summarizing his writings and lectures spanning 25 years, he described candidosis as "the local expression of a very bad state of the whole system". Other than oral thrush, skin thrush occurs characteristically in intertriginous areas, where locally moist conditions are caused by maceration and occlusion; lesions may also appear between the fingers or the toes, and some, such as barmaids or miners, may be at risk because of their occupations. English and her colleagues (1971) were able to isolate *C. parapsilosis* from the moist skin surrounding leg ulcers, although not from the ulcers themselves, demonstrating the propensity of candidas to colonize unhealthy skin. The clinical features and treatment of skin candidosis are discussed by Winner and Hurley (1964) and Odds (1979). Napkin rash is an important and common manifestation. In recent years, several cases of congenital cutaneous candidosis (Sonnenschein *et al.*, 1960; Gellis *et al.*, 1976) have been described. The postulated route of infection is intra-uterine, and the disease is benign and self-limiting, clearing spontaneously after a few weeks. The rash is widely

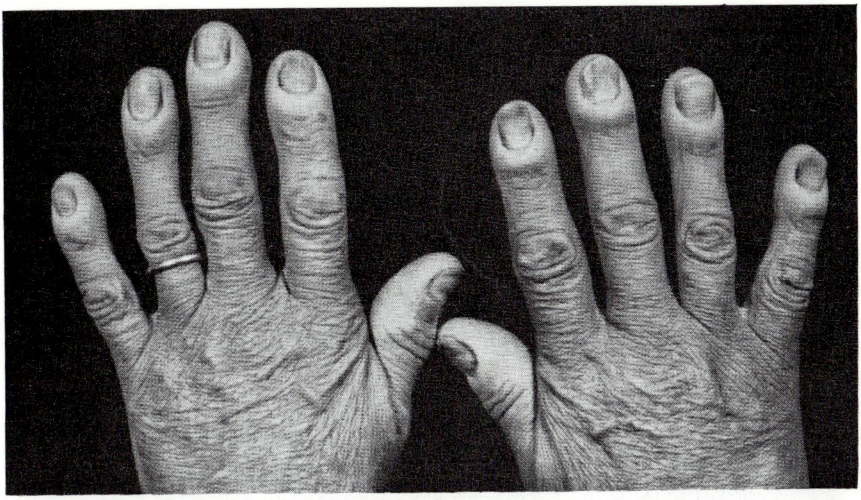

Fig. 5. Photograph showing chronic paronychia caused by *Candida albicans* affecting all digits. Note the bolster-like thickening of folds, pockets under the folds and loss of eponychium. From Winner and Hurley (1966).

distributed. Allergic "id" reactions to candida have been reported. Candida paronychia and onychia both occur as occupational diseases in housewives, fishmongers, fruit canners and others whose hands are macerated by prolonged immersion in water (Fig. 5). Unlike candida lesions elsewhere, species other than *C. albicans* predominate, especially *C. parapsilosis* and *C. guilliermondii. Candida albicans* causes otitis externa (Gregson and La Touche, 1961).

Most arresting of all the skin or mucous-membrane manifestations of candidosis is the syndrome known as chronic mucocutaneous candidosis (Higgs, 1973). The disease arises principally in children, usually in those under two years of age, occasionally presenting in the thirties or even in the fifties. It is extremely rare, often associated with genetic defect and endocrinological disorders, and almost certainly based on defects of cellular immunity, many of which have been classified in various ways, but that of Wells is widely used clinically (Wells *et al.*, 1972; Wells, 1973). The four groups comprise: those with familial chronic cutaneous candidosis with autosomal recessive inheritance, who have chronic oral thrush with minor lesions elsewhere; those with no obvious familial disposition but a general susceptibility to infection, including candidosis, in whom lesions are severe and extensive, and in whom candida granulomas occur (Fig. 6); those who have single or multiple endocrinopathy involving the thyroid, parathyroid, adrenal or pancreatic glands, and in whom there may be autosomal recessive inheritance of the endocrinopathy; chronic cutaneous candidosis of late

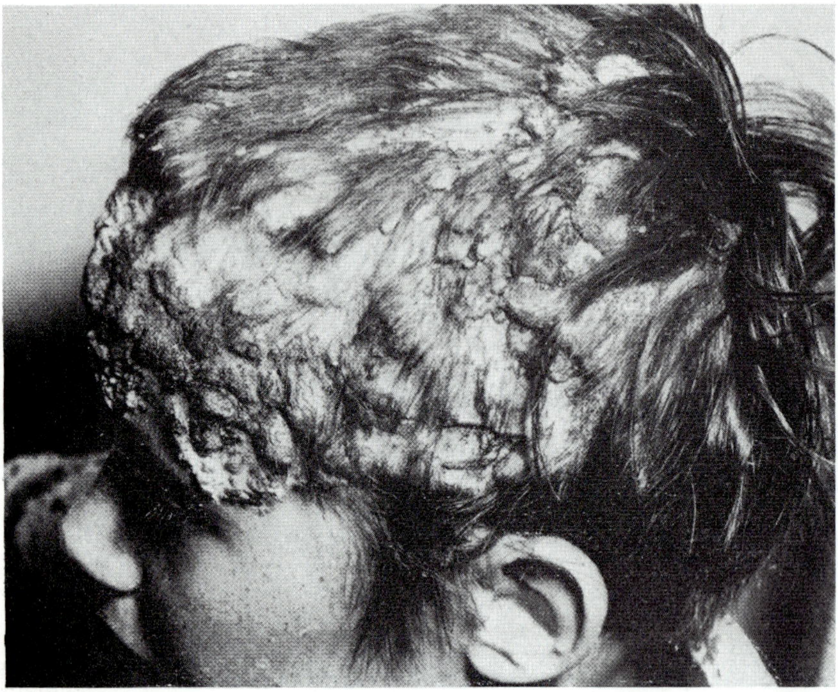

Fig. 6. Photograph showing Candida granuloma in a 14-year-old girl. From Winner and Hurley (1966).

onset. *Candida albicans* is the causative fungus in all cases. Dominant inheritance has been described (Sams *et al.*, 1979).

(ii) Deep-seated candidosis. Far less common but more severe than the superficial manifestations of candida mycosis are infections of a single viscus or widespread dissemination of candidas. Between 20 and 40 deaths attributable to systemic candidosis are reported to the Registrar General each year for England and Wales. Even in countries where other pathogenic fungi are endemic and cause serious disease, the death rate from candidosis is appreciable, the projected rate of cases requiring admission to hospital being a quarter of a million of the population in the U.S.A. in 1970 (Hammerman *et al.*, 1974). Candida accounted for 27 of 44 cases of fatal opportunistic fungal infections in women of child-bearing age revealed from scrutiny of the files of the Armed Forces Institute of Pathology (Purtilo, 1975).

The spread of candida lesions from the buccal cavity to involve other parts of the gastrointestinal tract, notably the fauces, oesophagus, stomach and intestines, has been recognized since the late eighteenth Century. Oesophageal candidosis is a well-recognized complication in patients with leukaemia and other blood dyscrasias, partly in consequence of the immunosuppressive agents used in therapy. Diarrhoea associated with candida is also well recognized, usually occurring in seriously ill patients who have been treated extensively with antibacterial antibiotics. In one case seen by us, virtually pure cultures of *C. albicans* were isolated from the cholera-like stools. Holti (1966) believed that allergy to yeasts underlay many cases of the "irritable colon syndrome".

Not only can candida spread from the mouth distally into the gastrointestinal tract proper, but it may also invade the larynx, and, more commonly, the main bronchi. Less well documented, though often reported and all too frequently diagnosed on less than satisfactory evidence, is involvement of the pulmonary parenchyma in the mycotic process, in the absence of disseminated infection. Pulmonary allergy to candida was described by Castellani (1912) as "teataster's cough", and enhancement by candida of allergy to threshing dust has been shown. So potent are the antigenic properties of *C. albicans* that it is widely used experimentally as an allergen.

Although lesions of bone or joints, and isolated organ or tissue lesions such as endophthalmitis, meningitis and peritonitis can occur, as well as relatively

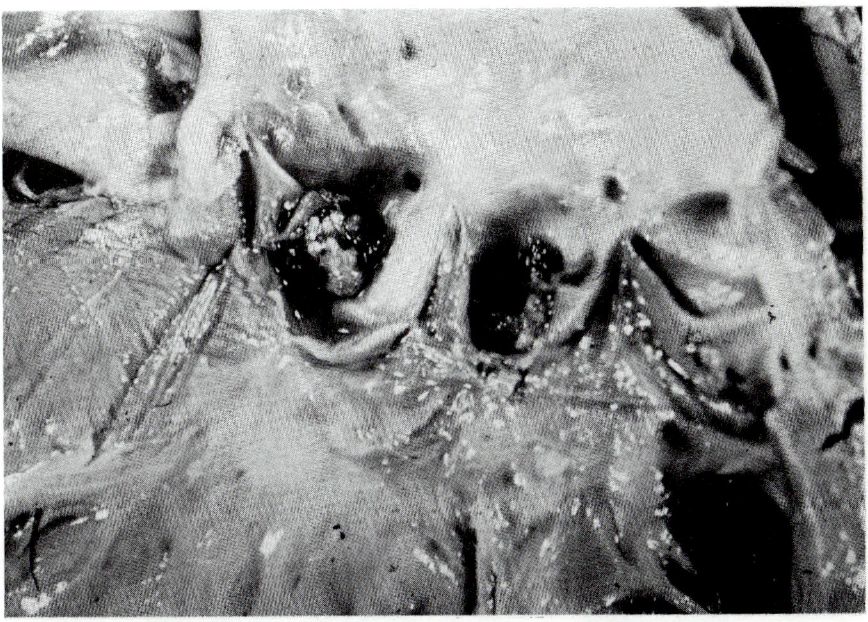

Fig. 7. Photograph showing Candida endocarditis. The aortic valve shows vegetations, and the cusps are thickened and ulcerated. From Winner and Hurley (1966).

Table XVII. Species of yeasts associated with cases of endocarditis in Man

Species	Percentage occurrence
Candida albicans	54.8
Candida guilliermondii	5.4
Candida krusei	5.4
Candida parapsilosis	25.8
Candida stellatoidea	2.2
Candida tropicalis	4.3
Candida glabrata	1.1
Rhodotorula pilimanae	1.1

minor infections of the urinary tract (cystitis, urethritis), the most serious infections caused by candidas are endocarditis, renal candidosis and acute disseminated (septicaemic) candidosis. These diseases share the common aetiology of being blood-borne. Few cases of candida endocarditis were recorded before the monograph of Winner and Hurley (1964) who reviewed all the cases reported up to 1962. Further reviews (Seelig *et al.*, 1974; Odds, 1979) have also been published. It has to be admitted that the majority of cases are iatrogenic, following open-heart surgery. Before the advent of this type of surgery, the disease occurred in main-line drug users who literally injected themselves with the pathogens contaminating the vehicles of addiction; some 20% of cases still arise in this way. The disease is an infective endocarditis arising primarily in patients with underlying disease of the valves, or in whom protheses have been fitted or homografts undertaken. The disease affects the aortic valve more than the mitral, and rarely involves the tricuspid (Fig. 7). The vegetations are large and friable and, consequently, embolic phenomena occur often involving the large arteries. The causal

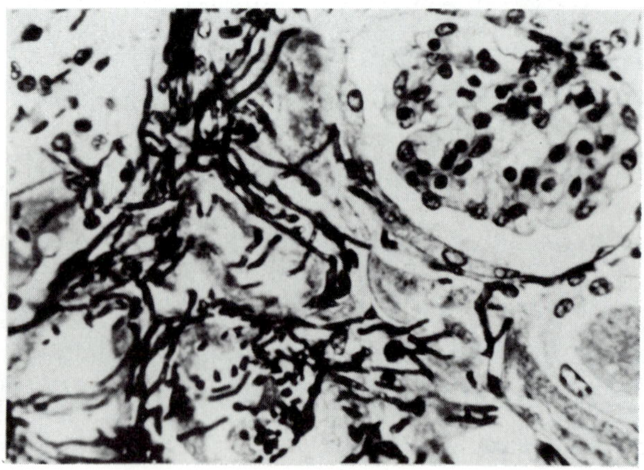

Fig. 8. High-power photomicrograph of a renal lesion in Man. The preparation was stained with Gomori stain. The hyphae are darkly staining.

yeasts are listed in Table XVII. The high incidence of *C. parapsilosis* is noteworthy and may indicate either a selective affinity of this yeast for endocardial tissue, as Odds (1979) suggests or, as we believe, reflect the propensity of *C. parapsilosis* to colonize damaged skin, and thus to have opportunity to gain ingress along intravascular lines, which are usually taped at the site of entry.

From study of isolated case reports and experimental work, Hurley and Winner (1963) adduced evidence of a variant form of the septicaemic disease, in which the renal parenchyma was the target organ, and the outcome a chronic fatal disease leading to hydronephrosis, sometimes with fungus balls in the pelvis of the ureter. The disease is uncommon. Experimentally, it can be induced by inocula smaller than that required to establish acute disseminated candidosis. This predilection of *C. albicans* for the kidney has long been known (Fig. 8).

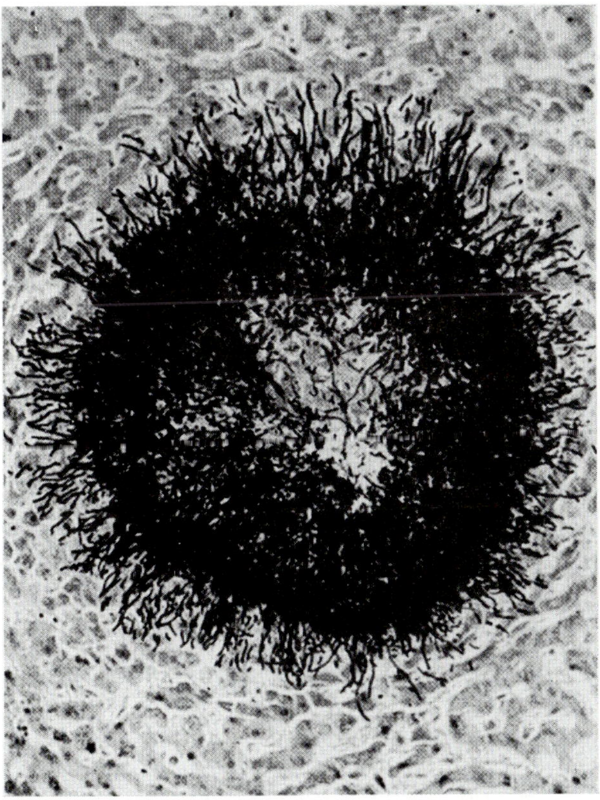

Fig. 9. A colony of fungi in human liver tissue. There is no tissue reaction. Magnification × 130. From Winner and Hurley (1966).

About 20 years ago, candida speticaemia and the syndrome of acute disseminated candidosis were unknown to all but a handful of experts, who admitted that on occasions candidas had been isolated from the bloodstream of seriously ill patients. Hurley (1964) located only 48 cases that she regarded as adequately documented. Now the disease is well known to hospital practitioners, and the regrettable (and lethal) error of ignoring blood cultures that yield yeasts is made far less frequently. On the subject of yeast-yielding blood cultures, Odds (1979) does not mince words; he advises practitioners to pay no heed to laboratories that describe them as contaminated. It seems likely that it is awareness of disease, rather than genuine increase in its frequency, that is responsible for the number of cases that are now recognized. The brain, the heart and the kidney are the primary target organs, but any viscus, particularly if already damaged by underlying disease, or chemical or physical poison, may provide lodgement for the fungi (Fig. 9). The untreated prognosis is poor, but less so in children under one year of age than in adults. In both it arises exclusively in debilitated hosts. The causal yeasts are listed in Table XVIII.

In the 1930s, disease in Man ascribable to *Candida* spp., even so common a disease as vulvovaginitis (Plass *et al.*, 1931), was regarded as a great rarity, and the deep-seated candida mycoses were virtually unknown. Now, 50 years later, all authorities agree that, numerically speaking, superficial candidosis ranks amongst the most common of all infectious diseases. Both superficial and deep candidosis are important public health hazards which are encountered throughout the world, not only in Man, but in wild and domestic animals.

(b) Candidosis in animals

Candida species, including those formerly classified as species of *Torulopsis* and members of the genus *Rhodotorula* have all been recovered from animals

Table XVIII. Species of yeasts associated with cases of acute disseminated candidosis in Man

Species	Percentage occurrence
Candida albicans	66.4
Candida guilliermondii	2.5
Candida krusei	0.9
Candida parapsilosis	3.1
Candida pseudotropicalis	0.3
Candida stellatoidea	1.6
Candida tropicalis	10.4
Candida glabrata	0.9
Rhodotorula species	1.3
Other (unidentified)	12.6

(see Section II. A, p. 213) and these yeasts cause both superficial and deep mycoses in many members of the Animal Kingdom.

Oral, genital and cutaneous lesions, including onychia and paronychia, have been reported in monkeys, and epidemic venereal disease has occurred amongst geese (Beemer *et al.*, 1973). Dogs, chickens, pigs and Central African hedgehogs have all been reported to suffer from cutaneous candidosis.

Lesions of the crop caused by *C. albicans* occur in wild and domestic birds; poultry crop candidosis is economically important, and epidemic candidosis is a major hazard of partridge rearing (Keymer and Austwick, 1961). A marked seasonal increase of the disease occurs between July and August (Austwick *et al.*, 1966); this seasonal incidence is also noted in vulvovaginitis in women (Herff, 1895). Candida oesophagitis occurs in pigs, caused mainly by *C. albicans* and *C. slooffii*, and similar infections have been reported in primates and in bovines.

Although the microbial aetiology of bovine mastitis is diverse, yeasts probably account for 2–3% of all cases. As well as *C. albicans, C. krusei, C. tropicalis* and *C. parapsilosis*, it is likely that *C. lusitaniae, Rh. mucilaginosa* and *Pichia farinosa* (an ascosporogenous yeast) are also implicated. *Candida lusitaniae* is not usually regarded as one of the pathogenic candidas, but more precise methods of speciation, including those of molecular taxonomy, led Holzschu *et al.* (1979) to record it as a cause of disseminated disease in Man. Pappagianis *et al.* (1979) also recorded a probable case.

Candida species, including *C. albicans, C. krusei, C. tropicalis, C. guilliermondii, C. lusitaniae, C. macedoniensis* and *Torulopsis* species have been reported to cause bovine abortion. Austwick *et al.* (1966) demonstrated transplacental passage of *C. tropicalis* and intrauterine infection in cows, experimentally.

Candida albicans and other candidas seem, therefore, to be able to produce the same spectrum of disease in animals as that which occurs in Man, as well as more specific types of disease, such as crop candidosis and mycotic abortion. Factors affecting the onset of disease are much the same as in humans, pigs in particular developing candidosis after nutritional disturbances. Vaccination, in Man, is not yet a possibility, although recent work with ribosomal preparations has been shown to modify murine candidosis.

III. *CRYPTOCOCCUS NEOFORMANS* AND CRYPTOCOCCOSIS

A. Taxonomy

Cryptococcus neoformans (Sanfelice) Vuillemin (1901) is the sole aetiological agent of cryptococcosis, and was the type species of the genus *Cryptococcus*

Kützing emend (Phaff et Spencer). Although a phylogenetic relationship seems to exist with the genera *Trichosporon* and *Candida*, Phaff and Fell (1970) excluded from the genus *Cryptococcus* those species that form pseudomycelium, true mycelium or arthrospores, and regarded the inositol utilization test as the major diagnostic criterion separating members of the genus *Cryptococcus* from species of *Rhodotorula* and *Torulopsis*. The genus included 17 species and six varieties (Phaff and Fell, 1970), and 39 synonyms were recorded for *Cryptococcus neoformans*. De Miranda (1984) accepts 19 species, retaining inositol assimilation as a feature of the genus which contains both ascomycetous and basidiomycetous species.

Discoveries of true basidia and basidiospores, prefaced by earlier observations of hyphal formation (Shadomy and Utz, 1966) and clamp connections (Shadomy, 1970), placed the sexual form of *Cr. neoformans* in a new genus *Filobasidiella*, as two species *Fil. neoformans* and *Fil. bacillispora* (Kwon-Chung, 1975, 1976a, b). The perfect (basidiomycete) states of other species of *Cryptococcus* have features that place them in the genus *Filobasidium* (Kwon-Chung, 1977); *Filobasidiella* is thus a genus of two species sharing a single asexual state.

B. Ecological considerations

Phaff and Fell (1970) studied 36 strains, commenting that nearly all isolations of *Cr. neoformans* have been from Man or warm-blooded animals, although Emmons (1951) reported it in soil, and Evenson and Lamb (1964) in the slime flux of the mesquite desert shrub. The fungus occurs in old dried accumulations of bird droppings (Littman and Schneierson, 1959; Emmons, 1960, 1962; Partridge and Winner, 1965), which are assumed to be an important source of human infection. Spontaneous cryptococcosis does not occur in the birds, their high body temperature probably protecting them from infection (Littman and Borok, 1968), but pigeons can be infected experimentally. Staib (1963) studied the conditions that favour growth of the yeast in bird faeces, noting that purines, uric acid, guanine and creatinine support good growth of *Cr. neoformans in vitro*. He advanced a plausible explanation for the association of *Cr. neoformans* with bird excreta, since it is the only yeast able to utilize creatinine as a nitrogen source. On the basis of agglutination and precipitation reactions (Evans, 1962), three serotypes of *Cr. neoformans* (A, B, C) can be recognized, a fourth (D) having been described by Wilson *et al.* (1968). Kwon-Chung (1976b) correlated the serotypes with sexuality in the fungus, noting that perfect forms were obtained by crossing suitable mating types within serotypes A and D. Crosses between strains of serotypes B and C were finally achieved, but the basidiospores, unlike the globose rough spores of *Fil. neoformans*, were smooth and rod-shaped. This feature characterized the second perfect species, *Fil. bacillispora*. Bennett *et al.* (1977) did not isolate B or C serotypes from 117 isolates from pigeon excreta and other saprophytic substrates. Saprophytic isolations of serotype C are rare, and none of serotype B is known. Campbell and Mackenzie (1980) suggest that, while pigeon excreta may be a frequent source of *Fil. neoformans*, it is not the habitat of *Fil. bacillispora*. The distribution of serotypes in human disease also shows a scarcity of B and C serotypes.

C. Pathogenicity of *Cryptococcus neoformans* and types of infection

Cryptococcosis (Busse-Buschke's disease, European blastomycosis, torulosis) is a deep visceral, systemic or generalized mycosis, caused by *Cr. neoformans*, having a worldwide distribution. With infections caused by *Candida* species, it is the most frequent systemic mycosis encountered in Europe, and there is some evidence that it is increasing in frequency. The Mycological Reference Laboratory in London reported a maximum of five cases per year (average 1.6) between 1963 and 1967, while from 1973 to 1977 they recorded three to nine cases per year (average 6.2). They analysed 69 cases, occurring over a 15-year period (Hay and Mackenzie, 1980). From 1965 to 1977, 1264 proven cases occurred in the U.S.A., compared with 300 cases recorded worldwide up to 1956 (Campbell and Mackenzie, 1980). The incidence, in association with malignant disease of the lymphoreticulo-endothelial system and its treatment, also seems to be on the increase (Salfelder, 1971). Tests for the disease have improved in the last two decades, so part of the increase may be more apparent than real. Formerly believed to be a rare and fatal disease, there is increasing evidence that benign and transient pulmonary infection, often subclinical, is not at all uncommon (Emmons *et al.*, 1977). Bone lesions occur in 5 to 10% of cases.

Mention has been made of the saprophytic source of *Cr. neoformans* in dry accumulations of pigeon droppings (Emmons, 1955). Inhalation of airborne yeasts is believed to initiate the infection, which spreads by the haematogenous route in certain individuals. In others, yeast may be seen at autopsy in single small sub-pleural nodules in persons dying of unrelated causes (Haugen and Baker, 1954). The nodes indicate the sub-clinical form of the infection. Larger lesions constitute cryptococcomas. Although many organs and tissues may be involved in disseminated disease, the yeast shows a strong neutropic tendency, lodging in the brain and, more especially, in the meninges. Brain and spinal fluid contain large amounts of thiamin, a vitamin stimulating growth of *Cr. neoformans*. Littman (1958) suggested that this may explain the neurotropism of the yeast. Such infection may begin in the nasal and pharyngeal cavities. Primary infection may, rarely, begin in the skin, especially after injury, although it is more usual for cutaneous lesions to follow haematogenous spread. Cryptococcosis is not confined to Man, having been reported in dogs, cats, pigs, horses, foxes, ferrets, cows, cheetahs, monkeys and sheep (Beneke and Rogers, 1970). The majority of bovine cases involve mammary tissue and draining lymph nodes while, in other animals, most reports have been of generalized, fatal disease. Infection of the central nervous system occurs, as in Man. The pathogenesis of the disease is extensively discussed by Salfelder (1971).

Zimmerman and Rappaport (1954) drew attention to the frequent association of lethal forms of cryptococcus with debilitating disease, particularly with malignant disease of the lymphoreticuloendothelial system. Salfelder (1971) listed some 19 conditions in which cryptococcosis had been reported as co-existent; they included sarcoidosis, tuberculosis, Hodgkin's disease and various forms of therapy, including steroid therapy. Pigeon breeders, as might be expected, are at special risk, and the disease is more frequent in males. Most cases occur between the ages of 40 and 60, but neonatal and infantile cryptococcosis has been reported (for a discussion of this aspect, see Salfelder, 1971). Hay and Mackenzie (1980) noted that 85% of 69 patients whose illness started at least three months after entry into the U.K. or Eire had serious underlying diseases. Principal amongst these were Hodgkin's disease, sarcoidosis, collagen disease, carcinoma and conditions requiring systemic corticosteroid therapy. The prognosis of the mycosis was adversely affected by diseases such as systemic lupus erythematosus, and tended to be poor if the values for the latex agglutination test for cryptococcal antigen in cerebrospinal fluid or serum were high, initially. The mortality in 45 cases was 52%.

Despite many studies attempting to equate both resistance to phagocytosis and virulence to capsulation of the yeast, there is no incontrovertible evidence of these relationships. Capsules do, however, tend to be lost if the yeast is kept in soil, and the capsulated form can be induced after animal passage.

D. Identification and laboratory diagnosis

Cryptococcus neoformans grows easily on media containing an organic source of nitrogen; thiamin, which is an essential growth factor, is present in sufficient quantity in peptone or malt extract. On Sabouraud's glucose agar, the yeast grows at 37°C or room temperature, forming moist mucoid white-to-pale colonies, composed of spherical cells 4–20 μm in diameter, that multiply by budding with one or more buds to each mother cell. Each cell is surrounded by a capsule, varying in size with the isolate, and seen best in nigrosin or India ink-stained preparations. Hyphal formation is observed extremely rarely (see above, p. 242). Specific identification is based on appropriate biochemical and physiological tests (Phaff and Fell, 1970), but urease is produced and inositol assimilated. Full details of identification are given by Beneke and Rogers (1970).

A special medium for isolation from heavily contaminated materials was developed by Shields and Ajello (1966). It contains glucose, creatinine and an extract of thistle seed. Colonies of *Cr. neoformans* develop a brown colour on this medium, whereas other cryptococci and candidas do not. Staib (1962, 1963) observed that *Cr. neoformans* produced brown-pigmented colonies if grown in extracts of bird feed or seeds of the Niger plant (*Guizotia abyssinica*). Added to agar, extracts of these seeds form the basis of a selective medium (Shields and Ajello, 1966; Botard and Kelley, 1968; Jennings *et al.*, 1968). Similarly pigmented colonies are observed if the yeast is grown in the presence of extracts of potato or carrot (Shaw and Kapica, 1972), caffeic acid (Korth and Pulverer, 1971) or dihydroxyphenylanaline (Chaskes and Tyndall, 1975). *Cryptococcus neoformans* thus produces a melanin-like pigment when grown in the presence of melanin precursors (*o*-diphenols); the chemical basis of the reaction is believed to be production of phenoloxidase by the fungus, catalysing synthesis of melanin from the diphenols. Chloramphenicol is often added to culture media to suppress bacterial growth; cycloheximide should not be used, as it retards growth of the fungus.

Specimens for culture include cerebrospinal fluid, sputum, urine, pus or biopsy material. The body fluids are examined directly, in nigrosin or India ink, for the presence of capsules, which are usually wide (sometimes more than twice the diameter of the cell) and refractile. In tissue sections, the yeasts can be seen in haematoxylin and eosin preparations, but periodic acid–Schiff, Gridley or the mucicarmine stains are superior. There is little tissue reaction round the capsules containing the cells. In tissues, the yeast occurs in the form of large unicellular cells, with a diameter of 4–10 μm. The capsule appears as a clear spherical zone, and may have a diameter five times that of the yeast cell. Lightly encapsulated strains may be confused with the tissue form of *Histoplasma capsulatum*, although careful search often reveals some definite capsules. Specific fluorescent antibody conjugates may be used to differentiate *Cr. neoformans* from morphologically similar yeast cells. Accounts of the histopathology of the lesions are given briefly by Chandler *et al.* (1980) and fully by Salfelder (1971).

Cryptococcus neoformans may be differentiated from other species by pathogenicity tests in mice. About 0.5 ml of infected material or a saline suspension of culture is injected by the intraperitoneal, intravenous or intracerebral route. Autopsy two to four weeks later shows gelatinous masses in the abdominal viscera, lungs and brain, the site of the lesions depending on the route of inoculation used. Negative staining demonstrates the encapsulated yeast.

Serological tests are most important in diagnosing cryptococcosis and they also have a role as a prognostic index. Three tests are used concurrently: the latex agglutination test for cryptococcal polysaccharide antigen, the tube whole-cell agglutination and indirect fluorescent tests for antibody. Using all three tests, cryptococcosis can be diagnosed serologically with an accuracy of 95% (Kaufman and Blumer, 1968). The realization that antigen, derived from capsular material which is formed abundantly *in vivo*, is more usually demonstrable than antibody led to development of the latex test (Bloomfield *et al.*, 1963) which can be quantified in cerebrospinal fluid and serum by titration. Successfully treated patients lose antigen while antibody appears and can be shown to rise; persistently high or rising titres of antigen indicate unsuccessful therapy and a poor prognosis.

Thus, diagnosis of cryptococcosis is based on isolation and culture of an encapsulated, budding yeast that grows at 37°C, does not assimilate nitrates, assimilates glucose, galactose, sucrose and inositol but not lactose, produces urease, is pathogenic to mice, and produces brown colouration in media containing creatinine and seed extract (*Guizotia abyssinica*), or colonies surrounded by red halos in Vogel's medium (Beneke and Rogers, 1970). A rapid test for urease production (Roberts *et al.*, 1978) has been described. The yeast may be seen and identified in biopsy material, and diagnosis can be established serologically.

IV. THE GENUS *PITYROSPORUM*

A. Incidence and distribution

The genus *Pityrosporum* in the family Cryptococcaceae was created by Sabouraud in 1904 to describe the fungus which Malassez (1874) had observed in cases of versicolor simplex. It included yeasts which are found only in close association with the skin of man and animals and currently contains three species, namely *Pityrosporum ovale* (the type species), *Pit. orbiculare* and *Pit. pachydermatis* (Faergemann and Bernander, 1981; Kreger-van Rij, Ch. 2 of this volume). Other strains of pityrosporum have been isolated from the ears of dogs (Gentles and La Touche, 1969).

Pityrosporum ovale and *Pit. orbiculare* are lipophilic yeasts and form part of the normal cutaneous human flora. *Pityrosporum orbiculare* is also found on the skin of patients with tinea versicolor (Gordon, 1951) to the extent that there is clearly a relationship between presence of the organism and the clinical condition (see p. 245). *Pityrosporum pachydermatis* does not need lipid for optimal growth and is found mainly in animals and only rarely on human skin (Noble and Sommerville, 1974). The justification for treating *Pit. ovale* and *Pit. orbiculare* as separate species has been questioned. Nevertheless, the present classification (see Kreger-van Rij, Ch. 2 of this volume) retains the two specific titles.

Pityrosporum ovale and *Pit. orbiculare* are found not only in the upper part of the stratum but also in deeper skin layers and hair follicles. Earlier reports that the organisms were only present above the acroinfundibulum (Pliwig and Kligman, 1975) could not be confirmed by Boyanowsky and Lischka (1977) who isolated them from the deepest part of the follicle. *Pityrosporum orbiculare* exists as an opportunistic pathogen as well as a skin saprophyte and it is now considered as the aetiological agent of tinea versicolor (Roberts, 1969; McGinley *et al.*, 1970; Potter *et al.*, 1973; Faergemann and Bernander, 1979). In this condition, *Pit. orbiculare* is found in the deepest areas of the stratum corneum where it grows in both yeast and mycelial forms. Recently filamentous forms of *Pit. ovale* and *Pit. orbiculare* have been grown *in vitro* (Porro *et al.*, 1977; Faergemann and Fredriksson, 1981). *Pityrosporum ovale* and *Pit. orbiculare* are able to survive and grow under anaerobic conditions and grow as well under micro-aerophilic atmospheric conditions as they do in air (Faergemann and Fredriksson, 1981). *Pityrosporum obiculare* is not found in the skin of infants (Faergemann and Fredriksson, 1980). It first appears before puberty and reaches its highest incidence at puberty or later. *Pityrosporum ovale* has been reported in the external ears of 7.5% of 212 neonates during the first three postnatal days, and in 70% of 132 specimens from adult female external genitalia but not from 46 intravaginal specimens (Randjandichi, 1981).

The first description of the yeast *Pit. ovale*, was by Rivolta (1873) under the name *Cryptococcus psoriasis*. More detailed descriptions followed (Malassez, 1874; Klamen, 1884; Sabouraud, 1904), but attempts to isolate the organism failed and, in 1904, Sabouraud concluded that previous claims to have isolated the organisms were mistaken. The review of Benham (1947) credits Ota and Huang (1933) with the first isolation of the lipid-dependent yeast and suggests that, as a result of this work, earlier reports of the isolation of *Pit. ovale* (Castellani, 1925; Acta and Panja, 1927) were authentic. Subsequent work by Benham (1947) using a synthetic medium showed conclusively the essential requirement that this organism has for a lipid substrate.

B. Isolation and cultural characteristics

1. *Pityrosporum ovale*

Provided that the surface of a solid medium is overlaid with a lipid, the composition of the basal medium used is not important for isolation of *Pit.*

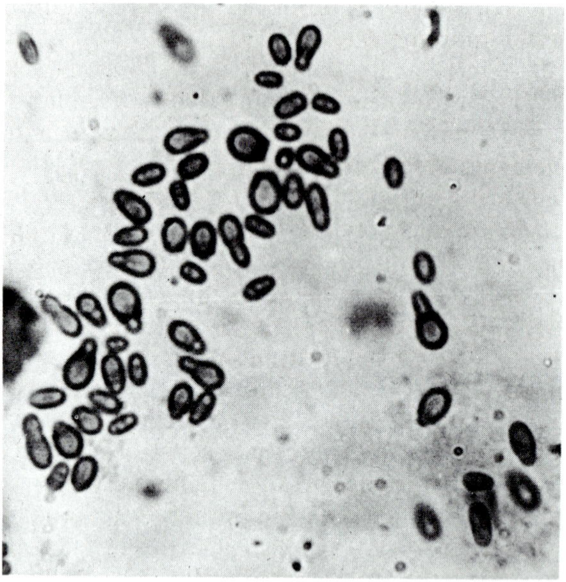

Fig. 10. Photomicrograph of *Pityrosporum ovale* isolated from a scraping of skin from a patient with seborrhoeic dermatitis. Magnification × 2000. From Gentles and La Touche (1969).

ovale, and a wide variety of media have been successfully used, including wort agar and a synthetic medium (Benham, 1947). Lipids in the form of lanolin, oleic acid or butter are most suitably added to the medium as an ether extract pipetted over the surface. When the ether evaporates, a thin film of lipid is left on the surface. After four or five days incubation at 37°C on fresh moist agar, *Pit. ovale* appears as creamy-to-orange coloured colonies which enlarge with further incubation and become wrinkled and dry.

In lipid medium, *Pit. ovale* grows as granules on the bottom of the culture vessel and also as a surface pellicle (Benham, 1947). Benham (1947) also demonstrated that oleic acid was the most suitable lipid substrate for growth, and that the organism had a requirement for asparagine or ethyl oxaloacetate. The optimum temperature for growth is 35–37°C and the organism grows at a wide range of pH values. Cells are pleomorphic, measuring 4–5 μm·2–3 μm. A proportion of them have a bottle form, consisting of a larger cell joined to a small one by a neck-like projection (see Fig. 10).

2. *Pityrosporum orbiculare*

Pityrosporum orbiculare was first described by Gordon (1951) who isolated the organism from the skin of 15 individuals and, in the absence of any previous report of the culture of this organism, proposed a new species of the genus *Pityrosporum*. *Pityrosporum orbiculare* shares with *Pit. ovale* a requirement for lipid, but differs from *Pit. ovale* in several morphological and physiological characteristics. It is probable that, although the organism had not been isolated previously, some of the earlier microscopic descriptions were of *Pit. orbiculare* (Bizzozero, 1884).

Pityrosporum orbiculare is a normal inhabitant of the skin (Gordon, 1951; Burke, 1961; Dorn and Roehnert, 1977; Faergemann and Bernander, 1979) where it occurs as spherical yeast-like cells. It is not primarily a contagious fungus but possibly becomes infectious as the result of yeast-to-mycelial conversion (McGinley *et al.*, 1970; Tosti *et al.*, 1972). Using appropriate culture media, it can be isolated invariably from the skin of patients with tinea versicolor (Faergemann, 1979). The characteristic microscopic appearance in skin scales from cases of tinea versicolor is of clusters of spore-like structures associated with short lengths of stout hyphae, and in this form, it has been known as *Malassezia furfur* (Fig. 11).

Germ tubes may be seen in some primary cultures of *Pit. orbiculare*, but these disappear on subculture (Gordon, 1951; Burke, 1961; Caprilli *et al.*,

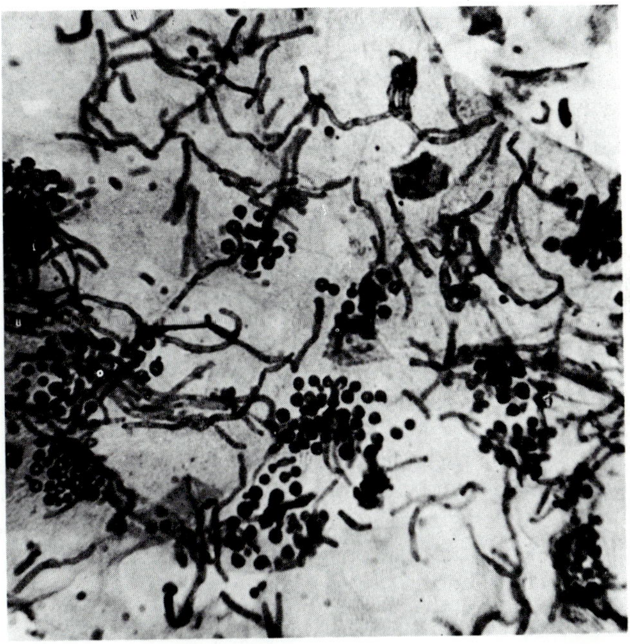

Fig. 11. Photomicrograph showing hyphae and spherical cells of *Malassezia furfur* from the pigmented area of skin from a patient with pityriasis versicolor. Magnification × 500. From Gentles and La Touche (1969).

1973). *In vitro* filamentous forms have been induced for *Pit. orbiculare* and *Pit. ovale* (Dorn and Roehnert, 1977; Porro *et al.*, 1977).

A suitable medium for primary growth of *Pit. orbiculare* has been described by Faergemann (1979) and consists of (l^{-1}, pH 5.6): neopeptone (Difco), 10 g; Bacto agar (Difco), 18 g; glucose, 40 g; yeast extract (Difco), 0.1 g; glycerol monostearate, 2.5 g; and Tween 80, 2.0 ml. After sterilization at 101 KPa (15 $lbf\,in^{-2}$) for 15 minutes, gentamicin and chloramphenicol are added to give final concentrations of 100 mg and 50 mg per litre respectively. After poured plates have been inoculated, they are overlaid with olive oil. Fine while granular colonies appear after four to five days incubation at 35–37°C. On further incubation, these spread until the agar surface is covered by a uniform creamy white layer. Cultures consist of pleomorphic smooth spherical cells 2.3–3.8 μm diameter with a double cell wall and featureless interior. There is no growth at 25°C after two weeks. Budding cells are spherical or slightly elongated and attached by a narrow isthmus, in contrast to *Pit. ovale*, the buds of which are attached by a broad stalk.

Oleic acid, which satisfies the lipid requirement of *Pit. ovale*, will not support growth of *Pit. orbiculare* for which olive oil, linseed oil, myristic acid or stearic acid are considered by Gordon (1951) to be most suitable. The lipid components essential for growth of *Pit. orbiculare in vitro* are fatty acids of chain length C_{12}–C_{24} (Porro *et al.*, 1976).

C. Tinea versicolor

Tinea versicolor is a fungus disease of the skin, characterized by development of flat or slightly raised scaly spots that gradually enlarge and coalesce. Versicolor lesions involving follicles occur, but only rarely. The lesions are light brown in colour and so may appear lighter or darker than the surrounding skin according to the level of natural pigmentation (Fig. 12). Infection is most common on the upper chest, back, arms and neck. The disease is usually chronic and lesions may persist, spread or reappear for years. Infection is usually asymptomatic and therefore only of cosmetic importance but, in severe cases, there may be itching. The condition is of worldwide distribution but is most prevalent in tropical and subtropical regions. The aetiological agent is not found in Nature, other than on human skin where it is part of the commensal flora. The reason that *Pit. orbiculare* becomes infectious in only a small proportion of the population is not understood, although it is apparent that high plasma cortisol concentrations (Boardman and Malkinson, 1962; Burke, 1961), climate (Roberts, 1969; Brinkman, 1974), poor nutrition (Burke, 1961) pregnancy and hereditary

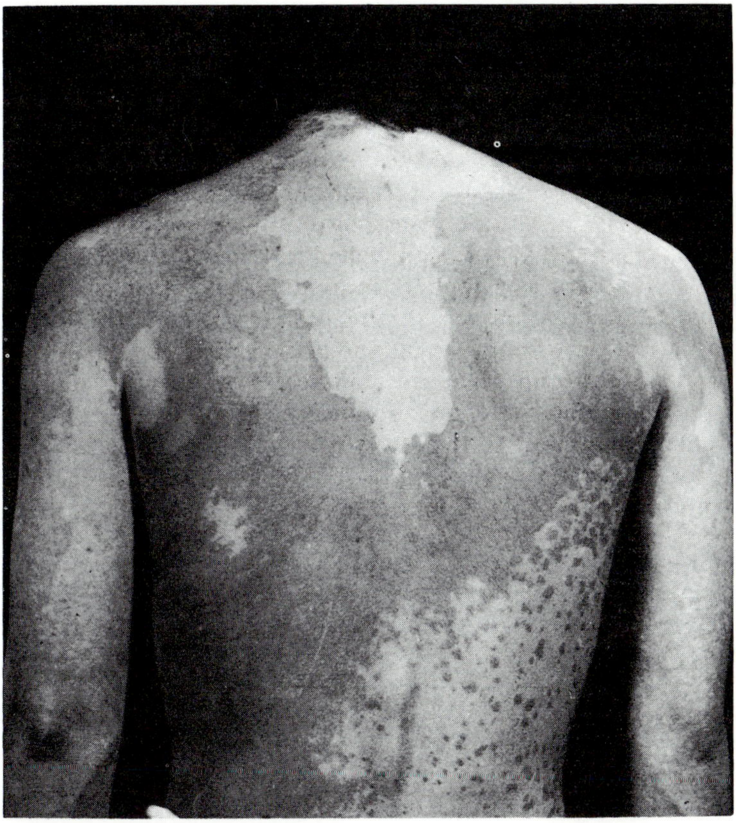

Fig. 12. Photograph showing pityriasis versicolor due to *Malassezia furfur* showing discrete and confluent pigmentation. From Gentles and La Touche (1969).

factors predispose to infection. The higher carbon dioxide concentration under occluded dressings also increases the susceptibility of the host to infection.

Cultural examination of skin scales is not necessary for diagnosis of tinea veriscolor. The characteristic macroscopic appearance and the yellow fluorescence of lesions under an ultraviolet lamp (Wood's light) are strongly indicative of tinea versicolor. Microscopic demonstration of the characteristic fungal elements in skin scales is diagnostic. *Pityrosporum orbiculare* (*Mal. furfur*) appears as short, curved, hyphal elements, 2.5–4.0 μm diameter, associated with clusters of oval or round, thick walled spores 3.0–8.0 μm in diameter. These cells are thought to be phialospores that are pushed out from the end of a hypha through a collarette. Secondary buds are produced in the same way.

Pityrosporum orbiculare can be cultured on routine mycological media overlaid with olive oil (Gordon, 1951) or by incorporation of sterile cream. Colonies are cream in colour and are yeast-like. In culture, yeast-like cells predominate although, under some conditions, hyphae may also be formed. The fungus has a nutritional requirement for lipid.

(a) Treatment

Traditionally, treatment consisted of topical application of 20% sodium hyposulphate in water or 2.5% sodium selenite suspension in a suitable detergent. More recently, considerable success has been reported following the use of the oral imidazole, ketoconazole, at a dose of 200 mg once a day for 4–6 weeks (Heel, 1982a, b).

(b) Histology

Histologically, tinea versicolor resembles tinea nigra. There is mild to moderate hyperkeratosis of the epidermis that may be associated with minimal mononuclear cell infiltration in the dermis. Fungal elements appear on short, curved or bent hyphal structures associated with clusters of round or oval thick-walled cells, some of which may show budding. These forms are usually found in the middle and deep layers of the stratum corneum. The fungal elements may be seen in sections stained with haematoxylin and eosin, but are best demonstrated by Giemsa or special fungal stains (Chandler *et al.*, 1980).

V. THE GENUS *TRICHOSPORON*

Three members of the genus *Trichosporon* are considered as potential human or animal pathogens. These are *Trichosporon cutaneum* (*Trichosporon beigelii*), causative agent of white Piedra, *Trichosporon capitatum* and *Trichosporon inkin*. A new species, *Trichosporon beemeri* sp. nov. is reported to infect crocodiles.

A. Habitat

Trichosporon capitatum has been isolated from sewage and wood pulp (Cooke *et al.*, 1960), whereas *Tr. cutaneum* has been found in soil (Polonelli *et al.*, 1981), beach sand and sea water (Anderson, 1979; Dabrowa *et al.*, 1964) and swimming pools (Dietz and Seeliger, 1977).

Carriage of *Trichosporon* sp. by man or animals may occur without any evidence of infection. *Trichosporon cutaneum* has been isolated from the oral cavity of man (Koch and Koch, 1981) and the digestive tract of animals (Taguchi *et al.*, 1979). The sputa of patients with bronchopulmonary conditions may contain *Tr. capitatum* (Ktanidou-Kartali *et al.*, 1979). As yet, *Tr. inkin* has only been found in association with human skin (Do Carmo-Sousa and van Uden, 1967).

B. Pathogenicity

1. Superficial infections

White piedra, an infection of the hair, is caused by *Tr. cutaneum*. It is characterized by the presence of nodules along the hair shaft formed by concretions of hyphae. These nodules may be microscopic, but larger ones give the hair a gritty feeling. The nodules of white piedra are softer and lighter in colour than those of *Piedraia hortai*, an ascomycete which is the causative agent of the more common black piedra. White piedra usually attacks the beard or moustache, but pubic hair may also be involved (Pankajalakshmi *et al.*, 1979). It occurs in temperate areas including South America, Central Europe, Western Europe and England. The thick hairs of the mane and tail of horses may also be infected with white piedra (Kaplan, 1959).

Sonck (1980) records 483 cases over a 16-year period of toe-nail or inter-digital skin infection with *Tr. cutaneum*, and simultaneous infection with a dermatophyte is not uncommon. Infection of human skin with *Tr. capitatum* has also been reported on rare occasions (Gründer, 1976). *Trichosporon inkin* is associated with a scrotal dermatosis most frequently observed in young Japanese men (Do Carmo-Sousa and van Uden, 1967). Both *Tr. inkin* and *Tr. cutaneum* have been isolated from penile ulcerations (Chapel *et al.*, 1978).

Piglets kept in conditions of high relative humidity may develop lesions over the head, abdomen and legs caused by *Trichosporon* sp. (Echevarria, 1977). A new species, *Tr. beemeri* sp. nov. recently described by Kuttin and Muller (1981) was isolated from a Nile crocodile with mucocutaneous mycosis.

2. Systemic infections

Certain species of *Trichosporon* may cause disseminated infection in Man, but only when the host's normal immune function has been depressed by use of cytotoxic drugs or by serious underlying disease. Evans *et al.* (1980) recorded two cases of systemic infection in patients with acute leukaemia who were receiving cytotoxic chemotherapy. *Trichsporon cutaneum* was isolated from the blood cultures of both patients, one of whom was successfully treated with amphotericin B. Fatal infection of a prosthetic heart valve with *Tr. cutaneum* has also been documented (Torok and Moravcsik, 1979). This species has been recovered from the contents of a brain abscess complicating a secondary deposit in a patient suffering from a primary bronchial carcinoma (Watson and Kallichurum, 1970). *Trichosporon capitatum* has been described as causing a fatal systemic infection involving virtually all of the organs of the body in a patient treated for aplastic anaemia by bone-marrow transplantation (Winston *et al.*, 1977). Although both *Tr. cutaneum* and *Tr. capitatum* may be isolated from sputa of patients with bronchopulmonary disease, their role as pathogens in this condition is unclear.

There is increasing evidence that *Trichosporon* sp. may be the aetiological agent in some cases of bovine mycological mastitis (Kadic, 1978; Biancardi *et al.*, 1976). They have been isolated also from the cervical mucus of infertile cows and buffaloes (Sinha *et al.*, 1980). Water-soluble sterile extracts of *Tr. capitatum* injected intravenously into cows can initiate a purulent mastitis and galactophoritis (Nicklas *et al.*, 1980).

C. Isolation and identification

A distinguishing feature of the genus *Trichosporon* is production of a true mycelium which frequently fragments to form an abundance of arthrospores.

1. *Trichosporon cutaneum*

After treatment with potassium hydroxide the nodules on the hair from cases of white piedra may be gently crushed and examined. Microscopically they are seen to consist of tightly packed hyphae, 4–8 μm in diameter. Culture is necessary to confirm the identity of the fungus. *Trichosporon cutaneum* grows rapidly on Sabouraud's agar, forming moist cream coloured colonies within a few days. Older colonies may become wrinkley, tough and hairy. Microscopic examination of corn-meal agar cultures reveals true mycelium, pseudomycelium and rectangular arthrospores. Less frequently, clusters of blastospores may be observed. Production of blastospores distinguishes *Trichosporon* sp. from *Geotrichum candidum* with which they are sometimes confused. Sugars are not fermented, but assimilation of glucose, galactose, sucrose, maltose and lactose occurs. Assimilation of other sugars may vary between strains and be lost in older cultures. Nitrate is not normally assimilated. *Trichosporon cutaneum* may be identified within 48 hours using the API 20C microtube system which contains 19 substrates for assimilation in a convenient strip form (de Louvois *et al.*, 1979).

2. *Trichosporon capitatum*

Whitish cream colonies of *Tr. capitatum* appear rapidly on Sabouraud's agar. Arthrospores, blastospores and true mycelium may all be observed on microscopic examination of corn-meal agar. *Trichosporon capitatum* assimilates only glucose and may thus be distinguished from *Tr. cutaneum*. Nitrate is not assimilated and no sugars are fermented.

3. *Trichosporon inkin* and *Trichosporon beemeri*

Physiologically *Tr. inkin* and *Tr. cutaneum* are very similar but the morphology of each is distinct. Culture of *Tr. inkin* on corn-meal agar produces thin hyphae and grape-like clusters of blastospores especially at the tip of the hyphae. Groups of cells dividing in different planes, with optically dense cytoplasm, also occur. *Trichosporon beemeri* sp. nov. also resembles *Tr. cutaneum*, but the blastospores of this new species are smaller and, unlike *Tr. cutaneum*, it is able to assimilate inulin.

VI. THE GENUS *GEOTRICHUM*

The genus *Geotrichum* consists of a single species *Geotrichum candidum* which is generally considered to be a saprophyte but some authenticated cases of infection in Man and animals have been reported. The nomenclature of *Geotrichum* is somewhat confused. *Geotrichum candidum* is referred to variously as *Endomyces lactis* (Do Carma Sousa, 1970) and *Endomyces candidus* (Chandler *et al.*, 1980). For the purpose of this description, the name *Geotrichum candidum* will be used.

A. Habitat

Geotrichum candidum has a worldwide distribution. It may be isolated from soil and sewage, and grows particularly well in sour-milk products and spoiled wine. The sputum and stools of healthy adults may be colonized by *Geotrichum candidum*, without any evidence of infection.

B. Pathogenicity

1. Infection in Man

Evidence for the pathogenicity of *Geotrichum candidum* in Man and animals is inconclusive. Chandler *et al.* (1980) were unable to find any histological material from a verified case of geotrichosis. Although *Geotrichum candidum* may be frequently isolated from clinical material, Chandler *et al.* (1980) cast doubt on the ability of *Geotrichum candidum* to cause tissue invasion. Geotrichoses of the mucous membranes of the alimentary tract and respiratory tract are most often reported but occasional instances of disseminated disease have been described. Geotrichosis is usually only manifested in patients with lowered resistance to infection. The number of cases reported is small, and histological studies have not been described. Although *Geotrichum candidum* may be isolated more frequently and in greater abundance from patients with varius underlying intestinal disease, for example ulcerative colitis (Ariyevich *et al.*, 1963), this may merely represent overgrowth of the normal flora, rather than a primary infection.

Between 1 and 6% of healthy persons harbour *Geotrichum candidum* in the respiratory tract (Péter *et al.*, 1967), and patients with pulmonary tuberculosis have a higher rate of carriage at 25% (Coudert *et al.*, 1957). Infrequent reports of infections believed to be pulmonary geotrichosis are described in the literature (Martins, 1928; Moore, 1934; Webster, 1959). Unfortunately, in many cases, other potential pathogens, for example *Klebsiella* spp., were also isolated and diagnosis was largely dependent on the success of antifungal therapy.

Castellani and Chalmers (1919) first described infection of the oral mucosa by *Geotrichum candidum*. The usual symptoms of this type of oral thrush are greyish yellow loosely attached patches developing on the mucous membrane of the mouth and throat. Chronic cutaneous infection by *Geotrichum candidum* may be superficial or involve deep areas of ulceration (Morenz, 1971). *Geotrichum candidum* may also occasionally cause corneal ulcers (Liesegang and Forster, 1980) usually as a result of traumatic injuries. Disseminated infection caused by *Geotrichum candidum* has been recorded in a diabetic (Bendove and Ashe, 1952) and a two-month-old child with miliary tuberculosis (Kaliski *et al.*, 1952).

2. Animal infections

Geotrichum candidum may be pathogenic for animals under certain conditions. It may cause adenitis in pigs (Morquer *et al.*, 1955) and mastitis in cows (Ainsworth and Austwick, 1955). Recently, *Geotrichum candidum* was isolated from aborted foetuses of pigs (Brömel and Zettl, 1979) and cows (Stuker *et al.*, 1979).

C. Isolation and identification

Geotrichum candidum appears in wet preparations or Gram-stained smears as rectangular, round or oval arthrospores ($5\,\mu m \cdot 10\,\mu m$) either singly or in chains, accompanied by hyphae 3–5 μm in diameter. Spherical forms may resemble the yeast-like cells of *Blastomyces dermatitidis*. Colonies forming on Sabouraud's dextrose agar are creamy white, flat and spreading. Microscopic examination of young cultures will reveal only yeast cells, but a submerged mycelium is apparent in older cultures. A short fluffy mycelium is characteristic of some strains. Microscopically, the mycelium is made up of coarse hyphae which become septate and readily break up into chains of typical arthrospores. *Geotrichum* spp. may be identified to the genus level using the API 20C microtube system (de Louvois *et al.*, 1979).

VII. HISTOPATHOLOGICAL DIFFERENTIATION OF YEAST-LIKE FUNGI IN TISSUES

Although not yeasts, since the dominant morphological form in ordinary culture is hyphal, the tissue phase of some fungi is yeast-like, and may give rise to confusion in histological diagnosis. These forms are considered briefly, here.

A. Sporotrichosis

This chronic, usually benign, infection is caused world wide by *Sporothrix* (*Sporotrichum*) *schenckii* and involves the skin and subcutaneous tissues. Occasionally, the musculoskeletal system, the viscera or the mucous membranes are also involved. Growth on Sabouraud's glucose agar at room temperature is mould-like but when grown on glucose/cystine blood agar or on brain–heart infusion agar at 37°C, the tissue or yeast phase is obtained. Morphologically, spindle-shaped elliptical or ovoid cells, reproducing by budding, are seen. The histology is fully described by Lurie (1971). The characteristic lesion is a granuloma. Most workers report that spores are never, or rarely, seen but three main types of fungal elements can sometimes be detected, namely cigar-bodies and their variants, asteroid bodies and hyphae. Yeast-like forms can be seen in pus.

Many variants of the cigar-body have been described but Lurie (1971), enumerating them, considers them all to represent conidia of the yeast phase of the fungus. All are Gram-positive when viable, and are positive when stained with periodic acid–Schiff reagent and can be stained by Gridley or Gomori techniques. They are difficult to see in haematoxylin and eosin preparations. Their true nature is ascertained from the clinical history and can be demonstrated by specific staining using fluorescent antibody techniques.

B. Histoplasmosis

Histoplasmosis has an almost worldwide distribution, being caused predominantly by *Histoplasma capsulatum* and, in the African continent, by *Histoplasma duboisii*. The portal of entry is the respiratory tract, and most infections (up to 95%) are benign and self-limiting, even subclinical. The frequency of infection in lower animals (mice, rats, dogs, cattle and horses) is similar to that in Man. Chronic pulmonary cavitary disease occurs in middle age, but pulmonary histoplasmosis occurs at all ages. There is a rare disseminated form of the disease, and complications of pulmonary histoplasmosis include meningoencephalitis, endocarditis and pericarditis.

The fungus is dimorphic, growing in tissues and *in vitro* at 37°C as a yeast, 2–5 μm in diameter. At room temperature, on laboratory media, hyphae develop. The latter shows septate hyphae, racquet hyphae, small conidia and large spores, either smooth or with tuberculate projections unique to the species. The yeast-like form is difficult to discern in tissues stained with haematoxylin and eosin where it may be confused with calcified particles, but is readily stained by the Gomori technique, and somewhat less reliably by Gridley and by the periodic acid–Schiff reagent. Specific fluorescent antibody staining is feasible, but often disappointing (Schwarz, 1971). The usual size of the tissue form is 1–5 μm, the shape varying from round to oval. Larger cells occur and can be differentiated from those of *Blastomyces dermatitidis* by their single nucleus. Schwarz (1971) discusses differential diagnosis of histoplasma-like structures in the tissues.

C. North American blastomycosis

North American blastomycosis is caused by the dimorphic fungus *Blastomyces dermatitidis* which grows as a mould in its saprophytic phase and as a yeast in its tissue phase. The disease is not entirely confined to the North American continent, but most cases reported from elsewhere have been in those who have had direct contact with North America, or which have been inconclusively demonstrated. The primary lesion appears to be in the lung, occasionally in the skin. There is slow progression with extension of the pulmonary lesion and/or spread to other organs. Asymptomatic infections may, conceivably, occur, as is the case with other endemic mycoses. Distribution of lesions is widespread, all systems and most organs having been involved. The pathology is extensively discussed by Chick (1971).

The yeast phase may be cultured *in vitro* at 37°C on brain–heart infusion blood agar, or other enriched medium. Colonies are composed of rounded thick-walled yeasts, ranging from 8–10 μm in diameter to 20 μm on occasion. The thick hyaline wall gives a doubly contoured appearance. Buds arise from a broad base on the parent cell, and some yeasts are elongated. Yeasts may also be seen in clinical specimens, such as pus, discharges, sputum, urine and exudates. Clearing with potassium hydroxide is often necessary. The histopathological appearances of the lesions will, often, not be sufficiently typical to differentiate blastomycosis from histoplasmosis or coccidioidomycosis, unless the fungi can be demonstrated. Haematoxylin and eosin staining is adequate for the purpose, but periodic acid-Schiff, Gridley and Grocott-Gomori stains are all used.

D. Paracoccidioidomycosis

Paracoccidioidomycosis is a chronic progressive granulomatous disease of the lungs, mucosa of the mouth and nose and the facial integument with spread to lymph nodes and viscera. It is confined to South America, is apparently limited to Man, and is caused by the dimorphic fungus *Paracoccidioides brasiliensis*. The fungus grows slowly as a mould at room temperature but rather faster, and as a yeast, at 35–37°C. The yeast cells are round to oval, with a rigid cell wall, up to 30 μm in diameter and sometimes 50–60 μm across. Small forms, 2–3 μm, are seen in actively growing cultures. Multiple budding occurs, daughter cells being attached by a narrow neck.

There is granulomatous response in tissues and, to verify the aetiology, the fungal cells should be demonstrated. In tissues, they appear as spherical or oval cells, between 10 and 60 μm in diameter, with multiple small buds on the periphery. This appearance is sometimes described as "airplane motor" or "ship's helm", and is characteristic of *Par. brasiliensis*. Fixation– retraction of the protoplasmic body in some cells causes the appearance of a non-staining space within some mother cells. Fragmented chromatin may also be seen. In fresh, active lesions *Par. brasiliensis* is recognizable in sections stained with haematoxylin and eosin while, in granulomas, Grocott-Gomori or Gridley stains, and periodic acid-Schiff reagent are required.

The appearance of the fungus is not always characteristic. Angulo and Pullar (1971) discuss the differential diagnosis. Small forms, especially when intracellular (3–4 μm), must be distinguished from *H. capsulatum*. Medium-sized forms may be confused with *Cr. neoformans*. Large empty forms without buds may resemble *Coccidioides immitis*. Confusion with *Blastomyces dermatitidis* seldom arises as the geographical distribution of the two mycoses is so distinct.

E. Coccidioidomycosis

Coccidioidomycosis, caused by *Coccidioides immitis*, may occur as an acute self-limiting primary infection of the respiratory tract or, more rarely, the skin, or as a chronic progressive often fatal disease involving the integument, the bones or the viscera. The disease is endemic to the arid Southwestern United States of America, and has also been reported from South America. It is not limited to Man, occurring also in cattle, dogs, sheep, rabbits, monkeys and desert rodents. In its pathogenic phase, the yeast is spherical in shape, reproducing by endosporulation, and growing as a saprophyte as aerial septate mycelium with arthrospores.

The round, thick-walled spherules (20–80 μm diameter) containing small endospores (2–5 μm diameter) can be seen in wet mounts of infected material, including sputum, gastric contents or cerebrospinal fluid. The spherules may be demonstrated in tissues, using haematoxylin and eosin, periodic acid-Schiff reagent, Gomori and Gridley's stains. Although the tissue phase is not really yeast-like, a brief description is included here, since, occasionally, the spherule structure may be confused histologically with aberrant forms of *Par. brasiliensis*. A full account of the histopathology is given by Huntingdon (1971).

VIII. ANTIFUNGAL CHEMOTHERAPY

A. Introduction

The chemotherapeutic era in the management of fungal infections began in 1903, when de Beurmann and Raymond reported the first successful treatment of a case of sporotrichosis with potassium iodide. Although the mode of action of potassium iodide remained obscure it was extensively used for treatment of sporotrichosis (Bennett, 1974) and subcutaneous phycomycosis (Koshi *et al.*, 1972). Griseofulvin, a metabolic product of *Penicillium griseofulvum* was the first antifungal antibiotic to be discovered (Oxford *et al.*, 1939; Grove and McGowan, 1947) but, since this compound has no useful activity against yeasts, it will not be discussed further. The first major advance in the search for antifungal drugs active against yeasts and yeast-like fungi was the discovery of two compounds produced by *Streptomyces noursei* (Hazen and

Brown, 1950, 1951). One of these, nystatin, was the first of more than 80 polyene antibiotics to be reported, and was the first to be used therapeutically. Since the introduction of nystatin, a large number of similar antifungal agents have been reported. Many have proved to be too toxic for clinical use and only those agents that are available for treatment of infections due to yeasts and yeast-like organisms will be discussed in this chapter.

B. Polyenes

Polyene antibiotics are the most commonly occurring single group of antimicrobial compounds produced by members of the genus *Streptomyces* and it is probable that, in addition to the 87 so far reported, there are others that await discovery. Although the list of polyenes is impressive, problems relating to their toxicity, solubility, stability and absorption have meant that only a few of these compounds are therapeutically useful, and only nystatin, natamycin, amphotericin B and candicidin are readily available for clinical use in the United Kingdom (Fig. 13). Candicidin is a heptaene produced by *Streptomyces griseus* with the empirical formula $C_{63}\,H_{85}\,N_2\,O_{19}$. Its complete chemical structure has not yet been delineated.

1. Chemical properties

All of the polyenes so far analysed have certain structural and chemical similarities. They are characterized by a macrolide ring of carbon atoms closed by formation of an internal ester or lactone (Fig. 13). The cyclic lactone is responsible for the highly characteristic peak on the infrared spectra of these compounds. The ring sizes of various polyenes range from 12–14 to 35–37 carbon atoms and the conjugated double bond is contained exclusively within the cyclic lactone. The second highly characteristic feature of the polyenes is the large number of hydroxyl groups on the molecule (6–14). They are usually distributed around the macrolide ring on alternate carbon atoms. The presence of polar hydroxyl groups and multiple hydrophobic double bonds confers on the polyene antibiotics their characteristic features.

These features probably play an important role in the mode of action of these compounds and the way in which they interact with cells. As a result of the amphipathic character of the polyenes they show limited solubility in water, ethanol or ether. In aqueous solvents they are known to produce micelles rather than true solutions. Polyenes are soluble in the polar organic solvents dimethyl sulphoxide and dimethylformamide. Solutions can be diluted further in water to produce a concentration of 50–100 $mg\,l^{-1}$. Stored

THE POLYENES

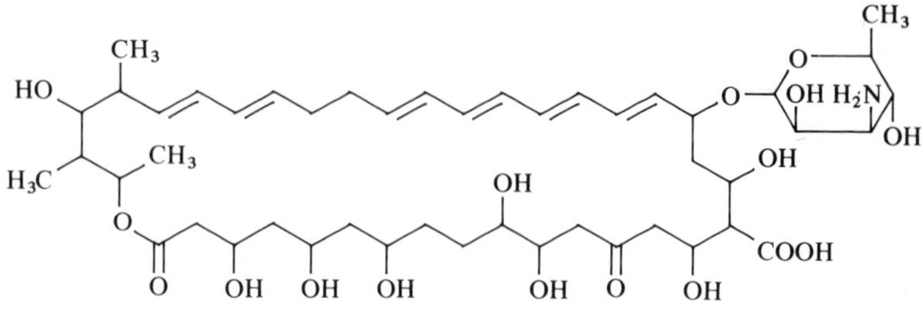

Nystatin

	Source:	*Streptomyces noursei*
	Proprietary names:	Fungicidin, Mycostatin, Nystan
	Chemistry:	$C_{47}H_{75}NO_{17}$
	Molecular weight:	926,100

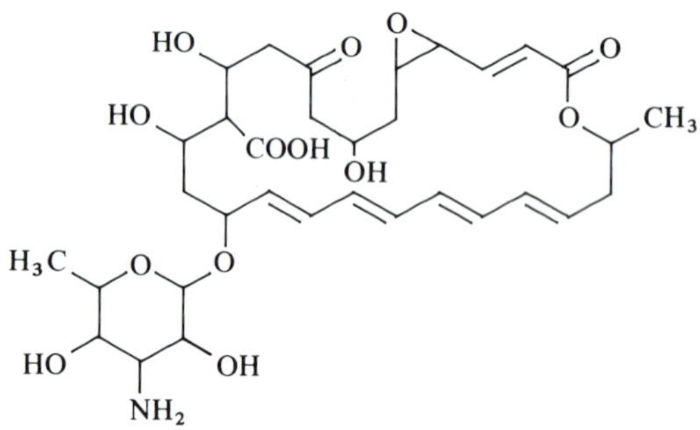

Natamycin

	Source:	*Streptomyces nataliensis*
	Proprietary names:	Natamycin, Pimafucin
	Chemistry:	$C_{33}H_{47}NO_{14}$
	Molecular weight:	665,700

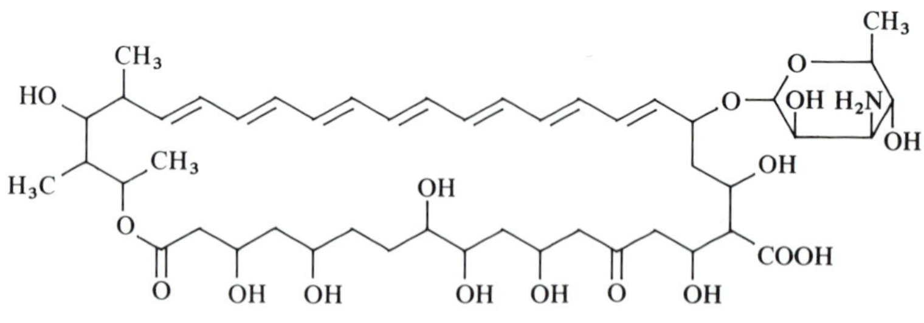

Amphotericin

	Source:	*Streptomyces nodosus*
	Proprietary names:	Fungizone, Fungitin
	Chemistry:	$C_{47}H_{73}NO_{17}$
	Molecular weight:	924,100

Fig. 13. Chemical structures and properties of some polyene antibiotics.

solutions in dimethyl sulphoxide or dimethyl formamide are stable provided that extreme care is taken to protect them from light or oxygen to which they are susceptible. Amphotericin B, which is produced from *Streptomyces nodosus*, is the only polyene for which the chemical structure and absolute configuration is known.

2. Physical properties

The most characteristic physical features of the polyenes are their ultraviolet absorption spectra, analysis of which has revealed several distinct classes of chromophores, namely tetraenes, pentaenes, hexaenes and heptaenes. Normal *et al.* (1972) related these ultraviolet absorption properties to the biochemical and biological mode of action of the polyenes.

3. Antifungal spectrum

The polyenes exert various degrees of activity against yeasts, dimorphic fungi, dermatophytes and moulds as well as being toxic to protozoa, but not to bacteria.

4. Mode of action

These compounds act by increasing the cell membrane permeability of sensitive fungi thus provoking leakage of amino acids, sugars and other metabolites from the cytoplasm leading to lysis and cell death (Kinsky, 1961a, b). Inhibition of aerobic and anaerobic respiration observed in cells treated with polyenes is a result of cytoplasmic leakage and is not due to inhibition of glycolysis or some other essential metabolic pathway. Inhibition of fungal growth by polyene antibiotics depends on the binding of the drug to the cell, and only cells that bind appreciable amounts of polyene are sensitive. There is a direct association between the sensitivity of an organism to a polyene and the presence of sterols in the plasma membrane of the cell. Thus bacteria that do not have sterols in their membranes are resistant, while all fungi sensitive to polyenes have sterol-containing plasma membranes. *Mycoplasma laidlawii* (*Acholeplasma laidlawii*) and fungal strains that are also unable to synthesize sterols *de novo* provide further proof of this association. When these organisms are grown in a sterol-deficient medium, they are resistant to polyenes, but, if grown in sterol-containing media, sterols are incorporated into the plasma membrane and the organisms are then sensitive to polyenes. Also, sensitive organisms can be protected from the fungicidal action of the drug by adding sterol to the culture medium which results in neutralization of the free polyene molecules. This effect is secondary to a physicochemical interaction between the polyene antibiotics and added sterol which prevents the antibiotics from interacting with cellular sterols. It has been shown by ultraviolet spectrum analysis and fluorimetric techniques that there is a direct interaction between sterols and polyene antibiotics. It has been shown also that amphotericin B and other polyenes interact with sterols in erythrocyte ghost cells, with sterols bound to liposomes and with membrane-bound cholesterol of *Acholeplasma laidlawii.* There is no evidence of interaction between polyenes and membranes lacking cholesterol.

It would appear that the effect of polyenes on plasma membranes of sensitive fungi is not to produce pores or holes but rather to interfere with the integrity of the membrane by binding to sterols thus decreasing the fluidity and permeability of the membrane and inhibiting movement of a carbohydrate carrier. It is currently considered (Medoff and Kobayashi, 1980) that the nature and severity of the membrane lesion induced by a polyene is dependent on the type of polyene and the kind of membrane under attack. Thus, yeasts are more sensitive to amphotericin B than to other polyenes because amphotericin B binds more avidly to the principal sterol in fungi, ergosterol. Polyene drugs that bind more avidly to cholesterol than to ergosterol (e.g. filipin) have increased activity against animal cells.

The cellular content of free sterol and the amount of polyene bound to the plasma membrane do not correlate directly with the differential sensitivity of fungal cells to the drug. Medoff and Kobayashi (1980) report that the sensitivity of fungal cells to a polyene depends on the cholesterol: phospholipid molar ratio in the cell membranes. It is probable that the lethal effects of polyenes can be separated from their effect on permeability. There are two types of binding of amphotericin B to fungal cells. The first is energy requiring and reversible and is probably responsible for the changes in cell permeability, while the second which is non-energy requiring and irreversible is responsible for lethality (Medoff and Kobayashi, 1980).

5. Effects *in vitro*

It is widely asserted that the antifungal activity of the polyenes increases as the number of double bonds rises although experimental data to support this view are hard to find. Antifungal activity increases progressively from the trienes through the tetraenes to the pentaenes (Utahara *et al.*, 1954; Gordee *et al.*, 1971; Armstrong *et al.*, 1965; Coronelli *et al.*, 1967). It has been shown also that haptaenes containing an aromatic moiety (e.g. candicidin) are more active than those without (e.g. amphotericin B; Pansy *et al.*, 1972). In general, the minimal fungicidal activity of the polyenes tends to be 2–10-fold higher than their minimal inhibitory concentrations. As with bacteria, inoculum size, temperature, duration of incubation and medium composition affect the results obtained (Kitahara *et al.*, 1976a). The relative activity of various

Table XIX. Minimal inhibitory concentrations of antifungal agents active against yeast and yeast-like fungi

	Concentration (mg l^{-1})			
Antifungal agent	*Candida albicans*	*Candida glabrata*	*Cryptococcus neoformans*	Dimorphic fungi
Nystatin	0.4–10	6	2	NA[b]
Natamycin	5	—	—	NA
Amphotericin B	0.5	1.6	0.01–1.6	0.04–1.6
Flucytosine	0.1–20[a]	0.1–3[a]	0.4–15[a]	R[c]
Clotrimazole	0.5–10	0.2–10	0.2–20	2–10
Miconazole and econazole	0.1–1	3–R	0.5–1	0.5–2
Ketoconazole	0.2–80	0.08–64	0.1–32	< 0.1–2

[a] indicates clinically sensitive strains

[b] indicates that the reagent was not active

[c] indicates that the micro-organism was resistant

polyene antibiotics against pathogenic yeasts and yeast-like organisms is shown in Table XIX. The presence of serum lowers the activity of polyenes partly through the binding effect of cholesterol and partly through the active repair processes of fungi that are stimulated by serum. The activity of nystatin is usually quoted in units and the International Standard contains 3000 units mg^{-1} (Lightbown *et al.*, 1963).

6. Interaction with other antimicrobial agents

In the presence of amphotericin B, the activity of flucytosine is increased against *Cr. neoformans* and *Candida* spp. (Medoff *et al.*, 1971). Tetracycline or rifampicin, which normally have no effect against fungal cells, inhibit synthesis of fungal protein and RNA respectively, if amphotericin B is also present. These *in vitro* effects have also been observed *in vivo*, and amphotericin B in combination with flucytosine has been shown to be more effective in treating cryptococcal meningitis than amphotericin B alone (Bennett *et al.*, 1978).

7. Resistance

Although fungi are intrinsically capable of giving rise to polyene-resistant variants, neither primary nor acquired resistance has been presented yet as a significant clinical problem. Laboratory-induced resistant variants of *Candida* spp. have lower levels of ergosterol in their membranes (Athar and Winner, 1971), grow more slowly and are less virulent than the parent strain. Polyene-resistant variants show no cross resistance with non-polyene antifungal drugs. Because of the absence of strains resistant to polyene antibiotics in the clinical situation, it is assumed that the defects observed in resistant organisms and the membrane changes responsible for resistance are detrimental to the overall survival of these organisms in the environment. Capek *et al.* (1970) reported that dermatophytes grown in the presence of nystatin or amphotericin B produced an enzyme capable of degrading nystatin which is analogous to the β-lactamase enzymes (penicillinases) produced by *Bacillus cereus* and *Staphylococcus aureus*.

8. Effects *in vivo* in experimental animals

The only member of the polyene group which is sufficiently non-toxic to be tolerated parenterally in experimental animals is amphotericin B. Experiments *in vivo*, using dosage regimens appropriate for animals, have usually confirmed the *in vitro* activity of amphotericin B against a variety of fungi and a number of animal studies have confirmed the *in vitro* evidence of synergy between amphotericin B and rifampicin, tetracycline and flucytosine (Titsworth and Gruneberg, 1973; Rabinovitch *et al.*, 1974; Arroyo *et al.*, 1977; Kitahara *et al.*, 1976b). Amphotericin B has also been used experimentally in combination with antitumour agents to treat successfully mice with murine leukaemia (Medoff *et al.*, 1974) under conditions in which the antitumour agent alone would be ineffective. Since these antitumour effects were achieved without intolerable polyene toxicity, it has been reasoned that tumour cells are more sensitive to polyene activity than normal cells (Medoff *et al.*, 1973). On the strength of experimental work in animals, it is proposed that amphotericin B has immunoadjuvant properties.

9. Clinical uses

Following its introduction, nystatin rapidly became the drug of choice in treatment of candidal infections of mucous membranes or skin. Its principal use has been in treatment of vaginal candidosis and it can be used safely during pregnancy (Kayser *et al.*, 1957). Used topically, it is virtually free from toxic or allergic side effects and resistance has not been reported even after prolonged usage. Nystatin has been used effectively in bladder washouts to treat candida cystitis in diabetic patients and orally to lower the yeast population of the gut. However, it is not absorbed from the gastrointestinal tract and cannot be given parenterally because it is unacceptably toxic.

Amphotericin B is the only polyene antibiotic available for parenteral administration. It is marketed as a bile salt complex (Fungizone from Squibb) that forms a colloidal suspension when hydrated. Several dosage regimens have been recommended to minimize the toxic effects of amphotericin B. The drug is normally given at a dose of 0.75–1.0 mg (kg body wt)$^{-1}$ day^{-1} in a four to six hour infusion. If clinically expedient, a 1 mg test dose (in 5% dextrose) is recommended initially increasing to the full dosage over five to seven days. Severely sick patients are given a 1 mg test dose followed four to six hours later by half the maximum dosage, increasing to the maximum dosage within 24 hours (Hay, 1982; Armstrong and Gold, 1980). Peak serum concentrations range from 0.5–5.5 mg l^{-1} (Medoff and Kobayashi, 1980; Winn, 1967). At optimal dosage, therapeutic serum concentrations will persist for 48 hours. Urine concentrations of amphotericin B are roughly parallel to serum concentrations, but penetration into cerebrospinal fluid, aqueous humour and haemodialysis solutions is poor (Feldman *et al.*, 1973). Intrathecal or intraventricular administration is necessary in cases of fungal meningitis. Most of an administered dose of amphotericin B is bound to cholesterol-containing membranes in a variety of tissues, but details of tissue distribution and catabolism are unknown.

In spite of the toxic side effects associated with amphotericin B chemotherapy, this drug has until recently remained the drug of choice for treatment of the majority of systemic fungal infections. In an attempt to minimize side effects a number of variations to the therapeutic regimen for amphotericin B have been proposed (Hay, 1982).

C. Flucytosine

Flucytosine was synthesized by Duschinsky *et al.* (1957) during a search for cytostatic compounds. Although flucytosine lacks any such activity, it was investigated further on the basis of a study by Berger and Duschinsky (1962) who demonstrated activity of the compound against yeasts *in vitro*. Grunberg *et al.* (1964) examined its therapeutic activity in experimental candidosis and crytococcosis in the mouse, and later reported on the first use of flucytosine to treat human mycoses (Grunberg *et al.*, 1967). More than 1000 cases of human infection have now been treated with parenteral or oral flucytosine, and there are a number of comprehensive reviews (Cartwright, 1975; Bennett, 1977; Scholer, 1980).

1. Chemical structure

Flucytosine (5-fluorocytosine or 4-amino-5-fluoro-2-pyrimidine) has a molecular weight of 129,100 and is an odourless, colourless, white cystalline solid which is stable at ordinary temperatures (Fig. 14). It is soluble in water up to 1.2% at 20°C and aqueous solutions are stable at pH 5.5–7.5.

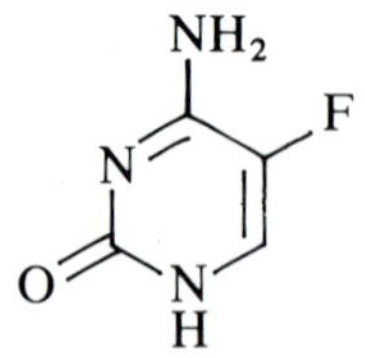

Fig. 14. Chemical structure of 5-fluorocytosine (4-amino-5-fluoro-2-pyrimidine). Molecular weight, 129,100.

2. Antifungal spectrum

The antifungal spectrum of flucytosine is limited, being most active against yeasts of the genera *Candida*, *Torulopsis*, *Cryptococcus* and *Geotrichum*, and slightly less active against *Aspergillus* spp. and dematiaceous fungi causing chromomycosis. The minimum inhibitory concentrations for these fungi range from 0.1 to 25 mg l^{-1}. The problem of resistant strains within essentially sensitive species is discussed on p. 268. Borderline activity of flucytosine (minimum inhibitory concentrations in the range 100 to 1,000 mg l^{-1}) is found with a range of micro-organisms including *Blastomyces dermatitidis* and *Sporotrichum schenckii*. The remaining fungi of medical importance (*Coccidioides* sp., *Histoplasma* sp. and the dermatophytes) are all resistant (Scholer, 1980).

3. Mode of action

Flucytosine exerts both fungistatic and fungicidal activity against candidas, cryptococci and *Torulopsis* spp. although a fungicidal effect requires a relatively high concentration and prolonged exposure (Shadomy *et al.*, 1973). The antifungal activity of flucytosine has two manifestations. It may disturb protein synthesis and/or the amino acid pool, resulting from extensive replacement of uracil by 5-fluorocytosine in fungal RNA; alternatively it may impair biosynthesis of fungal DNA following inhibition of thymidylate synthetase by 5-fluorodeoxyuridine monophosphate.

Flucytosine is taken up into sensitive cells by the cytosine permease and deaminated to fluorouracil by cytosine deaminase. The spectrum of activity of flucytosine is determined by the presence or absence of these enzymes. Fluorouracil cannot be used directly as an antifungal agent because, in low concentrations, its uptake into fungal cells is poor and, at high concentrations, it is too toxic to mammalian cells (Polak and Grenson, 1973).

The absence or low activity of cytosine deaminase in mammalian cells is considered the reason for the low toxicity of flucytosine to the mammalian host. Details of the intrafungal pathway and mode of action of flucytosine are now known, and are summarized by Scholer (1980). Recent work by Polak and Wain (1977) suggests that, in addition to incorporation of the deaminated drug into RNA there may also be an independent inhibition of DNA synthesis. This is supported by the finding that 5-fluorodeoxyuridine monophosphate which is present in cells of *C. albicans* treated with flucytosine produces a marked inhibition of thymidine synthetase which is consistent with blockage of DNA synthesis.

4. Resistance

Resistance to flucytosine may be produced by loss of an enzyme at any point on the pathway or by synthesis of sufficient normal components to compete with the fluorinated antimetabolites. Six methods of resistance to flucytosine have been described (Jund and Lacroute, 1970). To a greater or lesser degree, the majority of these mechanisms of resistance have been noted among clinical isolates. In all types of mutants studied so far, resistance has been genetically stable (Block *et al.*, 1973).

The overall resistance frequency of yeasts to flucytosine is 10^{-6} to 10^{-7} (Scholer, 1980). The frequency of resistant cells increases dramatically in *C. albicans* if the concentration of flucytosine drops below 25 mg l^{-1} and it is recommended that therapeutic serum concentrations be kept above this value (Normark and Schönebeck, 1972). Flucytosine-resistant mutants of *Cr. neoformans* and *C. albicans* occur spontaneously in the absence of the drug which, when present, provides the necessary selective pressure for them to dominate (Block *et al.*, 1973; Block and Bennett, 1974). Mutants that are resistant to flucytosine have unaltered sensitivity to polyenes and imidazoles. A certain degree of antifungal synergism has been reported between flucytosine and polyene antibiotics notably amphotericin B (Medoff *et al.*, 1971), and this combination has proved useful in treatment of patients.

5. Toxicity

The toxic manifestations associated with flucytosine therapy in Man are far less frequent (15%) and in general less severe than those resulting from therapy with amphotericin B. Gastrointestinal, hepatic and haematological signs of toxicity are usually mild and subside when the dosage is temporarily lowered. However, in certain cases, they may be severe. The risk of serious toxicity is dramatically increased if serum concentrations remain above 100 mg l^{-1} for long periods as might occur in patients with impaired renal function. In the majority of well documented cases, leucopenia, thrombocytopenia and disturbances in liver function develop slowly and are still reversible in the early stages. Hence, the grave consequences of flucytosine toxicity can be avoided by close observation (Scholer, 1980). The teratogenic effects of flucytosine observed in some animal experiments dictate that this drug should not be used in pregnancy other than in life-threatening situations.

D. Imidazoles

1. Introduction

The first report on the antifungal activity of an imidazole was by Woolley (1944) who also recognized the close structural similarity between benzimidazoles and purines. Thirty years later, Van den Bossche (1974) reported that the antifungal agent miconazole (phenethylimidazole) decreased uptake of purines by *Candida* spp. due to cell-membrane interference. The early report by Woolley (1944) and subsequent studies by Goldsworthy and Gertler (1949) stimulated a further study of the antimicrobial and pharmacological properties of approximately 100 benzimidazole derivatives (Holt, 1980). Of these, chlorabenzyl imidazole (chlormidazole) was the most active and well tolerated of these compounds and was selected for further investigation. The antifungal activity of chlormidazole (Myco–Polycid) is largely against yeasts and dermatophytes. It is active also against Gram-positive cocci and also has considerable antihelminthic activity.

The antifungal activity of chlormidazole is dramatically lowered in the presence of serum. Chlormidazole (5%) in cream, often with added antibiotics and steroids, has been successfully used to treat a large number of patients with dermatophytic eczema or other superficial mycoses. However, there are no reports of its use in the treatment of vaginal candidosis.

Following the development of chlormidazole a number of pharmaceutical companies in Europe and America undertook extensive research programmes which resulted in synthesis of a range of imidazole compounds with antifungal activity and led to two quite distinct imidazoles, clotrimazole (Canesten from Bayer) and miconazole (Daktarin from Janssen), being submitted for clinical trial. Although a large number of imidazole derivatives have now been described, only clotrimazole, miconazole, econazole and ketoconazole have been found to be clinically useful in treatment of infections due to yeasts and yeast-like fungi. The chemical structures of these four antifungal agents are shown in Fig. 15. These imidazoles are insoluble or only poorly soluble in water but will dissolve to give at least a 1% solution in a range of organic solvents. The dry powders are stable for very long periods and solutions in organic solvents are stable at 4°C if protected from air and light. These solutions are less stable when diluted with water and imidazoles may hydrolyse if the pH value is below 7. Aqueous solutions are opalescent but, with the exception of miconazole, show no loss of activity. Solutions of miconazole should be prepared fresh for use.

THE IMIDAZOLES

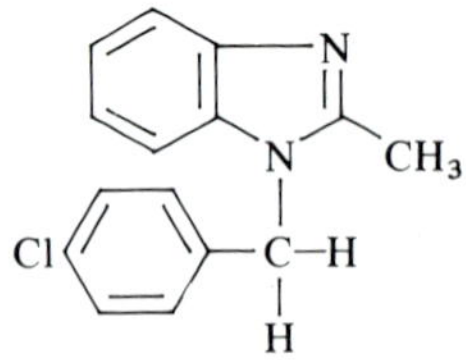

Chlormidazole
1-*p*-chlorobenzyl-2-methyl-benzimidazole
Proprietary name: MYCO-POLYCID

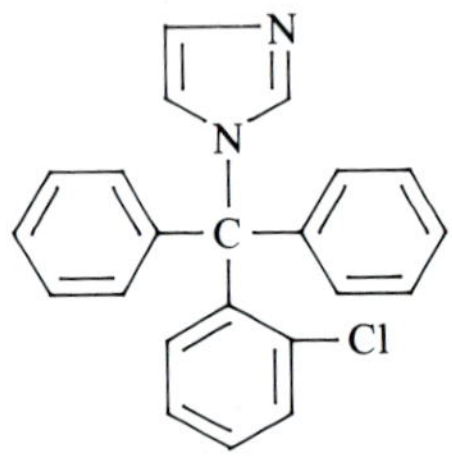

Clotrimazole
*bis*phenyl(2-chlorophenyl)-1-imidazolyl methane
Proprietary name: Canesten

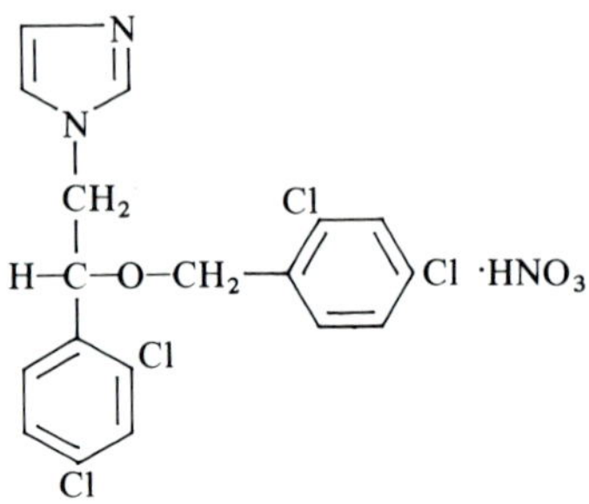

Miconazole
1-2,4-dichloro-β-(2,4-dichlorobenzyloxy)-phenethyl imidazole nitrate
Proprietary name: Daktarin

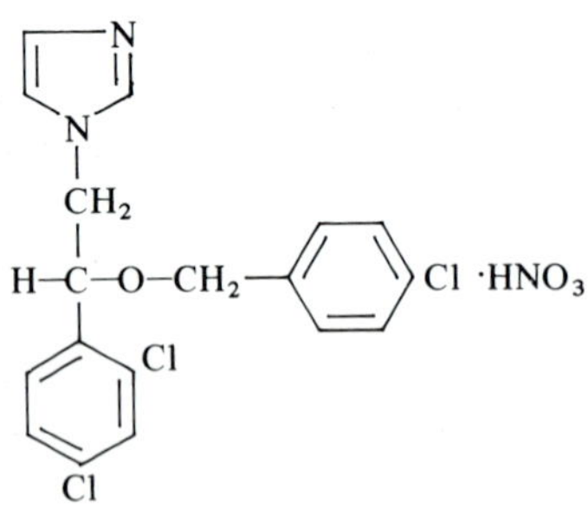

Econazole
1-2,4-dichloro-β-(*p*-chlorobenzyloxy) phenethyl imidazole nitrate
Proprietary name: Pevaryl

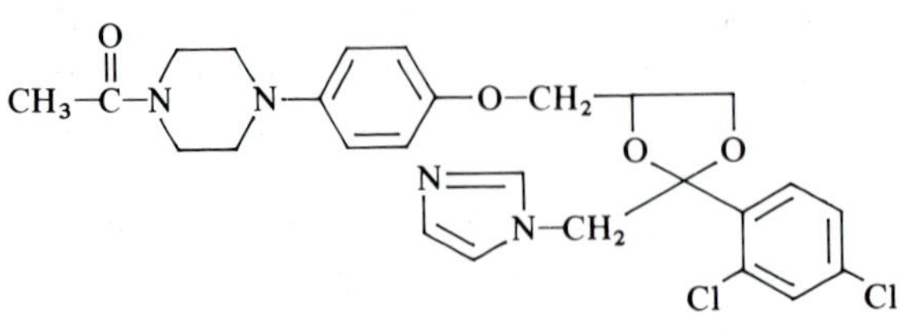

Ketoconazole
Cis-1-acetyl-4-(4[2-(2,4-dichlorophenyl)-2-(1H-1-imidazolyl-methyl)-1,3-dioxolan-4-yl] methoxy phenyl) piperazine
Proprietary name: Nizoral

Fig. 15. Structural formulae and proprietary names of imidazoles used to treat yeast diseases.

2. Antifungal spectrum

Almost all fungi of medical importance show *in vitro* sensitivity to these imidazoles. Clotrimazole is inhibitory to all of these organisms at a concentration of 10 mg l^{-1} in serum-free media and for *Madurella mycetomatis* and the dimorphic fungi the minimum inhibitory concentration is below 2 mg l^{-1} (Drouhet, 1973; Mahgoub, 1972). Miconazole is fungistatic at concentrations of 2 mg l^{-1} and fungicidal at 10 mg l^{-1}. Econazole at concentrations of 10–100 mg l^{-1} completely inhibits growth of medically important fungi in Sabouraud's broth (Holt, 1976). Ketoconazole is fungistatic to pathogenic fungi in the range 0.02–80 mg l^{-1}. The activity of imidazoles is lowered in the presence of serum due to binding to plasma proteins and, as a result, the minimum inhibitory concentrations in serum are 4–6-fold higher than in serum-free medium (Holt, 1980). The wide variation in imidazole activity noted in published reports is probably due to differences in the methodology used.

The antifungal imidazoles, notably miconazole, are unusual in that they also show some *in vitro* activity against Gram-positive bacteria notably *Staphylococcus aureus, Streptococcus agalactiae* and enterococci (de Louvois, 1981). There is no activity against Gram-negative organisms. The clinical value, if any, of this antibacterial activity has yet to be determined.

3. Mode of action

Studies of Iwata *et al.* (1973a, b) show that clotrimazole inhibits synthesis of various cellular constituents especially protein and RNA, with DNA, lipid and cell-wall polysaccharides less affected. This inhibition is brought about by clotrimazole-induced membrane leakage, with resultant loss of potassium and phosphate ions. There is evidence to suggest that, in addition to the direct action of clotrimazole on fungal cells, it may be capable also of stimulating host leucocyte myeloperoxidase activity. This enzyme is responsible for intracellular killing following phagocytosis by polymorphonuclear leucocytes and any deficiency in the enzyme results in increased cellular destruction by *C. albicans* (Lehrer and Cline, 1969a, b). In a controlled trial, Renz *et al.* (1974) observed that, in 15 out of 17 infants treated with clotrimazole for oral thrush, there was a marked and sometimes progressive increase in myeloperoxidase activity compared to a similar group of infants who received nystatin.

Unlike the polyene antibiotics, the effect of clotrimazole and other imidazole compounds on fungal plasma membranes is not associated with, or dependent on, the presence of sterols within the membrane. Fungicidal concentrations of miconazole alter the permeability of the cell membrane of

C. albicans with a resultant leakage of sodium and potassium ions and loss of protein, amino acids and phosphorus-containing compounds (Swamy *et al.*, 1974). Cytochemical studies suggest that, at fungicidal concentrations, intracellular peroxidase and catalase activities are abolished and that resultant cell death and necrosis are due to peroxide intoxication (De Nollin and Borger, 1974). In addition to a similar effect on the plasma membrane of fungi, ketoconazole inhibits sterol synthesis by sensitive organisms and De Brabander *et al.* (1980) have reported a synergistic action between ketoconazole and host defence cells. It appears that ketoconazole has different activity against yeasts and pseudohyphal cells (Borgers and Van den Bossche, 1982). Econazole is similar in action to miconazole.

4. Resistance

Among fungi of clinical importance wild strains resistant to imidazoles are exceedingly rare, although wide variation in inhibitory and fungicidal activity has been observed. This inability of most fungi to produce resistant variants is also reflected in the lack of success in producing resistant variants in the laboratory. No stable laboratory-induced resistant variants have been reported and laboratory strains which are resistant to polyenes are all sensitive to imidazoles. There is a single report (Holt and Azmi, 1978) of a clinical isolate of *C. albicans* which was resistant to flucytosine and miconazole with cross-resistance to clotrimazole, but which remained sensitive to nystatin and amphotericin B.

Chlormidazole, clotrimazole, miconazole and econazole have many of the characteristics required of an ideal antifungal drug for topical use. They are highly active against all of the pathogens involved; they are fungicidal; wild resistant strains are exceedingly rare; resistant mutants do not occur during treatment; they penetrate some distance into the cutaneous substrate without systemic absorption; adverse side effects are infrequent and many successful clinical studies have been published.

5. Clinical uses

The role of clotrimazole, miconazole and econazole in the treatment of systemic mycoses is less clear. Blood and body fluid levels after oral administration are unacceptably low and, with clotrimazole, the risk of short or longer term side effects is considerable. Oral miconazole is less toxic than clotrimazole but the resulting serum levels are low (Holt, 1976). Intravenous miconazole has been used by a number of groups in the U.S.A. to treat systemic infections successfully. In these cases, the considerable biochemical and haematological side effects have been considered justifiable. Much of the toxicity associated with systemic imidazole therapy is due to the solvents used, and alternative vehicles are currently being investigated. The most promising recent advance has been the development of ketoconazole, an orally absorbed imidazole. Therapeutic serum levels are achieved one to two hours after oral administration, and there are a number of reports of successful treatment of systemic fungal infections with oral ketoconazole (Heel, 1982a,b). Although the number of patients reporting side effects is lower (17%) than with other imidazoles, ketoconazole has been associated with fatal hepatotoxicity. The potential benefits of therapy should therefore be weighed against the risk of liver damage and patients should be carefully monitored both biochemically and clinically.

References

Acta, H.W. and Panja, G. (1927). *Indian Medical Gazette* **62**, 603.

Ainsworth, G.C. (1958). *In* "The Chemistry and Biology of Yeasts" (A.H. Cook, ed.), pp. 587–602. Academic Press, New York.

Ainsworth, G.C. and Austwick, P.K.C. (1955). *Transactions of the British Mycological Society* **38**, 369.

Aly, R., Britz, M.B. and Maibach, H.I. (1979). *British Journal of Dermatology* **101**, 445.

Anderson, J.H. (1979). *Sabouraudia* **17**, 1.

Angulo, A. and Pullar, A. (1971). *In* "The Pathologic Anatomy of Mycoses" (R.D.Baker, ed.), pp. 507–576. Springer-Verlag, Berlin.

Ariyevich, A.M., Krylov, L.M. and Stepanishcheva, Z.G. (1963). *Klinische Medica (Moskva)* **41**, 14.

Armstrong, D. and Gold, J. (1980). *In* "Antifungal Chemotherapy" (D.C.E. Speller, ed.), pp. 333–364. John Wiley, New York.

Armstrong, J.J., Grove, J.F., Turner, W.B. and Ward, G. (1965). *Nature, London* **260**, 399.

Arroyo, J., Medoff, G. and Kobayashi, G.S. (1977). *Antimicrobial Agents and Chemotherapy* **11**, 21.

Ashcraft, K.W. and Leape, L.L. (1970). *Journal of the American Medical Association* **212**, 454.

Athar, M.A. and Winner, H.I. (1971). *Journal of Microbiology* **4**, 505.

Audrey, C. (1887). *Revue Medicale* **7**, 586.

Austwick, P.K.C., Pepin, G.A., Thomson, J.C. and Yarrow, D. (1966). *In* "Symposium on *Candida* Infections" (H.I. Winner and R. Hurley, eds), pp. 89–100. Livingstone, London.

Bacle, J.R. and Bigland, C.H. (1962). *Journal of the American Veterinary and Medical Association* **141**, 476.

Band, J.D. and Maki, D.G. (1979). *American Journal of Medicine* **67**, 735.

Batista, A.C., Fischman, O. and Da Silva, J.O. (1961a). Publication of the Institute of Micologia et University of Recife, No.326.

Batista, A.C., Fischman, O., De Vasconcelos, C.T. and Da Rocha, I.G. (1961b). Publication of the Institute of Micologia et University of Recife, No.327.

Batista, A.C., De Vasconcelos, C.T., De Lima, J.A. and Shome, S.K. (1961c). Publication of the Institute of Micologia et University of Recife, No.329.

Beemer, A.M., Kuttin, E.S. and Katz, Z. (1973). *Avian Disease* **17**, 639.

Beheshti, F., Smith, A.G. and Krause, G.W. (1975). *Journal of Clinical Microbiology* **2**, 345.

Beijernick, M.W. (1889). *Archives Néerlandaises des Sciences Exactes et Naturelles* **23**, 367.

Bendove, R.A. and Ashe, B.I. (1952). *Archives of Internal Medicine* **89**, 107.

Beneke, E.S. and Rogers, A.L. (1970). "Medical Mycology Manual", 3rd edn. Burgess Publishing Co., Minneapolis.

Benham, R.W. (1931). *Journal of Infectious Diseases* **49**, 183.

Benham, R.W. (1947). *In* "Biology of Pathogenic Fungi" (W.J. Nickerson, ed.), pp. 63–70. Chronica Botanica Co., Waltham, Mass.
Bennett, J.E. (1974). *New England Journal of Medicine* **30**, 320.
Bennett, J.E. (1977). *Annals of Internal Medicine* **86**, 319.
Bennett, J.E., Kwon-Chung, K.J. and Howard, D.H. (1977). *American Journal of Epidemiology* **105**, 582.
Bennett, J.E., Plempel, M., Stevens, D.A. and Scholer, H.J. (1978). *In* "Current Chemotherapy". Proceedings of the Tenth International of Congress Chemotherapy, Zurich, 1977 (W. Wiegenthaler and R. Luthy, eds), vol. 1, pp. 53–55. American Society for Microbiology, Washington.
Berg, F.T. (1846). "Om Torsk Hos Barn". Hjerta, L.J., Stockholm.
Berger, J. and Duschinsky, R. (1962). *United States Patent Application Series* **181**, 822.
Berkhout, C.M. (1923). Dissertation: University of Utrecht.
Biancardi, G., Binaghi, C., Giorgi, G.C., Milani, P., Morganti, L. and Baudini, U. (1976). *Atti Della Societa Italiana di Buiatria.*
Bisping, W. (1961). *Mykosen* **4**, 137.
Bizzozero, J. (1884). *Virchows Archiv fur Pathologische Anatomie und Physiologie und fur Klinische Medizin* **98**, 441.
Block, E.R. and Bennett J.E. (1974). *American Review of Respiratory Diseases* **109**, 723.
Block, E.R., Jennings, A.E. and Bennett, J.E. (1973). *Antimicrobial Agents and Chemotherapy* **3**, 649.
Bloomfield, N., Gordon, M.A. and Elmendorf, F. (1963). *Proceedings of the Society for Experimental Biology and Medicine* **114**, 64.
Boardman, C.R. and Malkinson, F.D. (1962). *Archives of Dermatology* **85**, 44.
Bodey, G.P. (1966). *Journal of Chronic Disease* **19**, 667.
Borgers, M. and Van den Bossche, H. (1982). *In* "Ketaconazole in the Management of Fungal Disease" (H.B. Levine, ed.), pp. 25–47. ADIS Press, New York.
Botard, R.W. and Kelley, D.C. (1968). *Applied Microbiology* **16**, 689.
Boyanowsky, A. and Lischka, G. (1977). *Hautarzt* **28**, 409.
Brinkman, A. (1974). *Parasitologia* **25**, 350.
Brömel, J. and Zettl, K. (1979). *Praktischer Tierarzt* **60**, 882.
Buck, J.D. (1980). *Journal of Wildlife Diseases* **16**, 141.
Budtz-Jorgensen, E., Stenderup, A. and Grabowski, M. (1975). *Community Dental and Oral Epidemiology* **3**, 115.
Burke, R.C. (1961). *Journal of Investigative Dermatology* **36**, 389.
Campbell, C.K. and Mackenzie, D.W.R. (1980). *In* "Biology and Activities of Yeasts" (F.A. Skinner, S.M. Passmore and R.R. Davenport, eds), pp. 249–258. Academic Press, London.
Capek, A., Simek, A., Leiner, J. and Weichet, J. (1970). *Folio Microbiologica* **15**, 314.
Caprilli, F., Mercantini, R., Nazzaro, P.M., Passi, S. and Torolo, A. (1973). *Mycopathologia et Mycologia Applicata* **51**, 171.
Carroll, C.J., Hurley, R. and Stanley, V.C. (1973). *Journal of Obstetrics and Gynaecology of the British Commonwealth* **80**, 258.
Cartwright, R.Y. (1975). *Journal of Antimicrobial Chemotherapy* **1**, 141.
Castellani, A. (1912). *Lancet* **i**, 13.
Castellani, A. (1925). *Journal of Tropical Medicine and Hygiene* **28**, 217.
Castellani, A. (1963). *Mycopathologia* **20**, 1.
Castellani, A. and Chalmers, A.J. (1919). "Manual of Tropical Medicine", Balliere Tindall and Cox, London.
Cawson, R.A. (1966). *In* "Symposium on Candida Infections" (H.I. Winner and R. Hurley, eds), pp. 138–142. Livingstone, London.
Černíck, L., Hô Tân-Phi, Doan, Hông-Hoa, Šery, V. and Pham Vân-Nông (1960). *Časopis Lêkařu Českých* **99**, 1023.
Chandler, F.W., Kaplan, W. and Ajello, L. (1980). *In* "Histopathology of Mycotic Disease", pp. 120–121. Wolfe Medical Publications, London.
Chapel, T., Brown, W.J., Jeffries, C. and Stewart, J.A. (1978). *Journal of Infectious Diseases* **137**, 50.
Chaskes, S. and Tyndall, R.L. (1975). *Journal of Clinical Microbiology* **1**, 509.

Chick, E.W. (1971). *In* "The Pathologic Anatomy of Mycoses" (R.D. Baker, ed.), pp. 465–506. Springer-Verlag, Berlin.
Clarke, R.T.J. and di Menna, M.E. (1961). *Journal of General Microbiology* **25**, 113.
Clayton, Y.M. and Noble, W.C. (1966). *Journal of Clinical Pathology* **19**, 76.
Collin, J., Collin, C., Constable, F.L. and Johnston, I.D.A. (1975). *Lancet* **ii**, 150.
Cooke, W.B., Phaff, H.J., Miller, M.W., Shifrine, M. and Knapp, E.P. (1960). *Mycologia*, **52**, 210.
Coronelli, C., Pasqualucci, R.C., Thiemann, J.E. and Tamoni, G. (1967). *Journal of Antibiotics* **20A**, 329.
Coudert, J., Despierres, G., Saëz, H. and Hollard, J. (1957). *Semaine des Hospitaux de Paris* **33**, 2978.
Dabrowa, N., Landau, J.W., Newcomer, V.D. and Plunkett, O.A. (1964). *Mycopathologia et Mycologia Applicata* **24**, 137.
Dawson, C.O. (1962). *Sabouraudia* **1**, 214.
Davies, R.R. and Denning, T.J.V (1972). *Sabouraudia* **10**, 301.
de Beurmann, O. and Raymond, D. (1903). *Annals de Dermatologie et Syphilie* **4**, 678.
De Brabander, M., Aerts, F., Van Cutsem, J., Van den Bossche, H. and Borgers, M. (1980). *Sabouraudia* **18**, 197.
Deeb, E.N. and Natsios, G.A. (1972). *New England Journal of Medicine* **286**, 613.
Delafond, M. (1858). *Gazette Hebdomadaire de Médicine et de Chirurgie* **5**, 90.
de Louvois, J. (1981). *Clinical Research Reviews* No. 2.
de Louvois, J., Mulhall, A. and Hurley, R. (1979). *Journal of Clinical Pathology* **32**, 715.
de Miranda, L.R. (1984). *In* "The Yeasts, A Taxonomic Study" (N.J.W. Kreger-van Rij, ed.), pp. 845–872. Elsevier Biomedical Press, Amsterdam.
De Nollin, S. and Borgers, M. (1974). *Sabouraudia* **12**, 341.
de Stoechlin, H. (1898). *Archiv für Experimentelle Medizin* **10**, 1.
Dietz, R. and Seeliger, H.P.R. (1977). *Ernst Rodenwaldt Archiv* **4**, 65.
Do Carmo Sousa, L. (1969). *In* "The Yeasts" (A.H. Rose and J.S. Harrison, eds), vol. 1, pp. 79–105. Academic Press, London.
Do Carmo Sousa, L. (1970). *In* "The Yeasts — A Taxonomic Study" (J. Lodder, ed.), 2nd edn., pp. 1309–1352. North Holland Publishing Company, Amsterdam.
Do Carmo Sousa, L. and Van Uden, N. (1967). *Mycologia* **59**, 653.
Dorn, M. and Roehnert, K. (1977). *Journal of Investigative Dermatology* **69**, 244.
Drouhet, E. (1973). *Clinical Excerpts* **35**, 1219.
Drouhet, E. and Vieu, M. (1957). *Semaine des Hospitaux de Paris* **33**, 793.
Duschinsky, P., Pleven, E. and Heidelberger, C. (1957). *Journal of American Chemical Society* **79**, 4599
Echevarria, L.I. (1977). *Gaceta Vetinaria* **39**, 462.
Emmons, C.W. (1951). *Journal of Bacteriology* **62**, 685.
Emmons, C.W. (1955). *American Journal of Hygiene* **62**, 227.
Emmons, C.W. (1960). *Public Health Reports* **75**, 362.
Emmons, C.W. (1962). *Laboratory Investigations* **11**, 1026.
Emmons, C.W., Binford, C.W., Utz, J.P. and Kwon-Chung, K.J. (1977). "Medical Mycology", 3rd edn. Lea and Febiger, Philadelphia.
English, M.P., Smith, R.J. and Harman, R.R.M. (1971). *British Journal of Dermatology* **84**, 567.
Evans, E.E. (1962). *In* "Fungi and Fungus Diseases" (G. Daldorf, ed.), pp. 67–81. Charles Thomas Springfield, Illinois.
Evans, H.L., Kletzel, M., Lawson, R.D., Frankel, L.S. and Hopfer, R.L. (1980). *Cancer* **45**, 367.
Evenson, A.E. and Lamb, J.W. (1964). *Journal of Bacteriology* **88**, 542.
Faergemann, J. (1979). *Acta Dermato-Venereologica, Supplement*, 86.
Faergemann, J. and Bernander, S. (1979). *Sabouraudia* **17**, 171.
Faergemann, J. and Bernander, S. (1981). *Sabouraudia* **19**, 117.
Faergemann, F. and Fredriksson, T. (1980). *Acta Dermato Venereologica (Stockholm)* **60**, 531.
Faergemann, F. and Fredriksson, T. (1981). *Journal of Investigative Dermatology* **77**, 314.
Fahlberg, W.J., Dukes, C.D. and Guthrie, R.K. (1957). *Texas Reports on Biology and Medicine* **15**, 65.

Feldman, H.A., Hamilton, J.D. and Gutman, R.A. (1973). *Antimicrobial Agents and Chemotherapy* **4**, 302.

Fell, J.W., Tallman, A.S. and Ahearn, D.G. (1984). *In* "The Yeasts, A Taxonomic Study" (N.J.W. Kreger-van Rij, ed.), pp. 893–905. Elsevier Biomedical Press, Amsterdam.

France, D.R. (1968). *New Zealand Medical Journal* **67**, 552.

Gaines, J.D. and Remington, J.S. (1972). *Surgery* **72**, 730.

Gellis, S.S., Feingold, M., Kozinn, P.J., Tariq, A.A., Reale, M.R. and Rudolph, N. (1976). *American Journal of Disease in Childhood* **130**, 291.

Gentles, J.C. and La Touche, C.J. (1969). *In* "The Yeasts" (A.H. Rose and J.S. Harrison, eds.), vol. 1, pp. 108–182. Academic Press, London.

Gilbert, W.R. and Fetter, B.F. (1962). *Journal of Bacteriology* **84**, 961.

Goldman, H.J., Littman, M.L., Oppeheimer, G.D. and Glickman, S.I. (1960). *Journal of the American Medical Association* **174**, 359.

Goldstein, E., Grieco, M.H., Finkel, G. and Louria, D.B. (1965). *Journal of Infectious Diseases* **115**, 293.

Goldsworthy, M.C. and Gertler, S.J. (1949). *Chemical Abstracts* **43**, 6351.

Gordee, R.S., Butler, T.F. and Narasimhachari, N. (1971). *Journal of Antibiotics* **24**, 561.

Gordon, M.A. (1951). *Journal of Investigative Dermatology* **17**, 267.

Goulon, M., Babinet, P., Got, C., Gajdos-Preuss, A., Raphael, J.C. and Jaillon, P. (1973). *La Nouvelle Presse Médicale* **2**, 2033.

Gregson, A.E.W. and La Touche, C.J. (1961). *Journal of Laryngology and Otology* **75**, 45.

Griffin, J.R., Pettit, T.H., Fishman, L.S. and Foos, R.Y. (1973). *Archives of Ophthalmology* **89**, 450.

Grove, J.F. and McGowan, J.C. (1947). *Nature, London* **160**, 574.

Grunberg, E., Titsworth, E. and Bennett, M. (1964). *In* "Antimicrobial Agents and Chemotherapy — 1963", pp. 566–568. American Society of Microbiology, Ann Arbor.

Grunberg, E., Prince, H.N. and Utz, J.P. (1967). *Proceedings of the Fifth International Congress on Chemotherapy* **4**, 69.

Gründer, K. (1976). *Hautarzt* **27**, 422.

Hajsig, M., Setinski, Z. and Topolko, S. (1962). *Veterinarski Archiv* **32**, 232.

Hammerman, K.J., Powell, K.E. and Tosh, F.E. (1974). *Sabouraudia* **12**, 33.

Harmon, G.A., Slutzker, B., Davis, A. and Edmonds, P. (1962). *Archives of Dermatology* **85**, 510.

Hasenclever, H.F. (1959). *Journal of Bacteriology* **78**, 105.

Hasenclever, H.F. and Mitchell, W.O. (1961a). *Sabouraudia* **1**, 16.

Hasenclever, H.F. and Mitchell, W.O. (1961b). *Journal of Bacteriology* **82**, 578.

Hasenclever, H.S. and Mitchell, W.O. (1963). *Sabouraudia* **2**, 87.

Haugen, R.K. and Baker, R.D. (1954). *American Journal of Clinical Pathology* **24**, 1381.

Hausmann, D. (1875). "Parasites des Organes Sexuels Femeilles de l'Homme et de Quelques Animaux avec une Notice sur Developpement de *l'Oidium albicans Robin*". J.B. Baillière, Paris.

Hay, R.J. (1982). *Hospital Update*, 223.

Hay, R.J. and Mackenzie, D.W.R. (1980). *Journal of Infection* **2**, 13.

Hazen, E.L. and Brown, R. (1950). *Science* **112**, 423.

Hazen, E.L. and Brown, R. (1951). *Proceedings of the Society of Experimental Biology and Medicine* **76**, 93.

Heel, R.C. (1982a). *In* "Ketaconazole in the Management of Fungal Disease" (H.B. Levine, ed.), pp. 89–93. ADIS Press, New York.

Heel, R.C. (1982b). *In* "Ketaconazole in the Management of Fungal Disease" (H.B. Levine, ed.), pp. 151–153. ADIS Press, New York.

Herff, O. von (1895). *Sammlung Klinischer Vortrage in Verbindung mit Deutschen Klinikern* **137**, 493.

Hesseltine, H.C. (1933). *Journal of the American Medical Association* **100**, 177.

Higgs, J.M. (1973). *Transactions of St. John's Hospital Dermatological Society* **59**, 175.

Holt, R.J. (1976). *In* "Chemotherapy, Proceedings of the Ninth International Congress of Chemotherapy (J.D. Williams and A.M. Geddes, eds), vol. 6, pp. 119–125. Plenum Press, London.

Holt, R.J. (1980). *In* "Antifungal Chemotherapy" (D.C.E. Speller, ed.), pp. 107–147. John Wiley, New York.
Holt, R.J. and Azmi, A. (1978). *Lancet* **i**, 50.
Holti, G. (1966). *In* "Symposium on Candida Infections" (H.I. Winner and R. Hurley, eds), pp. 73–81. Livingstone, London.
Holzschu, D.L., Presley, H.L., Miranda, M. and Phaff, H.J. (1979). *Journal of Clinical Microbiology* **10**, 202.
Huntingdon, R.W. (1971). *In* "The Pathologic Anatomy of Mycoses" (R.D. Baker, ed.), pp. 147–210. Springer-Verlag, Berlin.
Hurley, R. (1962). M.D. Thesis: University of London.
Hurley, R. (1964). *Postgraduate Medical Journal* **40**, 644.
Hurley, R. (1965). *Journal of Pathology and Bacteriology* **90**, 351.
Hurley, R. (1966). *In* "Symposium on Candida Infections" (H.I. Winner and R. Hurley, eds), pp. 13–24. Livingstone, London.
Hurley, R. (1967). *Review of Medical and Veterinary Mycology* **6**, 159.
Hurley, R. (1975). *Practitioner* **215**, 753.
Hurley, R. and de Louvois, J. (1980). *Zentralblatt für Backteriologie, Parasitenkunde, Infektionskrankheiten und Hygiene, Supplement* **8**, 67.
Hurley, R. and Stanley, V.C. (1969). *Journal of Medical Microbiology* **2**, 63.
Hurley, R. and Winner, H.I. (1962). *Journal of Pathology and Bacteriology* **84**, 33.
Hurley, R. and Winner, H.I. (1963). *Journal of Pathology and Bacteriology* **86**, 75.
Hurley, R., Leask, B.G.S., Faktor, J.A. and de Fonseka, C.I. (1973). *Journal of Obstetrics and Gynaecology of the British Commonwealth* **80**, 252.
Hurley, R., Stanley, V.C., Leask, B.G.S. and de Louvois, J. (1974). *In* "The Normal Microbial Flora of Man" (F.A. Skinner and J.G. Carr, eds), pp. 155–183. Academic Press, London.
Iwata, K., Yamaguchi, H. and Hiratani, T. (1973a). *Sabouraudia* **11**, 158.
Iwata, K., Kanda, Y., Yamaguchi, H. and Osumi, M. (1973b). *Sabouraudia* **11**, 205.
Jarowski, C., Fialk, M.A., Murray, H.W., Gottlieb, G.J., Coleman, M., Steinberg, C.R. and Silver, R.T. (1978). *Archives of Internal Medicine* **138**, 544.
Jennings, A., Bennett, J.E. and Young, V. (1968). *Mycopathologia et Mycologia Applicata* **35**, 256.
Jund, R. and Lacroute, F. (1970). *Journal of Bacteriology* **102**, 607.
Kadic, S. (1978). *Veterinarski Archiv* **48**, 101.
Kaliski, S.R., Bennes, L.M. and Mattmann, L. (1952). *Journal of the American Medical Association* **148**, 1207.
Kandhari, K.C., Pasricha, J.S., Duggal, S.K. and Pandhi, R.K. (1969). *Indian Journal of Medical Research* **57**, 1295.
Kaplan, W. (1959). *Journal of the American Veterinary Medical Association* **134**, 113.
Kaufman, L. and Blumer, S. (1968). *Applied Microbiology* **16**, 1907.
Kawakita, S. and Van Uden, N. (1965). *Journal of General Microbiology* **39**, 125.
Kayser, J.L., Eisenberg, G.M., Shapiro, J. and Flippin, H. (1957). *American Journal of Obstetrics and Gynecology* **74**, 167.
Keymer, I.F. and Austwick, P.K.C. (1961). *Sabouraudia* **1**, 22.
Kinsky, S.G. (1961a). *Biochemical and Biophysical Research Communications* **4**, 353.
Kinsky, S.G. (1961b). *Journal of Bacteriology* **82**, 889.
Kitahara, M., Seth, V.K., Medoff, G. and Kobayashi, G.S. (1976a). *Antimicrobial Agents and Chemotherapy* **9**, 908.
Kitahara, M., Kobayashi, G.S. and Medoff, G. (1976b). *Journal of Infectious Diseases* **133**, 663.
Klamen, J. (1884). *Allgemeine Medizinische Zentralzeitung*, p. 23.
Knight, L. and Fletcher, J. (1971). *Journal of Infectious Diseases* **123**, 371.
Knudtson, W.U., Wohlgemuth, K., Kirkbride, C., Robl, M. and Kieffer, M. (1973). *Sabouraudia* **11**, 175.
Koch, Y. and Koch, H.A. (1981). *Mykosen* **24**, 197.
Korth, H. and Pulverer, G. (1971). *Applied Microbiology* **21**, 541.

Koshi, G., Kurien, T., Sudarsanam, D., Selvephandian, A.J. and Mammen, K.E. (1972). *Sabouraudia* **10**, 237.

Kozinn, P.J., Galen, R.S., Taschdhian, C.L., Goldberg, P.L., Protzman, W. and Kozinn, M.A. (1976). *Journal of the American Medical Association* **235**, 628.

Krause, W., Matheis, H. and Wulf, K. (1969). *Lancet* **i**, 598.

Kreger-van Rij, N.J.W. (ed.) (1984). "The Yeasts, A Taxonomic Study." Elsevier Biomedical Press, Amsterdam.

Ktanidou-Kartali, S., Jimakas, C., Franjidou, F., Stavropoulus, K. and Papapanagoitou, J. (1979). *Deltion Ellenikes Mikrobiologikes Etaireias* **24**, 136.

Kuttin, E.S. and Muller, J. (1981). *Antonie van Loeuwenhoek* **47**, 255.

Kwon-Chung, K.J. (1975). *Mycologia* **67**, 1197.

Kwon-Chung, K.J. (1976a). *Mycologia* **68**, 821.

Kwon-Chung, K.J. (1976b). *Mycologia* **68**, 942.

Kwon-Chung, K.J. (1977). *International Journal of Systematic Bacteriology* **27**, 293.

Langenbeck, B. (1839). *Neue Notizen auf dem Gebiete der Naturkund Heilkunde (Froriep)* **12**, 145.

Lehner, T. (1966). *In* "Symposium on Candida Infections" (H.I. Winner and R. Hurley, eds), pp. 73–81. Livingstone, London.

Lehrer, R.I. and Cline, M.J. (1969a). *Journal of Clinical Investigation* **48**, 1478.

Lehrer, R.I. and Cline, M.J. (1969b). *Journal of Bacteriology* **98**, 996.

Leisegang, T.J. and Forster, R.K. (1980). *American Journal of Opthalmology* **90**, 38.

Lightbown, J.W., Kogut, M. and Vemura, K. (1963). *Bulletin of the World Health Organisation* **29**, 87.

Littman, M.L. (1958). *Transactions of New York Academy of Sciences* **20**, 623.

Littman, M.L. and Borok, R. (1968). *Mycopathologia et Mycologia Applicata* **35**, 329.

Littman, M.L. and Schneierson, S. (1959). *American Journal of Hygiene* **69**, 49.

Lodder, J. (ed.) (1970). "The Yeasts–A Taxonomic Study", 2nd edn. North-Holland Publishing Company, Amsterdam.

Lodder, J. and Kreger-van Rij, N.J.W. (1952). "The Yeast–A Taxonomic Study". North-Holland Publishing Company, Amsterdam.

Lurie, H.I. (1971). *In* " The Pathologic Anatomy of Mycoses" (R.D. Baker, ed.), pp. 614–675. Springer-Verlag, Berlin.

Mackenzie, D.W.R. (1962). *Journal of Clinical Pathology* **15**, 563.

Mackenzie, D.W.R. (1966). *In* "Symposium on Candida Infections" (H.I. Winner and R. Hurley, eds), pp. 26–43. Livingstone, London.

Mahgoub, El Sheikh (1972). *Sabouraudia* **10**, 210.

Maiback, H.I. and Kligman, A.M. (1962). *Archives of Dermatology* **85**, 233.

Malassez, L. (1874). *Archives of Physiology, Series II* **1**, 451.

Martins, C. (1928). *Comptes Rendus de la Societé de Biologie* **98**, 1162.

McGinley, K.J., Lantis, L.R. and Marpless, R.R. (1970). *Archives of Dermatology* **102**, 168.

Medoff, G., and Kobayashi, G.S. (1980). *In* "Antifungal Chemotherapy" (D.C.E. Speller, ed.), pp. 3–33. John Wiley, New York.

Medoff, G., Comfort, M. and Kobayashi, G.S. (1971). *Proceedings of the Society for Experimental Biology and Medicine* **138**, 571.

Medoff, G., Kwan, C.M., Schlessinger, D. and Kobayashi, G.S. (1973). *Antimicrobial Agents and Chemotherapy* **3**, 441.

Medoff, G., Valeriote, F., Lynch, R.G., Schlessinger, D. and Kobayashi, G.S. (1974). *Cancer Research* **34**, 974.

Mehnert, B. and Koch, U. (1963). *Zentralblatt fur Bakteriologie, Parasitenkunde, Infecktionskrankheiten und Hygiene (Abteilung 1)* **188**, 103.

Meyer, S.A., Ahearn, D.G. and Yarrow, D. (1984). *In* "The Yeasts, A Taxonomic Study" (N.J.W. Kreger-van Rij, ed.), pp. 585–844. Elsevier Biomedical Press, Amsterdam.

Moore, M. (1934). *Annals of Missouri Botanical Garden* **21**, 349.

Morenz, J. (1971). *In* "The Pathological Anatomy of Mycoses" (R.D. Baker, ed.), pp. 919–952. Springer-Verlag, Berlin.

Morquer, R., Lombard, C. and Berthelon, M. (1955). *Comptes Rendus Academie des Sciences (Paris)* **240**, 378.

Munoz, M., Estes, G., Kilpatrick, M., Di Salvo, A. and Virella, G. (1980). *Mycopathologia* **72**, 47.

Murphy, N., Buchannon, C., Damjanovic, V., Whittaker, R., Hart, C.A. and Cooke, R.W.I. (1985). Communicable Disease Reports 85/34.

Nickerson, W.J. (1963). *Bacteriological Review* **27**, 305.

Nicklas, W., Suschka, C., Weigt, U. and Bohm, K.H. (1980). *Berliner und Münchener Tierarztliche Wochenschrift* **93**, 328.

Noble, W.C. and Sommerville, D.A. (1974). *In* "Microbiology of Human Skin" (W.C. Noble and D.A. Sommerville, eds), vol. 2, pp. 206–210. W.B. Saunders, London.

Norman, A.W., Demel, R.A., Dekruijff, B. and van Deenan, L.L.M. (1972). *Journal of Biological Chemistry* **247**, 1918.

Normark, S. and Schönebeck, J. (1972). *Antimicrobial Agents and Chemotherapy* **2**, 114.

Odds, F.C. (1979). "Candida and Candidosis." Leicester University Press, Leicester.

Odds, F.C. and Abbott, A.B. (1980). *Sabouraudia* **18**, 301.

Ota, M. and Huang, P.T. (1933). *Annales de Parasitologie Humaine et Comparée* **11**, 49.

Oxford, A.E., Raistrick, H. and Simonart, P. (1939). *Biochemical Journal* **33**, 240.

Pagano, J., Levin, J.D. and Trejo, W. (1958). *Antibiotic Annals* 1957–58, 137.

Pankajalakshmi, V.V., Taralakshmi, V.V., Paramasivan, C.N. and Subramanian, S. (1979). *Indian Journal of Dermatology, Venereology, and Leprology* **45**, 136.

Pansy, F.E., Jambor, W.P., Wilbur-Murphy, J. and Gadebusch, H.H. (1972). *Journal of Antibiotics* **25**, 405.

Pappagianis, D., Remingston, J., Collins, M.S. and Hector, R. (1979). *Antimicrobial Agents and Chemotherapy* **16**, 123.

Parrot, J. (1877). "Clinique des nouveau-nés l'athrepsie." Leçons recuellies par le Dr. Troisier. G. Masson, Paris.

Partridge, B.M. and Winner, H.I. (1965). *Lancet* **i**, 1060.

Partridge, B.M., Athar, M.A. and Winner, H.I. (1971). *Journal of Clinical Pathology* **24**, 645.

Péter, M., Horvath, G. and Domokos, L. (1967). *Medicina Interna* **19**, 875.

Phaff, H.J. and Ahearn, D.G. (1970). *In* "The Yeasts–A Taxonomic Study" (J. Lodder, ed.), 2nd edn., pp. 1187–1223. North Holland Publishing Company, Amsterdam.

Phaff, H.J. and Fell, J.W. (1970). *In* "The Yeasts–A Taxonomic Study" (J. Lodder, ed.), 2nd edn., pp 1088–1166. North Holland Publishing Company, Amsterdam.

Plass, E.D., Hesseltine, H.C. and Borts, I.H. (1931). *Amercian Journal of Obstetrics and Gynecology* **21**, 320.

Pliwig, G. and Kligman, A.M. (1975). *In* "Acne: Morphogenesis and Treatment", pp. 38–41. Springer-Verlag, Berlin.

Polak, A. and Grenson, M. (1973). *European Journal of Biochemistry* **32**, 276.

Polak, A. and Wain, W.H. (1977). *Chemotherapy (Basel)* **23**, 243.

Pollack, J.D. and Benham, W. (1957). *Journal of Laboratory and Clinical Medicine* **50**, 313.

Polonelli, L., Morace, G., Barcaroli, B.M. and Cossu, A.L. (1981). *Mycopathologia* **73**, 161.

Porro, M.N., Passi, S., Caprilli, F., Nazzaro, P. and Murpurgo, G. (1976). *Journal of Investigative Dermatology* **66**, 178.

Porro, M.N., Passi, S., Caprilli, F. and Mercantini, R. (1977). *Journal of Investigative Dermatology* **69**, 531.

Potter, B.S., Burgoon, C.F. and Johnson, W.C. (1973). *Archives of Dermatology* **107**, 388.

Prophet, K. (1963). *Tierarzt liche Hoehschule, Hannover* **35**, 1277.

Purtilo, D.T. (1975). *Gynaecology* **22**, 607.

Rabinovitch, S., Shaw, B.D. and Bryant, T. (1974). *Journal of Infectious Diseases* **130**, 28.

Ragab, A.H., Hazlett, B. and Cowan, D.H. (1972). *Diabetes* **21**, 906.

Randjandichi, M. (1981). *Sabouraudia* **19**, 143.

Redaelli, P. (1924). *Journal of Tropical Medicine and Hygiene* **27**, 211.

Renz, M., Farquhar, J.W., Cohen, M. and Harkness, R.A. (1974). *Postgraduate Medical Journal* **50**, *Supplement* **1**, 30.

Rivolta, S. (1873). Dei parassiti vegetali, Torino.

Robbins, J.S., Mittemeyer, B.T. and Borski, A.A. (1974). *Urology* **4**, 332.

Roberts, G.D., Horstmeir, C.D., Land, G.A. and Foxworth, J.H. (1978). *Journal of Clinical Microbiology* **7**,584.
Roberts, S.O.B. (1969). *British Journal of Dermatology* **81**, 315.
Rogers, A.L. and Beneke, E.S. (1964). *Mycopathologia et Mycologia Applicata* **22**, 15.
Sabouraud, R. (1894). "Les Trichophyties Humanines", pp. 49–55. Rueff et Cie, Paris.
Sabouraud, R. (1904). "Maladies du Cuir Chevelu." Mason et Cie, Paris.
Saez, H. (1979). *Mycopathologia* **69**, 179.
Salfelder, K. (1971). *In* " The Pathologic Anatomy of Mycoses" (R.D. Baker, ed.), pp. 383–464. Springer-Verlag, Berlin.
Sams, W.M., Jorizzo, J.L., Snyderman, R., Jegosothy, B.V., Ward, F.E., Weiner, M., Wilson, J.G., Yount, W.J. and Dillard, S.B. (1979). *American Journal of Medicine* **67**, 948.
Sandford, G.R., Merz, W.G., Wingard, J.R., Charache, P. and Saral, R. (1980). *Journal of Infectious Diseases* **142**, 503.
Sandhu, R.S., Randhawa, H.J. and Gupta, J.M. (1965). *Sabouraudia* **4**, 37.
Schaar, G., Long, I. and Widra, A. (1974). *Mycopathologia et Mycologia Applicata* **52**, 203.
Scherr, G.H. (1954). *Nebraska State Medical Journal* **39**, 342.
Schmorl, G. (1890). *Zentralblatt fur Bakteriologie* **7**, 329.
Scholer, H.J. (1980). *In* "Antifungal Chemotherapy" (D.C.E. Speller, ed.), pp. 35–106. John Wiley, New York.
Schönebeck, J. (1972). *Scandinavian Journal of Urology and Nephrology* **6**, 136.
Schönebeck, J. and Ånsehn, S. (1972). *Scandinavian Journal of Urology and Nephrology* **6**, 123.
Schwarz, J. (1971). *In* "The Pathologic Anatomy of Mycoses" (R.D. Baker, ed.), pp. 67–130. Springer-Verlag, Berlin.
Seelig, M.S., Speth, C.P., Kozinn, P., Taschdjian, C.L., Toni, E.F. and Goldberg, P. (1974). *Progess in Cardiovascular Disease* **17**, 125.
Segal, E., Romano, A., Eylan, E., Stein, R. and Ben-Tovim, T. (1975). *Mykosen* **18**, 107.
Shadomy, H.J. (1970). "Spectrum Monograph Series in the Arts and Sciences," vol. 1, pp. 67–72. Georgia State University, Atlanta.
Shadomy, H.J. and Utz, J.P. (1966). *Mycologia* **58**, 383.
Shadomy, S., Kirchoff, C.B. and Ingroff, A.E. (1973). *Antimicrobial Agents and Chemotherapy* **3**, 9.
Shaw, C.E. and Kapica, L. (1972). *Applied Microbiology* **24**, 824.
Shelmire, B. (1925). *Archives of Dermatology and Syphilology, Chicago* **12**, 789.
Sherry, J.P., Kuchma, S.R. and Dutka, B.J. (1979). *Canadian Journal of Microbiology* **25**, 1036.
Shields, A.B. and Ajello, L. (1966). *Science* **151**, 208.
Shome, S.K. (1963). *Review of the Faculty of Medicine of University of Ceará* **3**, 79.
Sinha, B.K., Sharma, T.S. and Mehrotra, V.K. (1980). *Veterinary Record* **106**, 177.
Skinner, C.E. (1947). *Bacteriological Reviews* **11**, 227.
Skinner, C.E. and Fletcher, D.W. (1960). *Bacteriological Reviews* **24**, 397.
Skoglund, R.W. (1971). *Lancet* **ii**, 1431.
Sonck, C.E. (1980). *Mykosen* **23**, 107.
Sonnenschein, H., Clark, H.L. and Taschdjian, C.L. (1960). *American Journal of Disease in Childhood* **99**, 81.
Sonnenschein, B., Weiss, R. and Bringewatt, W. (1978). *Deutsche Tierarzt liche Wochenschrift* **85**, 389.
Speller, D.C.E. (1975). *Journal of Antimicrobial Chemotherapy* **1**, 253.
Staib, F. (1962). *Zeitschrift für Hygiene und Infectionskrankheiten* **148**, 466.
Staib, F. (1963). *Mycopathologia et Mycologia Applicata* **19**, 143.
Stanley, V.C. and Hurley, R. (1967). *Journal of Pathology and Bacteriology* **94**, 301.
Stenderup, A., Karuru, E., Mataika, J. and Tuivaga, V.P. (1978). *Fiji Medical Journal* **6**, 334.
Stone, H.H., Kolb, L.D., Currie, C.A., Geheber, C.E. and Cuzzell, J.Z. (1974). *Annals of Surgery* **179**, 697.
Stuker, G., Ehrensperger, F., Pohlenz, J. and Troll, C. (1979). *Zentralblatt für Veterinärmedizin, B,* **26**, 184.
Swamy, K.H.S., Sirsi, M. and Rao, G.R. (1974). *Antimicrobial Agents and Chemotherapy* **5**, 420.

Taguchi, M., Tsukiji, M. and Tsuchiya, T. (1979). *Sabouraudia* **17**, 185.
Taschdjian, C.L., Burchall, J.J. and Kozinn, P.J. (1960). *American Journal of Disease of Children* **99**, 212.
Taschdjian, C.L., Seelig, S. and Kozinn, P.J. (1973). *Critical Reviews in Clinical and Laboratory Science* **4**, 19.
Tewari, R.P. (1963). *Agra University Journal of Research Science* **12**, 163.
Titsworth, E. and Gruneberg, E. (1973). *Antimicrobial Agents and Chemotherapy* **4**, 306.
Torok, I. and Moravcsik, E. (1979). *Mykosen* **22**, 345.
Tosti, A., Villardita, S. and Fazzini, M.L. (1972). *Journal of Investigative Dermatology* **59**, 233.
Trousseau, A. (1869). Lectures on Clinical Medicine, delivered at the Hotel Dieu, Paris. Translated from the edition of Cormach, J.R. (1868) volume 2. New Sydenham Society, London.
Utahara, R., Okami, Y., Nakamura, S. and Umezowa, H., (1954). *Journal of Antibiotics* **7a**, 120.
Van den Bossche, J. (1974). *Biochemical Pharmacology* **23**, 887.
Van Uden, N. and Buckley, H.R. (1970). *In* "The Yeasts–A Taxonomic Study" (J. Lodder, ed.), 2nd edn., pp. 893–1087. North Holland Publishing Company, Amsterdam.
Van Uden, N. and do Carmo-Sousa, L. (1962). *Journal of General Microbiology* **27**, 35.
Van Uden, N. and Vidal-Leiria, M. (1970). *In* "The Yeasts–A Taxonomic Study" (J. Lodder, ed.), 2nd edn., pp. 1235–1428. North Holland Publishing Company, Amsterdam.
Warnock, D.W., Speller, D.C.E., Day, J.K. and Farrell, A.J. (1979). *Journal of Applied Bacteriology* **46**, 571.
Watson, K.C. and Kallichurum, S. (1970). *Journal of Medical Microbiology* **3**, 191.
Webster, B.H. (1959). *Disease of the Chest* **35**, 273.
Weidman, F.D. (1925). Cited in Fox, H. (1925). Report of the Museum of Comparative Pathology and Zoological Society, Philadelphia.
Wells, R.S. (1973). *Proceedings of the Royal Society of Medicine* **66**, 801.
Wells, R.S., Higgs, J.M., Macdonald, A., Valdimarsson, H. and Holt, P.J.L. (1972). *Journal of Medical Genetics* **9**, 302.
Wickerham, L.J. and Burton, K.A. (1948). *Journal of Bacteriology* **56**, 363.
Wilkinson, J.S. (1849). *Lancet* **ii**, 448.
Wilson, D.E., Bennett, J.E. and Bailey, J.W. (1968). *Proceedings of the Society for Experimental Biology and Medicine* **127**, 820.
Winn, W.A. (1967). *In* "Coccidioidomycosis" (A. Ajello, ed.), pp. 55–61. University of Arizona Press, Tucson.
Winner, H.I. and Hurley R. (1964). "*Candida albicans*". J. and A. Churchill Ltd., London and Little Brown and Company, Boston.
Winner, H.I. and Hurley, R. (1966). "Symposium on Candida Infections." E. and S. Livingstone, Edinburgh and London.
Winston, D.J., Balsey, G.E., Rhodes, J. and Linne, S.R. (1977). *Archives of Internal Medicine* **137**, 1192.
Woolley, D.W. (1944). *Journal of Biological Chemistry* **152**, 225.
Yarrow, D. (1978). *International Journal of Systematic Bacteriology* **28**, 608.
Yarrow, D. and Ahearn, D.G. (1984). *In* "The Yeasts, A Taxonomic Study" (N.J.W. Kreger-van Rij, ed.), pp. 882–885. Elsevier Biomedical Press, Amsterdam.
Yarrow, D. and Meyer, S. (1978). *International Journal of Systematic Bacteriology* **28**, 611.
Zenker, W. (1862). Jahresbericht der gesellsch fur naturkund heilkunde, Dresden, p. 51.
Zimmerman, L.E. and Rappaport, H. (1954). *American Journal of Clinical Pathology* **24**, 1050.

8 Biology of the Cell Cycle in Yeasts

A.E. Wheals

Microbiology Group, School of Biological Sciences, University of Bath, Bath, U.K.

The Yeasts Vol. 1, 2nd edition
ISBN 0-12-596411-0

I. INTRODUCTION

Two species of yeast have proved themselves useful for studies of the cell cycle. *Schizosaccharomyces pombe* was chosen because of its fission mode of division, its unicellular simple form, ease of culture and fast growth rate (Mitchison, 1971). In the case of *Saccharomyces cerevisiae*, with its unusual budded mode of growth, the cell-cycle studies originated as by-products of general studies on the biology of a yeast of commercial importance. This picture was transformed about 20 years ago as it became clear that yeasts were relatively simple but *bona fide* eukaryotes which were amenable to both biochemical and genetical analysis. With the isolation of conditional lethal cell-division cycle (*cdc*) mutants in *Sacch. cerevisiae* (Hartwell, 1974; Simchen, 1978) began an era of studies which now uses the full array of molecular biological techniques (Struhl, 1983). Studies on *Schiz. pombe* have followed a similar pattern (Nurse, 1981a) and the two yeasts provide fascinating examples of similarities and differences in their biology, sufficient to prevent rash extrapolations being made from results on one species to eukaryotes as a whole. More recently a number of studies on *Candida albicans* (Soll, 1984; Chaffin, 1984a), *Candida utilis* (Thomas *et al.*, 1980b) and *Wangiella dermatitidis* (Roberts and Szaniszlo, 1980; Roberts *et al.*, 1979, 1980) have complemented this work.

Contemporary reviews on particular aspects of cell cycles are available, including those on yeast cell cycles (Elliott and McLaughlin, 1983; Lloyd, 1984), yeast-cell proliferation (Carter *et al.*, 1982) and its relation to sporulation (Dawes, 1983), the *Sacch. cerevisiae* cell cycle (Pringle and Hartwell, 1981), cell-cycle mutants (Simchen, 1978; Nurse, 1981a, 1985a; Pringle, 1981), yeast cytology (Byers, 1981a; Streiblová and Hašek, 1985) and many articles to be found in John (1981) and Nurse and Streiblová (1984) on the cell cycles of *Sacch. cerevisiae* (Reed, 1984), *Schiz. pombe* (Fantes, 1984) and

C. albicans (Soll, 1984) and on yeast cell walls (Streiblová, 1984). This review is divided into three parts. First, there is a review of the most widely used techniques for studying yeast cell cycles, secondly a survey of the major cellular components during the cell cycle, and thirdly a discussion of how this knowledge can be interpreted to give a comprehensible picture of how a cell cycle works.

A. Techniques for studying the cell cycle

Most of the work on yeast cell cycles can be conveniently classified under one of three headings – cytological, biochemical and genetical – although now it is the combination of these techiniques that is proving the most valuable, for example in cytochemistry or biochemistry of cloned gene products (Pringle *et al.*, 1984; Thomas *et al.*, 1984). The cell cycle is a dynamic process involving the growth and subsequent division of cells (Lloyd *et al.*, 1982). The original terminology of Howard and Pelc (1953) is still useful, dividing the cell cycle into a nuclear division or mitotic (M) phase, a gap period (G1) before the onset of DNA synthesis (S phase) and a post-synthetic G2 phase. Growing populations contain cells at all stages of the cell cycle and thus any biochemical analysis (which may require substantial amounts of material) forces the experimenter to resort to one of two strategies; either to synchronize the population, so that all cells will perform the stage-specific events at the same time, or to study exponential growing populations and attempt to deduce something about the cell-cycle stage-specific events of individual cells. The cytologist relies more on studying individual cells, either growing or of a known cell age, which can in *Sacch. cerevisiae* be approximately deduced from the size of the bud. The geneticist requires the isolation of mutants with specific defects in the cell cycle, for example cell-division-cycle (*cdc*) mutants, the methodology behind this being described in Section III.A.

B. Synchronous cultures

Two quite different approaches are available for synchronizing populations. Induction synchrony methods rely on there being a reversible and differential sensitivity of some cell-cycle stage to a specific physical or chemical environmental effector. Growing cells are treated with the agent, they gradually accumulate at a point in the cycle, the agent is withdrawn and the cells recommence the cycle in phase. Sometimes repeated treatment is needed to fully effect synchrony. In contrast, the aim of selection synchrony is to take an exponentially growing population and by some means sort the cells with respect to cell age through the cell cycle. It is then possible either to use one of the fractions to initiate a new culture or alternatively to analyse the fractions as being representative of cells of different ages (age fractionation).

1. Induction synchrony

(a) Starvation/refeeding

Starved *Sacch. cerevisiae* cells accumulate in the unbudded G1 phase of the cell cycle and when reinoculated into fresh medium rapidly initiate a cycle, which is signalled by a quasi-synchronous production of buds (Williamson and Scopes, 1960, 1962). Under ideal circumstances good synchrony can be obtained for at least two cycles (Fig. 1a). A noticeable feature, however, is that there is a pronounced lag before bud emergence, and the first cycle is longer than the second. The degree of variability in individual cycle times is large during the first cycle and much smaller, and more characteristic of exponentially growing cells, during the second cycle (G.C. Johnston *et al.*, 1977a). A similar procedure has been used with *Schiz. pombe* (Tauro and Halvorson, 1966).

(b) Entrainment

Dawson (1965) has adapted the starvation/refeeding technique to include repetitive cycles (phasing or fed-batch system), which can eventually produce a population of *C. utilis* that synchronously undergoes events during each entrainment cycle. Only certain media and certain periodicities produce good synchrony but these rhythms do not necessarily reflect natural endogenous periodicities (Thomas *et al.*, 1980a, b).

(c) Inhibitors of deoxyribonucleic acid (DNA) synthesis

Both 2′-deoxyadenosine (cordycepin) and hydroxyurea (HU) have proved effective, non-toxic, specific and reversible inhibitors (Mitchison and Creanor, 1971; Slater, 1973; Miyata *et al.*, 1978b, 1981a; Hereford and Hartwell, 1974; Mitchison and Carter, 1975) of DNA synthesis (Fig. 1b). The mode of action of 2′-deoxyadenosine is via inhibition of 2′-deoxycytidine triphosphate (dCTP), limiting the supply of 2-deoxycytosine. Hydroxyurea works by inhibiting ribonucleotide reductase (Timson, 1975), but it also prevents linking-up of short DNA fragments (L.H. Johnston, 1980) and this may be the direct cause of inhibition. Hydroxyurea will inhibit nuclear, mitochondrial and plasmid DNA synthesis in *Sacch. cerevisiae* (Slater, 1973).

(d) Chelators, ionophores and metabolic inhibitors

It has proved possible to produce synchronized cultures of *Schiz. pombe* with ionophore A23187 (Duffus and Paterson, 1974), ethylene diamine tetra-acetate (EDTA) (Ahluwalia *et al.*, 1978), sodium pyrophosphate and citric acid (Walker and Duffus, 1979) and in *Sacch. cerevisiae* with hydroxyquinoline and phenanthroline (G.C. Johnson and Singer, 1978), the *S*-adenosyl-methionine analogue, Sinefungin (Li *et al.*, 1984), and the glycosylation inhibitor tunicamycin (Arnold and Tanner, 1982). The action of some of these compounds is thought to be via magnesium ion concentration (Walker and Duffus, 1980, 1983; see Fig. 1c). In the case of *Sacch. cerevisiae*, ribonucleic acid (RNA) processing has been implicated (Singer and Johnston, 1982; personal communication).

(e) Pheromones

Cells of α mating-type of *Sacch. cerevisiae* secrete a dodecapeptide pheromone, α-factor, that acts on mating-type **a** cells, arresting them in the unbudded G1-phase (Bücking-Throm *et al.*, 1973). To a lesser extent a reciprocal action occurs with **a**-factor (Betz and Duntze, 1979). Removal of α-factor (or degradation via the action of extracellular proteases) allows cells to proceed into a cell cycle (Hartwell, 1973a; Chan, 1977; Shulman, 1978), but not with a high degree of synchrony (Shilo *et al.*, 1976, 1977) unless the preincubation with α-factor is long enough and the wash-out is at an acid pH (2.7) when a very high degree of synchrony can be obtained (Moore, 1984). The cells still grow while arrested, producing a characteristic elongated shape or shmoo. The buds that eventually appear on release normally arise from the tips of the shmoo.

(f) Mitotic inhibitors

The traditional mitotic inhibitor, colchicine (or its synthetic analogue colcemid) does not affect yeast cells to any great extent because its normal substrate, β-tubulin, seems to be modified in yeasts and so microtubules assemble and function normally. However, the anti-tubulin drug methyl benzimidazol-2-ylcarbamate (MBC), the active ingredient of benomyl, has a reversible effect on microtubule assembly (Quinlan *et al.*, 1980), arresting *Sacch. cerevisiae* cells as doublets with a short nuclear spindle, and *Schiz. pombe* cells as uninuclear cells without septa (Fantes, 1982; see Fig. 1d). Some loss of viability occurs if the cells are kept in MBC for extended periods.

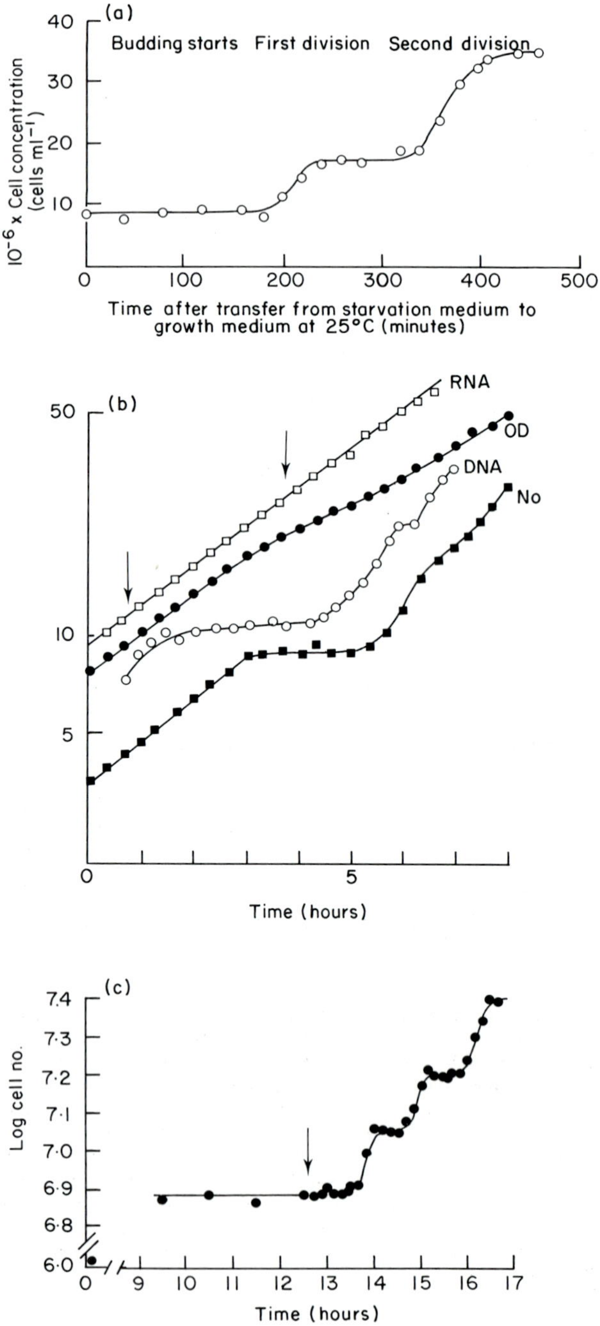
(a)
Budding starts
First division
Second division
10⁻⁶ x Cell concentration (cells ml⁻¹)
Time after transfer from starvation medium to growth medium at 25°C (minutes)
(b)
RNA
OD
DNA
No
Time (hours)
(c)
Log cell no.
Time (hours)

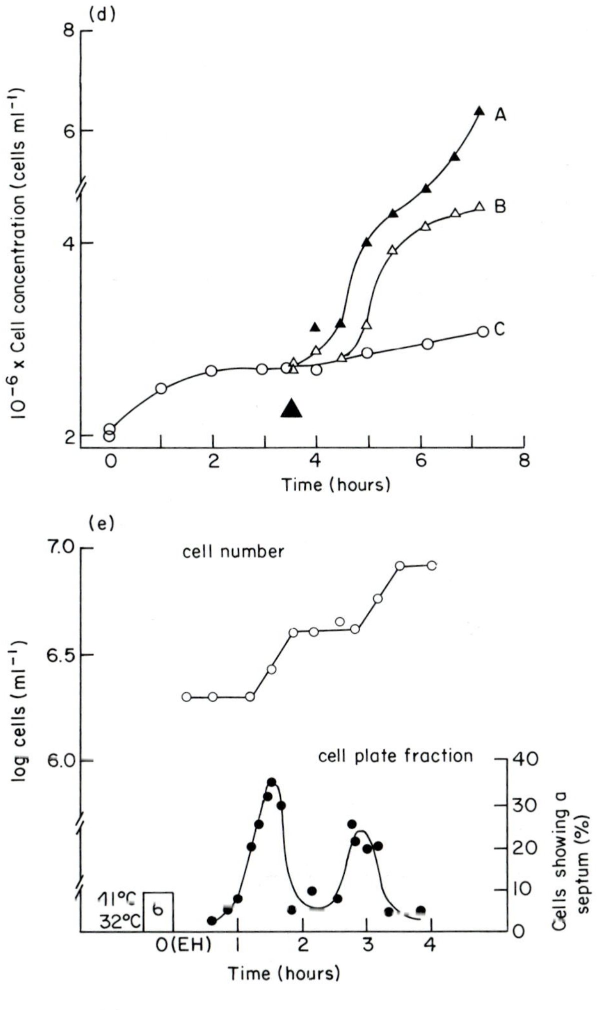

Fig. 1. Synchronous cultures produced by different induction synchrony methods, showing increase in cell concentration with time. (a) Starvation/refeeding of *Saccharomyces cerevisiae*. Reproduced with permission from Williamson and Scopes (1962). (b) Start of (first arrow) and release from (second arrow) DNA inhibition by 2′-deoxyadenosine of *Schizosaccharomyces pombe*. Reproduced with permission from Mitchison and Carter (1975). OD, optical density (absorbance); No., cell number. (c) Magnesium ion depletion (at start) and supplementation (at arrow) of culture of *Kluyveromyces fragilis*. Reproduced with permission from Walker and Duffus (1980). (d) Release (at 3.5 hours) from benomyl inhibition (applied at time 0) of *Schizosaccharomyces pombe* showing release at 35°C (0), at 25°C (△) and in the continued presence of benomyl (▲). Reproduced with permission from Fantes (1982). (e) Synchrony after six 30-minute 41°C heat shocks from a base of 32°C on *Schizosaccharomyces pombe*. The cell plate index (fraction of cells showing a septum) is also shown (●) as is cell number (0). Reproduced with permission from Zeuthen (1974).

(g) Heat shock

Alternating periods at 41°C and 32°C have produced synchronous cultures (Kramhøft and Zeuthen, 1975) of *Schiz. pombe* (Fig. 1e) but unusual proteins are synthesized in response to this procedure (Iida and Yahara, 1984a).

(h) Cell-division cycle (cdc) mutants

A large number of temperature-sensitive *cdc* mutants (both heat and cold sensitive) have been isolated in *Sacch. cerevisiae* and *Schiz. pombe*. Arrest of these strains is achieved by raising (or lowering) the temperature from the permissive to the non-permissive (or restrictive) conditions and maintaining that temperature for about two population doubling times, during which period all the cells (at least for some of the mutants) will accumulate at a particular terminal phenotype (King and Hyams, 1982b). Return to the permissive temperature results in recommencement of the cycle. Although good synchrony can be obtained, results may be difficult to interpret because of the simultaneous release from the block and change in temperature which will affect many biological reactions but to different extents. Furthermore, many *cdc* mutants rapidly lose viability at the restrictive temperature.

2. Selection synchrony

(a) Gradient separation

In this procedure cells are loaded on the top of a centrifuge tube or into a zonal rotor. In equilibrium centrifugation this medium is of increasing density and separation is achieved by cells banding at their own buoyant density. In velocity sedimentation, the separation occurs because of differences in size and density of the cells and the main purpose of the gradient is to prevent mixing during acceleration and deceleration and to assist in cell loading. After centrifugation, which is ideally as short as possible, a portion of the population – frequently the topmost fraction consisting of the smallest and/or lightest cells – is removed and used to inoculate fresh media and produce a synchronous culture. The technique was pioneered using sucrose (Mitchison and Vincent, 1965) as the buoyant medium (Fig. 2a), but other sugars such as lactose, glucose and glycerol have been used where non-utilization of the sugar has been thought desirable, or where an attempt has been made to avoid starvation or the induction of catabolic repression (Mitchison and Carter, 1975).

The density method has employed a wide range of compounds, usually of high molecular weight, which avoids the complications of osmotic pressure. The agents have included Renografin (Hartwell, 1970), percoll (Dwek *et al.*, 1980), colloidal silica (Fig. 2b) or Ludox (Shulman, 1978; Shulman *et al.*, 1973) and dextran (Wiemken *et al.*, 1970). The methods have been devised for *Sacch. cerevisiae* (Hartwell, 1970) using Renografin and polyvinylpyrollidone, for *C. utilis* (Nurse and Wiemken, 1974) using dextrin, and for *Schiz. pombe* (Poole and Lloyd, 1973) using dextran.

In most cases, claims have been made that 10^{10} cells can be successfully loaded onto gradients. Since only 5% of the population is often used for setting up the culture during a further 1 in 10 dilution, then initial densities of $5 \cdot 10^7$ are possible. However, all the published curves seem to have values of about $2 \cdot 10^6$ cells ml^{-1} which is nevertheless quite sufficient for modern biochemical assays. Starting with 10^{10} cells does imply a prior concentration step of exponentially growing cells and such a procedure is both time-consuming and biologically undesirable. Such larger scale cultures (Mitchison and Carter, 1975) can be made with zonal rotors (Fig. 2c).

(b) Continuous- flow centrifugation

This method utilizes a continuous-flow centrifuge running at suboptimal speeds for pelleting the cells so that a proportion of the most slowly sedimenting cells are continually washed out in the effluent (Lloyd *et al.*, 1975, 1977). These cells will be the smallest and can be used to initiate synchronous cultures at a density of about 10^7 ml^{-1} (Fig. 2d). It is possible to use the same medium for culturing, harvesting and producing the synchronous culture (Lloyd, 1984). The method is quick, with the cells in the rotor often for less than a minute, and can be scaled up by running the harvesting for longer periods. The synchrony with cells of *Schiz. pombe* (Lloyd *et al.*, 1975; Edwards and Lloyd, 1977; Bullock and Coakley, 1979; Walker and Duffus, 1980) and *C. utilis* cells (Lloyd *et al.*, 1975; Kader and Lloyd, 1979) can be good, but the results are rather variable (Creanor and Mitchison, 1979).

(c) Elutriation

This method is based on a counter-flow principle with cells being centrifuged to the bottom of a tube while experiencing a flow of medium up the tube (Gordon and Elliott, 1977). By a judicious choice of centrifuge speed and flow rate it is possible to pump out successive fractions of cells, which, like the continuous-flow method, will usually start with the smallest. These can then be used to initiate a new culture (Fig. 2e). Like the continuous-flow method, the cells remain in growth medium throughout the separation

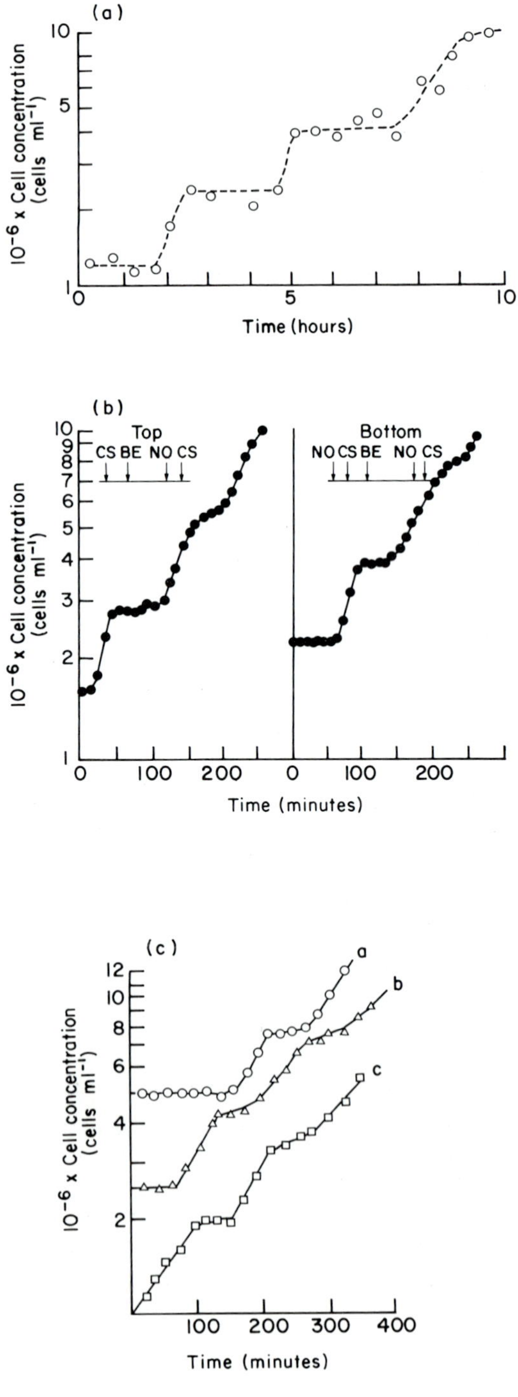

(a)
10^-6 x Cell concentration (cells ml^-1)
10
5
1
0
5
10
Time (hours)
(b)
Top
CS BE NO CS
Bottom
NO CS BE NO CS
10
9
8
7
6
5
4
3
2
1
0
100
200
0
100
200
Time (minutes)
(c)
12
10
8
6
4
2
a
b
c
100
200
300
400
Time (minutes)

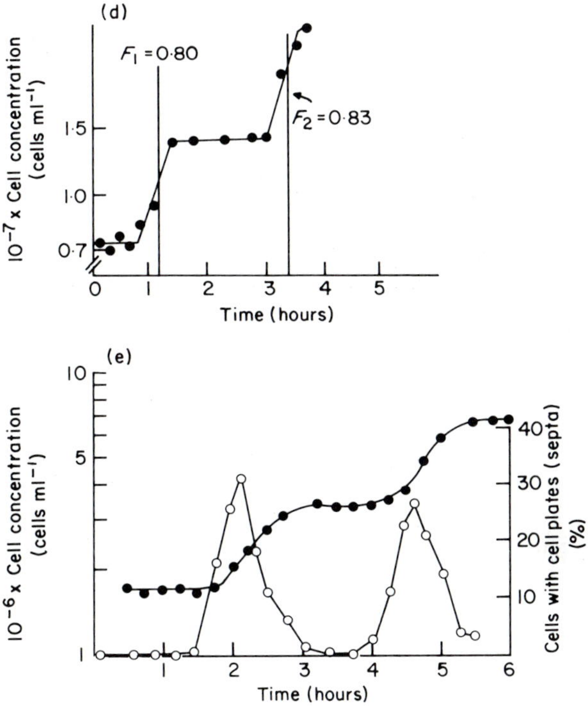

Fig. 2. Synchronous cultures produced by different selection synchrony methods. (a) Top fraction of a sucrose gradient of *Schizosaccharomyces pombe* cells inoculated into fresh medium. Reprinted by permission from Nature, vol. 205, p. 987. Copyright © 1965 MacMillan Journals Limited. (b) Top and bottom fractions of a Ludox gradient of *Saccharomyces cerevisiae* cells inoculated into fresh medium. The timings of bud emergence (BE), cell (bud) separation (BS) and nuclear division (ND) are shown. Reproduced with permission from Shulman (1978). (c) Zonal centrifugation of *Saccharomyces cerevisiae* cells and inoculation of early (0), middle (△) and late (□) fractions into fresh medium. Reproduced with permission from Sebastian *et al.* (1973). (d) Continuous flow centrifugation of *Candida utilis* cells kept in the same medium. *F* is the synchrony index of Blumenthal and Zahler (1962). Reproduced with permission from Lloyd *et al.* (1975). (e) Elutriation of *Schizosaccharomyces pombe* cells using the smallest cells to inoculate a fresh culture (●). The cell plate index (percentage of cells with a septum) is also shown (O). Reproduced with permission from Creanor and Mitchison (1979).

procedure. This method has proved highly successful for use with *Sacch. cerevisiae* (Lorincz *et al.*, 1982) and *Schiz. pombe* (Creanor and Mitchison, 1979, 1982), with good synchrony and yields ($5 \cdot 10^6$ cells ml^{-1}). Evidence has been obtained for minimal perturbation of at least some cell-cycle processes in *Schiz. pombe* (Creanor and Mitchison, 1982). The method has mostly been used to fractionate cell populations (see Section I.C.4). A detailed theoretical critique of this method has been made (Fraser and Barnes, 1983; see Section II.F.).

3. Selection and induction synchrony compared

None of the published methods give perfect synchrony. This is partly because of imperfections in the method, but equally because of the momentary variation in size or density of cells of identical cell age and the natural variability in cycle times which can readily generate a coefficient of variation of cycle times of 15% for cells synchronized at birth (Lord and Wheals, 1981). The imperfect synchrony will thus decay rapidly. It is customary to measure the population doubling time (τ) from the mid-points of successive increases in cell numbers. Few papers report the synchrony index, but an F value of 0.7 or better (using the index of Blumenthal and Zahler, 1962) would be regarded as satisfactory although this does signify an instantaneous event taking the whole population one-third of a mean cycle time to complete.

The distinct advantage of induction synchrony is the simplicity of the method and the production of large quantities of material. However, the strategy has been frequently criticized on several grounds (Mitchison, 1981). First, only some cell-cycle events may become synchronized – particularly those dependent on the stage that has become inhibited – and thus the cells examined are "mixtures" of synchronous and asynchronous material (Mitchison, 1971). Secondly, the effects that are seen may not be part of the normal cell cycle but may be produced by the synchronizing procedure itself. This kind of artefact has been most noticeably attributed to studies on the rate of enzyme synthesis (Mitchison, 1977). Thirdly, the procedure may alter the order or timing of events in normal cell cycles. For example, it is quite often seen that induction procedures produce artificially long cell cycles (Williamson and Scopes, 1960), perhaps due to recovery from the procedure, or alternatively, they give abnormally short cell cycles (Kramhøft and Zeuthen, 1975), as delayed events are accelerated up to their normal rate.

These potential drawbacks can be turned to advantage by regarding selection synchrony as an analytical tool rather than as a preparative procedure. For example, synchrony by release from 2′-deoxyadenosine should lead to synchrony of events directly dependent on DNA synthesis and asynchrony of DNA-independent events, and thus is should be possible to deduce dependency relationships of cell-cyle events. Selection synchrony methods are ostensibly the best for studying cells since, in principle, only those cells of interest from an exponential population are being studied. However, it has gradually become apparent that even "gentle" procedures can subject the cells to mild heat shock, oxygen deprivation and nutrient depletion, sufficient to perturb the normal metabolic flux in the cell and produce artefacts (Walker *et al.*, 1980; Mitchison, 1977). It is usually possible to obtain a good internal control by remixing a fractionated population and confirming that an asynchronous population is regenerated (Bostock *et al.*, 1966; Kader and Lloyd,

1979), but even though such controls have been included from the outset, some misleading conclusions have been made.

In evaluating published data, a number of questions should be asked. (i) What was the synchronizing procedure and what controls were performed? (ii) Did the population double in number and the item under study double in activity/amount? (iii) Did the event occur at the same stage in two successive cycles? – a question unanswerable if the synchrony was poor or if age fractionation was used. (iv) What was the synchrony index? An observed event more synchronous than cell-number synchrony warrants a cautious approach.

At present, the method of choice is probably elutriation (although expensive and with a low yield), but a prudent approach would be to attempt a study using two or three independent methods. If the same result were obtained in each case, confidence could be put on the conclusion.

C. Asynchronous cultures

1. Analysis of steady-state cultures

Exponentially growing cultures in balanced growth are, by definition, unperturbed and it is possible to measure the fraction of cells possessing a certain character (e.g. percentage of budded cells, or cells in S phase). Knowing the population doubling time (τ) and the timing of the initiation or completion of the event, it is possible to use the cell age distribution to calculate the duration of that event (Tyson *et al.*, 1979). In an idealized population there are always twice as many newborn cells as cells about to divide and the frequency of cells of intermediate age shows an exponential distribution (Hartwell and Unger, 1977). In practice, variation in cycle times of individual cells skews the distribution markedly (Lord and Wheals, 1981). In the case of *Sacch. cerevisiae*, the cycle times of daughters are, on average, longer than that of parents, and hence the cell age distribution is more complicated (Fig. 3).

This kind of analysis has been limited to morphological events (bud emergence, nuclear division) but, with the advent of flow microfluorimetry, can be extended to biochemical measurements, such as G1, S and G2 phases (Slater *et al.*, 1977; Hutter *et al.*, 1978; Vanoni *et al.*, 1983) but probably not to more subtle rate changes. However, flow microfluorimetry data must be treated carefully since the modest amounts of material present in individual cells means that the instruments are generally working at the limits of their resolution and there is poor discrimination in the read out. Proper analysis requires computer programs (Zietz and Nicolini, 1978; Fried and Mandel,

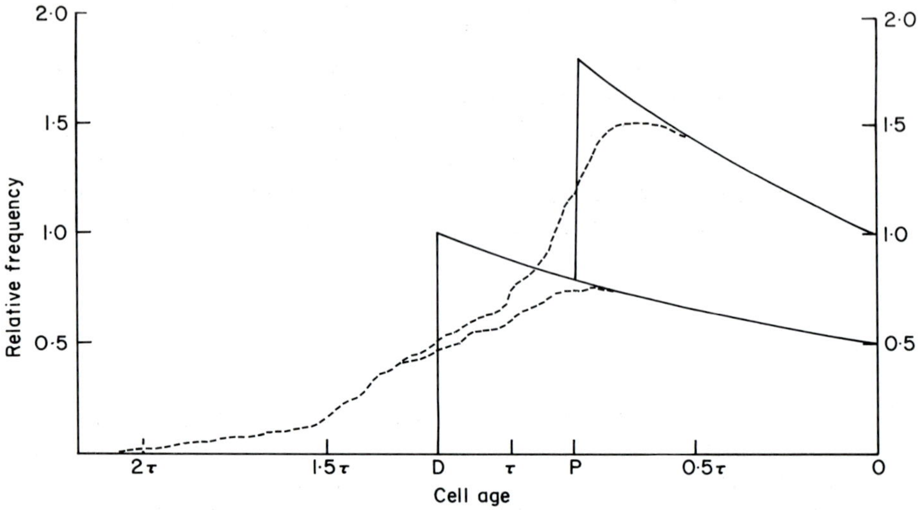

Fig. 3. Cell age distribution of *Saccharomyces cerevisiae* cells dividing asymmetrically with mean parent cycle time being approximately 0.8 τ and mean daughter cycle time approximately 1.2 τ, where τ is the mean population doubling time. ———, idealized cell age distribution; – – – –, actual cell age distribution calculated from Fig. 3 of Wheals (1982). Cell age is measured in units of τ to division. D, daughter; P, parent.

1979) which, unfortunately, have not always been used (Slater *et al.*, 1977; G.C. Johnston *et al.*, 1980).

2. Transition point studies

Certain chemicals, often used for synchronizing cells, and temperature-sensitive *cdc* mutants both have characteristic points at which they act in the cell cycle. This point can be determined with exponentially growing cultures by shifting a *cdc* mutant to the non-permissive temperature or by adding the inhibitor. The increase in residual cell numbers is carefully measured and if it reaches a plateau it is possible to calculate the time of action of the inhibitor (Slater, 1973; Fantes, 1982) or *cdc* allele (Hartwell, 1974) from the age distribution. This point is called the transition or execution point of the *cdc* allele or agent, the rationale behind the analysis being that cells that have passed the transition point will continue on for a further cycle, whereas those cells before the transition point will arrest in that cycle (Hartwell *et al.*, 1974).

The problems with this kind of analysis, especially with *cdc* mutants, are that many are leaky (slowly escaping the block), many do not show first-cycle arrest, the transition point is allele, not gene, specific (Pringle, 1975, 1978, 1981) and is affected by environmental and genetic background (Richmond and Williamson, 1983). Despite these potential problems, this kind of analysis with *cdc* mutants has proved highly informative. The production of cells all arrested at the same point has also proved useful for the isolation (King *et al.*, 1982a) and study (Byers and Goetsch, 1974) of cellular components.

3. Step and shift-up experiments

The timing and control of events can sometimes be deduced by studying the transition period between two different steady states. Changing from one growth rate (usually the slower) to a faster rate is called a shift-up experiment (Carter *et al.*, 1978), and changing one DNA synthesis rate (usually the slower) to another, without altering the growth rate, is called a step-up experiment (P.W. Thompson and A.E. Wheals, unpublished work). So far this technique has only been applied to *Sacch. cerevisiae.*

4. Age fractionation

Exponentially growing cells can be "fixed" instantaneously by a non-destructive method (i.e. cold) and then fractionated according to cell size and thus to cell age. Each fraction can then be analysed for any particular parameter (Elliott and McLaughlin, 1978; Ludwig *et al.*, 1982). The advantage of this method is that the cells should be unperturbed and the yield high. For example, measurements of rates of synthesis can be made by pulse-labelling the exponential population and then fractionating the cells. The main disadvantages are that only one cycle can be investigated, and the resolution of the technique is not good for events that occur near the beginning or end of the cycle (Creanor *et al.*, 1983). Age fractionation by elutriation looks particularly effective (Elliott and McLaughlin, 1978, 1979a,b; Elliott *et al.*, 1979; Ludwig and McLaughlin, 1982; Ludwig *et al.*, 1982).

5. Individual cells

It is possible to grow exponential populations of cells under the microscope and observe the timing of morphological events in individual cells (Mitchison, 1971; Lord and Wheals, 1981; Chaffin, 1984b; Herman and Soll, 1984). More recently, it has become possible to control the surrounding environment of the cells under study so that, for example, specific chemicals can be added or removed from the culture (Lord and Wheals, 1983). This technique, although limited in range, is particularly useful for studies on the co-ordination of growth and division, and is the only one available for analysis of differences between individual cells.

D. Genetic analysis

If biochemical analysis of the cell cycle was the major advance of the 1960s, it has been the genetic dissection of the cell cycle that has revolutionized progress since the 1970s. The work was pioneered by Hartwell and his colleagues for *Sacch. cerevisiae* (Hartwell, 1978; Hartwell *et al.*, 1973) and by Nurse and his collaborators for *Schiz. pombe* (Nurse *et al.*, 1976; Nurse, 1981a). Initially all the mutants were isolated as temperature (heat)-sensitive cell-division cycle (*cdc*) mutants that could not complete a specific cell-cycle step, but now there is an array of mutants including cold-sensitive mutants, suppressors, mutants with altered timing of events, those with an altered response to cell-cycle arrest and *in vitro* mutagenized genes with a phenotype determined only after they are returned to the parent cell (Thomas *et al.*, 1984). They are fully described in Section III.A.

II. SYNTHESIS OF CELLULAR COMPONENTS

Mitchison (1971) made the useful distinction between those cellular processes that occurred discontinuously, the set of events comprising a "DNA-division cycle", and those processes that looked as if they were concerned with overall growth, "the growth cycle". Although evidence for a DNA-division cycle has increased enormously since its proposal, there is little evidence for a separate growth *cycle*. This is not to say that all growth processes are continuous, some clearly are not, but rather that they do not constitute an ordered array of events whose sequence and integrity must be maintained for the completion of a normal cell cycle. Growth and division are co-ordinated with each other and are discussed in Section III, p. 326. The evidence on the rates and timing of synthesis of the major cellular components, both macromolecules and organelles, are presented in the following subsections.

A. Summary of events

The major cell-cycle events of *Sacch. cerevisiae* and *Schiz. pombe* are shown in Figs. 4 and 5. Newborn daughter cells of *Sacch. cerevisiae* are normally ovoid in shape, with a slight raised area visible under the scanning electron microscope, showing the position of the birth scar, the site of previous attachment to the parent cell. The cell has the usual array of contents with one unusual structure, the spindle pole body or spindle plaque, a disc-like structure embedded in the nuclear membrane and important in mitotic development (Byers, 1981a; King and Hyams, 1982a). Growth occurs by budding, which is first signalled by the appearance of microtubules emanating from the spindle pole body through the cytoplasm, to a site on the wall where a ring of chitin is deposited. Shortly after bud emergence a ring of microfilaments is detectable below the cell membrane at the bud isthmus (Byers and Goetsch, 1976a). In haploid cells at faster growth rates the bud

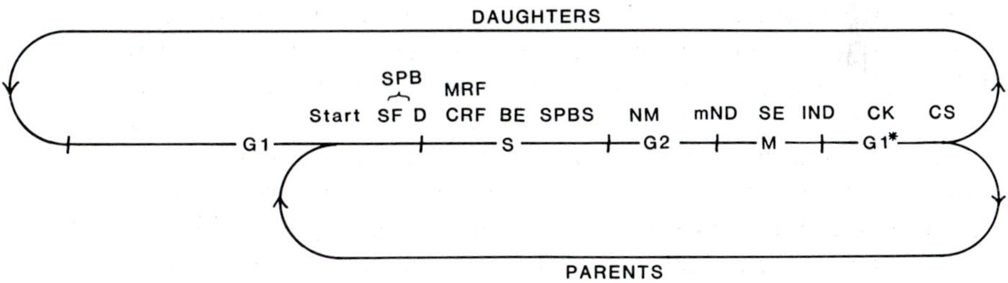

Fig. 4. Major landmark events of the *Saccharomyces cerevisiae* cell cycle. The diagram indicates the temporal order of events, but distances between events in the diagram are only approximately proportional to the time intervals between these events. In some cases uncertainty exists about the temporal order. Abbreviations: BE, bud emergence; CK, cytokinesis; CRF, formation of the chitin ring; CS, cell separation; IND, late stage of nuclear division; mND, medial stage of nuclear division; MRF, formation of the microfilament ring; NM, nuclear migration; SE, spindle elongation; SPBD, spindle-pole-body duplication; SPBS, spindle-pole-body separation; SPBSF, spindle-pole-body satellite formation. G1, S, G2, M are the normal cell cycle phases, with G1* indicating a pre-division G1 period. The fate of daughters and parents is indicated since they rejoin the cycle at different places.

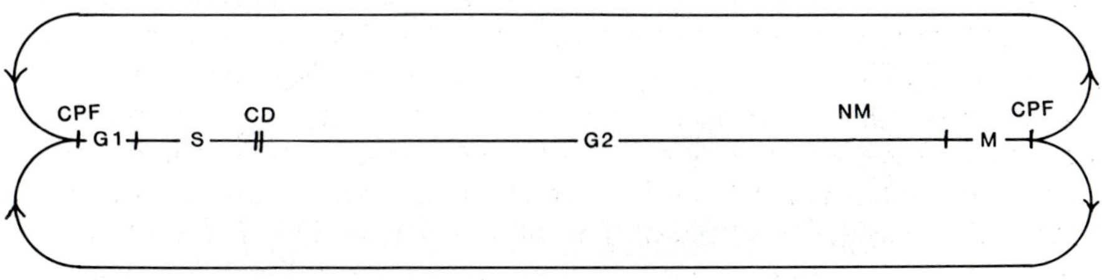

Fig. 5. Major landmark events of the *Schizosaccharomyces pombe* cell cycle. The scale is as in Fig. 4., CD, cell division; CPF, cell plate (septum) formation; NM, nuclear migration; G1, S, G2, M, normal cycle phases. The cycle is drawn from successive cell plate formations (when the cells are physiologically separate) in order to align the conventional cell cycle phases.

occurs in a polar position close to the site of the previous mother/bud junction, and subsequent buds appear adjacent to the previous one (Fig. 6). In diploid cells equatorial or polar bud location, at the opposite pole, is more common (Freifelder, 1960). Before bud emergence occurs, the spindle pole body produces a small satellite structure which grows to duplicate the whole structure which then separates into two distinct spindle pole bodies. A spindle develops between them as the nucleus elongates and moves to a central location between parent and daughter. There is no clear metaphase stage but nuclear division is distinct, the intact nuclear membrane dividing into two. As a septum grows between the cells cytokinesis is rapidly followed by cell separation (Molano *et al.*, 1980). The chitin ring is left as a bud scar on the parent, and with appropriate fluorescent stains (Primulin or Calcofluor) can

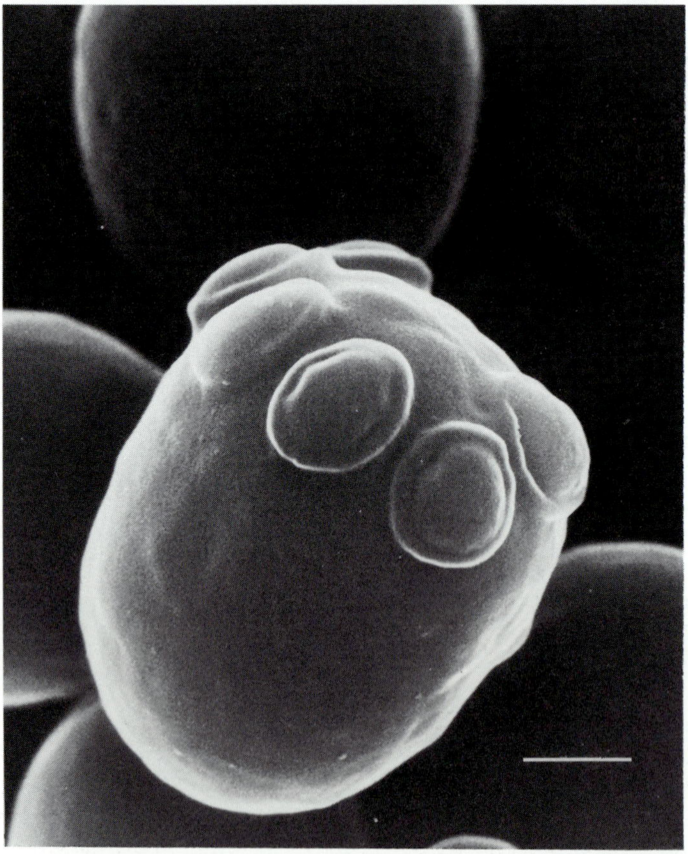

Fig. 6. Scanning electron micrograph of *Saccharomyces cerevisiae* parent cell. This haploid parent cell has produced at least six daughters as shown by the polar contiguous array of circular bud scars. The bar represents 1 μm. Photograph by Anna Gosney and John Forsdyke.

be visualized and hence the genealogical age of the cell determined (Lord and Wheals, 1980).

The cell cycle of *Schiz. pombe* has fewer events characterized (Fig. 5). The cell is cylindrical in shape with hemispherical ends (Streiblová, 1982). Growth is by elongation from both ends (Creanor and Mitchison, 1984) and there is a modest increase in width (Johnson *et al.*, 1979). The two clear cytological events are mitosis, which is intranuclear (King and Hyams, 1982c) and occurs about three-quarters of the way through the cycle (measured from cell division), and cell-plate (septum) formation which occupies the last 0.15 of the cycle and is completed by cell separation (Johnson *et al.*, 1982). Each division produces a stainable scar but the development of the pattern is more complex than in *Sacch. cerevisiae* (Streiblová, 1982; Calleja *et al.*, 1980). At a superficial level, the other yeasts mentioned in this review are similar to *Sacch. cerevisiae*, but no comparably detailed cytological investigation has been made on them.

B. Volume

Saccharomyces cerevisiae and *Schiz. pombe* behave quite differently in their mode of volume increase. Volume (V), calculated from the measurements of the major (a) and minor (b) axes and assuming $V = ab^2/6$, is continuous in *Sacch. cerevisiae* throughout the unbudded and budded phases. During the budded phase most of the volume increase is in the bud, the parent cell initially decreasing in volume (Johnson, 1965a) and then showing a net increase (Lord and Wheals, 1981). There are indications of a reduction in the rate of volume increase during cytokinesis (Bayne-Jones and Adolph, 1932; Mitchison, 1958). Under the unusual conditions of phasing, *C. utilis* cells show a different pattern (Thomas *et al.*, 1980a,b). Daughter cells which, like *Sacch. cerevisiae* cells, are born smaller than parents, show a rapid increase in volume to reach the same size as parents, both of which can then bud. There is no further net volume increase until division, so increase in bud volume is at the expense of the mother cell. Mechanisms whereby a rigid yeast cell can dramatically decrease in size have not been suggested. In the case of *Schiz. pombe*, there are two distinct phases; during the first three-quarters of a cycle cell volume increases continuously at two different rates, followed by a complete cessation during the remainder of the cycle (Mitchison, 1971). Although this view has been challenged after an analysis of asynchronous chemostat cultures (Barford, 1985), direct measurements have confirmed the original claim (Mitchison and Nurse, 1985). *Candida albicans* increases in volume continuously through the cell cycle, with a slight pause before bud emergence (Herman and Soll, 1984). The rate of volume increase is the same for yeast and mycelial forms even though the surface areas are quite different.

C. Mass/dry weight

Interferometry of individual cells of both *Schiz. pombe* and *Sacch. cerevisiae* revealed that dry mass increased continuously through the cycle, the detailed pattern varying with species and temperature (Mitchison, 1957). Since the rate of increase is not the same as volume, density fluctuations through the cycle are predicted and seem to be in agreement with observation for *Sacch. cerevisiae* (Hartwell, 1970; Baldwin and Kubitschek, 1984). However, careful measurements of *Schiz. pombe* have revealed constant bouyant density through the cell cycle, but an increase during stationary phase (Kubitschek and Ward, 1985). This result perhaps indicates why density-separation techniques have been unsuccessful with *Schiz. pombe*. The soluble pool also fluctuates through the cycle of both yeasts, reaching a maximum during the mid-point of the cycle (Mitchison, 1971).

D. Deoxyribonucleic acid synthesis

1. Nuclear deoxyribonucleic acid

There have been numerous studies on nuclear DNA synthesis in *Sacch. cerevisiae* (Table I). There is basic agreement that the S phase (see Section I.A.) is initiated early in the cell at about the time of bud emergence, requires protein synthesis for initiation but not continuation (Hereford and Hartwell, 1973; Williamson, 1973), and lasts for no more than half the cycle even in fast-growing cells. However, there has been little consensus over the finer details. Some of these differences are due to problems of technique. For example, standard errors are rarely quoted but can be quite large; flow microfluorimetry requires computer-assisted analysis for correct interpretation, and this has often not been done; autoradiography needs Poisson analysis correction of grain-count distributions (Rivin and Fangman, 1980a) and even this implies assumptions about labelled and unlabelled classes of cells that are not always met (D.H. Williamson, personal communication); and altering the growth rate by temperature or by limitation of carbon or nitrogen sources probably has different effects on the duration of the S phase (L.H. Johnston, 1980).

Autoradiographic methods tend to give values of about 30 to 40 minutes for the duration of the S phase. Flow microfluorimetric methods consistently give lower values. Unfortunately, the instrument is working at the limits of its resolution and the clearly separated G1 and G2 peaks which are found in the analysis of genomes with high DNA content are decreased in yeast to two overlapping distributions that need complex analysis and not a little faith to produce convincing answers. Resolution may become easier with the advent of more powerful lasers.

Most of the results showed an increase in the duration of the S phase at slower growth rates. The most pronounced effect was when the growth rate was lowered via nitrogen limitation (Rivin and Fangman, 1980a). Since it was demonstrated that the increase in S phase could be solely accounted for by a decrease in replication fork velocity, the clear implication in these experiments was that the method of growth-rate limitation was also directly increasing S phase (Rivin and Fangman, 1980b). When growth rate was decreased by lowering the temperature, S phase remained relatively unaffected until the temperature dropped below 20°C (Vanoni *et al.*, 1984).

The general conclusion that S phase starts at the same time as bud emergence is in clear disagreement with the observation that all budded cells can complete cell division in the presence of hydroxyurea (Slater, 1973; Hartwell, 1976). Since hydroxyurea abolishes DNA synthesis, and the completion of DNA synthesis is a prerequisite for division, the data imply that all budded cells have already completed DNA synthesis. This interpretation agrees better with more recent data, all of which suggest that the initiation of DNA synthesis occurs shortly before bud emergence, even at fast growth rates (Rivin and Fangman, 1980a; Vanoni *et al.*, 1983; Brewer *et al.*, 1984; Fingerhut *et al.*, 1984; P.W. Thompson and A.E. Wheals, unpublished work).

The observation that different genes exhibit different times of maximum mutagenesis by nitrosoguanidine (acting preferentially at the replication fork) during S phase suggests that there is a specific temporal programme of replication, as in other eukaryotes (Dawes and Carter, 1974; Burke and Fangman, 1975; Kee and Haber, 1975). Using plasmids containing ARS sequences (which are putative chromosome replication origins (Williamson, 1985)) as molecular probes, Fangman *et al.* (1983) have shown that (i) plasmids containing ARS1 replicate early in S phase, and (ii) the time of replication of these chromosomal sequences is in the order ARS1, ARS2, 10Z (where ARS1 and ARS2 are characterized ARS sequences (Zakian and Scott, 1982; Tschumper and Carbon, 1982) and 10Z is an uncharacterized sequence replicated under the control of a third ARS sequence). The difference in timing is presumably related specifically to differences in nucleotide sequence.

Candida albicans and the basidiomycetous yeast *Rhodosporidium toruloides* are similar to *Sacch. cerevisiae* in that the onset of DNA synthesis in the yeast-like form occurs early in the cycle very shortly after bud emergence (Soll *et al.*, 1978, 1981b; Soll and Bedell, 1978; Abe *et al.*, 1984). The situation in *Schiz. pombe* is quite different, with DNA synthesis occurring synchronously in the two halves of a cell recently separated by the appearance of a cell plate (Nasmyth, 1979). DNA synthesis only takes about 0.1 of a cycle for completion, and thus cell separation occurs at about the same time as the

Table I. Deoxyribonucleic acid synthesis in *Saccharomyces cerevisiae*

Method of analysis	Temperature (°C)	Population doubling time (minutes)	Time of initiation[a]	Duration	Comment	Reference
Autoradiography	25	120	BE	0.27 τ	Age by bud size	Williamson (1965)
Continuous label	23	135	0.1 τ	0.3 τ	Isopycnic synchrony	Hartwell (1970)
$^{14}N/^{15}N$ ratio	30	120	BE	–	Age fractionation	Sena *et al.* (1975)
Autoradiography	30	100 187 436	5 min 30 min 283 min	45 min 46 min 48 min	τ varied by different carbon source	Barford and Hall (1976)
Flow microfluorimetry	30	84 130 195 258	17 min 39 min 101 min 146 min	20 min 22 min 20 min 19 min	τ varied by different carbon source	Slater *et al.* (1977)
Diphenylamine	28	130	BE	0.2 τ	Sucrose gradient synchrony	Matur and Berry (1978)
Flow microfluorimetry	23	144 156 240 336 384 456	112 min 114 min 149 min 133 min 141 min 221 min	19 min 24 min 18 min 44 min 53 min 73 min	Doubling time controlled by varying nitrogen source	G.C. Johnston *et al.* (1980)
Autoradiography	30	110 to 915	0.25 τ	0.5 τ	Nitrogen limitation α-factor release (13 experiments)	Rivin and Fangman (1980a)
Autoradiography	36	140	32 min	41 min	Carbon limitation (three experiments)	P. Thompson and A. Wheals (unpublished work)

Autoradiography	30	93 80	37 min 21 min	26 min 31 min	Mother[b] Daughter	Brewer *et al.* (1984)
Flow microfluorimetry	30	75 to 314	Before bud emergence	0.14τ to 0.20τ	τ varied by different carbon and nitrogen sources	Vanoni *et al.* (1983)
Flow microfluorimetry	21–37 18 15	104–190	Before bud emergence	20–30 min 40 min 90 min	τ varied by temperature	Vanoni *et al.* (1984)

[a]Expressed as time of bud emergence (BE), as proportion of population doubling time (τ), or in minutes from birth.
[b]Population analysed separately for parents and daughters.
[-]Not measured

onset of G2. In cells that are born very small, either because of growth on poor medium or because of outgrowth from spores, or because they contain a *wee* mutation, a different pattern is seen, with DNA synthesis occurring later in the cycle, the smaller the cell, the later the S phase (Nurse, 1975; Nurse and Thuriaux, 1977). This size-related phenomenon is discussed in Section III.

A number of *cdc* mutants defective in some aspects of DNA synthesis have been detected, and five have been characterized. In *Sacch. cerevisiae CDC21* is allelic to *TMP1* (Game, 1976), the structured gene for thymidylate synthetase (Bisson and Thorner, 1977), *CDC9* is the structural gene for DNA ligase (L.H. Johnston and Nasmyth, 1978; Barker and Johnston, 1983) and *CDC8* codes for thymidylate kinase (Sclafani and Fangman, 1984; Jong and Campbell, 1984). In *Schiz. pombe, CDC19* is the structural gene for DNA ligase (Nasmyth, 1977) and *CDC22* was thought to be the gene for nucleoside diphosphokinase (J.R. Dickinson, 1981), but this is now in doubt (P. Fantes, personal communication).

2. Mitochondrial deoxyribonucleic acid (mtDNA)

No work has been done on mitochondrial DNA synthesis except in *Sacch. cerevisiae*. Earlier claims for a discontinuous replication of mtDNA (Wells, 1974) have been countered with other reports pointing to continuous synthesis through the cell cycle (Williamson and Moustacchi, 1971; Sena *et al.*, 1975; Cottrell, 1979, 1981, 1982) using cells either synchronized or sorted with respect to cell age, although "petites" are still claimed to have discontinuous synthesis (Del Giudice *et al.*, 1981). Additionally, mtDNA synthesis continues in *cdc* arrested cells (in contrast to nuclear DNA synthesis) except in those mutants specifically defective in DNA metabolism (e.g. *cdc 8, 9* and *21*), indicating that although the basic biochemistry of replication may be the same, the control of replication is organized differently (Petes and Fangman, 1973; Newlon and Fangman, 1975). It is uncertain whether each molecule (there are about 50 per diploid cell) replicates once, or whether there is a random replication (as in high-copy-number bacterial plasmids). The limited data suggest the former (Cottrell and Lee, 1981; Cottrell, 1982), but this will be difficult to determine unequivocally because the appearance of intermediate density DNA, the usual marker of replication, is rapidly created in mtDNA by the proficient recombination and repair processes.

3. Plasmid deoxyribonucleic acid

The 2 μm circular plasmid in *Sacch. cerevisiae*, despite earlier claims to the contrary, probably has a nuclear location, at least for most of the cell cycle

(Broach, 1981). Not surprisingly, the same mutants that affect nuclear DNA synthesis directly or indirectly (e.g. *cdc4, 7* and *28*) also abolish 2 μm DNA synthesis (Petes and Williamson, 1975; Livingston and Kupfer, 1977). Each molecule of the 2 μm circle is replicated once per cycle during early S phase (Zakian *et al.*, 1979), but on introduction into a cell lacking 2 μm circle it seems that additional rounds of replication are possible, raising the copy number to the usual 30–50 per haploid cell (Broach, 1981) within a few generations. The same control over artificially produced plasmids also seems to be in operation (Zakian and Scott, 1982; Fangman *et al.*, 1983). Over-replication of the 2 μm circle in strains carrying the *nib* allele has been suggested as the cause of cell-cycle arrest and cell death in "nibbled" colonies (Holm, 1982).

E. Ribonucleic acid synthesis

In *Sacch. cerevisiae* the data now clearly point to bulk RNA synthesis (including ribosomal RNA, some tRNA species and polyA-RNA) occurring continuously throughout the cycle, probably at an exponentially increasing rate (Elliott and McLaughlin, 1979b; Hynes and Phillips, 1976). The measurements, made with elutriation techniques, used pulse-labelling, dual-labelling and age fractionation to avoid artefacts due to pools and perturbation of the cycle. When the cell cycle is blocked using *cdc* mutants (Culotti and Hartwell, 1971; Hartwell, 1971a, 1973b) or inhibitors (Slater, 1973) or α-factor (Throm and Duntze, 1970; Wilkinson and Pringle, 1974), RNA synthesis continues unabated.

There are now becoming available some detailed studies of the synthesis of individual RNA species based on the study of cloned genes. The *Schiz. pombe* genes *CDC2* and *CDC10* show no changes in the level of transcription through the cell cycle (Aves *et al.*, 1985; Paul Nurse, personal communication) whereas the histone genes of both budding yeast (Hereford *et al.*, 1981, 1982) and fission yeast (Aves *et al.*, 1985) and the thymidylate synthase gene of *Sacch. cerevisiae* (Storms *et al.*, 1984) all show a peak in late G1/S phase. It seems likely that only cellular proteins required in relatively large amounts for the completion of a cell-cycle specific stage (in this case chromosome replication) will be regulated periodically at the transcriptional level. The actual level of histone mRNA is controlled post-transcriptionally (Osley and Hereford, 1981). One clear exception is the HO endonuclease which is transcribed periodically after the execution of the *CDC28* gene step early in the *Sacch. cerevisiae* cell cycle (Nasmyth, 1983), and a sequence has been identified in the 5′-flanking region conferring this property on the gene (Nasmyth, 1985).

PolyA-mRNA synthesis has been studied in *Schiz. pombe* and the data supported the view that there was a doubling in rate at some point during the cycle (Fraser and Carter, 1976; Fraser and Moreno, 1976; Fraser and Nurse 1978; Barnes *et al.*, 1979). This point in time of doubling of the rate correlated closely with the size of the cell rather than with any other cell-cycle event or even cell ploidy. Such a size control over the rate of synthesis would readily explain how cells of different sizes could ensure that the overall rate of synthesis was commensurate with their size (Fraser and Nurse, 1978, 1979). More recent results from the same laboratory have produced a somewhat different picture for ribosomal RNA. Elutriation was used to set up synchronous cultures. This method seemed to give minimal perturbations when examined in asynchronous controls (Elliott, 1983a,b). Measurements of the rate of ribosomal RNA synthesis have confirmed a step-like doubling during the cell cycle. Attempts were made to correlate this step with other cell-cycle features such as cell size, DNA synthesis or septation. The last two events were clearly ruled out (using *cdc* mutants) and cell size only poorly correlated with the rate change. The best candidate was an event associated with nuclear division, and a control over transcription maxima via a nuclear-limited regulator has been suggested (Elliott, 1983b).

Many yeast species are host to double-stranded (ds) RNA virus-like particles (killer particles) encoding for, amongst other things, a toxin that kills sensitive cells. Synthesis of dsRNA in *Sacch. cerevisiae* is unaffected by *cdc* mutants which inhibit nuclear-related DNA synthesis (but not mtDNA synthesis). There is some dispute over the timing of dsRNA synthesis. Experiments based on the elutriation rotor suggested continuous exponential synthesis (Newman *et al.*, 1981). Experiments based on selection synchrony cultures suggest synthesis at all times except during nuclear S phase (Zakian and Fangman, 1979).

F. Protein and enzyme syntheses

Some of the earliest cell-cycle measurements made included those on the amount of protein and those on variation in enzyme activity, but the conclusions reached have remained some of the most problematic during the ensuing 25 years (Mitchison, 1971, 1977, 1981). It is only recently that a clearer picture is emerging of the pattern of protein synthesis, though this picture is not without controversy and some remaining contradictions. The confusion in the literature stems from three quite different sources. Firstly, there are artefacts due to the synchronizing procedure. Induction-synchrony techniques lead to metabolic perturbations which can produce artificial periodicities in the pattern of protein synthesis. It has only been recently appreciated that similar criticisms apply to many selection synchrony methods as well. More subtle methods and abundant controls are needed to minimize this source of error. Secondly, measurements were usually made either of amount of a protein or activity of an enzyme rather than its *de novo* rate of synthesis. Pulse-labelling techniques and resolution of individual proteins mean direct measurement of rates are now possible. Thirdly, there are problems with precursor pools. Changes in the size of internal pools or compartmentalization of pools can give erroneous measurements of rates.

The bulk of the measurements in *Sacch. cerevisiae* were made by Halvorson and his colleagues using various kinds of synchronized cells including starved cells, and material from zonal rotors. What was usually measured was the total enzyme activity of a culture which, when calculated on a "per cell" basis, seemed to show clear periodicities in activity through the cell cycle consistent with a view of a "quantum" of protein being synthesized once per cycle at a particular time. Further, tests were made in an attempt to relate this periodic synthesis to the predictions of the oscillatory repression and linear reading models of control (Carter and Halvorson, 1973). Over 30 enzymes were at one time thought to be periodic (Halvorson *et al.*, 1970). Re-assessment of the position became necessary because of the difference in results found with *Schiz. pombe* (Mitchison, 1977), the lack of adequate internal "remixing" controls, and because activity rather than synthesis was measured. This last aspect would not distinguish between periodic synthesis, periodic activation or a two-fold change in rate of synthesis of a continuously produced unstable enzyme even if the periodic patterns were not artefacts.

Recently, McLaughlin and his colleagues, using quite different techniques, have presented a contrasting view. They have chosen to measure the rate of synthesis directly, using a short-term (about 10 minutes) label to measure instantaneous rate, and a different long-term (one cycle) label to measure total average rate, the ratio of the two being a good measure of the differential rate of synthesis. They have done these experiments with exponentially growing asynchronous cultures and then metabolically "frozen" the cells by cooling rapidly to 0°C, having already shown the stability of proteins under this regime. The cells were loaded on to an elutriator and fractionated into age classes (usually between six and nine). Each sample was then treated appropriately to isolate the proteins, which were then further separated by two-dimensional gel electrophoresis (O'Farrell, 1975). This resolves the proteins into a large number of spots (revealed by autoradiography) and the rate of synthesis can be assayed as the ratio of the two labels at different temporal locations through the cell cycle, although until recently the actual nature of each protein was unknown.

Initially about 550 proteins were distinguished, 111 of which were studied in more detail. Virtually all showed continuous exponential increase through the cell cycle (Elliott and McLaughlin, 1978). They have gradually extended the work to show that all of the ribosomal proteins are synthesized exponentially (Elliott *et al.*, 1979) and that there are no periodic changes in the rate of modification or degradation of the proteins (Elliott and McLaughlin, 1979a). They have also looked at heat-shock and heat-stroke proteins, that is, proteins which are known to be very rapidly synthesized or degraded with respect to the cell cycle in response to an external stimulus (heat). These proteins are also synthesized exponentially (Ludwig *et al.*, 1982). Some individual protein spots have now been identified, including structural proteins (e.g. actin), enzymes (hexokinase), virus protein (VLP protein V), the *CDC19* gene product, pyruvate kinase, and the enzyme α-glucosidase, which had been reported to be periodic (Tauro and Halvorson, 1966). All show an exponentially increasing rate of synthesis through the cycle and, in the case of α-glucosidase, the same for the specific activity.

The work has also been extended by Geiduschek and his colleagues (Lorincz *et al.*, 1982). They have used computer-assisted quantification of the radioactivity incorporated in proteins separated on a two-dimensional gel, which has enabled them to look carefully at up to 900 proteins. In addition to age fractionation, they have also used synchronous cultures and the effect of α-factor and hydroxyurea on the labelling patterns. They confirm that the vast majority of proteins are synthesized at a constant differential rate throughout the cell cycle. However, they have detected a few periodic proteins. In addition to histones H2A, H2B and H4, there were eight proteins whose rates of synthesis were modulated between 3- and 25-fold in the cycle, and nine proteins of which five were unstable, and possibly related, with half-lives of 10–15 minutes, which were regulated up to 20-fold in the cell cycle either by synthesis, modification or degradation. None of these proteins would be included in a survey of the 200 most abundant spots in a two-dimensional gel. Clearly there are some periodic proteins, but none have been associated with the enzymes previously described as periodic.

Using Percoll gradients to fractionate cells, Gulløv *et al.* (1981) arrived at different conclusions about bulk protein synthesis, claiming that it increased non-exponentially during the cell cycle at faster growth rates and reached a maximum during the late unbudded period at slower growth rates. The most likely cause of the discrepancy is the difference in technique used, and both have their drawbacks. The distinct advantage of elutriation seems to be a better resolution of fractions and a proven ability to detect known periodic activities.

There are two other reports of periodicity of enzyme activity of budding yeasts. Investigations of the 1, 3-β-glucanases of *Sacch. cerevisiae* showed periodicities of activity in both endo- and exo- forms (Del Rey *et al.*, 1979). The use of a zonal rotor with sucrose gradients for the preparation of fractions of different cell age should have avoided major artefacts, but the twofold change in activity could be due to activation and not *de novo* synthesis. One additional report (Yashphe and Halvorson, 1976) concerns the level of activity of β-galactosidase in individual cells of *Kluyveromyces lactis*, assigned to a particular cell age by the relative size of the mother cell and bud. The data show a very sharp doubling in activity at the end of the cycle. Besides the problems of interpretation discussed by Creanor *et al.* (1983) concerning momentary age variation and the threefold change in volume during one cell cycle, there is the added anomaly that the mid-point of rate doubling in this paper is 0.82 whereas previously it had been reported to be almost half a cycle earlier at 0.26 (Carter and Halvorson, 1973). Furthermore, measurements of β-galactosidase activity in cultures obtained by elutriation in which there were no serious perturbations (Creanor *et al.*, 1983) failed to show periodicities but did show a continuous rise in activity through the cell cycle, as was also evident in α-glucosidase from *Sacch. cerevisiae*.

With an increase in the number of cloned genes have come reliable estimates of the rates of synthesis of their gene products. Four examples are now known where there is periodic rise in rate of synthesis at one point during the cell cycle. In all cases this occurs during late G1/early S phase, involves a concomitant rise in RNA synthesis and is associated with DNA metabolism/chromosome replication. Hereford and her colleagues (Hereford *et al.*, 1982; Osley and Hereford, 1981) have studied the histones H2A, H2B and H3 and the results have been confirmed using two-dimensional gels (Ludwig and McLaughlin, 1982). Thymidylate synthase (Storms *et al.*, 1984) and DNA ligase (Peterson *et al.*, 1985) both affect DNA replication, and the HO endonuclease (Nasmyth, 1983) affects G1-specific mating-type switching.

Two-dimensional gel electrophoresis seems to be the best method of analysis. The technique is direct, sensitive (twofold differences can be detected), reproducible, quantitative, and an increasing number of the spots can be assigned to specific proteins. The limitations are that activity cannot be measured for proteins and very basic, very acidic or insoluble proteins will not appear. Nevertheless, the conclusions from this new technique, and the careful application of older ones, show most proteins in *Sacch. cerevisiae* to be continuously synthesized, the few periodic proteins are less abundant and the only characterized ones are the histones whose requirement is associated with a known periodic event, DNA synthesis. The implication of an exponentially increasing rate is that gene dosage is not the rate-limiting step for synthesis and is consistent with a given mRNA species being translated on ribosomes at a rate proportional to its own relative concentration.

The situation in *Schiz. pombe* has also been recently clarified, though with different conclusions from *Sacch. cerevisiae*. Bulk protein synthesis occurs in two phases (Creanor and Mitchison, 1982): there is a period of rising rate of synthesis starting from an "acceleration point" at about 0.9 of a cycle before reaching a plateau for the remaining quarter of a cycle. Using a number of *cdc* and *wee* mutants it has been established that although the "acceleration point" is unaffected by a block in DNA synthesis, it is affected by nuclear division blocks, reaching a plateau rate at a characteristic protein/DNA ratio (Creanor and Mitchison, 1982, 1984).

An analysis of the rates of sythesis of individual proteins using two-dimensional gel electrophoresis revealed no major differences in the pattern of spots in wild-type and various *cdc* mutants blocked in the cell cycle, suggesting that dependent sequences of gene expression do not appear to make a significant contribution to total polypeptide synthesis during the cell cycle, or to the occurrence of periodic cell-cycle events (D.P. Dickinson, 1981).

The pattern of individual enzyme activity shows a more complex pattern. Originally six enzymes were thought to show periodic patterns, and a further 13 showed continuous synthesis (Mitchison, 1977). Careful re-examination using gentler non-perturbing techniques, largely based on selection synchrony with controls or elutriation, showed that periodic patterns were largely artefacts. Of the 19 different enzymes studied in detail, 18 showed a continuous pattern of synthesis (Mitchison, 1977). Some enzymes, such as sucrase, were rarely perturbed by synchronizing procedures, some, such as arginase, showed initial perturbations which rapidly disappeared and others, such as maltase, showed prolonged perturbations with a fortuitous periodicity close to the mean population doubling time (see Tyson, 1983). The enzyme showing step synthesis was deoxythymidine monophosphate kinase, an enzyme associated with DNA synthesis. Of the continuous patterns, some showed an approximately exponential increase, whereas others showed a linear increase in activity with a doubling in rate at a characteristic time in the cycle, a pattern different from that of bulk protein synthesis.

It should be emphasized that all of these enzyme measurements were of enzyme activity and no estimates of the rate of synthesis or degradation were made directly. An interesting observation was that the time of rate-change was not directly related to the doubling in gene dosage, which was the obvious rate-limiting step for rate of synthesis for linear increasing enzymes. This was confirmed by an examination of the enzyme potential (i.e. the maximum rate of increase of enzyme activity in an induced sample) of two inducible genes, sucrase and arginase (Benitez *et al.*, 1980). Not only did their potential double 0.2 of a cycle after DNA synthesis (which could be explicable as unavailability of newly replicated genes), but doubling in potential was unaffected for at least one generation after inhibition of DNA synthesis. Gene dosage may present the ultimate limit on potential but it is not the immediate control. For these last two enzymes it was also shown that cell size was not the determining step in potential doubling.

The conclusion for *Schiz. pombe* must be that, like *Sacch. cerevisiae*, most proteins are synthesized continuously. There is good evidence with some enzymes for a linear increase in activity during the cell cycle, as has also been suggested for acid phosphatase of *Sacch. cerevisiae* (Creanor *et al.*, 1983), but the cause of the rate change is unknown. It could be related to the rate change in mRNA synthesis (Elliott, 1983,b). There is a good evidence for at least one periodic "step" enzyme, nucleoside diphosphokinase (Dickinson, 1983). The disputes over continuous and periodic synthesis have recently been joined by a new contender. A number of measurements of mitochondrially located enzymes of *Sacch. cerevisiae* and *C. utilis* have shown regular oscillations in activity in one cycle (Lloyd *et al.*, 1981; Edwards and Lloyd, 1977). The periodicities have varied from about 50 minutes for *Schiz. pombe* with a three-hour doubling time, to 15 to 20 minutes for *C. utilis* with a 100-minute doubling time. The measurements have been extended to sensitivity of the enzymes to various inhibitors, and in one case to the direct assay of enzyme protein using antibody. In both cases similar oscillations were seen.

Although a number of controls have been done, including the remixing experiments, the work has been criticized, largely for three reasons. First, instantaneous events (such as cell separation) took a third of a cycle to complete and, since some of the oscillations seen have periodicities of a third to a sixth of the cycle, it is implied that the real oscillations are much sharper than the observed ones, since the experimenter is seeing a sharp peak smeared by asynchrony. This means that there must be changes in rates of synthesis and degradation of sufficient magnitude and rapidity to be readily detected by pulse-labelling. No such oscillatory rate changes have been seen by two-dimensional gel analysis (e.g. Elliott and McLaughlin, 1978) and Lloyd and his colleagues have done no rate measurements. Secondly, since the cycle times of individual *Sacch. cerevisiae* cells show no periodicities and can take any value (Lord and Wheals, 1981; Wheals, 1982) and yet the cultures synchronized with respect to birth and division do show periodicities, the implication is that either cells reset a clock at birth or they entrain each other. No explanation of these possibilities has been offered or tested. Thirdly, examination of the published data suggests that cells make far more protein per cycle than needed, a sevenfold excess being seen in one case. Some rationale needs to be suggested for this energetically wasteful strategy.

In conclusion, my prejudice is that the case for periodic synthesis of most RNA and protein during the cell cycle is not proven. The ineluctable trend in the data over the years has been towards continuous synthesis. Whether this is truly exponential or periodic rate-doubling cannot always be decided because elutriation techniques may not be sufficiently sensitive to discriminate between these alternatives (Fraser and Barnes, 1983). All the good exceptions relate to the known periodicity of DNA synthesis. The periodic events of the cell cycle, such as mitosis, DNA synthesis, division etc., cannot be ascribed simply to the periodic activity or synthesis of a body of proteins (D.P. Dickinson, 1981). They do not seem to exist. What is more likely is that the cell is competent to perform most events at most times, but there are just a few rate-limiting steps for their initiation. This idea is fully consistent with the important observation of Byers and Sowder (1980) who showed that most of the *CDC* gene products are present in excess, sufficient to complete several cell cycles. One of the few exceptions was *CDC4* whose function is necessary for spindle pole body separation. The corollary of this observation, that the amounts or activities of enzymes are not crucial to complete a cell cycle, is seen when "normal" cell cycles are achieved in cells showing distinct perturbations of enzyme activity. Something as important as the replication of the cell is robust enough to resist modest changes in cellular content. The somewhat depressing implication of this conclusion is that an explanation of the driving force behind cell-cycle periodic events is still no nearer.

G. Membranes

The only work on membrane composition has been done on *Sacch. cerevisiae* and on cultures that were either phased synchronized or starvation synchronized. Modest changes in phospholipid composition were detected (Hossack *et al.*, 1979; Cottrell *et al.*, 1981) but could well have been caused by the synchronizing procedure. Incorporation of leucine and phosphate occurs at an exponentially increasing rate, and the fluidity of the membrane ensures a uniform distribution over mother cell and bud (Galdiero, 1973). A mutant, defective in the synthesis of phosphatidylinositol, arrests as cell doublets (Nikawa and Yamashita, 1982), suggesting a role for this compound in cytokinesis or cell separation. A corresponding depletion of phosphatidylinositol has been seen in *cdc28* arrested cells (Dudani *et al.*, 1983).

Entry of cells into stationary phase does increase both the proportion of cells with paracrystalline arrays and the depth and density of the groove (invagination) in the plasma membrane (Willison and Johnston, 1985). During the cell cycle an area of plasma membrane free of invaginations is seen associated with the bud isthmus in *Sacch. cerevisiae* and the plane of septum formation in *Schiz. pombe* (Takeo, 1984).

H. Walls

Most work has been done on *Sacch. cerevisiae* (Cabib *et al.*, 1982; Slater, 1982). Unbudded daughter cells probably incorporate new material over the whole cell surface. The onset of budding is anticipated by the appearance of vesicles below the cell wall (Matile *et al.*, 1969) and the formation of an array of microtubules extending from the spindle pole body on the nucleus (Byers and Goetsch, 1974) to the presumptive site. The switch to localized growth means that there is now only a limited incorporation into the mother cell wall, the bulk of the material being deposited in the bud. Since there is negligible relocation of material this means that mother cell walls are largely composed of older material (Chung *et al.*, 1965; Skutelsky and Bayer, 1979). There is a burst of chitin synthesis shortly before bud emergence, and a ring of chitin is laid down at the presumptive bud emergence site (Hayashibe and Katohda, 1973). The location of the chitin has been proposed to be under the control of the *CDC24* gene since mutants at the non-permissive temperature insert chitin over the entire cell surface, fail to bud, and continue to enlarge all over (Sloat and Pringle, 1978). However, many other *cdc* mutants seem to produce a similar effect (Roberts *et al.*, 1983) which leads the authors to speculate that activation of chitin synthetase zymogen may be a non-specific response to cell-cycle arrest. Furthermore, the dependency relationship between chitin ring formation, budding and septation can be broken since bud emergence can occur in the absence of chitin ring formation (Cabib and Bowers, 1975) and aberrant septa can be made away from the mother/bud junction (Slater *et al.*, 1985). The switch from generalized to localized growth represents one of the most important uninvestigated cellular events.

Arrest of the cells at the unbudded stage using mating pheromone leads to the formation of a shmoo (Fig. 7). Mannan and glucan synthesis, which are normally continuous (Katohda *et al.*, 1976; Hayashibe *et al.*, 1977), continue unabated (Lipke *et al.*, 1976). Chitin synthesis, which is normally periodic and related to bud emergence, still results in the appearance of chitin in the shmoo tip (Schekman and Brawley, 1979). New material, including mannan and glucan, is mainly deposited at the bud tip (Johnson and Gibson, 1966; Farkaš *et al.*, 1974; Field and Schekman, 1980; Tkacz and Lampen, 1972; Adams and Pringle, 1984). The ellipsoid shape of the bud is probably the result of hydrostatic pressure on the elastic properties of the wall, growing mainly at the tip but to some extent over the whole surface (Linnemans *et al.*, 1977).

During bud enlargement a filamentous ring (Fig. 8) appears at the neck (Byers and Goetsch, 1976a). This seems crucial for cytokinesis because cytokinesis mutants (*cdc3, cdc11* and *cdc12*) all lack the ring. In wild-type cells a second burst of chitin synthesis produces a ring around the neck on the mother-cell side. Activation of a latent chitin synthetase is not brought about by proteinase B as was formerly supposed (see Fernandez *et al.*, 1982). Fusion of vesicles in the neck produces a double membrane separating mother and daughter, and this is shortly followed by the deposition of wall material and ingrowth of a septum (Cabib, 1975; Linnemans *et al.*, 1977). Following cell separation, daughters are left with a birth scar, and parents

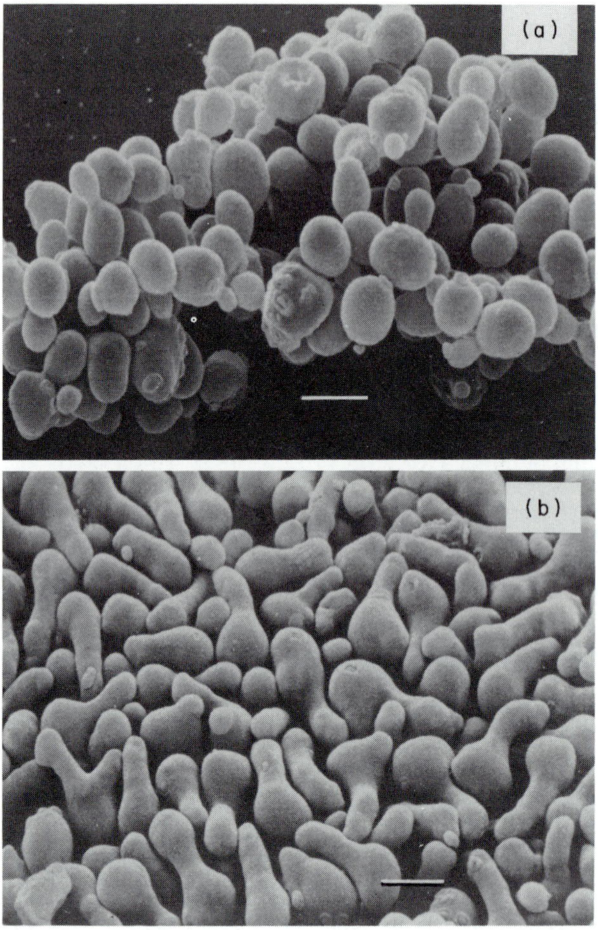

Fig. 7. Effect of α-factor on **a** cells. The scanning electron micrograph shows (a) untreated cells, and (b) **a** cells treated with 4 units ml^{-1} of α-factor for four hours at 30°C. Bar, 5 μm. Photographs provided by Jeremy Thorner and Michael Davis.

with a bud scar composed of a ring of chitin and central core of mannan, glucosamine and glucan. The increase in scarred wall in each generation is more than compensated for by an overall increase in surface area of 10–20% per generation (Johnson and Lu, 1975; Hayashibe and Katohda, 1973).

Growth of the *Schiz. pombe* cell wall, largely α- and β-glucan, occurs as extension at a polar growth zone (May, 1963; Johnson, 1965b; Streiblová and Wolf, 1972). Initially extension is at the new pole only but later in the cycle extension can occur at both poles. The switch from unipolar to bipolar

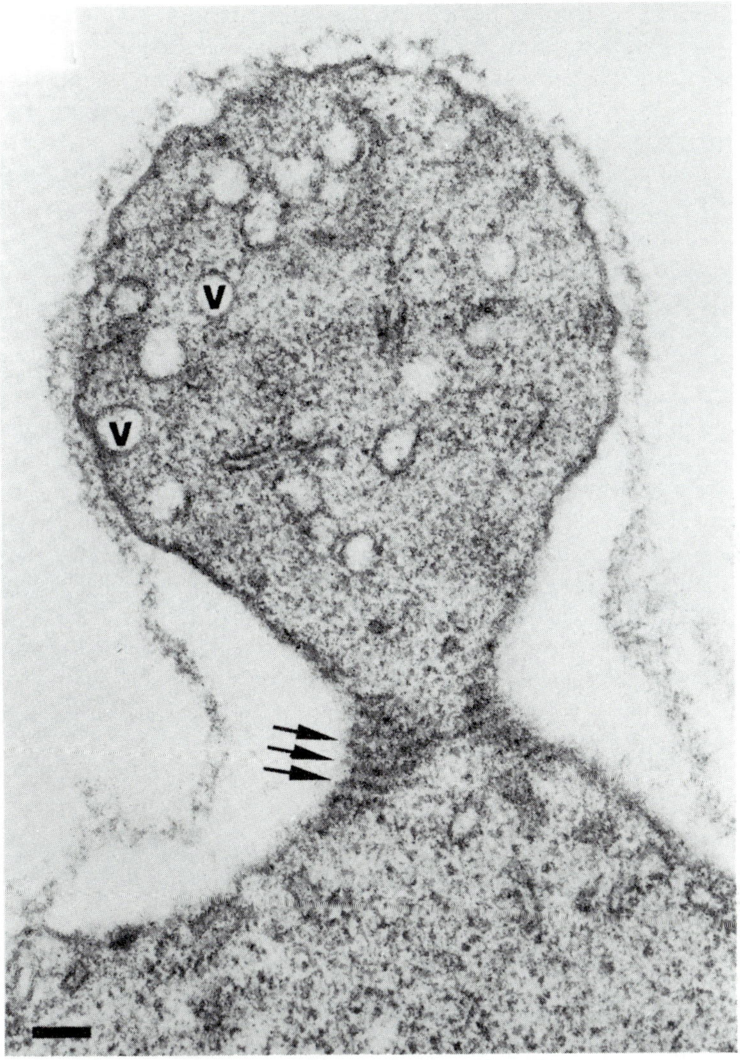

Fig. 8. Filamentous ring of *Saccharomyces cerevisiae*. The tangential section at the bud isthmus indicates the presence of seven or more rings. V indicates vesicles. Bar, 0.1 μm. Reproduced with permission from Byers and Goetsch (1976a).

growth is dependent on (i) cell length being greater than a critical value (of about 9 μm), and (ii) the cell having traversed the first one-third of a cycle and passed early G2 (Mitchison and Nurse, 1985). Cessation of extension, which occurs about three-quarters of the way through a cell cycle, is followed shortly afterwards by septation. The septum is normally centrally located and is initiated by centripetal growth of the cell membrane, which forms the primary septum. Appearance of cell-wall material creates both the secondary septum and the fission scar which is left as an annulus on the divided cells (Streiblová, 1982; Johnson *et al.*, 1982). A number of *cdc* mutants produce delocalized septa and there is a suggestion that this is due to a disruption of the microfilament organization (Streiblová *et al.*, 1984). The patterns and number of bud scars gives information on the genealogical age of a cell (Calleja *et al.*, 1980; Streiblová, 1982).

Studies on the dimorphic yeast *C. albicans* have concentrated on the factors that determine whether the cell will assume a budded or filamentous mode of growth (Soll, 1985a). Using zinc-starved stationary-phase cells at 25°C (Soll *et al.*, 1981a; Soll, 1985a), it was shown that these pluripotent cells will form buds if inoculated into fresh medium at 37°C and pH 4.5, and mycelia if inoculated at 37°C and pH 6.5 (Soll and Bedell, 1978; Mitchell and Soll, 1979a). Commitment to either one form or another occurs 20–30 minutes after cell-wall evagination (Fig. 9), and interestingly commitment to the mycelial form seems to be only possible from stationary phase whereas buds can form from growing mycelia (Mitchell and Soll, 1979a). There seems to be much more chitin (Chattaway *et al.*, 1968) and higher chitin synthetase activity (Braun and Calderone, 1978) in mycelial forms, but this has not been

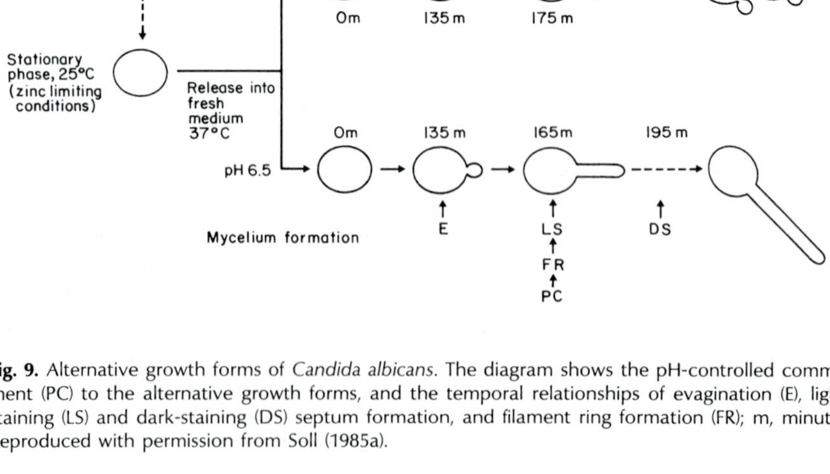

Fig. 9. Alternative growth forms of *Candida albicans*. The diagram shows the pH-controlled commitment (PC) to the alternative growth forms, and the temporal relationships of evagination (E), light-staining (LS) and dark-staining (DS) septum formation, and filament ring formation (FR); m, minutes; Reproduced with permission from Soll (1985a).

measured when pH value has been used to give the alternative types. Since it has been possible to produce filamentous forms of *Sacch. cerevisiae* (Thompson and Wheals, 1981) and branched forms of *Schiz. pombe* (Johnson *et al.*, 1982) simply by growing cells in chemostats at very fast growth rates or by carbon/nitrogen limitation, it has been suggested (Thompson and Wheals, 1981) that the morphological differences seen may be simply due to differences in the timing of cellular events, rather than in the composition of cells *per se*. This view has been echoed in *C. albicans* (Mitchell and Soll, 1979b; Soll, 1985a) because of the lack of difference in proteins synthesized during the dimorphic switch (Brummel and Soll, 1982), and the discovery that a filament ring found at the mother–bud junction in budded forms (as in *Sacch. cerevisiae*) appears 30 minutes later and 2 μm down the evaginated tube in mycelial forms. In both cases the ring indicates the site of future septum formation (Soll and Mitchell, 1983). Although new buds arise adjacent to old bud scars at lower temperatures, at higher temperatures and in germ-tube formation site selection seems more random (Chaffin, 1984a).

A switch from budded to a multicellular form can also be achieved in the polymorphic zoopathogenic fungus *Wangiella (Phialophora) dermatitidis* (Szaniszlo *et al.*, 1976) by inoculating cells into medium at pH 2.5. A similar switch is seen at pH 6.5 by incubating temperature-sensitive *Mc* (multicellular) mutants at the non-permissive temperature (Roberts and Szaniszlo, 1978; Roberts *et al.*, 1979). The switch to the multicellular form is prevented by the microtubule inhibitor nocodazole (Jacobs and Szaniszlo, 1982).

J. Cytoskeletal proteins

An elegant series of experiments using immunofluorescence techniques (Kilmartin, 1984, Kilmartin and Adams, 1984; Adams and Pringle, 1984; Pringle *et al.*, 1984) has revealed much of the detail of the organization of tubulin and actin polymers through the cell cycle of the budding yeasts *Sacch. cerevisiae* and *Sacch. uvarum* (which is now regarded as being the same as *Sacch. cerevisiae*). Using monoclonal antitubulin it was shown that in virtually all cells with short and medium length spindles, cytoplasmic microtubules extended from the nucleus to, or into, enlarging buds. The spatial relationship between cytoplasmic microtubules and the binding site established early in the cycle is maintained until the nucleus itself migrates into the neck. However, in nocodazole-treated cells (which have lost their microtubules), bud emergence, bud enlargement, bud tip growth and chitin ring formation are all normal suggesting that microtubules are not directly involved in these processes. In contrast, the nuclei of nocodazole-treated cells seemed to be randomly positioned and correct spatial positioning of the nucleus may be the major cytoplamic role for microtubules. In addition, prominent microtubules would also extend from the other pole. The rest of the data broadly confirmed the electron microscopic results.

Using purified polyclonal antiactin and fluorescent phalloidin (that binds to F-actin), actin was revealed as a set of cortical dots or patches and as distinct fibres. During the cell cycle the actin dots are associated with areas of cell-wall deposition (the base of small buds, in the bud itself, particularly at the tip, and in the neck region at the time of cytokinesis). Actin dots are also found associated with the formation of the chitin ring which signals bud emergence, and since the dots sometimes seem to occur in the absence of this ring they presumably anticipate its formation. This is the only site where there is any appreciable association between microtubules and actin.

K. Nucleus

The nucleus and its contents undergo major transformations associated with mitosis (Byers, 1981a; King and Hyams, 1982a). There is little intranuclear structure except for the appearance of somewhat condensed chromatic bodies (Gordon, 1977; Petersen and Ris, 1976). In *Sacch. cerevisiae* the appearance of a satellite structure on the spindle pole body is the earliest known morphological event in the cell cycle (Fig. 10). As duplication and separation of the spindle pole bodies occurs, they act as microtubule organizing centres for extranuclear microtubules (Quinlan *et al.*, 1980), defining the location of the bud, and intranuclear microtubules which form the mitotic spindle (King *et al.*, 1982a). The short intranuclear spindle, which forms early in the cycle, remains in this state until the nucleus has moved to the bud

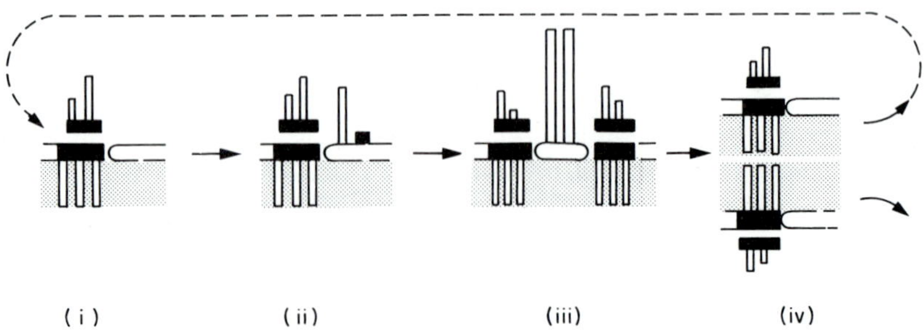

Fig. 10. Diagram of spindle plaque development: (i) single plaque, with outer plaque and spindle microtubules; (ii) spindle plaque with satellite, (iii) double plaque, with cytoplasmic microtubules, and (iv) complete spindle with a single plaque at each pole. Reproduced, with modifications and with permission from Byers and Goetsch (1975).

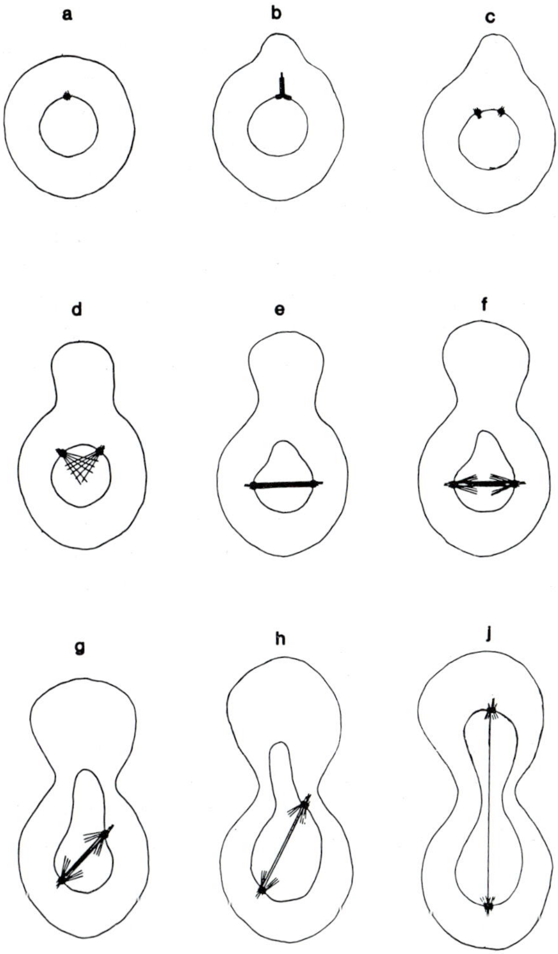

Fig. 11. Diagrammatic representation of nuclear-related events during the cell-cycle of *Saccharomyces cerevisiae*. a, Interphase cell with single spindle pole body; b, duplicate spindle pole body with microtubules pointing towards emerging bud; c, spindle pole body separation and intranuclear microtubule formation; d, interdigitating intranuclear microtubules; e, continuous microtubule bundle; f, start of nuclear migration; g–j, successive mitotic stages with nucleus preceding microtubules into neck of bud (h) and microtubules reducing to one (j). Reproduced with permission from King and Hyams (1982a).

isthmus; it then elongates to its final length (Byers and Goetsch, 1974). It has been suggested (Nurse, 1985a) that, since bud formation is effectively the first stage of a process that leads to cell division and mitosis is normally initiated in eukaryotes before cell division, the development of a short spindle is the first step of mitosis. The spindle is then "frozen" in this state and the chromosomes remain decondensed (to allow cell-cycle transcription) until the cell is ready for nuclear separation. The microtubules from each spindle pole body gradually interdigitate and eventually connect the two poles (Fig. 11). Discontinuous microtubules develop from each pole, the same number as there are linkage groups, though direct chromosome/microtubule contact has not been shown, except in the related yeast *Lipomyces lipofer* (Tanaka, 1976). During this process the nucleus elongates and moves towards the neck, a change in shape not produced by microtubule extension. The gradual elongation of both the nucleus and intranuclear spindle to fill almost the whole

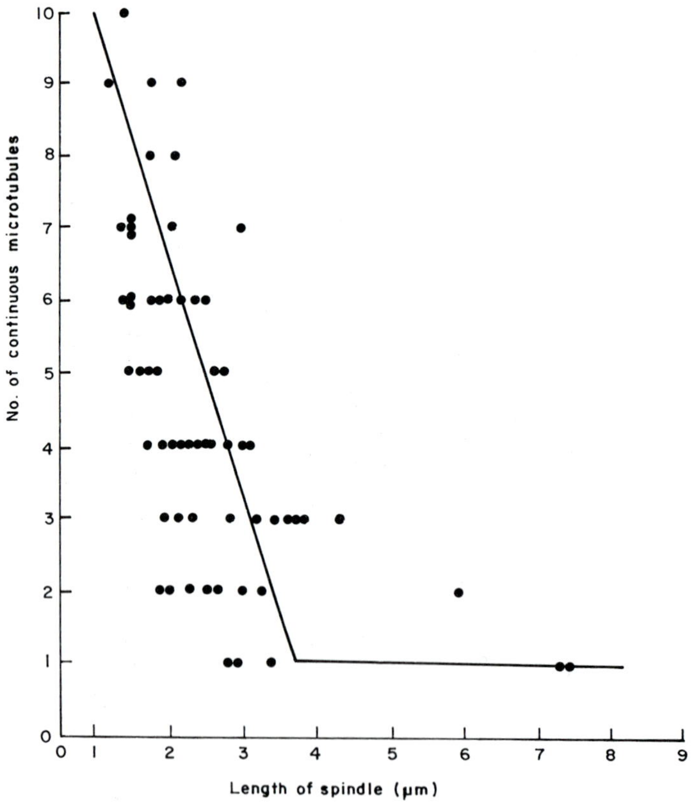

Fig. 12. Relationship between spindle length and number of continuous microtubules. The data are taken from transmission electron micrographs of intact spindles isolated from a strain of *Saccharomyces cerevisiae* carrying the *cdc6* mutations kept at the non-permissive temperature. Reproduced with permission from King *et al.* (1982a).

length of the two cells is accompanied by gradual loss of microtubules to leave only one (Fig. 12) during the last stages of anaphase (King *et al.*, 1982b).

The *Schiz. pombe* nucleus prepares for mitosis late in the cycle with the nucleus first elongating then becoming rectangular in longitudinal section (Fig. 13). At the same time the spindle pole body duplicates, separates and a continuous array of microtubules extends between them (King and Hyams, 1982c). The rapid elongation of the nucleus produces a dumb-bell-shaped structure (Fig. 14) which extends the length of the cell (McCully and Robinow, 1971; Toda *et al.*, 1981). The separated chromatic bodies show sequential changes in shape, from a hemisphere to a U-shape, to two daughter hemispheres, with short chromatic bodies protruding from the flat face (Toda *et al.*, 1981). In a few thiabendazole-treated cells it is possible to see the three pairs of chromosomes. The whole process of nuclear changes including mitosis only takes 5–10% of the cycle compared with up to 50% in *Sacch. cerevisiae*.

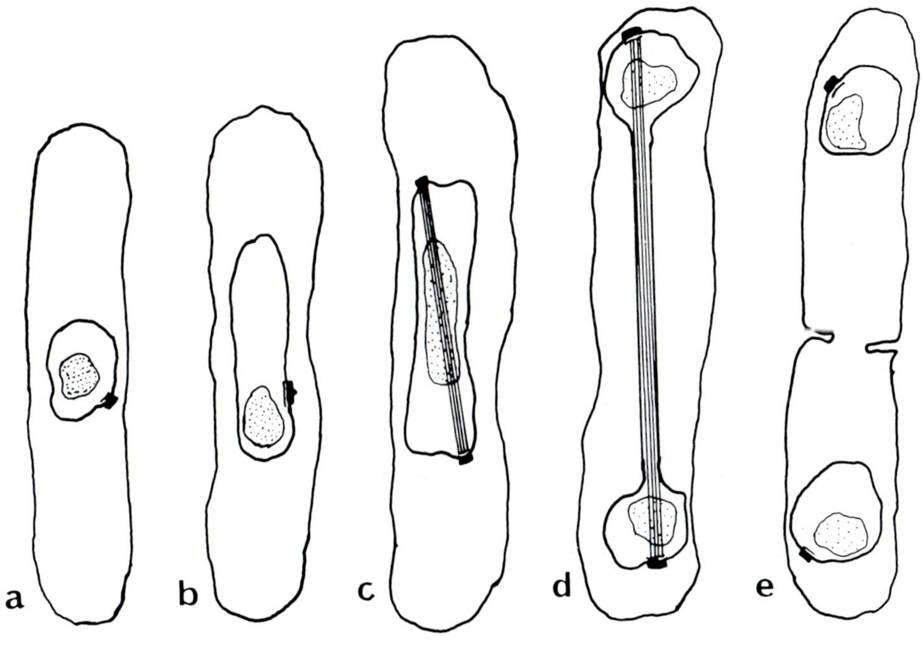

Fig. 13. Main features of mitosis in *Schizosaccharomyces pombe*. a, Interphase with a nucleolus (stippled) in a spherical central nucleus containing a single spindle pole body; b–d, successive stages of mitosis involving spindle pole body duplication (b), separation and spindle elongation (c and d); e, cytokinesis and onset of septum formation. Reproduced with permission from King and Hyams (1982c).

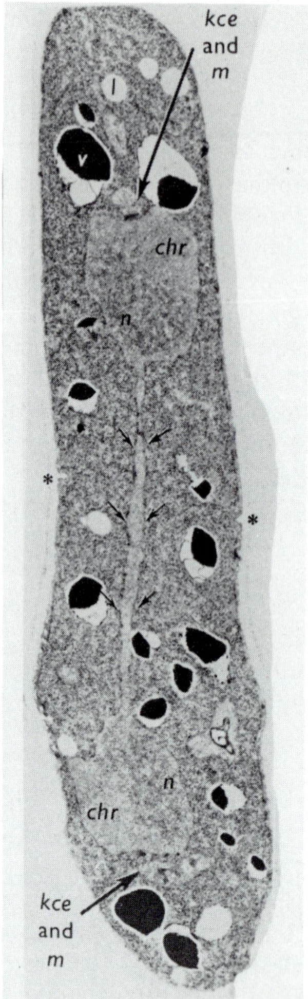

Fig. 14. Transmission electron micrograph of *Schizosaccharomyces pombe* during mitosis. The arrows indicate the spindle channel, and the asterisks show where the cell plate is forming. The structures labelled are spindle pole body (Kce or kinetochore equivalent), mitochondrion (m), lipid droplet (L), vacuole (v), nucleolus (n) and chromatin (chr). The distance from Kce to Kce is 9.25 μm. Reproduced with permission from McCully and Robinow (1971).

The analysis of nuclear movement will be facilitated using the benomyl-resistant mutants of *Schiz. pombe* (see Appendix 2D) recently isolated (Yamamoto, 1980; Roy and Fantes, 1982; Toda *et al.*, 1983; Umesono *et al.*, 1983b). Since one of these is allelic to the cell-cycle mutant *nda3*, and benomyl is thought to act on tubulin (Quinlan *et al.*, 1980), the mutant provides a direct link between structure (tubulin) and function (nuclear movement).

Some measurements have been made on the frequency and size of nuclear pores through the cell cycle and G1-arrested cells. The number of pores per unit area of nuclear surface oscillates by a small amount, the highest value being during S phase (Jordan *et al.*, 1977). The size of nuclear pores remains relatively constant but enlarged pores are seen in G1-arrested cells (Willison and Johnston, 1978).

L. Mitochondria

It is now clear that the mitochondria in *Sacch. cerevisiae* are extremely plastic in form and number (Stevens, 1977), with the possibility of fusions and fissions regularly occurring and mitochondria moving back and forth between parent and daughter (Stevens, 1981). Changes in overall form and amount are primarily associated with changes in growth rate, and there are no suggestions of similar changes occurring regularly during the cell cycle (Cottrell, 1982). In contrast, *C. albicans* has a regular cycle (Tanaka *et al.*, 1985), the few mitochondria fusing into a single organelle at the time of entry into the bud. During mitosis it fragments and then reforms before equal partition to the two daughter cells. Similar events were seen during germ-tube formation.

M. Vacuoles

There is a fairly regular vacuolar cycle in *Sacch. cerevisiae*. The large vacuoles in the bud and parent of a late budded cell shrink and then fragment during division, and subsequently enlarge and fuse during the ensuing bud emergence (Wiemken *et al.*, 1970; Lord and Wheals, 1981).

N. Metabolic activity

Metabolic activity, as measured by oxygen uptake and carbon dioxide evolution, has shown as many patterns as there are investigators (Elliott and McLaughlin, 1983; Creanor, 1978b). However, when attention is reserved for those experiments in which adequate controls were done to show that perturbations were not present, then the variety is greatly decreased. A linear increase with doubling in activity was seen in *Schiz. pombe* using three different approaches (Hamburger *et al.*, 1977), whereas a stepwise doubling in overall rate was seen with selection synchronized cells, one step for carbon dioxide (Creanor, 1978a), two steps for oxygen (Creanor, 1978b). There is no easy way to resolve these differences, but it does look as if the metabolic state of cells can be very sensitive to external conditions.

O. Recombination and repair

A number of reports have indicated that *Sacch. cerevisiae* cells are more sensitive to DNA damage during the G1 phase of the cell cycle and most resistant during S/G2 phases (Bacchetti *et al.*, 1967; Fabre, 1973; Hatzfield and Williamson, 1974; Chanet *et al.*, 1975; Brunborg and Williamson, 1978; Kelly *et al.*, 1983; Dubeau and Chung, 1984). The basis of the difference is unclear since it could be due to differences in shielding, in DNA content or in DNA repair. There is some evidence in favour of the last interpretation (Brunborg *et al.*, 1980) but it is also clear that mitotic recombination and gene conversion can occur at all stages of the cell cycle (Roman and Fabre, 1983; L.H. Johnston and Johnson, 1984) and some of the excision repair gene products are expressed at all stages of the cell cycle and are not induced by DNA-damaging agents (Nagpal *et al.*, 1985).

III. INTEGRATION OF THE CELL CYCLE

Section II recounted the types of events and processes that occur during a cell cycle of yeasts, and this description forms the basis of any cell-cycle study. The observation of sets of periodic events raises some difficult questions which are now being extensively tackled. For example, why do events occur in a particular order; how are superficially unrelated events kept in phase; how are growth and division co-ordinated; how does the cell know where to locate new structures? More generally, these are all aspects of asking how the integration of the cell cycle is achieved, and it was the isolation of temperature-sensitive cell-division-cycle mutants that provided the major impetus in an attempt to answer these kinds of questions.

A. Cell-division-cycle mutants

By definition, a *cdc* (cell-division cycle) mutation leads to a defect in a particular stage-specific function of the cell cycle (Hartwell *et al.*, 1970). Ideally, the *cdc* mutant cells should all accumulate at the same stage, known as the terminal phenotype, and an initially heterogeneous asynchronous

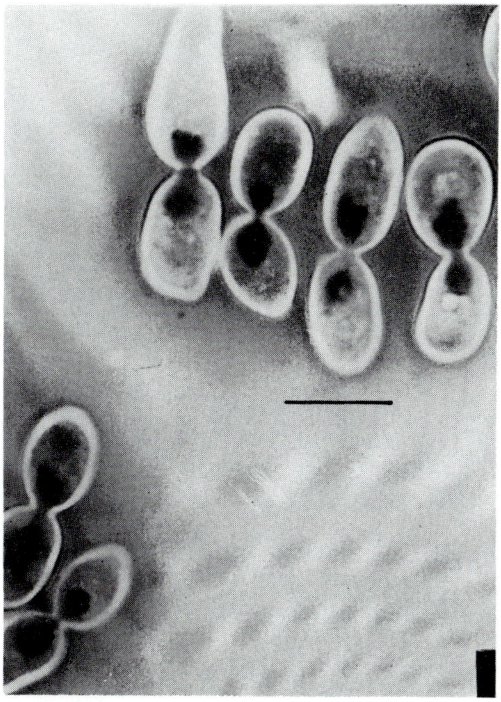

Fig. 15. Terminal phenotype of a *cdc* mutant of *Saccharomyces cerevisiae*. Strain 428DI carrying a *cdc9* mutation was incubated at the non-permissive temperature until it reached the terminal phenotype. The cells were fixed and stained with Giemsa to reveal the nucleus lodged in the bud isthmus. Bar, 5 μm. Reproduced with permission from Culotti and Hartwell (1971).

culture should become synchronous and homogeneous (Fig. 15). Since growth usually remains unaffected, at least for a few doubling times, the terminal phenotype of the arrested cells may take on a bizarre form which is not part of the normal cell cycle (Fig. 16). It is important to distinguish between the primary defect, which is usually unknown, and the diagnostic landmark, which is the first event known to be unaffected by the mutation. As more landmarks become recognizable so the diagnostic landmarks change. For example, some cytokinesis mutants (Hartwell, 1971b) are now known to be defective in the earlier event of microfilament ring formation (Byers and Goetsch, 1976b).

About 70 genes with a *cdc*-related phenotype have been detected in *Sacch. cerevisiae* (Hartwell *et al.*, 1973; Wolfner *et al.*, 1975; Kassir and Simchen, 1978; L.H. Johnston and Game, 1978; Reed, 1980a; Bedard *et al.*, 1981; Moir *et al.*, 1982; L.H. Johnston and Thomas, 1982a; Piggott *et al.*, 1982; Dumas *et al.*, 1982; Huffaker and Robbins, 1982; Klebl *et al.*, 1984) and are described

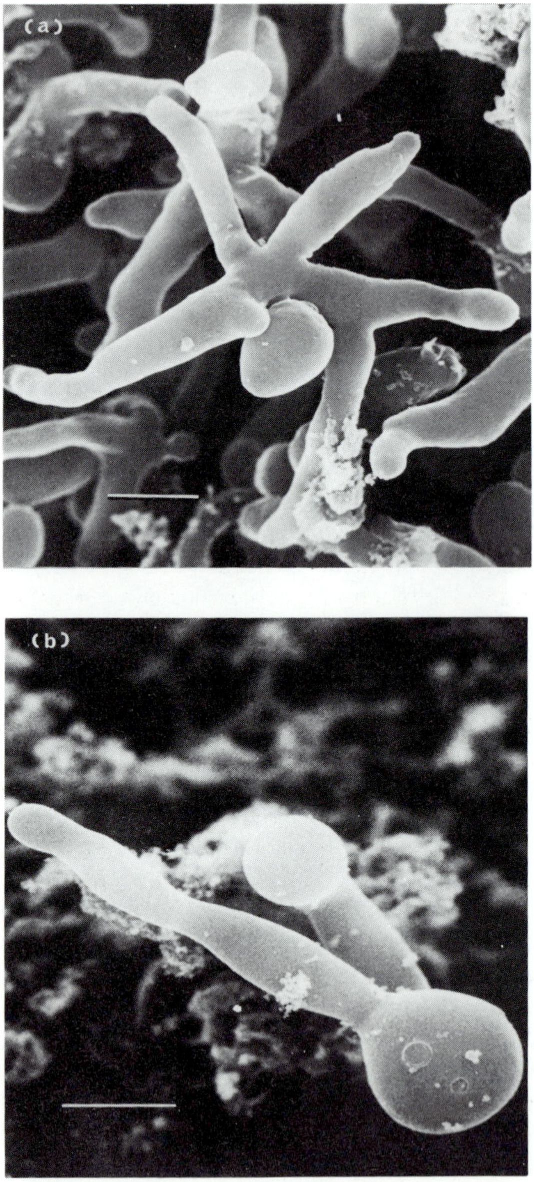

Fig. 16. Terminal phenotype of two *cdc* mutants of *Saccharomyces cerevisiae*. Strains derived from A364A carrying the *cdc4* (a) and *cdc10* (b) mutations were incubated at the non-permissive temperature for 10 and 4 hours respectively. The scanning electron micrographs reveal in (a) multiple abnormal buds from the original spherical cell, and in (b) a pseudomycelial form. Bar, 5 μm. Photographs from Anna Gosney and John Forsdyke.

in Appendix 1. Over 30 have been mapped (Hartwell *et al.*, 1973; Pringle and Hartwell, 1981; Mortimer and Schild, 1985; Hanic-Joyce, 1985). Thirty-two genes with a *cdc* phenotype (see Appendix 2) have been detected in *Schiz. pombe* (Nurse *et al.*, 1976; Nasmyth and Nurse, 1981; Roy and Fantes, 1982; Toda *et al.*, 1983), thirteen of which have been mapped over all three chromosomes (Gygax and Thuriaux, 1984).

The initial detection of 32 *cdc* genes revealed that the frequency of alleles did not show a Poisson distribution, suggesting that the selection protocol was enriching for certain kinds of *cdc* mutants. This is emphasized by the current distribution of alleles (see Appendices 1A and 2A) where certain phenotypes have been specifically selected, for instance, for dumb-bell-shaped terminal phenotype, or certain classes have been selected, for example, for "start" mutants, and different protocols have produced different kinds of mutant (Reed, 1980a; Bedard *et al.*, 1981). In addition to the characterized *cdc* strains, there are a large number of mutants in both *Sacch. cerevisiae* (Appendix 1B) and *Schiz. pombe* (Appendix 2B) awaiting assignment, and in *Sacch. cerevisiae* at least there could be in excess of 100 *cdc* genes. Since the selection protocol is designed to detect only those genes with a cell-cycle-stage specific function it comes as a surprise to find that *CDC19* in *Sacch. cerevisiae* codes for pyruvate kinase and it may well be that many *cdc* genes have general as well as specific cell-cycle functions.

Three kinds of analysis have been employed with the *cdc* mutants; execution (transition)-point analysis, double-mutants and order-of-function mapping. Execution-point analysis aims to map the mutations temporally by shifting an asynchronous culture to the non-permissive temperature and determining the fraction of cells that completes a cycle. Those cells before the execution point will arrest in that cycle; cells after the execution point will arrest in the next cycle. However, some mutants do not arrest immediately, making the execution-point analysis invalid. These differences between mutants probably reflect the different kinds of mutations possible. Tight conditional-for-function mutants should become blocked immediately, whereas leaky conditional-for-synthesis mutants could cycle for many generations. Cell-cycle mutant phenotypes are thus allele specific rather than gene specific (Pringle, 1975, 1978).

Despite potential complications with leakiness, allele specificity, effect of genetic background and loss of viability (Richmond and Williamson, 1983) when held at the non-permissive temperature, two major conclusions can be drawn. Firstly, most mutants affecting the same diagnostic landmark temporally map over a small fraction of the cell cycle. Secondly, this temporal location is usually at, or shortly before, the event in which they are involved (Fig. 17). This latter conclusion has important implications for our understanding of the cell cycle since it implies that there are a few major rate-

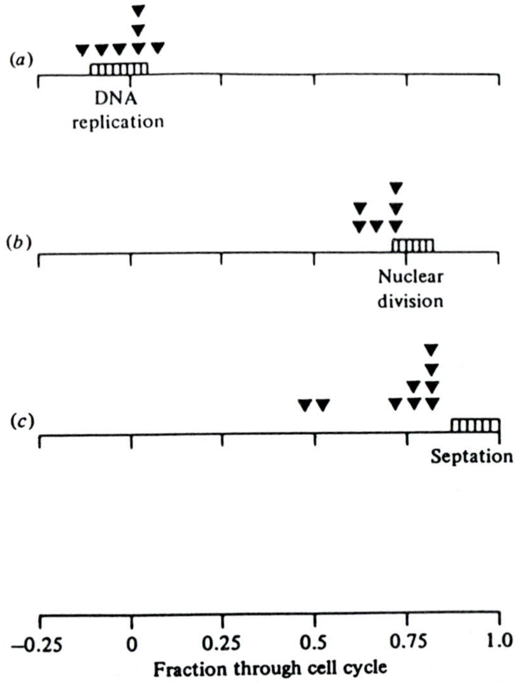

Fig. 17. Execution points of *cdc* mutants of *Schizosaccharomyces pombe*. (a) DNA replication genes, (b) nuclear division genes, (c) septation genes. Hatched boxes mark the timing of DNA replication, nuclear division and septation and ▼ the execution points. Reproduced with permission from Nurse (1981a).

controlling steps in the cell cycle and only when they are complete can a large number of other dependent events take place, hence the clustering of the execution points (Nurse, 1981a).

Double-mutant analysis combines two mutants with different terminal phenotypes in the same strain and looks at the phenotype of the strain when both lesions are expressed simultaneously (as in two heat-sensitive mutants), compared with expression on their own. The terminal phenotype of the double mutant reflects the functional relationship of the primary-defect event rather than the landmark event.

The third and most powerful method of analysis is order-of-function mapping by reciprocal-shift analysis, originally pioneered in bacteriophages (Jarvik and Botstein, 1973). The method relies on reversibly arresting cells by two different methods either of which can be applied independently of the other, as for example, by having a heat-sensitive *cdc* mutation and using hydroxyurea, or by having a heat- and a cold-sensitive double mutant. In the first experiment, cells are accumulated at a cycle block with agent A and then

Table II. Experimental determination of dependency relationship of two events, A and B

Experimental protocol		Conclusion
Block at A, then release and attempt to block at B	Block at B, then release and attempt to block at A	
+	−	A dependent on B
−	+	B dependent on A
+	+	A and B independent of each other
−	−	A and B interdependent

+, cell completes a cycle; −, cell does not complete a cycle

released and immediately arrested with agent **B**. The second experiment is the reciprocal of the first, using agent **B** before **A**. In both cases the fate of the cell population is monitored and conclusions can be drawn about the dependency relationships of the primary defect that the agents affect, as shown in Table II.

Utilizing these methods a functional sequence map can be drawn for the cell cycles of both *Sacch. cerevisiae* (Fig. 18) and *Schiz. pombe* (Fig. 19). In the latter case, the map is drawn strictly from order-of-function experiments, whereas in *Sacch. cerevisiae* various sorts of information have been used to deduce the map, which is based on Pringle and Hartwell (1981). (A map solely based on order-of-function mapping is shown in Fig. 27.) It should be emphasized that the diagrams are provisional. Some of the relationships are thoroughly tested (i.e. the dependency of most *cdc* mutations in *Sacch. cerevisiae* on *cdc28*), but many are based on experiments that are difficult to interpret, on single allele experiments, or had been done before the diagnostic landmarks had proliferated to 16. Nevertheless it is possible to draw a number of important conclusions for both diagrams. First, the cellular events of the cell cycle occur as a defined programme of discrete gene-controlled events. Second, most of the cellular events and their partitioning are covered by one or more mutations. Third, the programme contains dependent events, branch points and points of convergence. Fourth, there is a definite beginning and end to a normal programme. Fifth, the functional interrelationship between the events correlates well with the temporal map (see Figs. 4 and 11). Since there is a common start and end to the cycle, these events cannot get much out of phase. Sixth, some of the mutants have important spatial as well as temporal functions (Sloat *et al.*, 1981; Dutcher, 1982; Minet *et al.*, 1979). The circuitry in Fig. 18 suggests that new cell-cycle initiations can occur even in the absence of proper cytokinesis and cell separation, and this is confirmed

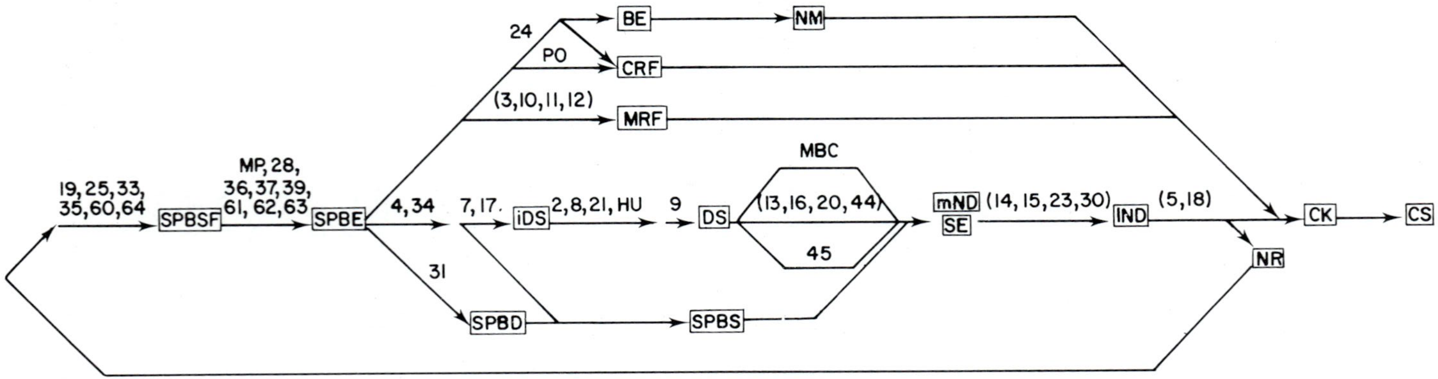

Fig. 18. Functional sequence map of *cdc* mutations of *Saccharomyces cerevisiae*. The numbers refer to the *cdc* mutation. Diagnostic landmark events are in boxes; dependent events are on consecutive arrows; independent events on parallel pathways; interdependent events on the same arrow. The positions of the landmark events across the figure are approximately to a correct temporal scale for the normal cell cycle. BE, bud emergence; CK, cytokinesis; CRF, chitin ring formation; CS cell separation; DS, DNA synthesis; iDS, initiation of DNA synthesis; HU, hydroxyurea; MBC, methyl benzimidazole-2-ylcarbamate (benomyl); MP, mating pheromone; MRF, microfilament ring formation; IND, late stage of nuclear division; mND, medial stage of nuclear division; NM, nuclear migration; NR, nuclear reorganization; PO, polyoxin D; SE, spindle elongation; SPBD, spindle pole body elongation; SPBE, spindle pole body enlargement; SPBS, spindle pole body separation; SPBSF, spindle pole body satellite formation; This figure is based on Fig. 2 of Pringle and Hartwell (1981) with revisions after Wood and Hartwell (1982), Moir and Botstein (1982) and Bedard *et al.* (1981).

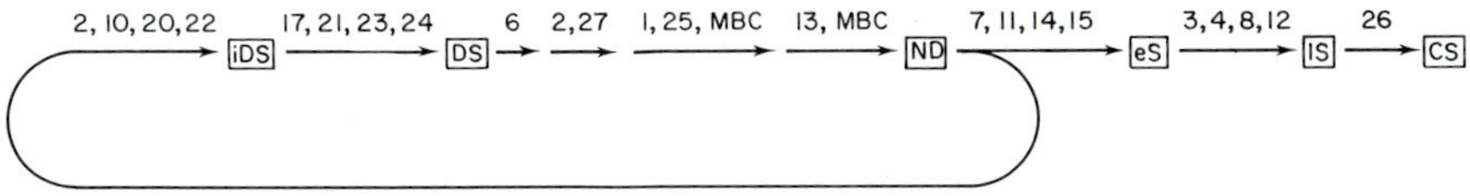

Fig. 19. Functional sequence map of *cdc* mutations of *Schizosaccharomyces pombe*. The numbers refer to the *cdc* mutation. CS, cell separation; DS, DNA synthesis; iDS, initiation of DNA synthesis; MBC, methyl benzimidazole-2-ylcarbamate (benomyl); ND, nuclear division, es, early septation; ls, late septation. This figure is based on the data of Nurse (1981a) and Fantes (1982).

in experiments with sphaeroplasts in which division is impossible but repeated rounds of DNA synthesis still occur (Doi and Doi, 1982; Murakami and Livingston, 1982).

"The events of the cell cycle occur in a prescribed order because the gene products that mediate these events are restricted to a prescribed sequence of function. This raises the question, how is the order of gene product function achieved. In principle there are two possibilities for the dependence of one gene product on another: (i) that the second product is dependent on the first for its synthesis or activation, (ii) that the second product is dependent on the former as its substrate" (Hartwell, 1976; and Fig. 20). The data from rates of synthesis of proteins (see Section II.F.), amount of *CDC* gene product (Byers and Sowder, 1980) and execution-point analysis (see Fig. 17) all point to possibility (ii) rather than possibility (i), but strong evidence on this point is generally lacking. Only in the case of DNA synthesis mutants can possibility (ii) be demonstrated since DNA cannot be synthesized without precursors made by *CDC21* (thymidylate synthetase) and DNA cannot be joined by DNA ligase until some DNA has been made.

In addition to *cdc* mutants that have been specifically searched for, several genes have been detected which can confer a *cdc* phenotype, such as *cal1-1*, which is a mutation conferring calcium dependency, lack of which leads to arrest of *Sacch. cerevisiae* as uninucleate cells with a small bud (Ohya *et al.*, 1984).

The *cdc* mutants all have a conditional block on completion of a cycle. Two other classes of mutants with an altered cell-cycle behaviour have been

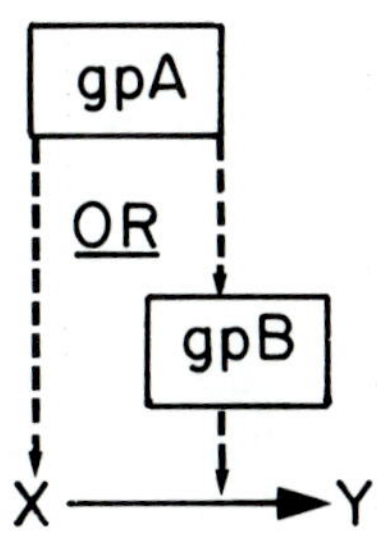

Fig. 20. Theoretical basis for cell-cycle-dependency relationships. Gene product B (gpB) controls the conversion of X into Y, which is necessary for a cell-cycle step, and the functioning of gpB is dependent on gene product A (gpA). This dependency is because either gpA acts (indirectly or directly) on gpB, or gpA (directly or indirectly) on the supply of reactant X.

selected. The first comprises cells which are usually non-conditional, but execute certain periodic events with altered timings, particularly with respect to cell size (Appendices 1C and 2C). These *wee* and *whi* mutants have proved particularly useful in investigation of the major rate-determining steps in the cycle (see Sections III. C., D. and E). The other major class of mutants isolated are those that are unresponsive to normal cell-cycle arrest, either by carbamates (Appendix 2D) or by mating pheromone (Appendix 1D). The latter were initially detected as being unable to mate, hence they are called sterile (*ste*) mutants. The intimate relationship between mating and cell-cycle control is described in Section III.B. and the particular value of these mutants with respect to cell cycle studies is that some *cdc* mutants can suppress some *ste* mutants (Section III.K.). Once *cdc, wee* or *ste* mutants are available it is possible to isolate suppressors of these mutations (Pringle *et al.*, 1984) and this is proving a highly profitable method for isolating other novel cell-cycle-related mutants, for example *cdc46* (Appendix 1A) and *sst1* (Appendix 1D).

B. Co-ordination of growth and division in *Saccharomyces cerevisiae*

So far the discussion has centered on the order of cellular events. In parallel with these events are the continuous processes of protein, RNA and wall synthesis. Since there is one set of discontinuous events per mass doubling, the question is raised as to how the discontinuous events are coupled to the continuous processes (Fantes and Nurse, 1981). In *Sacch. cerevisiae* this coupling occurs at the beginning of the cycle during a process that is called start (Hartwell *et al.*, 1974; Edwards *et al.*, 1978; Nurse, 1981b; Pringle and Hartwell, 1981; Reed, 1980a; Bedard *et al.*, 1981). The major features of the start concept are: first, all cellular events are dependent on a group of mutants (*cdc25, 28, 33, 35, 36, 37, 39, 60, 61, 62, 63, 64*) each of which acts approximately at the same point early in the cycle, and before which no other mutants have been discovered. They thus define the first event of the cell cycle which is under genetic control. Although spindle pole body duplication was originally described as the first event, it is possible to sort the start mutants into those arresting with a single spindle pole body and those with a spindle pole body plus satellite (see Fig. 10). Second, cells treated with mating pheromone reversibly arrest in G1 at a point interdependent with the start mutants. Growth continues but none of the morphological events associated with the cycle occur, although some biochemical changes are seen (Sumrada and Cooper, 1978b). Only cells accumulated at the *cdc28* or mating-pheromone step are competent to conjugate. Third, cells deprived of nutrient, as for example during entry into stationary phase, arrest as uninucleate unbudded cells in the G1 period at or before start (with an unusual chromatic configuration (Piñon, 1978, 1979; Piñon and Pratt, 1979)). Fourth, once vegetative cells have traversed the start event they are committed to a mitotic cell cycle and this can continue even in the absence of further growth (G.C. Johnston *et al.*, 1977a, b; G.C. Johnston, 1977). Only severe nutrient deprivation (Cooper *et al.*, 1979; Sumrada and Cooper, 1978a) or inhibition of protein synthesis (Slater, 1974; Unger and Hartwell, 1976; G.C. Johnston *et al.*, 1977b) will arrest cells between start and cell separation. Fifth, cells need to be sufficiently large before they have a high probability of traversing start (Hartwell and Unger, 1977; G.C. Johnston *et al.*, 1977a; Carter and Jagadish 1978a; Wheals, 1982). Sixth, cells at start are at a "decision point" (see Fig. 19) and can undertake one of four developmental pathways: (i) the mitotic cycle (Hartwell *et al.*, 1974); (ii) conjugation (Reid and Hartwell, 1977); (iii) the meiotic cycle (Simchen, 1974) after completing the *CDC4* step (Hirschberg and Simchen, 1977); (iv) non-proliferation or G0 (Hartwell, 1974). The actual pathway chosen will depend on the nature of the cell and the environment (Fig. 21).

The crucial role of the start process in co-ordinating growth on the one hand with the mitotic cycle on the other was investigated by G.C. Johnston *et al.* (1977a). They made two important conclusions. First, growth was the rate-limiting step for cell proliferation rather than progress through the programme of cell-cycle events. This meant that a cell could complete the set of events between start and cell separation in less time than it took to double cell mass. Secondly, they proposed that growth to a critical size was, in the absence of other constraints, the only requirement for the traverse of start and not a requirement for any other cell-cycle step. The rate of traverse of start thus determined the rate of division. The model also implied that there was no control of growth from the stage-specific events (Singer and Johnston,

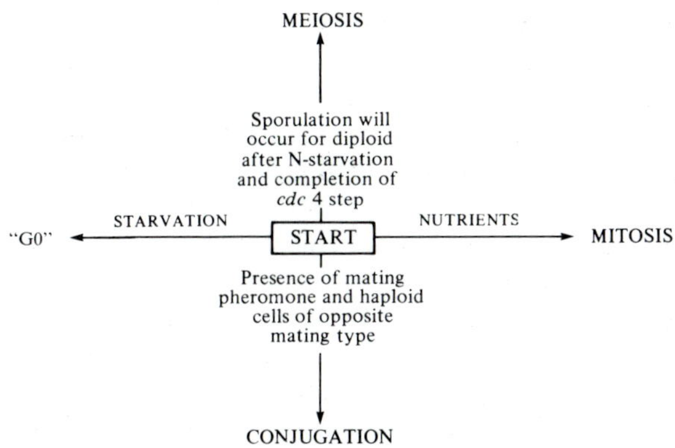

Fig. 21. "Start" as a decision point. Cells arrested in the G1 phase at start can, on release, develop in one of four different ways depending on the nature of the cell and its environment, as summarized in the diagram. "G0", non-proliferation.

1983), and it is a characteristic of *cdc* mutants that at the non-permissive temperature they will, within limits, continue growing. There is thus a kind of cell-cycle "clock", but the unit of time is biological and equal to mass doubling time, and thus the rate of "ticking" is proportional to growth rate.

Start thus represents a unique control point in the *Sacch. cerevisiae* cell cycle, and control of the traverse of start is tantamount to control of the cell cycle. These concepts have been tested by culturing cells under conditions in which the DNA-division sequence, and not cellular growth, was the rate-limiting factor for cell proliferation. Relief of these conditions, which caused an accelerated rate of cell division, did not result in increased rates of mass accumulation, suggesting that in proliferating *Sacch. cerevisiae* cells the DNA-division sequence does not affect cellular growth (Singer and Johnston, 1985).

Depletion of carbon, energy, ammonia, sulphate, biotin, phosphate or potassium leads to uniform arrest of a population of cells as budded G1 cells (Pringle and Mor, 1975; Unger and Hartwell, 1976; G.C. Jonhston *et al.*, 1977a; Piñon, 1978; Sumrada and Cooper, 1978b; Piñon and Pratt, 1979; Lillie and Pringle, 1980). This would be explained in terms of the start concept as a failure of these cells to grow to the critical size for reinitiating a new cycle, and thus they accumulate at or before start. A number of organic chemicals also specifically arrest cells at or before start, and these include *o*-phenanthroline and hydroxyquinoline (G.C. Johnston and Singer, 1978), Sinefungin (Li *et al.*, 1984), tunicamycin (Arnold and Tanner, 1982) and *Kluyveromyces lactis* toxin (Sugisaki *et al.*, 1983). A mutant, *whi2* (Appendix 1C), behaves phenotypically as if it does not recognize starvation conditions (Fig. 22), and it continues to proliferate without concomitant growth producing abnormally small cells (Sudbery *et al.*, 1980; Saul *et al.*, 1985). Reed (1980a) classified the start mutants into two classes. Class 1 mutants mimicked mating-pheromone arrest and thus had the phenotype of continued growth, "shmoo" formation, mating competence and an unduplicated spindle pole body. Class II mutants mimicked nutrient-starvation arrest with rapid cessation of growth, failure to conjugate and a spindle pole body with a satellite. The Class I mutants can be further subdivided into two. There are those (which I will call Class 1b), like *cdc36* and *cdc39*, that are suppressed by growth on non-fermentable carbon sources or by mating-type heterozygosity, arrest randomly through the cell cycle when made alpha-factor

Fig. 22. A *whi2* mutant of *Saccharomyces cerevisiae*. A wild-type strain, X4003-5B, and a strain, 15034, containing a *whi2* mutation, which was 87.5% isogenic to X4003-5B, were grown to stationary phase with vigorous aeration. The wild-type strain (A) consists largely of unbudded phase-bright (G0?) cells, whereas the *whi2* strain has small phase-dark cells at all stages of the cell cycle. Bar 5 μm. Photograph provided by Peter Sudbery.

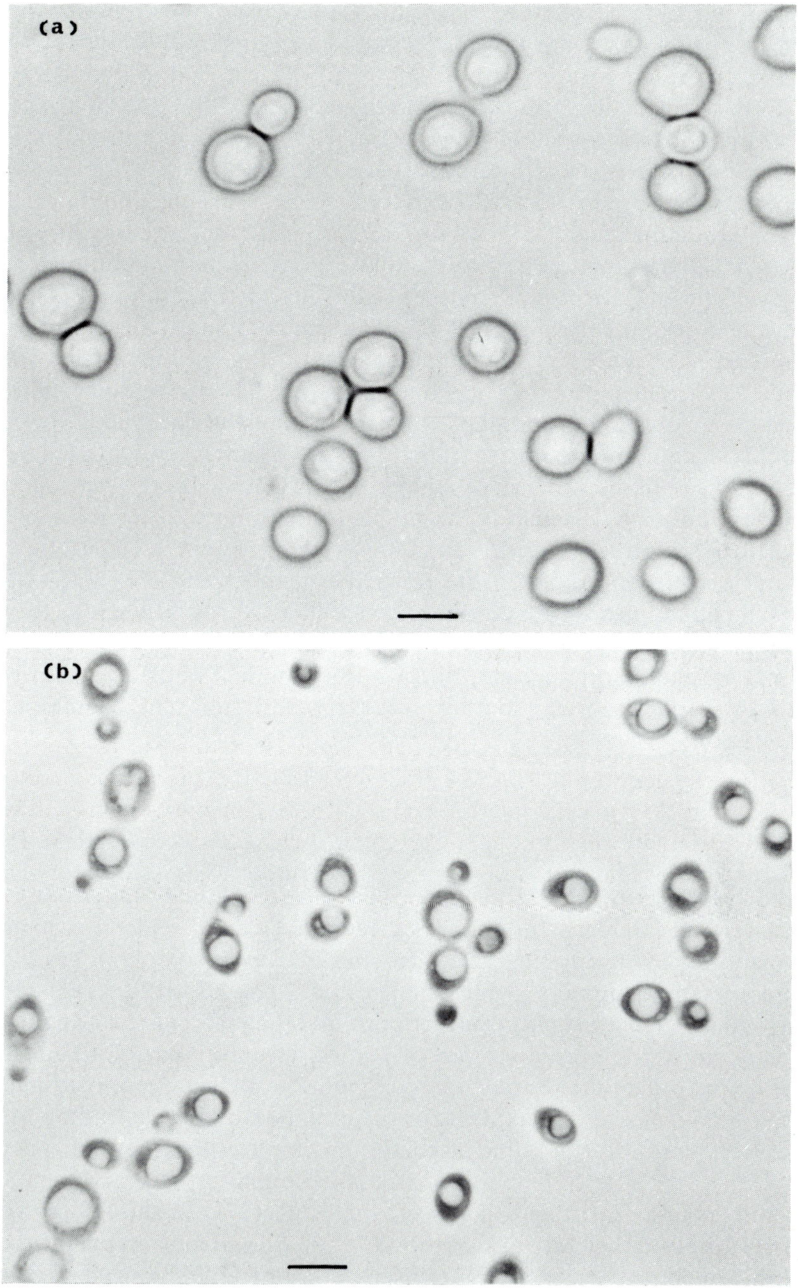
(a)
(b)

resistant and undergo normal karyogamy, and there are those (Class Ia), like *cdc28* and *cdc37*, that are not suppressed, have a normal G1 arrest when alpha-factor resistant and have a karyogamy defect (like *kar* cells). Subsequently new selection protocols led to the detection of five more start genes (Bedard *et al.*, 1981) some of which were like Class II mutants (*cdc60* and *cdc64*), and some of which were like Class Ib mutants (*cdc62* and *cdc63*) except that they did not shmoo. From their phenotypic behaviour it looks as if Class II mutants (*cdc25, cdc33, cdc35, cdc60, cdc64* and probably *cdc19* and *gcdl* (*tra3*)) are intimately involved with nutrient arrest recognition, a view supported by the discovery that *cdc35* is allelic to the structural gene for adenylate cyclase (*cyr1*) which is involved in mediating nutrient response (see Section III.N.). The Class I mutants appear to be operating further downstream in the start process (Iida and Yahara, 1984a) and some may be intimately involved in the mating response (Connolly *et al.*, 1983). However, it would be premature to regard Class I mutants as intrinsically more important than Class II mutants since mating response and size control are both dependent on growth which is the primary environmental response of the cell. Further, an allele of *cdc63* can show random arrest, Class I arrest or Class II arrest depending on the restrictive temperature (G.C. Johnston, personal communication). Start may be viewed as a process whereby the cell assesses its response to three "questions". (i) Are there nutrients present? (ii) Are mating pheromones absent? (iii) Am I big enough? Satisfactory answers to all three questions are both a necessary and sufficient condition for proliferation.

C. Asymmetrical division

A study of the growth of individual cells led Hartwell and Unger to extend the start concept ideas further. They rediscovered the older observation (Johnston and Gibson, 1966) that division in *Sacch. cerevisiae* was asymmetrical, with daughters being smaller than parents at cell separation (Hartwell and Unger, 1977). Furthermore, the time taken for a daughter to divide again was longer than that of parents (see also Burns, 1956). Both division and cycle time were regularly asymmetrical. They proposed that (i) there is a critical size requirement for the initiation of a cycle leading to cell separation, and (ii) the time needed to complete these events (largely S + G2 + M) is less than the population doubling time, and relatively constant, as in many other cell types (Mitchison, 1971). Although growth is almost entirely in the bud during the budded phase and most of the post-start period is budded, nevertheless at cell separation the daughter cell might not have achieved the same size as the parent from which it came. Furthermore, the mode of growth could mean that the parent cell was already above the critical size (achieved at the previous cell-cycle initiation) but the daughter might need some time to grow to the critical size, depending on the degree of asymmetry at division. The cycle times would thus be asymmetrical. The degree of asymmetry is greater the slower the doubling time (Hartwell and Unger, 1977; G.C. Johnston *et al.*, 1977a; Carter and Jagadish, 1978a; Lord and Wheals, 1980, 1981; Thompson and Wheals, 1980). The model also explains the observation of an increased fraction of unbudded cells at slower growth rates (von Meyenburg, 1968) since the pre-start period is unbudded. A critical test of the model (Lord and Wheals, 1980) confirmed its basic tenets, and there are now sets of equations enabling the duration of cell-cycle events (Hartwell and Unger, 1977), volumes at times during the cell cycle (Tyson *et al.*, 1979) and genealogical age distribution (Lord and Wheals, 1980) to be calculated from mean population data. Populations of yeast cells (including *C. albicans* (Chaffin, 1984b)) must be considered as a mixed population of biologically distinct mothers (parents) and daughters.

The asymmetry has an important molecular dimension. Homothallic strains of *Sacch. cerevisiae* regularly switch mating type (Herskowitz and Oshima, 1981). The rules governing the switch are (i) only mother cells can undergo switching, and (ii) both progeny of a switching cell change mating type. The latter rule implies that the switch occurs in the G1 phase. The initiation of interconversion is caused by a double-stranded endonuclease (Strathern *et al.*, 1982), encoded or regulated by the H0 gene. It has been shown (Nasmyth, 1983) that (i) the endonuclease activity only appears during the G1 phase of mother cells, (ii) it is unstable and disappears by the time of mitosis and (iii) activity is dependent only on a previous mitosis and the execution of the start (*CDC28*) function and on no other cell-cycle step (such *CDC4, CDC7, CDC31*). A repetitive eight base-pair sequence has been identified as conferring on the H0 gene *CDC28*-dependent transcription for both mothers and daughters (Nasmyth, 1985). The asymmetrical response of mothers and daughters could conceivably be due to size, except that the size distributions of mothers and daughters do overlap (Lord and Wheals, 1981), or due to regularly asymmetrical segregation of some unknown factor (Nasmyth, 1983).

Two features of the asymmetrical division model of Hartwell and Unger (1977) were not predicted. First, the "critical size" varies with growth rate. At slower growth rates there seems to be a constant value, rising rapidly at faster growth rates (Tyson *et al.*, 1979; Lorincz and Carter, 1979; G.C. Johnston *et al.*, 1979; Lord and Wheals, 1980). By performing shift-up and shift-down experiments (transferring cells from poor to rich medium and *vice versa*) it was shown that the point of commitment to a particular critical size occurs shortly before traverse of start, i.e. in late G1 (this point is discussed further in Section III). In chemostat cultures there is a change from an oxidative to a fermentative mode of growth (Beck and von Meyenburg, 1968) at the same growth rate at which there is a switch-over in the relationship between the duration of daughter, parent and budded periods (Thompson and Wheals, 1980) and the appearance of the minimum critical size (Lorincz and Carter, 1979; G.C. Johnston *et al.*, 1979). Investigations with non-respiring petite mutants showed this to be a coincidence since the effects are growth-rate dependent (Bugeja and Carter, 1982). Secondly, although the time from the beginning of G2 to cell separation (Jagadish and Carter, 1978), the time from the beginning of S phase to cell separation (Barford and Hall, 1976; Slater *et al.*, 1977), the time from start (as assayed by the *cdc28* step or the mating-factor step) to cell separation (Jagadish and Carter, 1977; Carter and Jagadish, 1978b) are all *relatively* constant, the budded period shows a clear increase in duration at slower growth rates in all published data (see Wheals, 1981); this effect is most noticeable in chemostat cultures (Thompson and Wheals, 1980; Vanoni *et al.*, 1983) indicating that there may be a growth-rate-dependent step after start.

The evidence for a size control over start has been reinforced by the isolation of a mutant (Appendix 1C) called *whi1* (Carter and Sudbery, 1980) which initiates the cycle at half the normal volume (Sudbery *et al.*, 1980; Lord and Wheals, 1984) resulting in a population that grows at the normal growth rate but has small cells. The *whi1* mutant still retains the growth-rate modulated effect over size at start that is shown by wild-type cells (B.L.A. Carter, personal communication).

D. Co-ordination of growth and division in *Schizosaccharomyces pombe*

At the same time as evidence was obtained for size control over start in *Sacch. cerevisiae*, Fantes (1977) showed that there was a size-regulated control of division in *Schiz. pombe*. By studying the cycle time and cell size in both normal and perturbed culture he showed that there was a homoeostatic regulation of cell size by change in cycle time. Large cells at birth had, on average, shorter cycle times and *vice versa*. Cells above a certain size at birth had a minimum cycle time and recovery took more than one generation (Fantes, 1977; Miyata *et al.*, 1978a,b, 1979a). The mean size of *Schiz. pombe* cells varies with growth rate, and the transition between two growth rates and its effect on proliferation rate and cell size has been investigated (Fantes and Nurse, 1977a; Miyata *et al.*, 1979b, 1981b). The major conclusion was that there is a size control over entry into nuclear division and septation, and that homoeostatic alteration of cycle time occurred in the G2 period. Cells wait to attain a critical size before they can enter mitosis, although this control is less evident in cells from a nitrogen-limited chemostat (Nasmyth, 1979). Cells that have been accelerated into division take rather longer than usual to complete the next cycle (Smith and Mitchison, 1976).

Although these data placed the co-ordination of growth and division by a nutrient-modulated size control as occurring in G2, evidence had already been presented for a size control over DNA synthesis occurring in G1. Two genes, *wee1* and *wee2*, had been detected (Nurse, 1975; Thuriaux *et al.*, 1978) which caused cells to divide at half the size of wild-type strains (Fig. 23; Appendix 2C). Strains carrying the *wee* mutations had a greatly lengthened G1 period and a comparably shortened S + G2+ M period and an S period that did not start at the physiological division of the cells and before cell septation was complete (Nasmyth, 1979), but about one-third into the cycle. A similar delay in the onset of S could be demonstrated in wild-type cells that had been made very small, either by nitrogen starvation or on germination from spores (Nurse and Thuriaux, 1977), showing that the delayed S phase was not a direct consequence of the *wee* mutation. In all three cases the S period was initiated at the same cell size (measured as weight of protein), a size below that normally seen in fast-growing wild-type cultures. The conclusion from these observations and detailed experimental tests (Fantes and Nurse, 1978) was that *wee* mutants had lost their size control over nuclear division, are thus accelerated into mitosis a minimum time after the completion of the S phase and so divide at a smaller cell size (Nurse and Thuriaux,

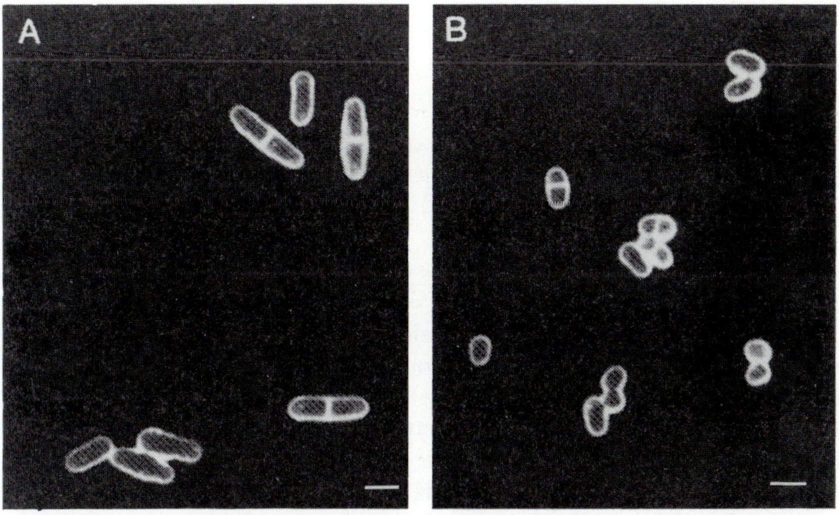

Fig. 23. Cell-size mutant of *Schizosaccharomyces pombe*. The micrographs show dark field optics of wild-type (A) and *wee1-50* cells (B) grown at 35°C. Bar, 5 μm. Reproduced with permission from Thuriaux et al. (1978).

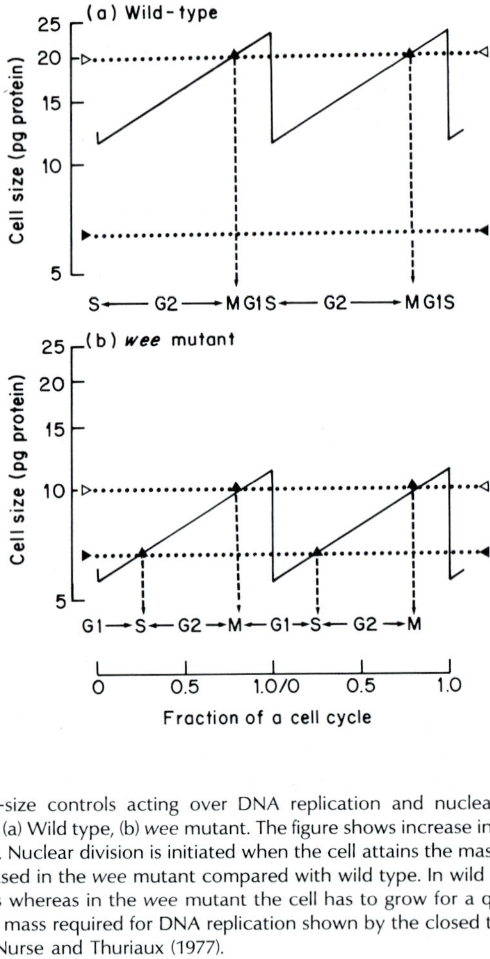

Fig. 24. Model of cell-size controls acting over DNA replication and nuclear division in *Schizosaccharomyces pombe*. (a) Wild type, (b) *wee* mutant. The figure shows increase in cell mass during two consecutive cell cycles. Nuclear division is initiated when the cell attains the mass shown by the open triangles. This is decreased in the *wee* mutant compared with wild type. In wild type, S phase follows rapidly on after mitosis whereas in the *wee* mutant the cell has to grow for a quarter of a cell cycle to attain the minimum mass required for DNA replication shown by the closed triangles. Reproduced with permission from Nurse and Thuriaux (1977).

1977; Fantes and Nurse, 1978). The delayed S period in small cells revealed a cryptic size control over DNA synthesis in wild-type cells, which are normally born at a size greater than the size control over DNA synthesis and are thus competent to initiate DNA synthesis immediately (Fig. 24). Small cells are born below the critical size for the initiation of DNA synthesis and thus there is a size-related delay before the onset of S phase (Nurse, 1975; Nurse and Thuriaux, 1977; Fantes and Nurse, 1978). An additional effect of a *wee1* mutation is to advance the transition points of those genes involved in mitosis, implying a role for *wee1* in a system that controls the timing and co-ordination of developmental steps controlled by three mitotic genes (Fantes, 1983). *Schizosaccharomyces pombe wee* mutants behave like wild-type *Sacch. cerevisiae*, with a size control over DNA synthesis and a constant S + G2 + M period (see also Nasmyth, 1979).

In *Sacch. cerevisiae* DNA-division sequence events are dependent on growth but not *vice versa* (G.C. Johnston *et al.*, 1977a; Singer and Johnston, 1985). However, the same does not appear to be true for *Schiz. pombe*. When constraints were applied to growing cells to the rate of completion of the DNA-division sequence (by using DNA synthesis inhibitors or *cdc* mutants), the rate of growth (mass accumulation) was limited in parallel with the rate of division. Moreover, when these constraints were relieved, growth showed an immediate and dramatic increase matching the increased rate of cell division (G.C. Johnston and Singer, 1984; G.C. Johnston and R.A. Singer, personal communication). This intriguing result implies that the DNA-division sequence modulates growth rate and that growth rate may be less than maximal in normally proliferating *Schiz. pombe* cells.

E. Comparison of start in *Saccharomyces cerevisiae* and *Schizosaccharomyces pombe*

Further similarities between control in these two yeasts were increased when it was discovered that *cdc2* and *cdc10* in *Schiz. pombe* both have an effect in G1 on commitment to conjugation (Nurse and Bissett, 1981), whereas all other *cdc* mutations, which act later in the cycle, prevent conjugation. Since *cdc2*, which is allelic to *wee2*, was originally isolated as a mutant of mitosis, leading to failure of septation and production of filaments, it seems to have a dual role both over commitment to a mitotic cycle and the initiation of DNA synthesis (like a *Sacch. cerevisiae* start mutant) and over mitosis. A reinvestigation of some alleles of *cdc28* of *Sacch. cerevisiae* has shown that it too blocks cells at mitosis as well as start (Piggott *et al.*, 1982), one allele blocking only at mitosis. The fraction of cells between start and mitosis is sufficiently small to confuse the observer into thinking that the small proportion of cells arrested with buds is due to leakiness past the *cdc28* block. The analogy between these two genes was further strengthened by an elegant cloning experiment in which the *CDC28* gene of *Sacch. cerevisiae* (Nasmyth and Reed, 1980) was used to transform *cdc2* mutants of *Schiz. pombe* (Beach *et al.*, 1982). Not only did the *CDC28* gene complement the *cdc2* mutation, but it also gave the cell the alternative phenotype of the *cdc2* gene, the *wee* or *cdc2-1w* phenotype. The reciprocal experiment was unsuccessful, probably because of an unspliced intron in the *CDC2* clone. Thus it looks as if the *Schiz. pombe CDC2* gene and the *Sacch. cerevisiae CDC28* gene perform exactly homologous functions (but see Hughes, 1983). Considering that these two species are phylogenetically distant from each other, the result suggests that there has been conservation of the structure of a gene crucial to cell division. This has been confirmed by sequencing studies. The sequence of *CDC28* shows that it codes for a protein of 298 amino acids (a relative molecular mass of 34 kDa) which has significant homologies to bovine cyclic AMP-dependent protein kinase and a number of tyrosine-specific protein kinase oncogenes (Lorincz and Reed, 1984). Sequence analysis of *CDC2* reveals it to be similar to *CDC28*, particularly in those parts of the molecule associated with protein kinase activity (Hindley and Phear, 1984). Furthermore, both genes have structures which makes them look like substrates for protein kinases. Using antisera raised against the *CDC28* gene product (Reed *et al.*, 1982), a phosphoprotein, p36, was precipitated from yeast cell lysates. These immune complexes contained a protein kinase activity (Reed *et al.*, 1985) as determined by the transfer of radio-labelled ATP to a coprecipitated protein (p40) of molecular weight 40,000. This activity was absent or thermolabile when using extracts from *cdc28* strains. The phosphate group was transferred specifically to serine and threonine residues and not to tyrosine (see also Section III.N.). The sequence of the *CDC10* gene shows no homology to any other known DNA sequence (Aves *et al.*, 1985).

F. Co-ordination of growth and division in other yeasts

The observation of asymmetry in the mode of growth of a number of other yeast species (Flegel, 1978) is suggestive of similar rules to *Sacch. cerevisiae* being applicable to them. Studies on *C. utilis* (Thomas *et al.*, 1980a, b; Vrana and Votruba, 1981) have shown some differences to *Sacch. cerevisiae*, but either the unusual conditions under which the cells were grown or the method of analysis preclude direct comparison.

In *C. albicans*, cells undergoing mycelial formation do not undergo nuclear division until the length of the mycelium is equal to or greater than the diameter of the mother cell (Soll *et al.*, 1978). A related effect was seen with budded cells when the majority of the 40% of cells that did undergo nuclear division were above a particular budded volume, rather than cell mass (Bedell *et al.*, 1980; Soll *et al.*, 1981b). When vacuolar volume is taken into account, daughter pseudomycelial cells are larger at bud initiation than their mother cells (Yokoyama and Takeo, 1983).

A comprehensive analysis of cycle times and cell sizes of parent and daughter *C. albicans* yeast-phase cells over a range of growth rates gave broadly similar results to that found in *Sacch. cerevisiae* (Chaffin, 1984b). The major difference was that cycle-time asymmetry could not be directly ascribed to cell-size asymmetry but it must be emphasized that measurements of external cell volume may be a poor indicator of effective internal volume (Yokoyama and Takeo, 1983).

G. Models of size control

The term size is deliberately vague but has been extensively used because although there is a lot of evidence that cells monitor how big they are, the method by which they do this is completely unknown. It is presumably some function of cell mass or protein or volume, rather than any of these three directly because it is possible to create conditions of unbalanced growth where any or all of these may be increasing without concomitant cell proliferation (Shilo *et al.*, 1978b; Singer *et al.*, 1978; Lillie and Pringle, 1980).

Although a large number of models are capable of giving size control (Fantes *et al.*, 1975), the asymmetrical mode of development in *Sacch. cerevisiae* puts a number of constraints on the range of possible models (Lorincz and Carter, 1979) and only those incorporating an unstable effector, whose concentration varies through the cell cycle and is titrated away by a nucleus-related function or structure, give a different response for small daughters and large parents (Fantes *et al.*, 1975; Wheals and Silverman, 1982).

Despite the homoeostatic behaviour of *Schiz. pombe* cells and the size-controlled traverse of start in *Sacch. cerevisiae*, claims have been made, arising from experiments with mammalian cells (Smith and Martin, 1973), that the initiation of events leading to division occurs at random, with a transition probability that is related to growth rate (Shilo *et al.*, 1976, 1977, 1978b; Brooks, 1981; Samokhin *et al.*, 1981a, b). The arguments have been refuted on several different grounds (Fantes and Nurse, 1977b; Wheals, 1977; Nurse and Fantes, 1981), but the clearest evidence against the concept has come from detailed analysis of cell-cycle kinetics. Cells of *Schiz. pombe* do not show the predicted kinetics (R. Brooks, personal communication) despite early indications that this was so (Fantes, 1977). In the case of *Sacch. cerevisiae*, simultaneous measurements of cycle time and cell size in steady-state (Lord and Wheals, 1981; Wheals, 1982) and perturbed (Lord and Wheals, 1983) cultures, have shown no agreement with the basic tenets of the model, providing that bud emergence is an adequate marker of recent traverse of start. Equally important, the data clearly show that there is not a defined critical cell volume at which cells initiate events. Either cell volume is only loosely coupled to what the cell measures, or the control has both probabilistic and deterministic elements which can be combined in a single sloppy size-control model (Wheals, 1982). The model proposes that cells monitor the concentration of unstable effector, and that an allosteric interaction between effector and monitor results in an increasing probability of a cell initiating a cycle as it increases in size, rising to a high plateau value, which is less than unit probability, in agreement with the data of Moore (1984).

H. The nature of the G1 period

Whatever the final outcome, most of the models applicable to *Sacch. cerevisiae* state that entry into G1 occurs after nuclear division, and exit from G1 can occur immediately after entry into G1, the time spent in G1 being a function of size. That is to say, there are virtually no G1-specific events that have to be accomplished before traverse of start. This is in full agreement with the genetic data and order-of-function mapping which locates no dependent events before start. What then is the nature of G1? Rather than comprising a set of events, it must be viewed as simply the time between two other events, mitosis and DNA synthesis, the duration being simply due to the amount of mass the cell has to synthesize after birth before it can reinitiate a cycle. Cooper (1979, 1982) has forcefully argued that this "continuum model" has direct analogies with the situation in bacteria (Fig. 25). In this view the cell cycle is not viewed as a cyclical process, which implies a fixed repetition of a set of events, but as a linear process – the DNA-division cycle (Mitchison, 1971) or cell-division sequence (Lord and Wheals, 1981) – comprising start, post-start G1, S, G2, M and G1*, which is repeated with a variable periodicity, the variation occurring in G1. Evidence consistent with this view comes from an examination of the duration of cell events in populations of cells, where there is a modest variation in the duration of the post-start events but a large variation in the unbudded period, the smaller cells (daughters) having a long unbudded period (G.C. Johnston *et al.*, 1977a; Lord and Wheals, 1981, 1983). This conclusion has been reinforced by experiments which show that artificial elongation of the cell-division sequence without affecting growth, results in cells that are larger at division and thus need a shorter period before a new cycle is initiated, showing that G1 is largely dispensable (Singer and Johnston, 1981, 1983; G.C. Johnston and Singer, 1983; P. Thompson and A. Wheals, unpublished observations; see Fig. 25). The duration of the cell-division sequence was greatly extended without appreciably affecting growth rate in a number of ways (using hydroxyurea, methyl benzimidazol-2-yl-carbamate, *cdc* mutants at intermediate temperatures, thymidylate limitation of auxotrophs), and the execution point of start, nuclear division and the *cdc4* and *cdc7* mutants was determined (G.C. Johnston and Singer, 1983). They discovered that start could be completed in the previous cycle, giving a negative execution point, before nuclear division had occurred (Fig. 26). The *cdc4* and *cdc7* execution

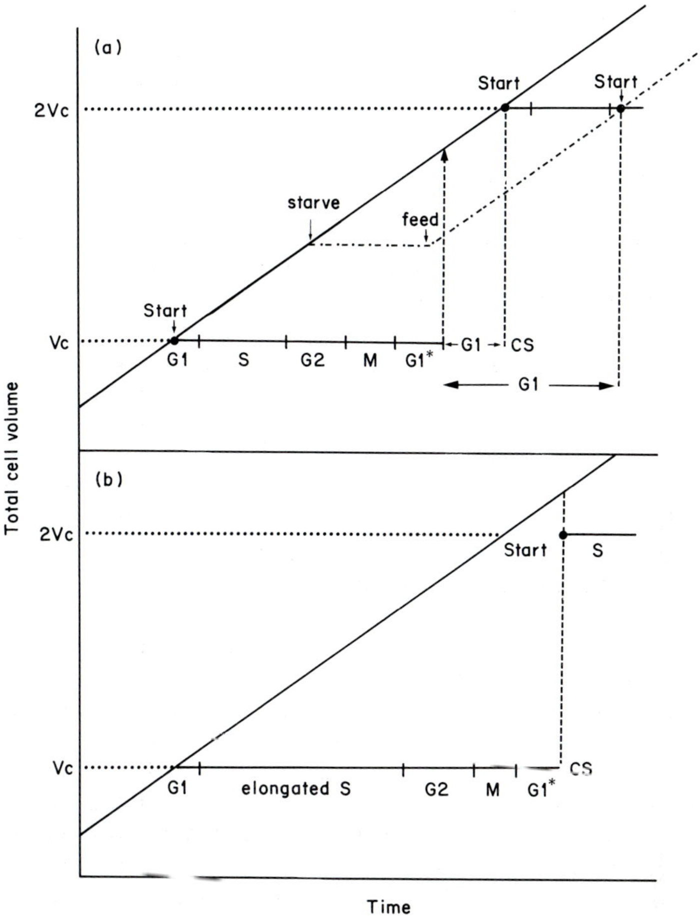

Fig. 25. Application of continuum model to *Saccharomyces cerevisiae*. The figures show (i) volume increase of a yeast cell (on an exponential scale) and its progeny, simplified to ignore volume decrease per cell at division, and (ii) cell-cycle events from "start". Diagram (a) shows how the duration of G1 is determined by the volume at cell division and the critical volume (Vc) at the next reinitiation. On starvation (which abolishes increases in cell volume but not the completion of cell-cycle events) and re-feeding, the duration of G1 is dependent on cell volume at starvation and Vc. Starved cells are heterogeneous with respect to cell volume and hence have a variable G1 on re-feeding. Diagram (b) shows how G1 can be abolished as a result of delays to the completion of cell-cycle events which result in 2Vc being achieved before cell separation.

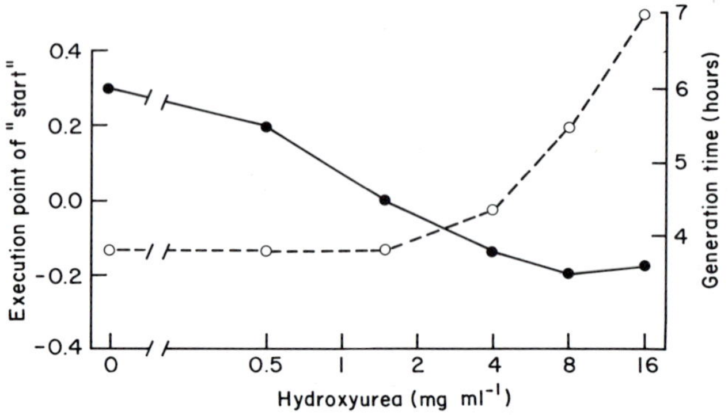

Fig. 26. Execution point of "start". As the concentration of hydroxyurea increases the generation time begins to increase (0) and eventually the execution point of "start", which occurs at about + 0.3 of a cycle in untreated cells, falls to a negative value indicating completion of the "start" event before cell separation. The execution points were measured using both time-lapse and residual-number increase of mating-pheromone treated or *cdc* start-mutant-containing cells. Data provided by Gerry Johnston and Rick Singer.

points, however, always occurred after nuclear division, even though they are normally executed immediately after start (Hereford and Hartwell, 1974). The time interval between start and *cdc4* increased as the execution point of start became more negative. They concluded that start and nuclear division are each necessary but insufficient for initiation of cycle, and it is only those cells that have completed start, and are in a "start complete" state, and have executed nuclear division which are capable of cell-cycle initiation. In a parallel analysis (Singer and Johnston, 1983) it was shown that when the enlarged cells caused by these experimental methods were relieved from these constraints a burst of rapid divisions ensued with the absence of a G1 period, leading to the production of a normal-sized population.

They concluded that start is independent of other periodic events and is only dependent on growth. Under normal conditions these other events would be completed before a mass doubling and so the independency of the relationship would be obscured. This important new concept is in full agreement with the concepts underlying the continuum model (Cooper, 1979, 1982) and is shown in Fig. 27. A corollary of these results is that for titration of cell growth at least one cellular component has to double in order to titrate away a twofold increase in volume and it is of interest that periodic budding in *cdc4* strains, as they complete start, is accompanied by the appearance of an enlarged spindle pole body (Byers and Goetsch, 1974).

Similar experiments have been done with *Schiz. pombe* (Singer and Johnston, 1984; R.A. Singer and G.C. Johnston, personal communication). Since the initiation of S phase in wild-type cells is already at the beginning of the cycle, *wee1* cells were used which have a start *cdc10* execution point of

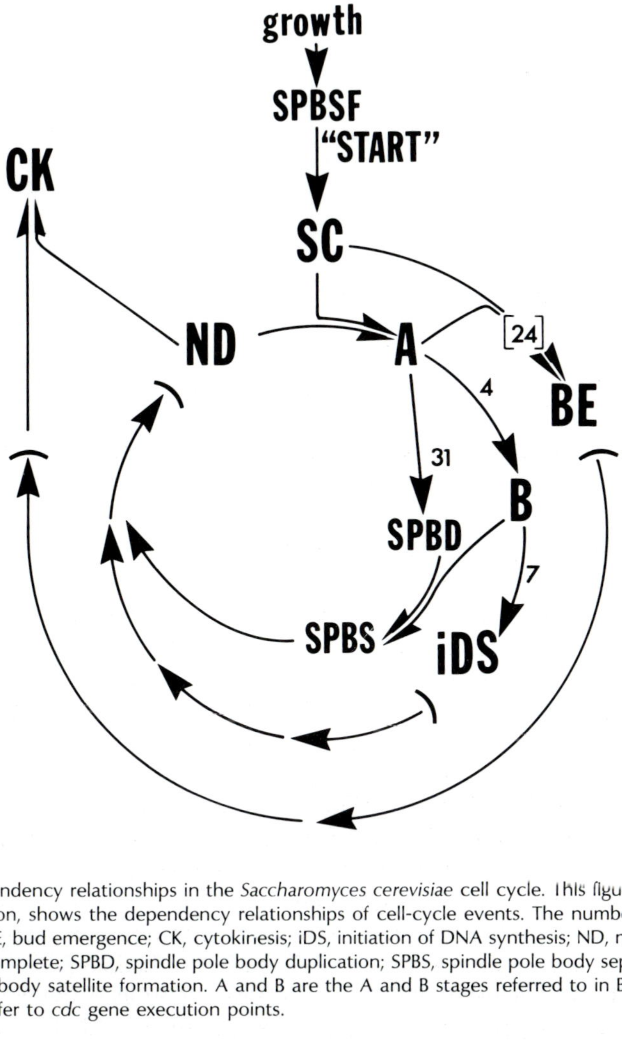

Fig. 27. Dependency relationships in the *Saccharomyces cerevisiae* cell cycle. This figure, provided by Gerry Johnston, shows the dependency relationships of cell-cycle events. The numbers refer to *cdc* mutations. BE, bud emergence; CK, cytokinesis; iDS, initiation of DNA synthesis; ND, nuclear division; Sc, "start" complete; SPBD, spindle pole body duplication; SPBS, spindle pole body separation; SPBSF, spindle pole body satellite formation. A and B are the A and B stages referred to in Byers (1981b). 4, 24 and 31 refer to *cdc* gene execution points.

+ 0.2. Prolonging S phase by using inhibitors of DNA synthesis or *cdc* mutants at semi-permissive temperatures again caused an increase in cell size at constant growth rate, so the *wee1* phenotype was suppressed, the G1 period was abolished and the *cdc10* execution point became − 0.1, results directly comparable with *Sacch. cerevisiae*. Analogous results have also been seen by Nasmyth (1979). In fast-growing wild-type cells DNA synthesis occurs synchronously in the two daughter cells just separated by the cell plate, and before cell separation is complete. At slower growth rates in nitrogen-limited chemostat culture, there is a delay before the initiation of

DNA synthesis. Nevertheless there is an apparent constancy of the size at which cells execute the *CDC10* function, and a constancy in the execution point to division, of 4–4.5 hours. Cells growing faster than this execute the function in the previous cycle, like *Sacch. cerevisiae* start genes, become DNA synthesis competent but delay initiating DNA synthesis because they are dependent on division for initiation to occur. The analogy with *Sacch. cerevisiae* becomes more apparent when it is realized that one of the two execution points of *CDC2* is the same as *CDC10*, under normal conditions, and *CDC2* is homologous to the *Sacch. cerevisiae* start gene *CDC28*.

J. G1-arrest and G0 in *Saccharomyces cerevisiae*

G1 is a transient step between nuclear division and DNA synthesis of a cell undergoing growth and proliferation. A number of treatments can arrest cells after completion of cell division and before the onset of DNA synthesis (see Section III.B. and Li *et al.*, 1984). Although they are all called G1 arrest, several criteria can be used to show that they are different. Arrest by starvation leads to cessation of growth and so the cells do not accumulate in a "cycling" G1 state. This is emphasized by the difference in appearance (Sudbery *et al.*, 1980), sphaeroplast formation (Deutch and Parry, 1973), nystatin treatment (Snow, 1966), heat resistance (Walton *et al.*, 1979), protein profile (Iida and Yahara, 1984a), stringent response (Iida and Yahara, 1984b) and state of the chromatin (Piñon, 1978). Arrest using Class II start mutants produces changes similar to starvation (Iida and Yahara, 1984a,b; Popolo and Alberghina, 1984). Arrest using mating pheromone or Class I start mutants seems to be approaching "cycling" G1 arrest because growth is initially unaffected (preventing induction of most of the changes above) and the protein profile is closer to untreated cells (Iida and Yahara, 1984a; J.R. Ludwig and L. Popolo, personal communications). Nevertheless, physiological changes (e.g. shmooing) and chromatin changes (Piñon, 1979; Petersen and Piñon, 1979) still occur and a general response to the arrest is the induction of proteins also induced in response to heat (heat-shock proteins (Iida and Yahara, 1984a)).

Should the non-proliferating state that the cells reach be regarded as a G0 phase? There is no clear definition of G0 (Cooper, 1982), but it is normally regarded as a specific differentiation from the G1 phase into a quiescent state which may or may not be reversible. The only stage in yeast corresponding to a quiescent state is the growth-arrested G1 state. To show that this is equivalent to a G0 phase it will be necessary to demonstrate two features. First, that there are a unique set of properties associated with stationary phase and secondly that it is possible to isolate mutants specifically defective in either entry into or exit from stationary phase. Some progress has been made in establishing both criteria: (i) starved cells have a unique chromatin structure (Piñon, 1978, 1979; Petersen and Piñon, 1979); (ii) sulphur-starved cells and Class II start mutants at the restrictive temperature produce a unique set of proteins (in addition to the heat shock proteins) (Iida and Yahara, 1984a,c; Popolo and Alberghina, 1984; Martegani *et al.*, 1984); (iii) there is at least one short-lived protein which is synthesized on recovery from arrest induced by Class II start mutants (Popolo and Alberghina, 1984); (iv) both *whi2* (Saul *et al.*, 1985) and *ard1* mutations (Whiteway and Szostack, 1985) show properties consistent with failure to enter stationary (or G0) phase on starvation; (v) after an initial failure to isolate specific G0 mutants (Bedard *et al.*, 1981), a strain has now been isolated (M. Drebot, R. Singer and G. Johnston, personal communication) that is temperature sensitive for exit from stationary phase. In addition, slow-growing chemostat cultures can show a proportion of cells that can reversibly enter a state indistinguishable from stationary phase (Bugeja *et al.*, 1982; N.F. Britton and A.E. Wheals, unpublished results).

K. Conjugation

Exponentially growing *Sacch. cerevisiae* cells constitutively produce mating pheromones (Stötzler *et al.*, 1976; Betz and Duntze, 1979) that synchronize the cycles of cells of opposite mating type as a prelude to cell fusion (Hartwell, 1973a) although mating-pheromone inactivation activity may vary during the cell cycle (Tsuchiya *et al.*, 1981). The cells are specifically arrested at start (Hereford and Hartwell, 1974; Hartwell *et al.*, 1974; Wilkinson and Pringle, 1974) and, after agglutination and cell fusion, the nuclei fuse at the spindle pole body (Byers and Goetsch, 1975; Delgado and Conde, 1984). Arresting cells at different stages of the cell cycle using *cdc* mutants revealed that only cells arrested at the start (*CDC28*) step could mate (Reid and Hartwell, 1977). A similar set of experiments performed on *Schiz. pombe* cells containing *cdc* mutations (Nurse and Bissett, 1981), revealed an analogous start control over conjugation. Only *cdc2* and *cdc10* mutants, which block during the G1 period, are able to conjugate; other mutants blocked during G1, S, G2 or mitosis are already committed to cell division and cannot mate. The *CDC2* gene also has a function during mitosis.

The intimate relationship between mating and G1 arrest is further revealed by the discovery that α-factor-resistant mutants of *cdc28* and *cdc37* show normal G1 arrest at 37°C whereas α-factor-resistant mutants of *cdc36* and *cdc39* show random arrest (Connolly *et al.*, 1983). This implies that (a) *cdc36* and *cdc39* need an active mating-pheromone response system to show G1 arrest, and (b) *cdc28*- and *cdc37*-induced arrest is not equivalent to α-factor arrest even though it occurs at a similar stage of the cell cycle.

The process of mating is now being analysed genetically using mutants that are insensitive to arrest by mating pheromone (*steriles*) or are super-sensitive to its effect (*sst* mutants), and are listed in Appendix ID. *STE13* encodes a dipeptidyl aminopeptidase involved in α-factor processing, *STE2* is the α-factor receptor, *STE4* and *SST2* control the activity of the mating-pheromone receptor and *STE8* regulates the expression of the *MAT* locus. It has been suggested, on the basis of suppressibility, that *CDC36* and *CDC39* interact directly with the *STE5* gene product (Shuster, 1982b) but other interpretations are possible (see Fig. 29).

L. Meiosis and sporulation

Since the process of meiosis and sporulation includes many processes such as DNA synthesis and nuclear segregation that are also common to the mitotic cycle, it was of great interest to see whether the same or different gene functions were utilized. The unambiguous answer for *Sacch. cerevisiae* (Simchen, 1974) was that nearly all genes known at that time to be involved in SPB duplication or segregation, DNA replication and chromosome segregation were also necessary for meiosis and/or sporulation. In contrast, four of the genes, involved in bud emergence and cytokinesis (*CDC3, CDC10, CDC11* and *CDC24*), were not necessary, suggesting that (a) the meiotic programme is in essence a modification of a mitotic programme, and (b) the process of bud emergence was a specific "yeast like" adaptation evolved at a later date than meiotic development (Hartwell *et al.*, 1974; Nurse, 1985a). With respect to (a) it is worth noting that there seems to be a minimum size requirement for entry into meiosis (Calvert and Dawes, 1984a).

The switch to meiotic development for a diploid cell, heterozygous at the *MAT* locus, is normally achieved by starvation for nitrogen on a non-fermentable carbon source (Varma *et al.*, 1985). Cells complete their mitotic cycle and initiate the meiotic cycle from the G1 phase (Hirschberg and Simchen, 1977). An analysis has been made as to whether it is possible for cells to initiate meiotic events from other than the G1 phase (Hirschberg and Simchen, 1977). This was achieved by accumulating diploid cells homozygous for a *cdc* mutation at their terminal phenotype (by raising the temperature) and then switching the cells to starvation conditions at the permissive temperature and asking whether the cells could undergo meiosis directly, or first had to complete the mitotic cycle. The answer was that all *CDC* steps subsequent to *CDC4* required prior completion of mitosis before meiotic development, therefore the point of commitment is at *CDC4* one step after start (see Fig. 21). This is one piece of evidence that has led Nurse (1981b) to regard the *CDC4* step as being within the "start area", but this undermines the strict functional definition that other workers have used to characterize start, particularly by order-of-function mapping (Hereford and Hartwell, 1974). Further aspects of sporulation are discussed in Section III.N.

M. Mechanisms of cell-cycle control

Models can reveal something of the properties of the control mechanisms for initiating events but eventually the hope is that the molecular species involved may be identified. Although this has not been achieved so far, there are a number of promising lines of enquiry leading to experimentally testable predictions.

1. *wee* Mutants in *Schizosaccharomyces pombe*

Genetic evidence (Fantes, 1981) shows that at least three genes are involved in size monitoring. Since *wee 1* alleles are easy to isolate, and loss of function mutants are the most common kind to produce, it is thought that *WEE1* codes for a repressor/inhibitor preventing premature entry into mitosis. Since *cdc2* mutants are easy to isolate and cannot enter mitosis, *CDC2* must have a function required for mitosis. The rare *cdc2-1w* (= *wee2*) allele occurs by a change of function (i.e. increase in activity) causing acceleration into mitosis. The gene *CDC25* is implicated (Fantes, 1979) since *wee1* mutants can be isolated as suppressors of *cdc25* suggesting an interaction between the two

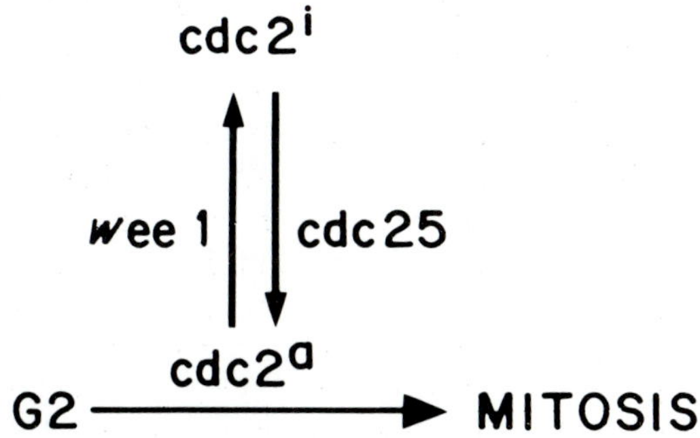

Fig. 28. Genetic model for control of mitosis in *Schizosaccharomyces pombe*. The *cdc2* gene encodes a product necessary for mitosis which is present in both active and inactive forms. The *cdc25* and *wee1* genes control the equilibrium between these two states. Figure provided by Paul Russell.

gene products. Cloning of the genes (Beach *et al.*, 1982) should facilitate analysis of the interactions between them (Nurse and Thuriaux, 1984) which has already led to a model of their relationship (Fig. 28) which is amenable to both *in vivo* and *in vitro* testing (Paul Russell and Paul Nurse, personal communication).

2. Magnesium ions in *Schizosaccharomyces pombe*

Both *Schiz. pombe* and *Kluyveromyces fragilis* are reported to show regular oscillations in the total amount of Mg^{2+} through the cell cycle (Walker and Duffus, 1980), although the selection-synchronized cells used for the experiments did not include proper controls. Return of Mg^{2+} ions to a magnesium-starved culture that had accumulated elongated cells led to synchronous rounds of division and sharp peaks of Mg^{2+} content in the cell. It is proposed that magnesium concentration is the transducer for size-related control of the cell cycle – a "fall in magnesium concentration, accompanying an increase in cell size eventually reaches a level which permits tubulin polymerisation and spindle formation. Once chromosome separation is complete, a rapid influx of magnesium causes breakdown of the spindle, and nuclear and cell division ensue" (Walker and Duffus, 1980). It is not possible to exclude the possibility that the steady-state oscillations are artefacts or effects of rather than the cause of cycle control, and that magnesium-induced synchrony is caused by addition of a cofactor which is now the rate-limiting step in many cellular processes. Perhaps the most intriguing result obtained was a pronounced acceleration of cells into division, obtained by adding the ionophore A23187 as a short pulse late in the cycle to calcium-starved cells. Accelerations have also been seen with the addition of 1% ether (Smith and Mitchison, 1976) in the middle of the cycle. Perhaps both ether and magnesium interact with products that are natural rate-limiting steps for progression (such as *WEE 1*).

3. Spindle pole body in *Saccharomyces cerevisiae*

The spindle pole body is the first structure to alter after entry into the cell-division sequence, undergoing duplication and then separation, locating it as an early cell-cycle event. The most persuasive evidence for relating start to the spindle pole body comes from studies on karyogamy. Nuclear fusion occurs shortly after cell fusion as a result of fusion between the spindle pole bodies of the two nuclei (Byers and Goetsch, 1975). Mutants defective in karyogamy have the property of being dominant to wild-type alleles in heterokaryons and recessive in heterozygotes, consistent with the mutant gene product being part of a nucleus-limited structure, such as the spindle pole body (Conde and Fink, 1976; Fink and Conde, 1977; Dutcher and

Hartwell, 1983b). An analysis of 45 *cdc* mutants (Dutcher and Hartwell, 1982; Bedard *et al.*, 1981) has shown that *cdc4, cdc28, cdc37, cdc60* and *cdc64*, which are involved in start or the first step after start, all show a karyogamy defect. Most interestingly, *cdc4* is dominant to *CDC4* in heterokaryons for the karyogamy defect, but recessive for the cell-division defect, an observation congruent with *CDC4* being a component of the spindle pole body (Dutcher and Hartwell, 1983b).

In addition, the spindle pole body of diploid cells is approximately twice the size of haploid cells, consistent with a "size titrating" function since diploid cells are approximately twice the size of haploid cells (Adams, 1977). Despite this circumstantial evidence that the spindle pole body is involved in an early step of the cycle, there are no data to show that initiation of spindle pole body duplication is the normal rate-determining step of cell proliferation.

4. Processing of ribosomal precursor RNA in *Saccharomyces cerevisiae*

Using compounds that specifically affect ribosomal RNA (rRNA) gene expression, it was shown (G.C. Johnston and Singer, 1978; Singer *et al.*, 1978; Singer and Johnston, 1979; Bedard *et al.*, 1980) that the initiation of new cell cycles could be prevented, leading to G1 arrest. The production of mature rRNA did not correlate with cell-cycle behaviour, rather the G1 arrest response was associated with ribosomal precursor RNA (rpreRNA) and its maturation (G.C. Johnston and Singer, 1980). Cells of *Sacch. cerevisiae* are also responsive to rates of protein synthesis, decreased rates leading to preferential accumulation in G1 (Popolo *et al.*, 1982). Since the stringent response in yeast (Warner and Gorenstein, 1978) leads to decreased rates of protein synthesis, which in turn leads to decreased rates of rpreRNA synthesis, it was of interest to see that G1 arrest can be achieved by impaired rpreRNA without concomitant impaired protein synthesis (G.C. Johnston and Singer, 1978). Therefore, decreased rpreRNA synthesis is a common feature of all metabolically induced G1 arrest.

It is hypothesized (G.C. Johnston and R.A. Singer, personal communication) that the non-conserved portion of the rpreRNA, termed a "coupon", may bind stoichiometrically to a cellular component and accumulation of one cell cycles worth of coupons triggers a new cycle. The attraction of such a model is that, since ribosomes represent a constant fraction of the cell components, titration of ribosomal components, via coupons, would be a convenient measure of growth and cell size. The rRNA genes have been cloned and it should be possible to test the hypothesis by alteration of rRNA production and accurate assay of the processed components.

N. Cyclic AMP in *Saccharomyces cerevisiae*

The latest potential candidate for control is the ubiquitous cyclic 3′,5′-adenosine monophosphate (cyclic AMP), and a persuasive case can be made for its importance in growth control (Thorner, 1982). (i) There is an entire cyclic AMP-based regulatory system in *Sacch. cerevisiae* (Thorner, 1982), the concentration of endogenous cyclic AMP being maintained by the equilibrium between synthesis, by adenylate cyclase (Liao and Thorner, 1980), and degradation, by cyclic nucleotide phosphodiesterase (Liao and Thorner, 1981). (ii) α-Factor inhibits adenylate cyclase activity *in vitro* (Liao and Thorner, 1980). (iii) Addition of exogenous cyclic AMP decreases G1 arrest (Liao and Thorner, 1980). (iv) Mutant *ste5*, which cannot mate and is insensitive to α-factor arrest at 34°C, has an adenylate cyclase activity that is insensitive to α-factor inhibition *in vitro* (Liao and Thorner, 1980). (v) Mutant *ste5* is suppressed by the start mutants *cdc36* and *cdc39* (D. Jenness, personal communication). (vi) Compounds that inhibit cyclic nucleotide phosphodiesterase *in vitro*, such as theophylline, aminophylline and isobutylmethylxanthine (IBMX), prevent α-factor arrest (Liao and Thorner, 1981). (vii) Isobutylmethylxanthine suppresses G1 arrest of *cdc36* (E. McGoldrick) and A. Wheals, unpublished work). (viii) Catabolite derepressing medium, which should increase the concentration of endogenous cyclic AMP (Shuster, 1982a), suppresses start mutants *cdc28, cdc36* and *cdc39* and decreases or abolishes G1 arrest by mating pheromone (Ruiz *et al*., 1984).

However, the most compelling direct evidence for a role for cyclic AMP in the regulation of the cell cycle has been provided by the work of Ishikawa and his colleagues who have genetically investigated the cyclic AMP regulatory system and the phosphorylation of proteins in yeast (Uno *et al*., 1982, 1984; Matsumoto *et al*., 1983a, 1984, 1985a,b). Cells that have a defective adenylate cyclase and cannot synthesize cyclic AMP (*cyr1* mutants, which are allelic to the Class II start mutant *cdc35* (Mortimer and Schild, 1985)) arrest as unbudded cells in the G1 phase of the cell cycle. This is because the regulatory subunit of cyclic AMP-dependent protein kinase, coded for by *CYR3*, binds to and inhibits the catalytic subunit, coded for by *CYR2*. The presence of cyclic AMP leads to the dissociation of catalytic and regulatory subunits and the activation of cyclic AMP-dependent protein kinase. Mutations in the *bcy1* gene produce a deficiency of the *CYR3* gene product leading to the constitutive activation of the catalytic subunit which is thus rendered independent of cyclic AMP. Thus *bcy1* mutants suppress *cyr1* mutants. The important conclusion from this is that the sole function of cyclic AMP in vegetative cells is to activate a cyclic AMP-dependent protein kinase.

This work has been complemented by the exciting discovery of genes in yeasts structurally and functionally homologous to mammalian oncogenes (see Wheals (1985) for review and references; Fukui and Kaziro, 1985;

Section III.E). *Saccharomyces cerevisiae RAS* genes (DeFeo-Jones *et al.*, 1983; Dhar *et al.*, 1984; Powers *et al.*, 1984) are about twice the molecular mass of the mammalian *H-ras* gene which codes for a protein known as p21 *H-ras*. The corresponding gene from *Schiz. pombe* (*SPRAS*) is, at the equivalent of 219 amino acids long, only slightly larger than *H-ras* (Fukui and Kaziro, 1985). For *RAS1, RAS2*, and *SPRAS*, the greatest homology is shown in the first 80 (*N*-terminal) amino acids which are 90% homologous to *H-ras*. It is the *N*-terminal part of the proto-oncogene which is associated with the oncogenic activation of the protein and where oncogenic modifications occur. Disruption of both *Sacch. cerevisiae RAS* genes is lethal (lack of germination or growth of haploid spores) but disruption of one is not, although *RAS2* disruptions lead to failure to utilize effectively non-fermentable carbon sources, hyperaccumulation of storage carbohydrate and sporulation on rich media (Kataoka *et al.*, 1984; Tatchell *et al.*, 1984; Fraenkel, 1985). Cells carrying disruptions in *RAS1* and *RAS2* are phenotypically like *cyr1* mutants in that the intracellular concentrations of cyclic AMP are low, sporulation occurs in nutrient media and they are suppressed by *BCY1*. Replacement of the yeast genes with the mammalian *H-ras* gene or a hybrid leads to spore germination (Kataoka *et al.*, 1985a; DeFeo-Jones *et al.*, 1985).

A reciprocal set of experiments investigating the ability of yeast genes to produce focal transformations of mammalian cells has been performed (DeFeo-Jones *et al.*, 1985). Two general conclusions have emerged from this work, namely(i) unaltered yeast genes do not transform mammalian cells and (ii) in order to produce a transformation, the yeast gene must have an "oncogenic type" amino-acid substitution at the right place and the gene must be shortened in length to that of *H-ras*. Thus, wild-type or mutant (oncogenic) human *ras* genes will function to a greater or lesser extent in yeast; and mutant yeast *RAS* genes will function as oncogenes in mammalian cells. The biochemical function of mammalian p21 protein has been sufficiently conserved for it to operate well in yeast and *vice versa*. It is not surprising therefore to find that the products of all three genes (*RAS1, RAS2* and *H-ras*) bind both GDP and GTP and have guanosine triphosphatase (GTPase) activity (Tamanoi *et al.*, 1984; Temeles *et al.*, 1985).

A mutant yeast *ras2* gene carrying valine at position 19 (*RAS2-val19*), an "oncogenic type" substitution, when put into yeast (which already has *RAS1* and *RAS2*), produces a pleiotropic response – a dominant deficiency in sporulation, failure to arrest uniformly at the beginning of the cell cycle, enhanced trehalase activity, failure to accumulate nutrient reserves like glycogen and trehalose, and cyclic AMP concentration elevated fourfold above the wild-type concentration (Kataoka *et al.*, 1984). The *RAS2-val19* gene produces analogous changes to a *bcy1* mutation (Matsumoto *et al.*, 1982).

The altered concentrations of cyclic AMP and membrane-bound adenylate cyclase in these mutants suggests that *RAS* is required for proper nutrient response, lack of *RAS* activity leading to a premature starvation response. The data are consistent with the hypothesis that strains lacking *RAS* activity cannot enter the cell cycle whereas strains expressing a hyperactive *RAS* gene are constitutive for entry into the cell cycle (Toda *et al.*, 1985). A third gene homologous to *H-ras*, *YP2*, has been discovered. It is a single-copy gene coding for a protein of similar molecular mass to p21 and of 196 amino acids (Gallwitz *et al.*, 1983). Again there is good homology at the *N*-terminal end sufficient to give essentially identical secondary and tertiary structure. Gene-disruption experiments show that *YP2* is required for spore viability and vegetative growth, but, unlike *RAS1* and *RAS2*, *YP2* is not suppressed by *bcy1* so that its mode of action is unlikely to operate through the adenylate cyclase system (Segev and Botstein, 1985).)

Three other genes in addition to the *RAS* genes and *CDC28/CDC2* have also been shown to have homologies with mammalian oncogenes and protein kinases. One is the *Sacch. cerevisiae* start gene *CDC36* (Peterson *et al.*, 1984), one is *CDC4*, which is not a start gene but is the next gene "downstream" from start in the functional sequence map (Peterson *et al.*, 1984) and the last is *CDC7*, which has homology to *CDC28* and thus to the *src* family of oncogenes (John Rosamond, personal communication). The homology is again restricted to the ATP-binding domain and the phosphoreceptor region implying that *CDC7* may also be a protein kinase. Interestingly, *CDC7* is the next gene "downstream" of *CDC4* in the functional sequence map.

These results have implications for growth control and the cell cycle. The start process seems to be the point where the cell monitors its internal and external environments before being committed to a new cycle. Amongst the parameters being measured are nutrient status, presence of mating factors (diffusible pheromones produced to arrest cells of opposite mating type in order to synchronize cell cycles at start) and cell size (the necessity for cells to be above a certain size before they have a high probability of entering the cell cycle). The behaviour of the *RAS* mutants is consistent with the *RAS* genes being involved in nutrient response by regulating the activity of adenylate cyclase. The mammalian adenylate cyclase regulatory system has at least three components: a surface receptor that binds hormones; a catalytic subunit that converts ATP into cyclic AMP, and a protein that couples the two. Binding of a hormone to the receptor stimulates GTP binding which stimulates the catalytic subunit and so cyclic AMP concentration rises. Slow hydrolysis of the GTP to GTPase activity returns it to the inactive form. It has been discovered that both oncogenic *H-ras* and "oncogenic" variants of *RAS1*, have lost their GTPase activity (Temeles *et al.*, 1985). If the direct substrate for both human and yeast *RAS* proteins is the catalytic subunit of adenylate cyclase then loss of the GTPase activity would leave *RAS* in a permanently activated state stimulating cyclic AMP production.

Evidence for a direct involvement of *RAS* proteins in adenylate cyclase activity has come from experiments in which purified *RAS* proteins have been added to membranes prepared from *bcy1 ras1 ras2* cells (Broek *et al.*, 1985). Both human and yeast *RAS* proteins restore the response of adenylate cyclase to magnesium ions and guanine nucleotides. Furthermore, there is a biochemical difference between *RAS2* and *RAS2-val19* proteins in their ability to activate adenylate cyclase after preincubation with GTP. It is possible that normal *RAS* genes are stimulated both by interaction with a surface receptor and concentration of internal GTP. This could imply that the primary internal mediator of nutrient status is GTP and that transmission of this signal to the cell cycle is via cyclic AMP.

If increased concentrations of cyclic AMP are necessary for activation of cyclic AMP-dependent protein kinase then a possible link with the cell cycle can be seen with *CDC28* since its sequence suggests it is a substrate for a protein kinase. These ideas also have implications for the *CDC* genes downstream of start. Although their relationships to protein kinases have not yet always been established, it is possible that the sequential operation of the cell cycle is by a phosphorylation cascade system. Since the actual amount of *CDC* gene products in the cell does not seem to vary much through the cell cycle it is unlikely that the dependency relationship through the cell cycle is due to the sequential appearance of different amounts of protein. The idea of sequential *activation* would be consistent with these observations. However, to maintain dependency of action, there would have to be a degree of specificity in each component of the cascade system.

These results with cyclic AMP mutants also reveal aspects of the decision process of diploid cells whether to sporulate or proliferate. When homozygous, strains of *Sacch. cerevisiae* carrying *cdc25* and *cdc35* (= *cyr1*) have the unusual property of sporulating when placed in acetate growth medium (Shilo *et al.*, 1978a). Both are still subject to glucose repression but are not repressed by nitrogen.

Although the mutants of Matsumoto and coworkers (such as *cyr1*) demonstrate a relationship between cyclic AMP and sporulation they are unable to show if the observed effects are direct or indirect. The conclusion that "cyclic AMP works as a negative effector on the initiation of meiosis via the activation of cyclic AMP dependent protein kinase" (Matsumoto *et al.*, 1983b) implies a direct role for cyclic AMP in the regulation of sporulation. One of the greatest difficulties encountered when working with mutants like *cyr1* is to differentiate between primary effects in which cyclic AMP has a direct regulatory role and secondary effects caused by the inability of *cyr1* cells to continue to grow.

The involvement of cyclic AMP and cyclic AMP-dependent protein kinase in a wide spectrum of metabolic effects is demonstrated by the highly pleiotropic nature of both *cyr1* and *bcy1* cells (Matsumoto *et al.*, 1982). The existence of mutations not implicated in cyclic AMP metabolism that lead to similar phenotypes should serve to exemplify the caution needed in differentiating primary and secondary effects. Dawes (1975) has isolated a series of mutants that lead to sporulation on rich media. These *spd* (sporulation deficient) mutants show many similarities to the *cyr1* mutation of Matsumoto *et al.* (1982). For both, haploids arrest in G1 as nutrient-starved cells. Both show similar patterns of sporulation on media containing non-fermentable carbon sources whilst retaining sensitivity to glucose repression of sporulation (Vezinhet *et al.*, 1979). Further evidence that *cyr1* and *spd1* are phenotypically, though not genetically, similar is demonstrated by the remarkable similarity of phenotypes shown by suppressor mutations of these genes. Both *bcy1*, and the *spd1* suppressing *spo* mutants (Calvert and Dawes, 1984b), are sporulation deficient; both suppress the ability to G1 arrest under nutrient starvation; both lose viability rapidly under starvation conditions and neither shows commitment to premeiotic DNA synthesis or intragenic recombination under restrictive conditions. The phenotype of the *spd1* mutation has been attributed to a defect in a "central metabolic function involving some, but not all, non-fermentable carbon sources" (Dawes, 1983), a secondary effect of which is to alter nitrogen repression of sporulation. Vezinhet *et al.* (1979) state that the *spd1* mutants are "probably not affected directly in events unique to the initiation of sporulation", indicating them to be secondary consequences of mutations that affect inputs to start via growth. The physiology of *spd1* and *spo1* mutants leads Calvert and Dawes (1984b) to suggest that they relate to the "fundamental and complex regulatory connection between respiration and sporulation". Considering the phenotypic similarities between *cyr1* and *spd1*, and *bcy1* and *spo1*, it is possible that the explanation offered to account for the sporulating properties of *spd1* and *spo1* can be extended to *cyr1* and *bcy1*.

Rather than a direct role, the evidence that Matsumoto and coworkers present supports the suggestion that the altered meiotic features of the *cyr1* and *bcy1* cells can be explained in terms of a secondary metabolic defect leading to nitrogen starvation deregulation. Wild-type diploid yeast cells undergo G1 arrest in response to failing nutrient supply. They remain in G2 until the nitrogen supply becomes inadequate. Thus it is the lack of nitrogen rather than G1 arrest that triggers commitment to meiosis.

A comparison of temperature-sensitive *cyr1-2/cyr1-2* diploids at permissive and restrictive temperatures confirms the low concentration of cyclic AMP associated with sporulation (Uno *et al.*, 1985). Mostly two-spored asci are formed from meioses with defective spindle pole bodies lacking the outer plaque. Intriguingly, partial GTP deprivation also initiates meiosis and sporulation (Varma *et al.*, 1985). There is thus mirror-image symmetry between high concentrations of cyclic AMP and GTP being associated with proliferation and low concentrations with sporulation (see Section III.N). Mutants of *Schiz. pombe* have also been isolated which can sporulate in nitrogen-containing growth medium. The independent isolates of *ran1* (Nurse, 1985b) and *pat1* (Iino and Yamamoto, 1985) are allelic. The difference between these *ran/pat* mutants and the *Sacch. cerevisiae* mutants is that the *ran* mutants can sporulate even if haploid or homozygous at the mating-type locus!

An explicit model in which cyclic AMP plays a major role in the integration of the various inputs to and outputs from start (McGolorick, 1986) is shown in Fig. 29 (see also Matsumoto *et al.*, 1985a; Jacquet and Camonis, 1985). Cyclic AMP, acting through cyclic AMP-dependent protein kinase, would be responsible for the regulation of specific proteins in a cascade type system. The basis of the model is that effectors decreasing the concentration of cyclic AMP lead to G1 arrest. However, although a high concentration of cyclic AMP is necessary to ensure traverse of start it is not sufficient. Thus cyclic AMP acts as a master switch capable of preventing start in the absence of favourable conditions. Cyclic AMP is a metabolite that has been implicated in the regulation of all pathways from start. Its ability to stimulate protein kinase accounts for the pleiotropic effects that such an "indicator" molecule must have. Figure 29 is a diagrammatic representation of the model. Centrally and on a vertical axis is placed cyclic AMP and changes in its concentration. Decreased concentrations lead to G1 arrest and increased concentrations lead to traverse of start via the activation of cyclic AMP-dependent protein kinase and consequential protein phosphorylation. A number of the enzymes involved in cyclic AMP metabolism (in its widest sense) are shown. On this framework we have also attempted to fit the data pertaining to many of the mutations affecting cell-cycle regulation, the majority of which can be unambiguously assigned a position. Figure 29 should serve as a useful reference since it indicates, where known, the nature of the mutations and may serve to allow predictions to be made concerning interactions between the different mutations.

The essence of the model is that the major (but not the only) determinant of cellular development is the *in vivo* concentration of cyclic AMP. External and internal inputs mediate their effects via alterations to the concentration of cyclic AMP. In turn the concentration of cyclic AMP affects the degree of activation of cyclic AMP-dependent protein kinase(s), which in turn activates other protein kinases downstream. One of these steps is probably involved in the completion of the start events leading to initiation of the cell cycle proper. The least well characterized stage is the link to the traverse of start.

The high/go, low/stop model is ultimately dependent, not on the concentration of cyclic AMP, but on the activity of cyclic AMP-dependent protein kinase. There is evidence to suggest that the activity of this phos-

The "High/Go Low/Stop Model"

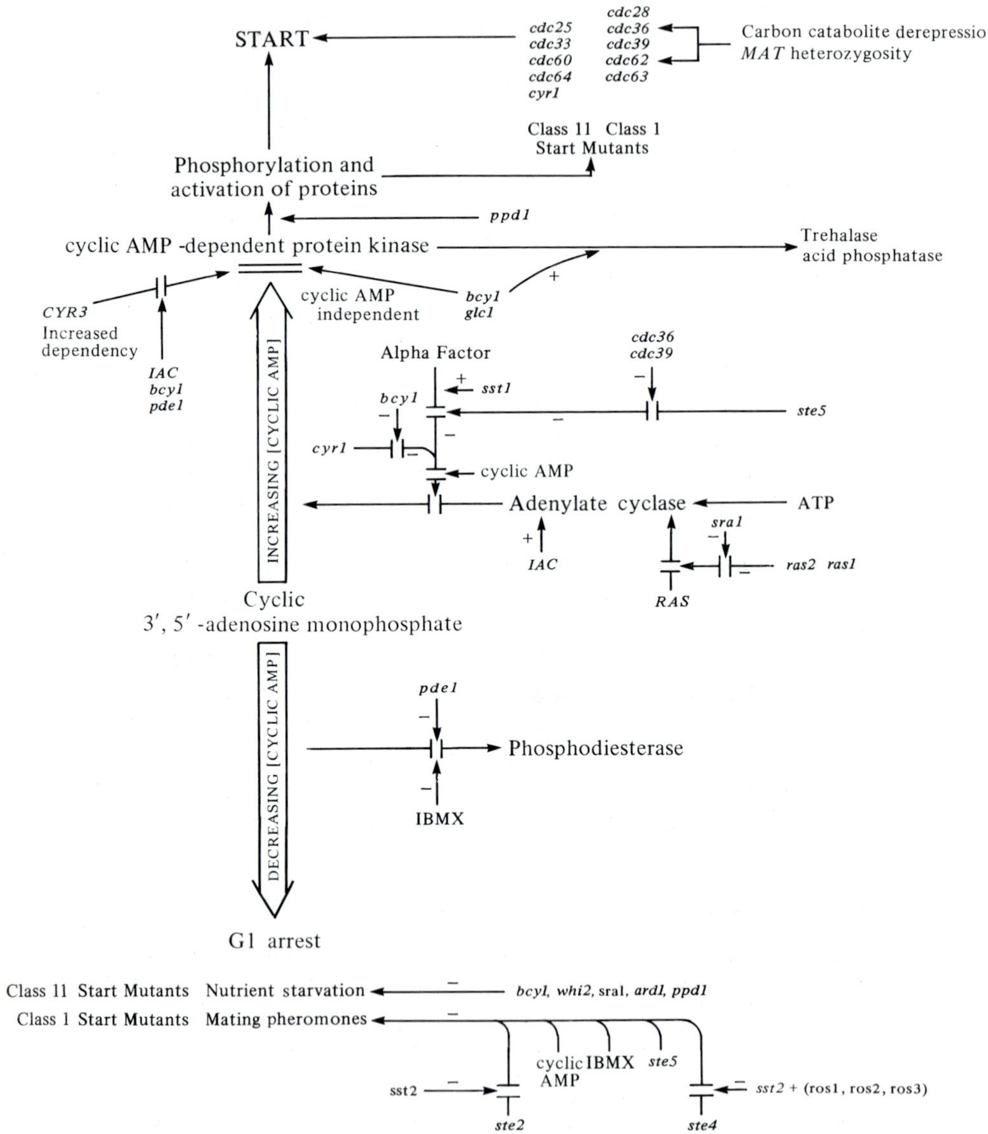

phorylating enzyme is affected by other metabolites. Uno *et al.* (1982), indicate that high concentrations of cyclic IMP are partially capable of supporting the growth of *CYR3* mutants in the absence of cyclic AMP, presumably by playing the role of a molecule capable of dissociating the regulatory and catalytic subunits of cyclic AMP-dependent protein kinase. Although both cyclic IMP and cyclic GMP are effective competitors of cyclic AMP for its site on protein kinase, the activation constant for cyclic IMP is two hundred times greater than for cyclic AMP (Uno *et al.*, 1982). These figures suggest that other cyclic nucleotides may act as effective inhibitors of cyclic AMP-dependent protein kinase by restricting cyclic AMP access to sites on the regulatory subunit. In this scheme it is the effective concentration of cyclic AMP, that is, its concentration relative to inhibitors rather than its absolute concentration, that is important. Regulation by two or more metabolites increases the resolution of the system, allowing the cell to inactivate cyclic AMP-dependent protein kinase by either decreasing the concentration of cyclic AMP or increasing its competitive inhibitors.

The clear differences between the G1 arrest imposed by nutrient starvation and that imposed by mating pheromones cannot be accommodated in a model in which a single compound has sole responsibility. Consequently, although cyclic AMP may determine the phosphorylating ability of cyclic AMP-dependent protein kinase, it is the nature of the substrates available to the kinase that determines both the form of arrest and the course taken on its release. To date, two enzymes have been identified as substrates for cyclic AMP-dependent protein kinase. Trehalase, the enzyme that catalyses the conversion of trehalose into glucose, is activated by phosphorylation (Uno *et al.*, 1984). This explains the failure of mutants with an overactive protein kinase to accumulate carbohydrate. Acid phosphatase is the second proven substrate (Matsumoto *et al.*,1985a); its ability to cleave phosphate groups make it part of the phosphorylation/dephosphorylation system regulated by glucose. The isolation of a strain deficient in phosphoprotein phosphatase 1

Fig. 29. The "High/Go, Low/Stop Model". The diagram depicts factors affecting cyclic AMP-dependent protein kinases and an enhanced level of protein phosphorylation of proteins, some of which are involved in subsequent traverse of start. Conversely, a decreased concentration of cyclic AMP leads to G1 arrest. Mutations affecting cyclic AMP metabolism or cell development are shown. Arrows marked with a positive sign indicate a stimulatory or synergistic effect; those marked with a negative sign have inhibitory or antagonistic action on arrows crossed with a double bar. See the text for further explanation. A brief description of most of the mutants is given in the Appendix. Those not included are: *ard1*, arrest deficient (Malcolm Whiteway, personal communication); *bcy1*, mutation causing a cyclic AMP-dependent protein kinase to become cyclic AMP independent (Matsumoto *et al.*, 1985a); *glc1*, deficient in glycogen and trehalose accumulation (Ortiz *et al.*, 1983); *IAC*, increased adenylate cyclase activity (Uno *et al.*, 1983); IBMX, isobutylmethylxanthine, an inhibitor of phosphodiesterase (Liao and Thorner, 1981); *pde*, reduced phosphodiesterase I activity (Uno *et al.*, 1983); *ppd1*, deficient in phosphoprotein phosphatase activity (Matsumoto *et al.*, 1985b); RAS, the yeast *RAS* proteins (Wheals, 1985); *ros1, 2, 3*, relaxation of sterility, *ros1* allelic to *cdc39* (D. Jenness, personal communication); *sra1*, suppressor of *ras*, (K. Tatchell, personal communication).

(*ppd1*) (Matsumoto *et al.*, 1985b), has allowed an investigation of the relative importance of phosphorylation and dephosphorylation in the cell cycle. Since mutants with decreased phosphorylation ability (*cyr1, cyr2* and *CYR3*) G1 arrest prematurely, whereas those with enhanced ability have deregulated proliferation (*bcy1, IAC, pde1* and *ppd1*), Uno and coworkers concluded that "phosphorylation of cellular proteins is indispensable for cell cycle, but dephosphorylation is not". Although this statement is correct if cell cycle is taken to mean "continued mitotic cycle" the characteristics of enhanced phosphorylators clearly indicate that control over both phosphorylation and dephosphorylation is necessary for the regulation and integration of the component pathways that define the cell cycle. Cloning of the relevant genes (Caspersson *et al.*, 1985) should enable a more precise molecular model to be built and tested. As a final cautionary tale it should be noted that mutations of the *SPRAS* gene of *Schiz.* pombe have no effect on adenylate cyclase activity (Fukui *et al.*, 1986).

O. Conclusions and speculations

A few very general conclusions may be made on the studies described above.

1. Most cellular processes are continuous.
2. The periodic events are associated with the replication and partitioning of the nuclear genetic material, and the division of a cell into two.
3. The periodic events occur in a defined order because the gene products associated with their execution are dependent for their function on the execution of previous events.
4. The cell cycle may be more accurately described as a periodically repeating linear set of events with one or two major rate-determining steps for their initiation and completion.
5. Growth and division are co-ordinated because the major rate-determining steps are dependent on the growth of the cell for their initiation.
6. There are spatial as well as temporal controls on cell-cycle development.

Possible future developments include:

1. Cloning of the *CDC* genes to determine something of the nature of their gene product and eventually the amount, activity and function of that gene product *in vivo*.
2. The development of *in vitro* systems to mimic the major *in vivo* events of DNA synthesis and mitosis, enabling assays to be made of *CDC* gene products (Broach, 1986).
3. Studies on the interaction of *CDC* gene products *in vitro* particularly from those genes that are known to interact *in vivo*.

4. The isolation of mutants specifically affecting morphogenesis.
5. Investigations of the biology of the cell using cloned genes with different gene dosage and variable gene product output *in vivo*.
6. The increased use of trans-specific complementation tests to determine gene homology.
7. Investigations on chromatin structure as a possible candidate for temporal control of gene activity.
8. The use of nuclear magnetic resonance to investigate the biochemistry of *CDC* gene products (Broach, 1986).
9. The increased use of "reverse genetics" to isolate novel "*CDC*" mutants and their function.

So far there is a reasonable natural history of the yeast cell cycle. The next decade should reveal something of the molecular biology of the process.

IV. ACKNOWLEDGEMENTS

I thank those colleagues, too numerous to mention individually, who sent me reprints and preprints. I especially thank Eamon McGoldrick, Gerry Johnston, John Ludwig and Murdoch Mitchison for useful discussions and Jeremy Thorner, Anna Gosney and Peter Sudbery for photographs. Many thanks also to Marie-Claire Oliver for her diligent typing.

References

Abe, K., Fukui, S. and Kusanagi, A. (1984). *Journal of General and Applied Microbiology* **30**, 257.
Adams, A.E. and Pringle, J.R. (1984). *Journal of Cell Biology* **98**, 934.
Adams, J. (1977). *Experimental Cell Research* **106**, 267.
Ahluwalia, B., Duffus, J.H., Paterson, L.J. and Walker, G.M. (1978). *Journal of General Microbiology* **106**, 261.
Arendes, J., Kim, K.C. and Sugino, A. (1983). *Proceedings of the National Academy of Sciences of the United States of America* **80**, 673.
Arnold, E. and Tanner, W. (1982). *FEBS Letters* **148**, 49.
Aves, S.J., Durkacz, B., Carr, A. and Nurse, P. (1985). *EMBO Journal* **4**, 457.
Bacchetti, S., Casandro, M. and Maro, F. (1967). *Experimental Cell Research* **46**, 292.
Baldwin, W.W. and Kubitschek, H.E. (1984). *Journal of Bacteriology* **158**, 701.
Baranowska, H., Zaborowska, D. and Zuk, H. (1985). *Current Genetics* **9**, 447.
Barford, J.P. (1985). *Journal of General and Applied Microbiology* **31**, 87.
Barford, J.P. and Hall, R.J. (1976). *Experimental Cell Research* **102**, 276.
Barker, D.G. and Johnston, L.H. (1983). *European Journal of Biochemistry* **134**, 315.
Barker, D.G., Johnson, A.L. and Johnston, L.H. (1985). *Molecular and General Genetics* **200**, 458.
Barnes, A., Nurse, P. and Fraser, R.S.S. (1979). *Journal of Cell Science* **35**, 41.
Bayne-Jones, S. and Adolph, E.F. (1932). *Journal of Cellular and Comparative Physiology* **1**, 387.
Beach, D., Durkacz, B. and Nurse, P. (1982). *Nature, London* **300**, 706.
Beck, C. and von Meyenburg, H.K. (1968). *Journal of Bacteriology* **96**, 479.
Bedard, D.P., Singer, R.A. and Johnston, G.C. (1980). *Journal of Bacteriology* **141**, 100.
Bedard, D.P., Johnston, G.C. and Singer, R.A. (1981). *Current Genetics* **4**, 205.
Bedell, G.W., Werth, A. and Soll, D.R. (1980). *Experimental Cell Research* **127**, 103.

Benitez, T., Nurse, P. and Mitchison, J.M. (1980). *Journal of Cell Science* **46**, 399.
Betz, R. and Duntze, W. (1979). *European Journal of Biochemistry* **95**, 469.
Birkenmeyer, L.G., Hill, J.C. and Dumas, L.B. (1984). *Molecular and Cellular Biology* **4**, 583.
Bisson, L. and Thorner, J. (1977). *Journal of Bacteriology* **132**, 44.
Blumenthal, L.K. and Zahler, S.A. (1962). *Science* **135**, 724.
Bostock, C.J., Donachie, W.D., Masters, M. and Mitchison, J.M. (1966). *Nature, London* **210**, 808.
Braun, P.C. and Calderone, R.A. (1978). *Journal of Bacteriology* **135**, 1472.
Brendel, M. and Fäth, W.W. (1974). *Zeitschrift für Naturforschung* **29c**, 733.
Brewer, B.J., Chlebowicz-Sledziewska, E. and Fangman, W.L. (1984). *Molecular and Cellular Biology* **4**, 2529.
Broach, J.R. (1981). *In* "The Molecular Biology of the Yeast, *Saccharomyces*" (J. Strathern, E. Jones and J. Broach, eds), vol. 1, pp. 445–470. Cold Spring Harbor Press, Cold Spring Harbor.
Broach, J.R. (1986). *Cell* **44**, 3.
Broek, D., Samiy, N., Fasano, O., Fujiyama, A., Tamanoi, F., Northup, J and Wigler, M. (1985). *Cell* **41**, 763.
Brooks, R.F. (1981). *In* "The Cell Cycle" (P.C.L. John, ed.), pp. 35–61. Cambridge University Press, Cambridge.
Brummel, M. and Soll, D.R. (1982). *Developmental Biology* **89**, 211.
Brunborg, G. and Williamson, D.H. (1978). *Molecular and General Genetics* **162**, 277.
Brunborg, G., Resnick, M. and Williamson, D.H. (1980). *Radiation Research* **82**, 547.
Bücking-Throm, E., Duntze, W., Hartwell, L.H. and Manney, T.R. (1973). *Experimental Cell Research* **76**, 99.
Bugeja, V.C. and Carter, B.L.A. (1982). *Journal of Bacteriology* **151**, 1051.
Bugeja, V.C., Piggott, J.R. and Carter, B.L.A. (1982). *Journal of General Microbiology* **128**, 2707.
Bullock, J.G. and Coakley, W.T. (1979). *Experimental Cell Research* **121**, 441.
Burke, W. and Fangman, W.L. (1975). *Cell* **5**, 263.
Burns, V.W. (1956). *Journal of Cellular and Comparative Physiology* **47**, 357.
Byers, B. (1981a). *In* "Molecular Biology of the Yeast, *Saccharomyces*" (J. Strathern, E. Jones and J. Broach, eds), vol. 1, pp. 59–96. Cold Spring Harbor Press, Cold Spring Harbor.
Byers, B. (1981b). *In* "Alfred Benzon Symposium; Molecular Genetics of Yeast" (D. von Wettstein, J. Friis, M. Kielland-Brandt and A. Stenderup, eds), vol. 16, pp. 119–133. Munksgaard, Copenhagen.
Byers, B. and Goetsch, L. (1974). *Cold Spring Harbor Symposium on Quantitative Biology* **38**, 123.
Byers, B. and Goetsch, L. (1975). *Journal of Bacteriology* **124**, 511.
Byers, B. and Goetsch, L. (1976a). *Journal of Cell Biology* **69**, 717.
Byers, B. and Goetsch, L. (1976b). *Journal of Cell Biology* **70**, 35a.
Byers, B. and Sowder, L. (1980). *Journal of Cell Biology* **87**, 6a.
Cabib, E. (1975). *Annual Review of Microbiology* **29**, 191.
Cabib, E. and Bowers, B. (1975). *Journal of Bacteriology* **124**, 1586.
Cabib, E., Roberts, R. and Bowers, B. (1982). *Annual Review of Biochemistry* **51**, 763.
Calleja, G.B., Zuker, M., Johnson, B.F. and Yoo, B.Y. (1980). *Journal of Theoretical Biology* **84**, 523.
Calvert, G.R. and Dawes, I.W. (1984a). *Nature, London* **312**, 61.
Calvert, G.R. and Dawes, I.W. (1984b). *Journal of General Microbiology* **130**, 615.
Carson, M.J. and Hartwell, L.H. (1985). *Cell* **42**, 249.
Carter, B.L.A. and Halvorson, H.O. (1973). *Experimental Cell Research* **76**, 152.
Carter, B.L.A. and Jagadish, M.N. (1978a). *Experimental Cell Research* **112**, 15.
Carter, B.L.A. and Jagadish, M.N. (1978b). *Experimental Cell Research* **112**, 373.
Carter, B.L.A. and Sudbery, P.E. (1980). *Genetics* **96**, 561.
Carter, B.L.A., Lorincz, A. and Johnston, G.C. (1978). *Journal of General Microbiology* **106**, 221.
Carter, B.L.A., Piggott, J.R. and Walton, E.F. (1982). *In* "Yeast Genetics" (J.F.T. Spencer, D.M. Spencer and A.R.W. Smith, eds), pp. 1–28. Springer-Verlag, New York.

Casperson, G.F., Walker, N. and Bourne, H.R. (1985). *Proceedings of the National Academy of Sciences of the United States of America* **82**, 5060.
Chaffin, W.L. (1984a). *Journal of General Microbiology* **130**, 431.
Chaffin, W.L. (1984b). *Canadian Journal of Microbiology* **30**, 192.
Chaleff, D.T. and Tatchell, K. (1985). *Molecular and Cellular Biology* **5**, 1878.
Chan, R.K. (1977). *Journal of Bacteriology* **130**, 766.
Chan, R.K. and Otte, C.A. (1982). *Molecular and Cellular Biology* **2**, 11.
Chan, R.K., Melnick, L.M., Blair, L.C. and Thorner, J. (1983). *Journal of Bacteriology* **155**, 903.
Chanet, R., Waters, R. and Moustacchi, E. (1975). *International Journal of Radiation Biology* **27**, 481.
Chattaway, F.W., Holmes, M.R. and Barlow, A.J.E. (1968). *Journal of General Microbiology* **51**, 367.
Chung, K.L., Hawirko, R.Z. and Isaac, P.K. (1965). *Canadian Journal of Microbiology* **11**, 953.
Conde, J. and Fink, G.R. (1976). *Proceedings of the National Academy of Sciences of the United States of America* **73**, 3651.
Connolly, B.M., Bugeja, V.C., Piggot, J.R. and Carter, B.L.A. (1983). *Current Genetics* **7**, 309.
Conrad, M.N. and Newlon, C.S. (1983a). *Molecular and Cellular Biology* **3**, 1000.
Conrad, M.N. and Newlon, C.S. (1983b). *Molecular and Cellular Biology* **3**, 1665.
Cooper, S. (1979). *Nature, London* **280**, 17.
Cooper, S. (1982). *In* "Cell Growth" (C. Nicolini, ed.), pp. 315–336. Plenum Press, New York.
Cooper, T.G., Britton, C., Brand, L. and Sumrada, R. (1979). *Journal of Bacteriology* **137**, 1447.
Cottrell, S.F. (1979). *Experimental Cell Research* **118**, 398.
Cottrell, S.F. (1981). *Experimental Cell Research* **132**, 89.
Cottrell, S.F. (1982). *Cell Biology International Reports* **6**, 125.
Cottrell, S.F. and Lee, L.H. (1981). *Biochemical and Biophysical Research Communications* **101**, 1350.
Cottrell, S.F., Getz, G.S. and Rabinowitz, M. (1981). *Journal of Biological Chemistry* **256**, 10973.
Creanor, J. (1978a). *Journal of Cell Science* **33**, 385.
Creanor, J. (1978b). *Journal of Cell Science* **33**, 399.
Creanor, J. and Mitchison, J.M. (1979). *Journal of General Microbiology* **112**, 385.
Creanor, J. and Mitchison, J.M. (1982). *Journal of Cell Science* **58**, 263.
Creanor, J. and Mitchison, J.M. (1984). *Journal of Cell Science* **69**, 199.
Creanor, J., Elliott, S.G., Bisset, Y.C. and Mitchison, J.M. (1983). *Journal of Cell Science* **61**, 339.
Culotti, J. (1974). PhD Thesis: University of Washington, Seattle.
Culotti, J. and Hartwell, L.H. (1971). *Experimental Cell Research* **67**, 389.
Dawes, I.W. (1975). *Nature, London* **255**, 707.
Dawes, I.W. (1983) *In* "Yeast Genetics" (J.F.T. Spencer, D.M. Spencer and A.R.W. Smith, eds), pp. 29–64. Springer-Verlag, New York.
Dawes, I.W. and Carter, B.L.A. (1974). *Nature, London* **250**, 709.
Dawes, I.W. and Calvert, G.R. (1984). *Journal of General Microbiology* **130**, 605.
Dawson, P.S.S. (1965). *Canadian Journal of Microbiology* **11**, 893.
DeFeo-Jones, D., Scolnick, E.M., Koller, R. and Dhar, R. (1983). *Nature, London* **306**, 707.
DeFeo-Jones, D., Tatchell, K., Robinson, L.C., Sigal, I.S., Vass, W.C., Lowy, D.R. and Scolnick, E.M. (1985). *Science* **228**,, 179.
Delgado, M.A. and Conde, J. (1984). *Molecular and General Genetics* **193**, 188.
Del Giudice, L., Wolf, K., Mann, F. and Pagliuca, N. (1981). *Molecular and General Genetics* **182**, 252.
Del Rey, F., Santos, T., Garcia-Acha, I. and Nombeta, C. (1979). *Journal of Bacteriology* **139**, 924.
Deutch, C.E. and Parry, J.M. (1974). *Journal of General Microbiology* **80**, 259.
Dhar, R., Nieto, A., DeFeo-Jones, D. and Scolnick, E.M. (1984). *Nucleic Acids Research* **12**, 3611.
Dickinson, D.P. (1981). *Journal of Cell Science* **51**, 203.
Dickinson, J.R. (1981). *European Journal of Biochemistry* **119**, 341.
Dickinson, J.R. (1983). *Journal of Cell Science* **60**, 355.
Dinardo, S., Voelkel, K. and Sternglanz, R. (1984). *Proceedings of the National Academy of Sciences of the United States of America* **81** 2616.

Doi, A. and Doi, K. (1982). *Experimental Cell Research* **137**, 95.
Dominguez, A., Varona, R.M., Villanueva, J.R. and Sentandreu, R. (1982). *Antonie van Leeuwenhoek* **48**, 145.
Dubeau, H. and Chung, Y.S. (1984). *Molecular and General Genetics* **197**, 120.
Dudani, A.K., Trivedi, A. and Prasad, R. (1983). *FEBS Letters* **153**, 34.
Duffus, J.H. and Paterson, L.J. (1974). *Nature, London* **251**, 626.
Dumas, L.B., Lassky, J.P., McFarland, E.J. and Shampay, J. (1982). *Molecular and General Genetics* **187**, 42.
Dutcher, S.K. (1982). *Genetics* **102**, 9.
Dutcher, S.K. and Hartwell, L.H. (1982). *Genetics* **100**, 175.
Dutcher, S.K. and Hartwell, L.H. (1983a). *Cell* **33**, 203.
Dutcher, S.K. and Hartwell, L.H. (1983b). *Molecular and Cellular Biology* **3**, 1255.
Dwek, R.D., Kobrin, L.H., Grossman, N. and Ron, E.Z. (1980). *Journal of Bacteriology* **144**, 17.
Edwards, D.R., Taylor, J.B., Wakeling, W.F., Watts, F.Z. and Johnston, I.R. (1978). *Cold Spring Harbor Symposium on Quantitative Biology* **43**(1), 577.
Edwards, S.W. and Lloyd, D. (1977). *Biochemical Journal* **162**, 39.
Elliott, S.G. (1983a). *Molecular and General Genetics* **192**, 204.
Elliott, S.G. (1983b). *Molecular and General Genetics* **192**, 212.
Elliott, S.G. and McLaughlin, C.S. (1978). *Proceedings of the National Academy of Sciences of the United States of America* **75**, 4384.
Elliott, S.G. and McLaughlin, C.S. (1979a). *Journal of Bacteriology* **137**, 1185.
Elliott, S.G. and McLaughlin, C.S. (1979b). *Molecular and General Genetics* **169**, 237.
Elliott, S.G. and McLaughlin, C.S. (1983). *Progress in Nucleic Acid Research and Molecular Biology* **28**, 146.
Elliott, S.G., Warner, J.R. and McLaughlin, C.S. (1979). *Journal of Bacteriology* **137**, 1048.
Fabre, F. (1973). *Radiation Research* **56**, 528.
Fangman, W.L., Hice, R.H. and Chlebowicz-Sledziewska, E. (1983). *Cell* **32**, 831.
Fantes, P.A. (1977). *Journal of Cell Science* **24**, 51.
Fantes, P.A. (1979). *Nature, London* **279**, 428.
Fantes, P.A. (1981). *Journal of Bacteriology* **146**, 746.
Fantes, P.A. (1982). *Journal of Cell Science* **55**, 383.
Fantes, P.A. (1983). *Nature, London* **302**, 153.
Fantes, P.A. (1984). *In* "The Microbial Cell Cycle" (P.M. Nurse and E. Streiblová, eds), pp. 109–125. CRC Press, Boca Raton, Florida.
Fantes, P.A. and Nurse, P.M. (1977a). *Experimental Cell Research* **107**, 377.
Fantes, P.A. and Nurse, P.M. (1977b). *Nature, London* **267**, 647.
Fantes, P.A. and Nurse, P.M. (1978). *Experimental Cell Research* **115**, 317.
Fantes, P.A. and Nurse, P.M. (1981). *In* "The Cell Cycle" (P.C.L. John, ed.), pp. 11–33. Cambridge University Press, Cambridge.
Fantes, P.A., Grant, W.D., Pritchard, R.H., Sudbery, P.E. and Wheals, A.E. (1975). *Journal of Theoretical Biology* **50**, 213.
Farkaš, V., Kovarik, J., Košinová, A. and Bamer, Š. (1974). *Journal of Bacteriology* **117**, 265.
Fernandez, M.P., Correa, J.U. and Cabib, E. (1982). *Journal of Bacteriology* **152**, 1255.
Field, C. and Schekman, R. (1980). *Journal of Cell Biology* **86**, 123.
Fields, S. and Herskowitz, I. (1985), *Cell* **42**, 923.
Fingerhut, R., Kiefer, J. and Otto, F. (1984). *Molecular and General Genetics* **193**, 192.
Fink, G.R. and Conde, J. (1977). *In* "International Cell Biology" (B.R. Brinkley and K.R. Porter, eds), pp. 414–419. Rockefeller University Press, New York.
Flegel, T.W. (1978). *Canadian Journal of Microbiology* **24**, 827.
Fraenkel, D. (1985). *Proceedings of the National Academy of Sciences of the United States of America* **82**, 4740.
Fraser, R.S.S. and Barnes, A. (1983). *Journal of Cell Science* **62**, 187.
Fraser, R.S.S. and Carter, B.L.A. (1976). *Journal of Molecular Biology* **164**, 827.
Fraser, R.S.S. and Moreno, F. (1976). *Journal of Cell Science* **21**, 497.
Fraser, R.S.S. and Nurse, P.M. (1978). *Nature, London* **271**, 726.

Fraser, R.S.S. and Nurse, P.M. (1979). *Journal of Cell Science* **35**, 25.
Freifelder, D. (1960). *Journal of Bacteriology* **80**, 567.
Fried, J. and Mandel, M. (1979). *Computer Programs in Biomedicine* **10**, 218.
Fukui, Y. and Kaziro, Y. (1985). *EMBO Journal* **4**, 687.
Fukui, Y., Kozasa, T., Kaziro, Y., Takeda, T. and Yamamoto, M. (1986). *Cell* **44**, 329.
Galdiero, F. (1973). *Experientia* **29**, 496.
Gallwitz, D., Donath, C. and Sander, C. (1983). *Nature, London* **306**, 704.
Game, J.C. (1976). *Molecular and General Genetics* **146**, 313.
Gillies, R.J. and Benoit, A.G. (1983). *Biochimica et Biophysica Acta* **762**, 466.
Gordon, C.N. (1977). *Journal of Cell Science* **24**, 81.
Gordon, C.N. and Elliott, S.G. (1977). *Journal of Bacteriology* **129**, 97.
Gulløv, K., Friis, J. and Bonven, B. (1981). *Experimental Cell Research* **295**, 304.
Gygax, A. and Thuriaux, P. (1984). *Current Genetics* **8**, 85.
Hagen, D.C. and Sprague, G.P. (1984). *Journal of Molecular Biology* **178**, 835.
Halvorson, H.O., Carter, B.L.A. and Tauro, P. (1970). *Advances in Microbial Physiology* **6**, 47.
Hamburger, K., Kramhøft, B., Nissen, S.B. and Zeuthen, E. (1977). *Journal of Cell Science* **24**, 69.
Hanic-Joyce, P.J. (1985). *Genetics* **110**, 591.
Hartwell, L.H. (1970). *Journal of Bacteriology* **104**, 1280.
Hartwell, L.H. (1971a). *Journal of Molecular Biology* **59**, 183.
Hartwell, L.H. (1971b). *Experimental Cell Research* **69**, 265.
Hartwell, L.H. (1973a). *Journal of Bacteriology* **115**, 966.
Hartwell, L.H. (1973b). *Experimental Cell Research* **76**, 111.
Hartwell, L.H. (1974). *Bacteriological Reviews* **38**, 164.
Hartwell, L.H. (1976). *Journal of Molecular Biology* **104**, 803.
Hartwell, L.H. (1978). *Journal of Cell Biology* **77**, 627.
Hartwell, L.H. (1980). *Journal of Cell Biology* **85**, 811.
Hartwell, L.H. and Smith, D. (1985). *Genetics* **110**, 381.
Hartwell, L.H. and Unger, M.W. (1977). *Journal of Cell Biology* **75**, 422.
Hartwell, L.H., Culotti, J. and Reid, B. (1970). *Proceedings of National Academy of Sciences of the United States of America* **66**, 352.
Hartwell, L.H., Mortimer, R.K., Culotti, J. and Culotti, M. (1973). *Genetics* **74**, 267.
Hartwell, L.H., Culotti, J., Pringle, J.R. and Reid, B.J. (1974). *Science* **183**, 46.
Hatzfield, J. and Williamson, D.H. (1974). *Experimental Cell Research* **84**, 431.
Hayashibe, M. and Katohda, S. (1973). *Journal of General and Applied Microbiology* **19**, 23.
Hayashibe, M., Abe, N. and Matsui, M. (1977). *Archives of Microbiology* **114**, 91.
Hereford, L.M. and Hartwell, L.H. (1973). *Nature New Biology* **244**, 129.
Hereford, L.M. and Hartwell, L.H. (1974). *Journal of Molecular Biology* **84**, 445.
Hereford, L.M., Osley, M.A., Ludwig, J.R. and McLaughlin, C.S. (1981). *Cell* **24**, 367.
Hereford, L.M., Bromley, S. and Osley, M.A. (1982). *Cell* **30**, 305.
Herman, M.A. and Soll, D.R. (1984). *Journal of General Microbiology* **130**, 2219.
Herskowitz, I. and Oshima, Y. (1981). *In* "The Molecular Biology of the Yeast, *Saccharomyces*" (J. Strathern, E. Jones and J. Broach, eds), vol. 1, pp. 181–209. Cold Spring Harbor Press, Cold Spring Harbor.
Hindley, J. and Phear, G.A. (1984). *Gene* **31**, 129.
Hirschberg, J. and Simchen, G. (1977). *Experimental Cell Research* **105**, 245.
Holm, C. (1982). *Cell* **29**, 585.
Hossack, J.A., Rose, A.H. and Dawson, P.S. (1979). *Journal of General Microbiology* **113**, 199.
Howard, A. and Pelc, S.R. (1953). *Heredity, London (Supplement)* **6**, 261.
Huffaker, T. and Robbins, P. (1982). *Journal of Biological Chemistry* **257**, 3203.
Hughes, T.E. (1983). *Nature, London* **302**, 286.
Hutter, K.J., Gortz, T. and Eipel, H.E. (1978). *FEMS Microbiology Letters* **3**, 291.
Hynes, N.E. and Phillips, S.L. (1976). *Journal of Bacteriology* **128**, 502.
Iida, H. and Yahara, I. (1984a). *Journal of Cell Biology* **99**, 199.
Iida, H. and Yahara, I. (1984b). *Journal of Cell Biology* **98**, 1185.
Iida, H. and Yahara, I. (1984c). *Journal of Cell Biology* **99**, 1441.

Iino, Y. and Yamamoto, M. (1985). *Molecular and General Genetics* **198**, 416.
Jacobs, C.W. and Szaniszlo, P.J. (1982). *Archives of Microbiology* **133**, 155.
Jacquet, M. and Camonis, J. (1985). *Biochimie* **67**, 35.
Jagadish, M.N. and Carter, B.L.A. (1977). *Nature, London* **269**, 145.
Jagadish, M.N. and Carter, B.L.A. (1978). *Journal of Cell Science* **31**, 71.
Jarvik, J. and Botstein, D. (1973). *Proceedings of the National Academy of Sciences of the United States of America* **70**, 2046.
Jenness, D., Burkholder, A. and Hartwell, L.H. (1983). *Cell* **35**, 521.
John, P.C.L. (1981). "The Cell Cycle." Cambridge University Press, Cambridge.
Johnson, B.F. (1965a). *Experimental Cell Research* **39**, 577.
Johnson, B.F. (1965b). *Experimental Cell Research* **39**, 613.
Johnson, B.F. and Gibson, E.J. (1966). *Experimental Cell Research* **41**, 580.
Johnson, B.F. and Lu, C. (1975). *Experimental Cell Research* **95**, 154.
Johnson, B.F., Calleja, G.B., Boisclair, I. and Yoo, B.Y. (1979). *Experimental Cell Research* **123**, 253.
Johnson, B.F., Calleja, G.B., Yoo, B.Y., Zuker, K. and McDonald, I.J. (1982). *International Review of Cytology* **75**, 167.
Johnson, L.M., Snyder, M., Chang, L., Davis, R.W. and Campbell, J.L. (1985). *Cell* **43**, 369.
Johnston, G.C. (1977). *Journal of Bacteriology* **132**, 738.
Johnston, G.C. and Singer, R.A. (1978). *Cell* **14**, 951.
Johnston, G.C. and Singer, R.A. (1980). *Molecular and General Genetics* **178**, 357.
Johnston, G.C. and Singer, R.A. (1983). *Experimental Cell Research* **149**, 1.
Johnston, G.C. and Singer, R.A. (1984). *In* "Growth, Cancer and the Cell Cycle" (P. Skehan and S.J. Friedman, eds), pp. 213–219. Humana Press, Clifton, NJ.
Johnston, G.C., Pringle, J.R. and Hartwell, L.H. (1977a). *Experimental Cell Research* **105**, 79.
Johnston, G.C., Singer, R.A. and McFarlane, E.S. (1977b). *Journal of Bacteriology* **132**, 723.
Johnston, G.C., Ehrhardt, C.W., Lorincz, A. and Carter, B.L.A. (1979). *Journal of Bacteriology* **137**, 1.
Johnston, G.C., Singer, R.A., Sharrow, S.O. and Slater, M.L. (1980). *Journal of General Microbiology* **118**, 479.
Johnston, L.H. (1980). *Current Genetics* **2**, 1975.
Johnston, L.H. (1983). *Molecular and General Genetics* **190**, 315.
Johnston, L.H. and Game, J.C. (1978). *Molecular and General Genetics* **161**, 205.
Johnston, L.H. and Johnson, A.L. (1984). *Genetical Research, Cambridge* **43**, 1.
Johnston, L.H. and Nasmyth, K. (1978). *Nature, London* **274**, 891.
Johnston, L.H. and Thomas, A.P. (1982a). *Molecular and General Genetics* **186**, 439.
Johnston, L.H. and Thomas, A.P. (1982b). *Molecular and General Genetics* **186**, 445.
Johnston, L.H., Johnson, A.L. and Game. J.C. (1982). *Experimental Cell Research* **141**, 63.
Jong, A.Y.S. and Campbell, J.L. (1984). *Journal of Biological Chemistry* **259**, 14394.
Jordan, E.G., Severs, N.J. and Williamson, D.H. (1977). *Experimental Cell Research* **104**, 446.
Julius, D., Blair, L., Brake, A., Sprague, G. and Thorner, J. (1983). *Cell* **32**, 839.
Kader, J. and Lloyd, D. (1979). *Journal of General Microbiology* **114**, 455.
Kassir, Y. and Simchen, G. (1978). *Genetics* **90**, 49.
Kassir, Y., Kupiec, M., Shalom, A. and Simchen, G. (1985). *Current Genetics* **9**, 253.
Kataoka, T., Powers, S., McGill, C., Fasano, O., Strathern, J.,Broach, J. and Wigler, M. (1984). *Cell* **37**, 437.
Kataoka, T., Powers, S., Cameron, S., Fasano, O.,Goldfarb, M., Broach, J. and Wigler, M. (1985a). *Cell* **40**, 19.
Kataoka, T., Broek, D. and Wigler, M. (1985b). *Cell* **43**, 493.
Katohda, S., Abe, N., Matsui, M. and Hayashibe, M. (1976). *Plant Cell Physiology* **17**, 909.
Kee, S.G. and Haber, J.E. (1975). *Proceedings of the National Academy of Sciences of the United States of America* **72**, 1179.
Kelly, S.L., Merrill, C. and Parry, J.M. (1983). *Molecular and General Genetics* **191**, 314.
Kilmartin, J.V. (1984). *In* "Molecular Biology of the Cytoskeleton" (G. Borisy, D. Cleveland and D. Murphy, eds), pp. 185–192. Cold Spring Harbor Press, Cold Spring Harbor.
Kilmartin, J.V.and Adams, A.E. (1984) *Journal of Cell Biology* **98**, 922.
King, S.M. and Hyams, J.S. (1982a). *Micron* **13**, 93.

King, S.M. and Hyams, J.S. (1982b). *Canadian Journal of Microbiology* **28**, 261.
King, S.M. and Hyams, J.S.(1982c). *Protoplasma* **110**, 54.
King, S.M. and Hyams, J.S. (1982d). *In* "Microtubules in Micro-organisms" (N.R. Morris and P. Cappucinelli, eds), pp. 143–156. Marcel Dekker, New York.
King, S.M., Hyams, J.S. and Luba, A. (1982a). *European Journal of Cell Biology* **28**, 98.
King, S.M., Hyams, J.S. and Luba, A. (1982b). *Journal of Cell Biology* **94**, 341.
Klebl, F., Huffaker, T. and Tanner, W. (1984). *Experimental Cell Research* **150**, 309.
Kramhøft, B. and Zeuthen, E. (1975). *In* "Methods in Cell Biology" (D. Prescott, ed.), vol. 12, pp. 373–380. Academic Press, New York.
Kubitschek, H.E. and Ward, R.A. (1985). *Journal of Bacteriology* **162**, 902.
Kuo, C. and Campbell, J.C. (1982). *Proceedings of the National Academy of Sciences of the United States of America* **79**, 4243.
Kuo, C., Huang, N. and Campbell, J. (1983). *Proceedings of the National Academy of Sciences of the United States of America* **80**, 6465.
Li, A.W., Singer, R.A. and Johnston, G.C. (1984). *FEMS Microbiology Letters* **23**, 243.
Liao, H. and Thorner, J. (1980). *Proceedings of the National Academy of Sciences of the United States of America* **77**, 1898.
Liao, H. and Thorner, J. (1981). *Journal of Bacteriology* **148**, 919.
Lillie, S.H. and Pringle, J.R. (1980). *Journal of Bacteriology* **143**, 1384.
Linnemans, W.A., Boer, P. and Elbers, P.F. (1977). *Journal of Bacteriology* **131**, 638.
Lipke, P.N., Taylor, A. and Ballou, C.E. (1976). *Journal of Bacteriology* **127**, 610.
Little, J.G. and Haynes, R.H. (1979). *Molecular and General Genetics* **168**, 141.
Livingston, D.M. and Kupfer, D.M. (1977). *Journal of Molecular Biology* **116**, 249.
Lloyd, D. (1984). *In* "Fungal Morphogenesis" (J.E. Smith, ed.), pp. 107–145. Marcel Dekker, New York.
Lloyd, D., John, L., Edwards, C. and Chagla, A.H. (1975). *Journal of General Microbiology* **88**, 153.
Lloyd, D., John, L., Hamill, M., Phillips, C., Kader, J. and Edwards, S.W. (1977). *Journal of General Microbiology* **99**, 223.
Lloyd, D., Edwards, S.W. and Williams, J.L. (1981). *FEMS Microbiology Letters* **12**, 295.
Lloyd, D., Poole, R.K. and Edwards, S.W. (1982). "The Cell Division Cycle: Temporal Organization and Control of Cellular Growth and Reproduction." Academic Press, London.
Lord, P.G. and Wheals, A.E. (1980). *Journal of Bacteriology* **142**, 808.
Lord, P.G. and Wheals, A.E. (1981). *Journal of Cell Science* **50**, 361.
Lord, P.G. and Wheals, A.E. (1983). *Journal of Cell Science* **59**, 183.
Lord, P.G. and Wheals, A.E. (1984). *Journal of General Microbiology* **130**, 2803.
Lorincz, A. and Carter, B.L.A. (1979). *Journal of General Microbiology* **113**, 287.
Lorincz, A. and Reed, S.I. (1984). *Nature, London* **307**, 183.
Lorincz, A.T., Miller, M.J., Xuong, N. and Geiduschek, E.P. (1982). *Molecular and Cellular Biology* **2**, 1532.
Ludwig, J.R. and McLaughlin, C.S. (1982). *In* "Proceedings of the Berkeley Workshop on Recent Advances in Yeast Molecular Biology". (M. Esposito, ed.), vol. 1, pp. 113–121. U.S. Department of Commerce, Springfield, Virginia.
Ludwig, J.R., Foy, J.J., Elliott, S.G. and McLaughlin, C.S. (1982). *Molecular and Cellular Biology* **2**, 117.
Mackay, V. and Manney, T. (1974). *Genetics* **76**, 273.
Martegani, E., Vanoni, M. and Baroni, M. (1984). *European Journal of Biochemistry* **144**, 204.
Matile, P., Moor, H. and Robinow, C.F. (1969). *In* "The Yeasts" (A.H. Rose and J.S. Harrison, eds), vol. 1, pp. 219–302. Academic Press, London and New York.
Matsumoto, K., Uno, I., Oshima, Y. and Ishikawa, T. (1982). *Proceedings of the National Academy of Sciences of the United States of America* **79**, 2355.
Matsumoto, K., Uno, I. and Ishikawa, T. (1983a). *Experimental Cell Research* **146**, 151.
Matsumoto, K., Uno, I. and Ishikawa, T. (1983b). *Cell* **32**, 417.
Matsumoto, K., Uno, I. and Ishikawa, T. (1984). *Journal of Bacteriology* **157**, 277.
Matsumoto, K., Uno, I. and Ishikawa, T. (1985a). *Yeast* **1**, 15.
Matsumoto, K., Uno, I., Kato, K. and Ishikawa, T. (1985b). *Yeast* **1**, 25.
Matur, A. and Berry, D.R. (1978). *Journal of General Microbiology* **109**, 205.

May, J.W. (1963). *Experimental Cell Research* **27**, 170.
McCready, S.J. and Cox, B.S. (1982). *Current Genetics* **6**, 29.
McCully, K.E. and Robinow, C.F. (1971). *Journal of Cell Science* **9**, 475.
McGoldrick, E.M. (1986). Ph.D. thesis: University of Bath, Bath, UK.
Minet, M., Nurse, P., Thuriaux, P. and Mitchison, J.M. (1979). *Journal of Bacteriology* **137**, 440.
Mitchell, L.H. and Soll, D.R. (1979a). *Experimental Cell Research* **120**, 167.
Mitchell, L.H. and Soll, D.R. (1979b). *Experimental Mycology* **3**, 298.
Mitchison, J.M. (1957). *Experimental Cell Research* **13**, 244.
Mitchison, J.M. (1958). *Experimental Cell Research* **15**, 214.
Mitchison, J.M. (1971). "The Biology of the Cell Cycle." Cambridge University Press, Cambridge.
Mitchison, J.M. (1977). *In* "Cell Differentiation in Micro-organisms, Plants, and Animals" (L. Nover and K. Mothes, eds), pp. 377–401. Gustav Fischer Verlag, Jena.
Mitchison, J.M. (1981). *In* "The Cell Cycle" (P.C.L. John, ed.),pp. 1–10. Cambridge University Press, Cambridge.
Mitchison, J.M. and Carter, B.L.A. (1975). *In* "Methods in Cell Biology" (D.M. Prescott, ed.), vol. 11, pp. 201–219. Academic Press, New York.
Mitchison, J.M. and Creanor, J. (1971). *Experimental Cell Research* **67**, 368.
Mitchison, J.M. and Nurse, P.M. (1985). *Journal of Cell Science* **75**, 357.
Mitchison, J.M. and Vincent, W.S. (1965). *Nature, London* **205**, 987.
Miyata, H., Miyata, M. and Ito, M. (1978a). *Cell Structure and Function* **3**, 39.
Miyata, H., Miyata, M. and Ito, M. (1978b). *Cell Structure and Function* **3**, 153.
Miyata, H., Miyata, M. and Ito, M. (1979a). *Cell Structure and Function* **4**, 81.
Miyata, H., Miyata, M. and Ito, M. (1979b). *Cell Structure and Function* **4**, 209.
Miyata, H., Miyata, M. and Ito, M. (1981a). *Cell Structure and Function* **6**, 269.
Miyata, H., Miyata, M. and Ito, M. (1981b). *Journal of General and Applied Microbiology* **27**, 277.
Moir, D. and Botstein, D. (1982). *Genetics* **100**, 565.
Moir, D., Stewart, S.E., Osmond, B.C. and Botstein, D. (1982). *Genetics* **100**, 547.
Molano, J., Bowers, B. and Cabib, E. (1980). *Journal of Cell Biology* **85**, 109.
Montelone, B.A., Prakash, S. and Prakash, L. (1981). *Journal of Bacteriology* **147**, 517.
Moore, C. (1982). *Journal of Bacteriology* **151**, 1617.
Moore, S. (1984). *Experimental Cell Research* **151**, 542.
Mortimer, R.K. and Schild, D. (1985). *Microbiological Reviews* **49**, 181.
Murakami, S. and Livingston, D. (1982). *Molecular and General Genetics* **185**, 506.
Nagpal, M.L., Higgins, D.R. and Prakash, S. (1985). *Molecular and General Genetics* **199**, 59.
Nasmyth, K. (1977). *Cell* **12**, 1109.
Nasmyth, K. (1979). *Journal of Cell Science* **36**, 155.
Nasmyth, K. (1983). *Nature, London* **302**, 670.
Nasmyth, K. (1985). *Cell* **42**, 225.
Nasmyth, K. and Nurse, P. (1981). *Molecular and General Genetics* **182**, 119.
Nasmyth, K. and Reed, S.I. (1980). *Proceedings of the National Academy of Sciences of the United States of America* **77**, 2119.
Newlon, C.S. and Fangman, W.L. (1975). *Cell* **5**, 423.
Newlon, C.S., Ludescher, R.D. and Walter, S.K. (1979). *Molecular and General Genetics* **169**, 189.
Newman, A.M., Elliott, S.G., McLaughlin, C.S., Sutherland, P.A. and Warner, R.C. (1981). *Journal of Virology* **38**, 263.
Nikawa, J. and Yamashita, S. (1982). *European Journal of Biochemistry* **125**, 445.
Njagi, F. and Kilbey, B. (1982). *Molecular and General Genetics* **186**, 478.
Nurse, P.M. (1975). *Nature, London* **256**, 547.
Nurse, P.M. (1981a). *Symposium of the Society for General Microbiology* **31**, 291.
Nurse, P.M. (1981b). *In* "The Fungal Nucleus" (K. Gull and S. Oliver, eds), British Mycological Society Symposium, vol. 5, pp. 331–345. Cambridge University Press, Cambridge.
Nurse, P.M. (1985a). *Trends in Genetics* **1**, 51.
Nurse, P.M. (1985b). *Molecular and General Genetics* **198**, 497.

Nurse, P.M. and Bisset, Y. (1981). *Nature, London* **292**, 558.
Nurse, P.M. and Fantes, P.A. (1981). *In* "The Cell Cycle" (P.C.L. John, ed.), pp. 85–98. Cambridge University Press, Cambridge.
Nurse, P.M. and Streiblová, E. (1984). "The Microbial Cell Cycle". CRC Press, Boca Raton, Florida.
Nurse, P.M. and Thuriaux, P. (1977). *Experimental Cell Research* **107**, 365.
Nurse, P.M. and Thuriaux, P. (1980). *Genetics* **96**, 627.
Nurse, P.M. and Thuriaux, P. (1984). *Molecular and General Genetics* **196**, 332.
Nurse, P.M. and Wiemken, A. (1974). *Journal of Bacteriology* **117**, 1108.
Nurse, P.M., Thuriaux, P. and Nasmyth, K. (1976). *Molecular and General Genetics* **146**, 167.
O'Farrell, P. (1975). *Journal of Biological Chemistry* **250**, 4007.
Ohya, Y., Ohsumi, Y. and Anraku, Y. (1984). *Molecular and General Genetics* **193**, 389.
Ortiz, C.H., Maia, J.L., Tenan, M.N., Braz Padrao, G.R., Mattoon, J.R. and Panek, A.D. (1983). *Journal of Bacteriology* **153**, 644.
Osley, M.A. and Hereford, L.M. (1981). *Cell* **24**, 377.
Petersen, J. and Piñon, R. (1979). *Carlsberg Research Communications* **44**, 395.
Petersen, J.B. and Ris, H. (1976). *Journal of Cell Science* **22**, 219.
Peterson, T.A., Yochem, J., Byers, B., Nunn, M.,F., Duesberg, P.H., Doolittle, R.F. and Reed, S.I. (1984). *Nature, London* **309**, 556.
Peterson, T.A., Prakash, L., Prakash, S., Osley, M.A. and Reed, S.I. (1985). *Molecular and Cellular Biology* **5**, 226.
Petes, T.D. and Fangman, W.L. (1973). *Biochemical and Biophysical Research Communications* **55**, 603.
Petes, T.D. and Williamson, D.H. (1975). *Cell* **4**, 249.
Piggott, J.R., Rai, R. and Carter, B.L.A. (1982). *Nature, London* **298**, 391.
Piñon, R. (1978). *Chromosoma* **67**, 263.
Piñon, R. (1979). *Chromosoma* **70**, 337.
Piñon, R. and Pratt, D. (1979). *Chromosoma* **73**, 117.
Polakowska, R., Prakash, L. and Prakash, S. (1983). *Molecular and General Genetics* **189**, 513.
Poole, R.K. and Lloyd, D. (1973). *Biochemical Journal* **136**, 195.
Popolo, L. and Alberghina, L. (1984). *Proceedings of the National Academy of Sciences of the United States of America* **81**, 120.
Popolo, L., Vanoni, M. and Alberghina, L. (1982). *Experimental Cell Research* **142**, 69.
Powers, S., Kataoka, T., Fasano, O., Goldfarb, M., Strathern, J., Broach, J. and Wigler, M. (1984). *Cell* **36**, 607.
Prakash, L., Hinkle, D. and Prakash, S. (1979). *Molecular and General Genetics* **172**, 249.
Pringle, J.R. (1975). *In* "Methods in Cell Biology" (D.M. Prescott, ed.), vol. 12, pp. 233–272. Academic Press, New York.
Pringle, J.R. (1978). *Journal of Cellular Physiology* **95**, 393.
Pringle, J.R. (1981). *In* "Mitosis/Cytokinesis" (A.M. Zimmerman and A. Forer, eds), pp. 3–28. Academic Press, New York.
Pringle, J.R. and Hartwell, L.H. (1981). *In* "Molecular Biology of the Yeast *Saccharomyces*" (J. Strathern, E. Jones and J. Broach, eds), vol. 1, pp. 97–142. Cold Spring Harbor Press, Cold Spring Harbor.
Pringle, J.R. and Mor, J. (1975). *In* "Methods in Cell Biology" (D.M. Prescott, ed.), vol. 11, pp. 131–168. Academic Press, New York.
Pringle, J.R., Coleman, K., Adams, A., Lillie, S., Haarer, B., Jacobs, C., Robinson, J. and Evans, C. (1984). *In* "The Molecular Biology of the Cytoskeleton" (G. Borisy, D. Cleveland and D. Murphy, eds), pp. 193–209. Cold Spring Harbor Press, Cold Spring Harbor.
Quinlan, R.A., Pogson, C.I. and Gull, K. (1980). *Journal of Cell Science* **46**, 341.
Rai, R. and Carter, B.L.A. (1981). *Molecular and General Genetics* **181**, 556.
Reed, S.I. (1980a). *Genetics* **95**, 561.
Reed, S.I. (1980b). *Genetics* **95**, 579.
Reed, S.I. (1984). *In* "The Microbial Cell Cycle" (P.M. Nurse and E. Streiblová, eds), pp. 89–107. CRC Press, Boca Raton, Florida.
Reed, S.I., Ferguson, J. and Groppe, J.C. (1982). *Molecular and Cellular Biology* **2**, 412.

Reed, S.I., Hadwiger, J.A. and Lorincz, A.T. (1985). *Proceedings of the National Academy of Sciences of the United States of America* **82**, 4055.
Reid, B.J. (1979). PhD Thesis: University of Washington, Seattle.
Reid, B.J. and Hartwell, L.H. (1977). *Journal of Cell Biology* **75**, 355.
Richmond, K.V. and Williamson, D.H. (1982). *Current Genetics* **6**, 167.
Richmond, K.V. and Williamson, D.H. (1983). *Journal of Cell Science* **64**, 307.
Rine, J.D. (1979). PhD Thesis: University of Oregon, Oregon U.S.A.
Rivin, C.J. and Fangman, W.L. (1980a). *Journal of Cell Biology* **85**, 96.
Rivin, C.J. and Fangman, W.L. (1980b). *Journal of Cell Biology* **85**, 108.
Roberts, R.L. and Szaniszlo, P.J. (1978). *Journal of Bacteriology* **135**, 622.
Roberts, R.L. and Szaniszlo, P.J. (1980). *Journal of Bacteriology* **144**, 721.
Roberts, R.L., Lo, R.J. and Szaniszlo, P.J. (1979). *Journal of Bacteriology* **137**, 1456.
Roberts, R.L., Lo, R.J. and Szaniszlo, P.J. (1980). *Journal of Bacteriology* **141**, 981.
Roberts, R.L., Bowers, B., Slater, M.L. and Cabib, E. (1983). *Molecular and Cellular Biology* **3**, 922.
Roman, H. and Fabre, F. (1983). *Proceedings of the National Academy of Sciences of the United States of America* **80**, 6912.
Roy, D. and Fantes, P.A. (1982). *Current Genetics* **6**, 195.
Ruiz, T., Villanueva, J.R. and Rodriguez, L. (1984). *Journal of General Microbiology* **130**, 337.
Samokhin, G.P., Lizlova, L.V., Bespalova, H.D., Titov, M.I. and Smirnov, V.N. (1981a). *Experimental Cell Research* **131**, 267.
Samokhin, G.P., Minin, A.A., Bespalova, H.D., Titov, M.I. and Smirnov, V.N. (1981b). *FEMS Microbiology Letters* **10**, 185.
Saul, D.J. and Sudbery, P.E. (1985). *Journal of General Microbiology* **131**, 1797.
Saul, D.J., Walton, E.F., Sudbery, P.E. and Carter, B.L.A. (1985). *Journal of General Microbiology* **131**, 2245.
Schekman, R. and Brawley, V. (1979). *Proceedings of the National Academy of Science of the United States of America* **76**, 645.
Schild, D. and Byers, B. (1978). *Chromosoma* **70**, 109.
Schild, D. and Byers, B.(1980). *Genetics* **96**, 859.
Schild, D., Ananthaswamy, H.N. and Mortimer, R.K. (1981). *Genetics* **97**, 551.
Sclafani, R.A. and Fangman, W. (1984). *Proceedings of the National Academy of Sciences of the United States of America.* **81**, 582.
Sebastian, J.,Carter, B.L.A. and Halvorson, H.O. (1973). *European Journal of Biochemistry* **37**, 516.
Segev, N. and Botstein, D. (1985). *Cellular Biochemistry (Supplement)* **9C**, 111.
Sena, E.P., Welch, J.W., Halvorson, H.O. and Fogel, S. (1975). *Journal of Bacteriology* **123**, 497.
Shilo, B., Shilo, V. and Simchen, G. (1976). *Nature, London* **244**, 767.
Shilo, B., Shilo, V. and Simchen, G. (1977). *Nature, London* **267**, 648.
Shilo, V., Simchen, G. and Shilo, B. (1978a). *Experimental Cell Research* **112**, 241.
Shilo, B., Simchen, G. and Pardee, A.B. (1978b). *Journal of Cell Physiology* **97**, 177.
Shimoda, C. and Nishi, K. (1982). *Plant and Cell Physiology* **23**, 1133.
Shulman, R.W. (1978). *In* "Methods in Cell Biology" (D.M. Prescott, ed.), vol. 20, pp. 35–43. Academic Press, New York.
Shulman, R.W., Hartwell, L.H. and Warner, J. R. (1973). *Journal of Molecular Biology* **73**, 513.
Shuster, J.R. (1982a). *Journal of Bacteriology* **151**, 1059.
Shuster, J.R. (1982b). *Molecular and Cellular Biology* **2**, 1052.
Simchen, G. (1974). *Genetics* **76**, 745.
Simchen, G. (1978). *Annual Review of Genetics* **12**, 161.
Simchen, G., Kassir, Y., Horesh-Cabilly, O. and Friedmann, A. (1981). *Molecular and General Genetics* **184**, 46.
Singer, R.A. and Johnston, G.C. (1979). *Molecular and General genetics* **176**, 37.
Singer, R.A. and Johnston, G.C. (1981). *Proceedings of the National Academy of Sciences of the United States of America* **78**, 3030.
Singer, R.A. and Johnston, G.C. (1983). *Experimental Cell Research* **149**, 15.

Singer, R.A. and Johnston, G.C. (1984). *In* "Growth, Cancer and the Cell Cycle," (P. Skehan and S.J. Friedman, eds), pp. 205–211. Humana Press, Clifton, NJ.

Singer, R.A. and Johnson, G.C. (1985). *Experimental Cell Research* **157**, 387.

Singer, R.A., Johnston, G.C. and Bedard, D.P. (1978). *Proceedings of the National Academy of Sciences of the United States of America* **75**, 6083.

Skutelsky, E. and Bayer, E.A. (1979). *Experimental Cell Research* **121**, 331.

Slater, M.L. (1973). *Journal of Bacteriology* **113**, 263.

Slater, M.L. (1974). *Nature, London* **247**, 275.

Slater, M.L. (1982). *In* "Yeast Cell Envelopes: Biochemistry Biophysics and Ultrastructure" (W.N. Arnold, ed.), vol. 2, pp. 65–77. CRC Press, New York.

Slater, M.L., Sharrow, S.O. and Gart, J.J. (1977). *Proceedings of the National Academy of Sciences of the United States of America* **74**, 3850.

Slater, M.L., Bowers, B. and Cabib, E. (1985). *Journal of Bacteriology* **162**, 763.

Sloat,, B.P. and Pringle, J.R. (1978). *Science* **200**, 1171.

Sloat, B.P., Adams, A. and Pringle, J.R. (1981). *Journal of Cell Biology* **89**, 395.

Smith, B.T.B. and Mitchison, J.M. (1976). *Experimental cell Research* **99**, 432.

Smith, J.A. and Martin, L. (1973). *Proceedings of the National Academy of Sciences of the United States of America* **70**, 1263.

Snow, R. (1966). *Nature, London* **211**, 206.

Soll, D.R. (1984). *In* "The Microbial Cell Cycle" (P.M. Nurse and E. Streiblová, eds), pp. 143–162. CRC Press, Boca Raton, Florida.

Soll, D. (1985a). *In* "Fungal Dimorphism: with Emphasis on Fungi Pathogenic to Humans" (P.J. Szaniszlo ed.), pp. 167–195. Plenum Press, New York.

Soll, D.R. (1985b). *In* "Current Topics in Medical Microbiology" (M.R. Ginnis, ed.), vol. 1, pp. 167–189, Springer-Verlag, New York.

Soll, D.R. and Bedell, G.W. (1978). *Journal of General Microbiology* **108**, 173.

Soll, D.R. and Mitchell, L.H. (1983). *Journal of Cell Biology* **96**, 486.

Soll, D.R., Stasi, M. and Bedell, G. (1978). *Experimental Cell Research* **116**, 207.

Soll, D.R., Bedell, G.W. and Brummel, M. (1981a). *Infection and Immunity* **32**, 1139.

Soll, D.R., Bedell, G., Thiel, J. and Brummel, M. (1981b). *Experimental Cell Research* **133**, 55.

Stevens, B. (1977). *Biologie Cellulaire* **28**, 37.

Stevens, B. (1981). *In* "The Molecular Biology of the Yeast, *Saccharomyces*" (J. Strathern, E. Jones and J. Broach, eds), vol. 1, pp. 471–504. Cold Spring Harbor Press, Cold Spring Harbor.

Storms, R.K., Ord, R.W., Greenwood, M.T., Mirdamadi, B., Chu, F.K. and Belfort, M. (1984). *Molecular and Cellular Biology* **4**, 2858.

Stötzler, D., Kiltz, H. and Duntze, W. (1976). *European Journal of Biochemistry* **69**, 397.

Strathern, J.N., Klar, A.J., Hicks, J.B., Abraham, J.A., Ivy., J.M., Nasmyth, K.A. and McGill, C. (1982). *Cell* **31**, 183.

Streiblová, E. (1982). *In* "Yeast Cell Envelopes: Biochemistry, Biophysics and Ultrastructure" (W.N. Arnold, ed.), vol. 2, pp. 79–92. CRC Press, New York.

Streiblová, E. (1984). *In* "The Microbial Cell Cycle" (P.M. Nurse and E. Streiblová, eds), pp. 127–141. CRC Press, Boca Raton, Florida.

Streiblová, E. and Girbardt, M. (1980). *Canadian Journal of Microbiology* **26**, 250.

Streiblová, E. and Hašek, J. (1985). *Microbiological Sciences* **5**, 139.

Streiblová, E. and Wolf, A. (1972). *Zeitschrift fur Allgemeine Mikrobiologie* **12**, 673.

Streiblová, E., Hašek, J. and Jelke, E. (1984). *Journal of Cell Science* **69**, 47.

Struhl, K. (1983). *Nature, London* **305**, 391.

Sudbery, P.E., Goodey, A.R. and Carter, B.L.A. (1980). *Nature, London* **288**, 401.

Sugisaki, Y., Gunge, N., Sakaguchi, K., Yamasaki, H. and Tamura, G. (1983). *Nature, London* **304**, 464.

Sumrada, R. and Cooper, T.G. (1978a). *Journal of General Microbiology* **108**, 45.

Sumrada, R. and Cooper, T.G. (1978b). *Journal of Bacteriology* **136**, 234.

Szaniszlo, P.J., Hsieh, P.H. and Marlowe, J.D. (1976). *Mycologia* **68**, 117.

Takeo, K. (1984). *FEMS Microbiology Letters* **22**, 97.

Tamanoi, F., Walsh, M., Kataoka, T. and Wigler, M. (1984). *Proceedings of the National Academy of Sciences of the United States of America* **81**, 6924.
Tanaka, K. (1976). *Journal of Electron Microscopy* **25**, 188.
Tanaka, K., Kanbe, T. and Kuroiwa, T. (1985). *Journal of Cell Science* **73**, 207.
Tatchell, K., Chafell, D.T., DeFeo-Jones, D. and Scolnick, E.M. (1984). *Nature, London* **309**, 523.
Tauro, P. and Halvorson, H.O. (1966). *Journal of Bacteriology* **92**, 652.
Temeles, G.L., Gibbs, J.B., D'Alonzo, J.S., Sigal, I.S. and Scolnick, E.M. (1985). *Nature, London* **313**, 700.
Thomas, J.H. and Bottstein, D. (1986). *Cell* **44**, 65.
Thomas, J.H., Novick, P. and Botstein, D. (1984). *In* "The Molecular Biology of the Cytoskeleton (G. Borisy, D. Cleveland and D. Murphy, eds), pp. 153–174. Cold Spring Harbor Press, Cold Spring Harbor.
Thomas, K.C., Dawson, P.S.S. and Gamborg, B.L. (1980a). *Journal of Bacteriology* **141**, 1.
Thomas, K.C., Dawson, P.S.S., Gamborg, B.L. and Steinhauer, L. (1980b). *Journal of Bacteriology* **141**, 10.
Thompson, P.W. and Wheals, A.E. (1980). *Journal of General Microbiology* **121**, 401.
Thompson, P.W. and Wheals, A.E. (1981). *Journal of General Microbiology* **122**, 151.
Thorner, J. (1982). *Cell* **30**, 5.
Throm, E. and Duntze, W. (1970). *Journal of Bacteriology* **104**, 1388.
Thuriaux, P., Nurse, P. and Carter, B. (1978). *Molecular and General Genetics* **161**, 215.
Thuriaux, P., Sipiczki, M. and Fantes, P.A. (1980). *Journal of General Microbiology* **116**,525.
Timson, J. (1975). *Mutation Research* **32**, 115.
Tkacz, J.S. and Lampen, J.O. (1972). *Journal of General Microbiology* **72**, 243.
Toda, T., Yamamoto, M. and Yanagida, M. (1981). *Journal of Cell Science* **52**, 271.
Toda, T., Umesono, K., Hirata, A. and Yanagida, M. (1983). *Journal of Molecular Biology* **168**, 251.
Toda, T., Adachi, Y., Hiraoka, Y. and Yanagida, M. (1984). *Cell* **37**, 233.
Toda, T.,Uno, I., Ishikawa, T., Powers, S., Kataoka, T., Broek, D., Cameron, S., Broach, J., Matsumoto, K. and Wigler, M. (1985). *Cell* **40**, 27.
Tschumper, G. and Carbon, J. (1982). *Journal of Molecular Biology* **156**, 293.
Tsuchiya, E., Tsuboshita, A., Kamimura, K., Miyakawa, T. and Fukui, S. (1981). *Agricultural and Biological Chemistry* **45**, 2395.
Tyson, C.B., Lord, P.G. and Wheals, A.E. (1979). *Journal of Bacteriology* **138**, 92.
Tyson, J.T. (1983). *Journal of Theoretical Biology* **103**, 313.
Uemara, T. and Yanagida, M. (1984). *EMBO Journal* **3**, 1737.
Umesono, K., Hiraoka, Y., Toda, T. and Yanagida, M. (1983a). *Current Genetics* **7**, 123.
Umesono, K., Toda, T., Hayashi, S. and Yanagida, M. (1983b). *Journal of Molecular Biology* **168**, 271.
Unger, M.W. and Hartwell, L.H. (1976). *Proceedings of the National Academy of Sciences of the United States of America* **713**, 1664.
Uno, I., Matsumoto, K. and Ishikawa, T. (1982). *Journal of Biological Chemistry* **257**, 14110.
Uno, I., Matsumoto, K. and Ishikawa, T. (1983). *Journal of Biological Chemistry* **258**, 3539.
Uno, I., Matsumoto, K., Adachi, K. and Ishikawa, T. (1984). *Journal of Biological Chemistry* **259**, 12508.
Uno, I., Matsumoto, K., Hirata, A. and Ishikawa, T. (1985). *Journal of Cell Biology* **100**, 1854.
Vanoni, M., Vai, M., Popolo, L. and Alberghina, L. (1983). *Journal of Bacteriology* **156**, 1282.
Vanoni, M., Vai, M. and Frascotti, G. (1984). *Cytometry* **5**, 530.
Varma, A., Freese, E.B. and Freese, E. (1985). *Molecular and General Genetics* **201**, 1.
Vezinhet, F., Kinnaird, J.H. and Dawes, I.W. (1979). *Journal of General Microbiology* **115**, 391.
von Meyenburg, H.K. (1968). *Pathologia Microbiologica* **31**, 77.
Vrana, D. and Votruba, J. (1981). *Folia Microbiologica* **26**, 382.
Walker, G.M. and Duffus, J.H. (1979). *Journal of General Microbiology* **114**, 391.
Walker, G.M. and Duffus, J.H. (1980). *Journal of Cell Science* **42**, 329.
Walker, G.M. and Duffus, J.H. (1983). *Magnesium* **2**, 1.
Walker, G.M., Thompson, J.C., Slaughter, J.C. and Duffus, J.H. (1980). *Journal of General Microbiology* **119**, 543.

Walton, E.F., Carter, B.L.A. and Pringle, J.R. (1979). *Molecular and General Genetics* **171**, 111.
Warner, J.R. and Gorenstein, C. (1978). *Nature, London* **275**, 891.
Wells, J.R. (1974). *Experimental Cell Research* **85**, 278.
Wheals, A.E. (1977). *Nature, London* **267**, 647.
Wheals, A.E. (1981). *In* "Fifth International Symposium on Yeasts" (G.G. Stewart and I. Russell, eds), pp. 549–554. Pergamon Press, London.
Wheals, A.E. (1982). *Molecular and Cellular Biology* **2**, 361.
Wheals, A.E. (1985). *Bioessays* **3**, 108.
Wheals, A.E. and Silverman, B. (1982). *Journal of Theoretical Biology* **97**, 505.
Whiteway, M. and Szostak, J. (1985). *Cell* **43**, 483.
Wiemken, A., Matile, P. and Moor, H. (1970). *Archiv fur Mikrobiologie* **70**, 89.
Wilkinson, L.E. and Pringle, J.R. (1974). *Experimental Cell Research* **89**, 175.
Williamson, D.H. (1965). *Journal of Cell Biology* **25**, 517.
Williamson, D.H. (1973). *Biochemical and Biophysical Research Communications* **52**, 731.
Williamson, D.H. (1985). *Yeast* **1**, 1.
Williamson, D.H. and Moustacchi, E. (1971). *Biochemical and Biophysical Research Communications* **42**, 195.
Williamson, D.H. and Scopes, A.W. (1960). *Experimental Cell Research* **20**, 338.
Williamson, D.H. and Scopes, A.W. (1962). *Nature, London* **193**, 256.
Willison, J.H. and Johnston, G.C. (1985). *Canadian Journal of Microbiology* **31**, 109.
Wolfner, M., Yep, D., Messenguy, F. and Fink, G.R. (1975). *Journal of Molecular Biology* **96**, 273.
Wood, J.S. and Hartwell, L.H. (1982). *Journal of Cell Biology* **94**, 718.
Yamamoto, M. (1980). *Molecular and General Genetics* **180**, 231.
Yashphe, J. and Halvorson, H.O. (1976). *Science* **191**, 1283.
Yokoyama, K. and Takeo, K. (1983). *Archives of Microbiology* **134**, 251.
Zakian, V.A. and Fangman, W.L. (1979). *Journal of Cell Biology* **83**, 9a.
Zakian, V.A. and Scott, J.F. (1982). *Molecular and Cellular Biology* **2**, 221.
Zakian, V.A., Brewer, B.J. and Fangman, W.L. (1979). *Cell* **17**, 923.
Zeuthen, E. (1974). *In* "Cell Cycle Controls" (G.M. Padilla, I.L. Cameron and A. Zimmerman, eds), pp. 1–30. Academic Press, New York.
Zietz, S. and Nicolini, C. (1978). *In* "Biomathematics and Cell Kinetics" (A. Valleron and P.D.M. Macdonald, eds), pp. 357–394. Elsevier/North Holland Biomedical Press, Amsterdam.

Appendix 1. Cell cycle mutants of *Saccharomyces cerevisiae*

A. Characterized genes with *cdc* (cell-division cycle) phenotype

Mutant gene	Number of alleles	Terminal phenotype	Diagnostic landmark	First cycle arrest	Required for sporulation	Remarks	References
cdc1	8	A	BE	Yes	No	Rapidly stops growing at N.P.T.	1, 2
cdc2	7	H	DS	Yes	Yes	Defective in initiation of DNA synthesis	1–6
cdc3	5	L	MRF	Yes	No	Initially described as CK mutant; many extragenic suppressors	1, 2, 4, 7–10
cdc4	7	F	SPBS	Yes	Yes	Can undergo meiosis directly; defective in karyogamy	1, 2, 4, 8, 11–13, 101
cdc5	1	K	NR	Yes	Yes	Defective in mitochondrial transmission	1–4, 14–16
cdc6	3	H	mND	Yes	No(?)	Arrests with short (1.5 μm) spindle	1–4, 17–20
cdc7	14	H	iDS	Yes	Yes	Essential for nuclear DNA synthesis	1, 2, 4, 8, 17, 18, 21, 22
cdc8	6	H	DS	Yes	Yes	Involved in error-prone DNA repair; structural gene for thymidylate kinase	1–4, 23–32
cdc9	14	H	DS	Yes	Yes	Structural gene for DNA ligase; increased mitotic recombination in an allele *mms8-1*	1–4, 33–39
cdc10	9	L	MRF	Yes	No	Initially described as CK mutant; many extragenic suppressors	1, 2, 7, 9, 10
cdc11	10	L	MRF	Yes	No	Four cold-sensitive alleles; many extragenic suppressors	1, 2, 7, 9, 10 17, 40
cdc12	4	L	MRF	No	—	Initially described as CK mutant; one extragenic suppressor	1, 2, 7, 9, 10
cdc13	3	H	mND	Yes	Yes	Sister chromatids not segregated (?)	1–4, 11
cdc14	24	J	lND	Yes	Yes	Sister chromatids probably segregated	1–4, 16–18, 21
cdc15	31	J	lND	Yes	No		1–4, 18, 21, 41
cdc16	26	H	mND	Yes	Yes(?)	Sister chromatids not segregated (?)	1–4, 17, 18
cdc17	2	H	mND	Yes	Yes(?)	Prevents telomere elongation	1–4, 42

cdc18	3	K	NR(?)	No	—		1, 2
cdc19	4	A	SPBSF	No	—	Structural gene for pyruvate kinase	1, 2, 41, 43–44
cdc20	4	H	mND	Yes	Yes(?)		1, 2, 4, 21,
cdc21	5	H	DS	Yes	Yes	Structural gene for thymidylate synthetase; allelic to *tmp1*; 45 non-*ts* mutants isolated	1–3, 32, 45–48
cdc22	4	B(?)	SPBE(?)	Yes	—	Originally isolated as double mutant in which *cdc22* alone did not have a *cdc* phenotype	1, 2, 17
cdc23	6	H	mND	Yes	Yes(?)	Sister chromatids not segregated(?)	1–3
cdc24	29	E	CRF	Yes	No	Defective in spatial localization of growth; many extragenic suppressors	1, 2, 8, 10, 41, 48, 49
cdc25	25	A	SPBSF	Yes/No	No	Diploids sporulate in acetate growth medium at N.P.T.	1, 2, 49–52
cdc26	1	H	mND(?)	No	—	Stops growth at N.P.T.	1, 2
cdc27	7	J(?)	IND(?)	No	—	Defective in mitochondrial transmission	1, 2, 14, 18, 53
cdc28	23	B	SPBE/ND	Yes	Yes	Can conjugate at N.P.T.; defective in karyogamy; two alleles are amber suppressible; one allele has a defect only in nuclear division, suppressed in carbon-catabolite derepressing medium.	1, 2, 6, 8, 11, 21, 49, 50, 54–61
cdc29	2	A	SPBSF	Yes	—		1, 2
cdc30	2	J	IND	No	—		1, 2
cdc31	2	G	SPBD	No	—	Forms a unipolar spindle; undergoes endomitosis	1, 2, 62–64
cdc32	2	B	SPBE	Yes	—	A cold-sensitive allele isolated as extragenic suppressor of *cdc51*	1, 2, 40
cdc33	1	A	SPBSF	Yes	—	Stops growing at N.P.T.	1, 49, 63
cdc34	2	F	iDS	No	—	Defective in karyogamy	1, 11, 63
cdc35	39	A	SPBSF	No	No	Diploids sporulate in acetate growth medium at N.P.T.; allelic to *cyr1*, structural gene for adenylate cyclase	1, 49–51, 63–65
cdc36	15	B	SPBE	Yes	—	Phenotype altered by heterozygosity at *MAT* locus; can conjugate at N.P.T.; suppressed in carbon-catabolite derepressing medium; can suppress some alleles of *ste4* and *ste5*.	1, 55, 58, 61, 65, 101

A. Characterized genes with *cdc* (cell-division cycle) phenotype continued

Mutant gene	Number of alleles	Terminal phenotype	Diagnostic landmark	First cycle arrest	Required for sporulation	Remarks	References
cdc37	4	B	SPBE	Yes	—	Two amber suppressible alleles known; can conjugate at N.P.T.; defective in karyogamy; like *cdc28*	1, 11, 12, 55 56, 61
cdc38	?	B(?)	—	—	—	Like *cdc28* (?)	1
cdc39	1	B	SPBE	Yes	—	Like *cdc36*	1, 55, 58, 61 65
cdc40	1	H	?	Yes	Yes	Makes defective DNA at N.P.T.	1, 66, 67
cdc41	?	C	?	—	—		1
cdc42	1	E	CRF(?)	—	—	Like *cdc24*; many extragenic suppressors	1, 10, 48
cdc43	5	E	CRF(?)	—	—	Like *cdc24*	1, 48
cdc44	1	H	mND	—	—	Cold-sensitive allele	1, 40
cdc45	1	H	mND	—	—	Cold-sensitive allele suppressible by certain alleles of *cdc46* and *cdc47*	1, 40
cdc46	6	H	?	—	—	Temperature-sensitive alleles isolated as extragenic suppressors of *cdc45*	1, 40
cdc47	3	H	?	—	—	Like *cdc46*	1, 40
cdc48	1 + 4	H	?	—	—	One cold-sensitive (*cs*) allele; four temperature-sensitive intragenic suppressors of the *cs* allele	1, 40
cdc49	7	D	?	—	—	Cold-sensitive alleles	1, 40
cdc50	1	D	?	—	—	Like *cdc49*	1, 40
cdc51	1	H	?	—	—	Cold-sensitive allele	1, 40
cdc60	1	C	?	Yes	—	Growth stops and cannot conjugate at N.P.T.	68, 69
cdc61	1	C	?	Yes	—	Reduced conjugation ability at N.P.T.; growth unaffected.	68, 69
cdc62	1	C	?	Yes	—	Can conjugate at N.P.T.	68, 69
cdc63	1	C	?	Yes	—	Like *cdc62*. Growth unaffected	68, 69
cdc64	1	C	?	Yes	—	Like *cdc60*	68, 69
cdc66	1	C	?	—	—	—	69
dds4	1	H	?	—	—	Depressed DNA synthesis; protein and RNA synthesis normal for a period; not allelic to *cdc1* to *cdc31*	17

dds5	1	H	?	—	—	DNA and RNA synthesis stops after 30 minutes at N.P.T.; protein unaffected; not allelic to *cdc1* to *cdc31*	17
dbf1	1	H	?	—	—	*dbf1-9* all temperature sensitive and not allelic to *cdc1* to *cdc31*; DNA synthesis normal	18
dbf2	2	H	?	—	—	Like *dbf1*; see *dbf1*	18
dbf3	2	H	?	—	—	Specifically defective in DNA synthesis; RNA synthesis normal; see *dbf1*	18, 70
dbf4	2	H	?	—	—	Like *dbf3*, execution point after *cdc7* and before final protein synthesis for iDS; see *dbf1*	18, 70
dbf5	4	H	?	—	—	Allelic to *rna3* (which does not have *dbf* phenotype); phenotype is late G2 rather than dumb bell; see *dbf1*	18
dbf6	2	H	?	—	—	DNA and RNA synthesis stop after a delay; see *dbf1*	18
dbf7	1	H	?	—	—	Like *dbf6*; see *dbf1*	18
dbf8	1	H	?	—	—	Like *dbf6*; see *dbf1*	18
dbf9	1	H	?	—	—	Like *dbf6*; see *dbf1*	18
gcd1(tra3)	1	C	?	Yes	Yes	Depressed for certain amino acids biosynthetic pathways; additionally like a 'start' mutant at N.P.T.	50, 71
alg1	1	A	?	—	—	Defective in *N*-glycosylation of proteins	72, 73
cal1	1	(D)	NM	—	—	Calcium dependent cell cycle arrest	74
tub1	?	M	NM	—	—	Structural gene for β-tubulin	75
ndc1	1	N	(NM)	—	—	Cold-sensitive; leaky; partitions diploid parents and haploid daughters; spindle pole body segregation normal	75, 102
pol1	1	H	DS	Yes	—	Structural gene for DNA polymerase I; haploids with gene disruption inviable	103
top2	6	(H)	ND	No	—	Structural gene for DNA topoisomerase II. Temperature sensitive mutants show mitotic-dependent loss of viability	104, 105

B. Additional strains

Phenotype	Number of complementation groups (alleles)	Remarks	References
Arrest at mND	2(2)	Cold-sensitive alleles	40
Arrest as unbudded G1 cells at α-factor sensitive step	5(5)	Temperature sensitive; four can conjugate, one cannot at N.P.T.; not allelic to *cdc1* to *cdc35*.	64
Defective in DNA synthesis but not protein synthesis	55(79)	Temperature-sensitive alleles; not allelic to *cdc1* to *cdc31*; RNA synthesis and *cdc* phenotype not measured	21
Defective in DNA synthesis	1(1)	Temperature sensitive	76
Dumb-bell or unbudded terminal phenotype	2?(6)	Harbour amber mutations suppressible by a *ts* amber suppressor	77
Dumb-bell or unbudded terminal phenotype	?(9)	Selected as having depressed DNA synthesis at N.P.T.	17
Dumb-bell terminal phenotype	?(53)	Temperature sensitive for colony formation; DNA synthesis normal over a seven-hour period	18
Cell-cycle-specific arrest	5(6)	Isolated as suppressors of *ste* mutants: two suppress *ste4* and *ste5*; one suppresses *ste2* only	78

C. Genes affecting control of proliferation

Mutant gene	Number of alleles	Remarks	References
whi1	3	Initiates cell-division cycle at half volume of parent; not temperature sensitive	79–81
whi2	1	Normal during exponential growth but continues to proliferate without growth on entering stationary phase producing small cells	79, 80, 82, 83
cyr1	24	Accumulates as unbudded G1 cells; dependent on exogenous cyclic AMP for growth; sporulates in rich media; deficient in membrane-bound adenylate cyclase activity; structural gene for catalytic subunit of adenylate cyclase; allelic to *cdc35*; suppressed by *bcy1*	65, 84–88, 105
cyr2	3	Accumulates as unbudded G1 cells; dependent on exogenous cyclic AMP for growth; sporulates in rich media; adenylate cyclase activity normal; suppressed by *bcy1*; probably structural gene of the catalytic subunit of cyclic AMP-dependent protein kinase	84, 86–88
CYR3	2	Dominant mutation; like *cyr2*: probably structural gene of the regulatory subunit of cyclic AMP-dependent protein kinase	84–88
ras1	(1)	Not essential but in combination with a gene disruption of *ras2* leads to lack of germination and growth. Lowered cyclic AMP concentrations. Suppressed by *bcy1*.	87, 88
ras2	(2)	Gene disruption leads to greatly decreased concentrations of cyclic AMP. Cannot grow on non-fermentable carbon sources; accumulates storage carbohydrate and sporulates on rich media. A *RAS2-Val19* allele has elevated concentrations of cyclic AMP and cannot sporulate (like *bcy1*). See also *ras1*	87, 88

D. Genes with altered response to cell cycle arrest by α-factor

Gene	Number of alleles	Phenotype	References
ste2	30	Sterile; not arrested by α-factor; specifically defective in *MAT***a** inducible functions–division arrest and agglutination; hyperproduction of **a** factor; normal degradation of α-factor; putative α-factor receptor	89–92
ste3		Sterile; not arrested by α-factor; only expressed in α-cells; induced by α-factor	89–93
ste4	7	Sterile; not arrested by **a** or α-factor; defective in constitutive and inducible *MAT* functions; suppressible by *sst2* in strains carrying *ros* (relaxation of sterility) mutants; allele-dependent suppression of sterility by *cdc36* and *cdc39*; putative control gene of mating pheromone receptor	65, 89–92
ste5	7	Sterile; not arrested by **a** or α-factor; defective in constitutive and inducible *MAT* functions; allele-dependent suppression of sterility by *cdc36* and *cdc39*; adenylate cyclase activity insensitive to α-factor inhibition at N.P.T.	89–92
ste6	7	Sterile; specifically defective in *MAT***a** functions	94
ste7	4	Like *ste5* but not suppressible expresses silent copies of mating type	65, 90–92, 95
ste8	5	Sterile; indistinguishable from **a**/α diploids; expresses silent copies of mating type	65, 90, 91, 96
ste9	5	Like *ste8*	65, 90, 91, 96
ste11	4	Like *ste7*	65, 90–92, 95
ste12	5	Like *ste7*; positively regulates *MAT* genes	65, 90, 91, 97
ste13	?	Sterile; *MAT* cells produce incompletely processed α-factor; putative dipeptidyl aminopeptidase	98, 99
ste14	?	Like *ste6*	94, 99
sst1	5	Supersensitive to α-factor arrest because it cannot be inactivated by proteolytic degradation; *MAT***a** specific	99, 100
sst2	4	Supersensitive to both **a**- and α-factor; some *MAT***a** cells self-shmoo; suppresses *ste4* in some genetic backgrounds	92, 99, 100

Notes

The tables in Appendix 1 are based on Table 1 of Pringle and Hartwell (1981), from which additional information and further references can be obtained.

Mutant genes. Dominant (usually wild type) alleles are upper case; recessive (mutant) alleles are lower case. Most *cdc* genes from *cdc1* to *cdc35* have been fully defined by complementation and linkage tests. Subsequently, characterization may have been limited to complementation with *cdc1* to *cdc31* and/or with mutants having similar phenotype.

Number of alleles. This continually changing number is based on references in Pringle and Hartwell (1981), with more recent additions as indicated in the references and is given as a guide to the relevant abundance of allelic types.

Terminal phenotype. At non-permissive temperature (N.P.T.): A, growth stops, unbudded uninucleate G1 cells with no spindle pole body (SPB) satellite; B, cells shmoo, unbudded uninucleate G1 cells with SPB satellites; C, growing non-shmooing unbudded uninucleate G1 cells; D, uninucleate growing cells with small buds; E, growth, *DNA* synthesis and nuclear division continue, no budding; F, multiply-budded growing uninucleate G1 cells with SPB, not separated; G, growing uninucleate G2 cells with large SPB and short, unipolar spindle; H, growing, uninucleate G1 or G2 cells with short spindle and nucleus in neck of mother and single bud; J, growing uninucleate G2 cells with complete spindle and nucleus extending through length of mother and single bud; K, growing, binucleate G2 cells with single bud; L, growing, multinucleate, multiply budded cells; M, large-budded doublet cells, nucleus not migrated to neck; N, anucleate daughters, proliferating polyploid parents. Most alleles of a gene have similar terminal phenotypes (but see *cdc28, cdc63* and *dbf5*.

Diagnostic landmark. BE, bud emergence; CRF, chitin ring formation; DS, DNA synthesis; iDS, initiation of DNA synthesis; MRF, microfilament ring formation; ND, nuclear division; NR, nuclear reorganization; lND, late stage of nuclear division; mND, medial stage of nuclear division; SPBD, spindle pole body duplication; SPBE, spindle pole body enlargement; SPBS, spindle pole body separation; SPBSF, spindle pole body satellite formation.

Sporulation. Data from Simchen (1974), Schild and Byers (1978, 1980), Shilo *et al.* (1978a).

Remarks. Except where indicated the mutants are wild-type at 23°C and mutant at 36°C, the non-permissive temperature (N.P.T.).

References

1, Pringle and Hartwell (1981); 2, Hartwell *et al.* (1973); 3, Wood and Hartwell (1982); 4, Hartwell and Smith (1985); 5, Conrad and Newlon (1983b); 6, Conrad and Newlon (1983a); 7, Dominguez *et al.* (1982); 8, Roberts *et al.* (1983); 9, Slater *et al.* (1985); 10, Pringle *et al.* (1984); 11, Dutcher and Hartwell (1982); 12, Dutcher and Hartwell (1983a); 13, Adams and Pringle (1984); 14, Dutcher (1982); 15, Simchen *et al.* (1981); 16, Schild and Byers (1980); 17, L.H. Johnston and Game (1978); 18, L.H. Johnston and Thomas (1982a); 19, King and Hyams (1982d); 20, King *et al.* (1982a); 21, Dumas *et al.* (1982); 22, Njagi and Kilbey (1982); 23, Prakash *et al.* (1979); 24, Kuo and Campbell (1982); 25, Polakowska *et al.* (1983); 26, Arendes *et al.* (1983); 27. Sclafani and Fangman (1984); 28, Jong and Campbell (1984); 29, Birkenmeyer *et al.* (1984); 30, Baranowska *et al.* (1985); 31, L.H. Johnston and Johnson (1984); 32, Newlon *et al.* (1979); 33, L.H. Johnston (1983); 34, L.H. Johnston and Nasmyth (1978); 35, Moore (1982); 36, L.H. Johnston *et al.* (1982); 37, McCready and Cox (1982); 38, Barker and L.H. Johnston (1983); 39, Barker *et al.* (1985); 40, Moir *et al.* (1982); 41; D. Kaback and J. Pringle (personal communication); 42, Carson and Hartwell (1985); 43, G. Kawasaki (personal communication); 44, Gillies and Benoit (1983); 45, Brendel and Fäth (1974); 46, Little and Haynes (1979); 47, Game (1976); 48, Pringle (1981); 49, Iida and Yahara (1984b); 50, Shilo *et al.* (1978a); 51, Reid (1979); 52, Popolo and Alberghina (1984); 53, Richmond and Williamson (1982); 54, Hartwell *et al.* (1974); 55, Reed (1980a); 56, Reed (1980b); 57, Piggott *et al.* (1982); 58, Shuster (1982a); 59, Nasmyth (1985); 60, Reed *et al.* (1985); 61, Connolly *et al.* (1983); 62, Byers (1981b); 63, Culotti (1974); 64, Schild *et al.* (1981); 65, Shuster (1982b), Caspersson *et al.* (1985); 66, Kassir and Simchen (1978); 67, Kassir *et al.* (1985); 68, Bedard *et al.* (1981); 69, Hanic-Joyce (1985); 70, L.H. Johnston and Thomas (1982b); 71, Wolfner *et al.* (1975); 72, Huffaker and Robbins (1982); 73, Klebl *et al.* (1984); 74, Ohya *et al.* (1984); 75, Thomas *et al.* (1984); 76, Kuo *et al.* (1983); 77, Rai and Carter (1981); 78, M.E. Katz, P. Kaynes and S.I. Reed (personal communication); 79, Sudbery *et al.* (1980); 80, Carter and Sudbery (1980); 81, Calvert and Dawes (1984a); 82, Saul and Sudbery (1985); 83, Saul *et al.* (1985); 84, Matsumoto *et al.* (1982); 85, Matsumoto *et al.* (1983b); 86, Uno *et al.* (1985); 87, Matsumoto *et al.*(1985a); 88, Wheals (1985); 89, Mackay and Manney(1974); 90, Hartwell (1980); 91, Liao and Thorner (1980); 92, Jenness *et al.* (1983); 93, Hagen and Sprague (1984); 94, I. Herskowitz (personal communication); 95, Chaleff and Tatchell (1985); 96, Rine (1979); 97, Fields and Herskowitz (1985); 98, Julius *et al.* (1983); 99, Chan *et al.* (1983); 100, Chan and Otte (1982); 101, Peterson *et al.* (1984); 102, Thomas and Botstein (1986); 103, L.M. Johnson *et al.* (1985); 104, Holm *et al.* (1985); 105, Dinardo *et al.* (1984); 106, Kataoka *et al.* (1985b).

Appendix 2 Cell cycle mutants of *Schizosaccharomyces pombe*

A. Characterized genes with *cdc* phenotype

Mutant Gene	Number of alleles	Diagnostic landmark	Terminal phenotype	Conjugation at N.P.T.	Remarks	References
cdc1	4	ND	A	No	Two alleles are homoallelic. Nucleus has rectangular or dumb-bell shape as terminal phenotype.	1–7
cdc2	10	ND	A	Yes	Some alleles continue proliferation at decreased cell size (= *wee2* mutants); required in G1 for commitment to cell cycle and in G2 for control of mitosis; two pairs of alleles probably homoallelic	1–10
cdc3	2	lCPF	D	–		1, 6, 11, 12
cdc4	2	lCPF	D	–	Two alleles show intragenic complementation	1, 11, 12
cdc5	1	(ND)	(A)	–	Leaky, probably not *cdc*	1, 2
cdc6	18	ND	A	No	One allele leaky	1–5, 7
cdc7	1	eCPF	C	–		1
cdc8	3	lCPF	D	–		1, 6, 12
cdc9 renamed *weel*				1		13, 14
cdc10	15	DS	A	Yes	Two alleles are homoallelic; nucleus has rectangular shape as terminal phenotype	1–5, 6, 7, 10, 15, 16
cdc11	3	eCPF	C	–		1, 6, 7, 10
cdc12	1	lCPF	D	–		1, 6, 11, 12
cdc13	1	ND	B	No		1–4, 7
cdc14	1	eCPF	C	–	Leaky	1
cdc15	2	eCPF	C	–		1, 4, 6
cdc16	1	CPF	E	–		6, 17, 18
cdc17	3	DS	A	No	Structural gene for DNA ligase	2, 3, 7, 19–21
cdc18	1	DS/ND?	A	–		2, 6
cdc19	1	DS/ND?	A	–		2
cdc20	3	iDS	A	–		2, 7

cdc21	1	DS	A	–		2, 7
cdc22	1	iDS	A	No		2, 3, 7
cdc23	2	DS	A	No		2, 3
cdc24	4	DS	A	No	Two alleles are homoallelic	2, 3
cdc25	3	ND	A	No	Suppressible by *wee* mutations; allalleles act as allosuppressors; allelic to *sal2*	2–7, 22–24
cdc26	1	CS	F	–		25
cdc27	3	ND	A	No	Microtubules emanate from SPB at N.P.T.	2, 3, 4, 5
cdc28	1	ND	A	–		2, 5
ndal	2	ND	A	–	Cold-sensitive alleles; defective in early stage of ND.	6, 26
nda2	5	ND	A	–	Structural gene for α1-tubulin; one cold-sensitive allele; all supersensitive to tubulin-binding drugs at permissive termperature; defective in nuclear segregation	6, 26–28
nda3 29	5	ND	A	–	Two cold-sensitive alleles, one also benomyl resistant, three supersensitive to thiabendazole; defective in nuclear segregation; allelic to *ben1*.	26, 27, 29
ben4	7	ND	A	–	Cold-sensitive alleles, selected for benomyl resistance; arrested cells have chromatin strung out along elongated nucleus; nucleolus at one end; six alleles super-sensitive to isopropyl *N*-(3-chlorophenyl)carbamate	30
top2	3	ND	G	–	Structural gene for DNA topisomerase II	36

B. Uncharacterized genes with *cdc* phenotype

Phenotype	Number of genes (alleles)	Remarks	References
Nuclear-division arrest	9(9)	Like *nda1* to *3*	25, 31

C. Genes affecting cell proliferation

Mutant gene	Number of alleles	Remarks	References
wee1	67[a]	Cell length, protein and dry weight at division half that of wild-type cells; one allele is opal nonsense suppressible; three alleles are temperature sensitive; two are partially *wee*; epistatic to *cdc25*	2–4, 6, 7, 9, 11, 32, 33
wee2	3	Like *wee1*; allelic to *cdc2*; alleles designed *cdc2-1w, 2w* and *3w*; epistatic to *cdc25*	2–4, 9, 11, 24
ran1/pat1	14	Temperature sensitive; when growth arrested, haploids sporulate in rich media	34, 35

D. Genes with altered response to cell-cycle arrest

Mutant Gene	Number of alleles	Arresting agent	Phenotype	References
ben1	9	Thiabendazole (TBZ)	Resistant to anti-mitotic drug TBZ; not temperature sensitive; also resistant to methyl benzimidazolecarbamate (MBC); allelic to *nda3*	6, 26, 27, 29
ben2	4	TBZ	Like *ben1* but all temperature sensitive	6, 29
ben3	2	TBZ	Like *ben2*	29

Notes

Number of alleles. The non-random distribution between different genes is largely due to certain types of mutant. (e.g. nuclear division arrest or *wee* mutants) being specifically sought for in particular experiments.
*[a] Several hundred *wee1* alleles were detected and discarded without further analysis (Fantes, 1981).

Diagnostic landmark. CPF, cell-plate formation; eCPF early cell-plate formation; lCPF, late cell-plate formation; CS, cell separation; DS, DNA synthesis; iDS, initiation of DNA synthesis; ND, nuclear division.

Terminal phenotype. A, Elongated with single nucleus and no cell plate; B, elongated with single nucleus and multiple cell plates; C, elongated with 4–16 nuclei per cell and no cell plate; D, swollen elongated cells with 2–4 nuclei and disorganized cell-plate material; E, either binucleate and multiseptate or uninucleate with an asymmetrical septum; F, cells have multiple septa; G, septum partitions single nucleus.

Conjugation at N.P.T. (non-permissive temperature). Based on percentage of cells, relative to wild-type, conjugating at 25°C,33°C and 36°C (Nurse and Bissett, 1981).

References

1, Nurse *et al.* (1976); 2, Nasmyth and Nurse (1981); 3, Nurse and Bissett (1981); 4, Fantes (1982); 5, King and Hyams (1982c); 6, Gygax and Thuriaux (1984); 7, Mitchison and Nurse (1985); 8, Nurse and Thuriaux (1980); 9, King and Hyams (1982b); 10, Creanor and Mitchison (1984); 11, Streiblová and Girbardt (1980); 12, Streiblová *et al.*(1984); 13, Nurse (1975); 14, Thuriaux *et al.* (1978); 15, Shimoda and Nishi (1982); 16, Aves *et al.* (1985); 17, Minet *et al.* (1979); 18, Fantes (1981); 19, Nasmyth (1977); 20, Nasmyth (1979); 22, Fantes (1979); 23, Thuriaux *et al.* (1980); 24, Nurse and Thuriaux (1984); 25, P.M. Nurse (personal communication); 26, Toda *et al.* (1983); 27, Umesono *et al.* (1983a); 28, Toda *et al.* (1984); 29, Yamamoto (1980); 30, Roy and Fantes (1982); 31, Toda *et al.* (1981); 32, Fantes (1983); 33, D.P. Dickinson (1981); 34, Nurse (1985a, b); 35, Iino and Yamamoto (1985); 36, Uemara and Yangida (1984).

Subject Index

I

K

N

O

P

R

S

W

X

Y

Z

Author Index

H

I

J

Q

R

S

T

Y

Z